XLV International Symposium on Multiparticle Dynamics (ISMD 2015)

EPJ Web of Conferences Volume 120 (2016)

Munich, Germany
4-9 October 2015

Editors:

T. Barillari
S. Kluth
S. Bethke
S. Menke

ISBN: 978-1-5108-2603-8

Printed from e-media with permission by:

Curran Associates, Inc.
57 Morehouse Lane
Red Hook, NY 12571

Some format issues inherent in the e-media version may also appear in this print version.

Printed by Curran Associates, Inc. (2016)

For additional information, please contact EDP Sciences – Web of Conferences at the address below.

EDP Sciences – Web of Conferences
17, Avenue du Hoggar
Parc d'Activité de Courtabœuf
BP 112
F-91944 Les Ulis Cedex A
France

Phone: +33 (0) 1 69 18 75 75
Fax: +33 (0) 1 69 28 84 91

contact-edps@webofconferences.org

Additional copies of this publication are available from:

Curran Associates, Inc.
57 Morehouse Lane
Red Hook, NY 12571 USA
Phone: 845-758-0400
Fax: 845-758-2633
Email: curran@proceedings.com
Web: www.proceedings.com

TABLE OF CONTENTS

edp sciences

EPJ Web of Conferences

Statement of Peer review

In submitting conference proceedings to EPJ Web of Conferences, I certify to the Publisher that I adhere to the **Policy on Publishing Integrity** of the journal in order to safeguard good scientific practice in publishing.

1. All articles have been subjected to peer review administered by the proceedings editors.
2. Reviews have been conducted by expert referees, who have been requested to provide unbiased and constructive comments aimed, whenever possible, at improving the work.
3. Proceedings editors have taken all reasonable steps to ensure the quality of the materials they publish and their decision to accept or reject a paper for publication has been based only on the merits of the work and the relevance to the journal.

Title, date and place of the conference XLV International Symposium on Multiparticle Dynamics
Wildbad Kreuth, Germany
4-9 Oct 2015

Proceedings editor(s): S. Bethke, T. Barillari, S. Kluth, S. Menke

Date and editor's signature 13.5.16

17, avenue du Hoggar - PA de Courtabœuf – BP 112 - 91944 Les Ulis Cedex A (France)
Tél. : 33 (0)1 69 18 75 75 - Fax : 33(0)1 69 07 45 17 – www.edpsciences.org

EPJ Web of Conferences **120**, 01001 (2016) DOI: 10.1051/epjconf/201612001001
ISMD 2015

Bose-Einstein correlations and results on minimum bias interactions, underlying event and particle production from ATLAS

Yuri Kulchitsky[1,2,a] (on behalf of the ATLAS Collaboration)

[1] *Institute of Physics, National Academy of Sciences, Minsk, Belarus*
[2] *JINR, Dubna, Russia*

Abstract. The report on the recent results of soft-QCD with the ATLAS experiment at the LHC is presented. The effect of space-time geometry in the hadronization phase has been studied in the context of Bose-Einstein correlations between charged particles, for determining the size and shape of the source from which particles are emitted. Bose-Einstein correlation parameters are investigated in proton-proton collisions at 0.9 and 7 TeV, up to very high charged particle multiplicities. Measurements of the properties of charged particle production are presented from proton-proton collisions at different centre-of-mass energies in the range of 0.9 to 13 TeV and compared to various Monte Carlo event generator models. Furthermore, particle distributions sensitive to the underlying event have been measured and are compared to theoretical models. The production properties of mesons and baryons are presented and compared to predictions.

1 Introduction

The focus of the ATLAS experiment [1] at the LHC is high-p_T physics, and also provides a window into important softer QCD processes. These are of intrinsic interest but also their understanding underpins searches for new physics. The selected topics of this paper, including first 13 TeV results, are Bose Einstein correlations [2], charged particledistributions [3], underlying events for leading tracks at 13 TeV [4], Z-boson [5] and jets [6] at 7 TeV, and particleproduction [7, 8] at 7 TeV. The ATLAS Collaboration possess now data from 0.9 to 13 TeV in the same detector, allowing scale evolution to be probed.

2 Bose-Einstein correlations at 0.9 and 7 TeV

The studies of Bose–Einstein Correlations (BEC) for pairs of like-sign charged particles measured in the kinematic range $p_T > 100$ MeV and $|\eta| < 2.5$ in proton–proton collisions at centre-of-mass energies of 0.9 and 7 TeV with the ATLAS detector are presented [2]. The integrated luminosities are approximately 7 μb^{-1}, 190 μb^{-1} and 12.4 nb^{-1} for 0.9 TeV, 7 TeV minimum-bias (MB) and 7 TeV high-multiplicity (HM) data samples, respectively. Bose–Einstein correlations are measured in terms of a two-particle correlation function, $C_2(p_1, p_2) = \rho(p_1, p_2)/\rho_0(p_1, p_2) = C_0[1 + \lambda \cdot \exp(-RQ)](1 + \varepsilon Q)$, where p_1 and p_2 are the four-momenta of two identical bosons in the event, ρ is the two-particle density function, and ρ_0 is a two-particle density function (known as the reference function) specially

[a] e-mail: Iouri.Koultchitski@cern.ch

constructed to exclude BEC effects, $Q^2 = -(p_1 - p_2)^2$, and C_0, R, λ, ε are parameters. The unlike-charge reference sample is used. To account for the effects of resonances, the two-particle correlation function $C_2(Q)$ is corrected using Monte Carlo simulation without BEC effects via a double-ratio correlation function defined as $R_2(Q) = C_2(Q)/C_2^{\mathrm{MC}}(Q)$. There are no the anti-correlations in the $R_2(Q)$ correlation functions. The multiplicity and k_T dependences are shown in the Fig. 1 for the R (top) and λ (bottom) parameters. The R parameter (Fig. 1 (a)) increases with multiplicity up to about $n_\mathrm{ch} \simeq 50$ independently of the center of mass energy. For $n_\mathrm{ch} \leq 82$ at 0.9 TeV and $n_\mathrm{ch} \leq 55$ at 7 TeV, the n_ch dependence on R is fitted with the function $R(n_\mathrm{ch}) = \alpha\sqrt[3]{n_\mathrm{ch}}$. The n_ch dependence on R at 7 TeV is fitted for $n_\mathrm{ch} > 55$ with a constant and $R(n_\mathrm{ch}) = 2.28 \pm 0.32$ fm. A saturation effect in the multiplicity dependence of the correlation source size parameter is observed. The k_T dependence of the R parameter (Fig. 1 (b), (c)) is found to follow an exponential decrease. The shapes of the k_T dependence are similar for the 7 TeV and the 7 TeV HM data. The λ parameter exponentially decreases with multiplicity (Fig. 1 (d)) and k_T (Fig. 1 (e), (f)).

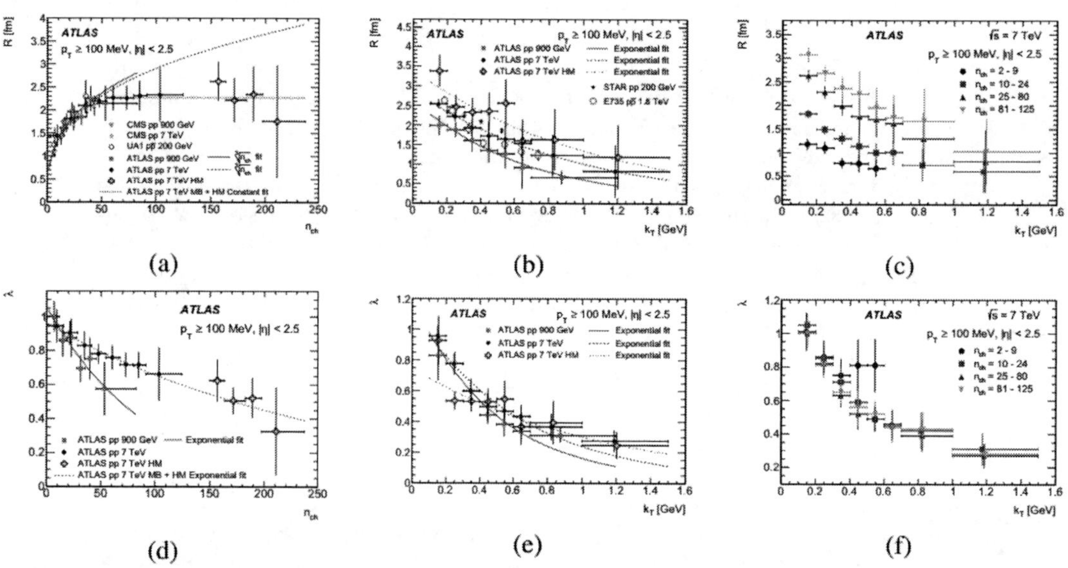

Figure 1. Dependences of the R (top) and λ (bottom) BEC parameters on (a, d) multiplicity, n_ch, and (b, e) pair transverse momentum, $k_\mathrm{T} = |\mathbf{p}_{\mathrm{T},1} + \mathbf{p}_{\mathrm{T},2}|/2$, obtained from the exponential fit to the two-particle double-ratio correlation functions $R_2(Q)$ at 0.9 and 7 TeV and 7 TeV HM events (taken from Ref. [2]). The solid and dashed curves are the results of (d) the exponential and (a) $\sqrt[3]{n_\mathrm{ch}}$ for $n_\mathrm{ch} \leq 55$ fits. The dotted line in (a) is a result of a constant fit to MB and HM events data at 7 TeV for $n_\mathrm{ch} > 55$. The solid, dashed and dash-dotted curves on (b) and (c) are results of the exponential fits for 0.9 TeV, 7 TeV and 7 TeV HM data, respectively. The results on the (b) are compared to the corresponding measurements by the E735 experiment at the Tevatron [9], and by the STAR experiment at RHIC [10]. The k_T dependence of the fitted parameters (c) R and (f) λ obtained from the exponential fit to the $R_2(Q)$ correlation function at 7 TeV for the different multiplicity regions: $2 \leq n_\mathrm{ch} \leq 9$ (circles), $10 \leq n_\mathrm{ch} \leq 24$ (squares), $25 \leq n_\mathrm{ch} \leq 80$ (triangles) and $81 \leq n_\mathrm{ch} \leq 125$ (inverted triangles).

3 Charged particle distributions at 13 TeV

Measurements of charged particle distributions in pp collisions at 13 TeV are presented, using a data sample of nearly 9 million events recorded by the ATLAS detector, corresponding to an integrated luminosity of ≈ 170 μb^{-1} [3]. The corrected distributions for primary charged particles for events with $n_\mathrm{ch} \geq 1$ in the kinematic range $p_\mathrm{T} > 500$ MeV and $|\eta| < 2.5$ are shown in Fig. 2. In most bins of all distributions the largest uncertainty comes from the track reconstruction efficiency. Fig. 2 (a) shows the charged particle multiplicity as a function of pseudorapidity. The mean particle density is roughly constant at 2.9 for $|\eta| < 1.0$ and decreases at higher values of $|\eta|$. Fig. 2 (b) shows the charged particle multiplicity as a function of transverse momentum. Fig. 2 (c) shows the charged particle multiplicity distribution. Fig. 2 (d) shows the mean transverse momentum versus the charged particle multiplicity. The $\langle p_T \rangle$ rises with n_{ch}, from 0.8 to 1.2 GeV. This increase is expected due to colour-reconnection effects. The results are compared to various Monte Carlo event generator models. The EPOS LHC [11] and the Pythia 8 [12] tunes describe the data the most accurately. The EPOS reproducing the η, p_T, and $\langle p_T \rangle$ distributions best and Pythia 8 A2 describing the multiplicity best in the low and mid-n_{ch} regions.

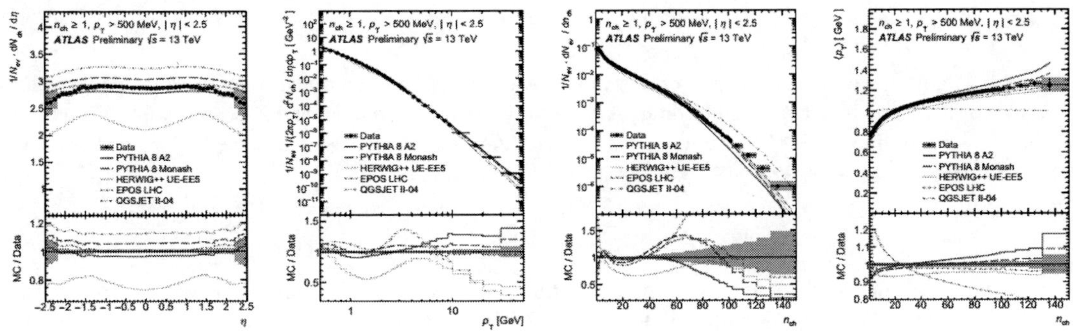

(a) (b) (c) (d)

Figure 2. Charged particle densities as a function of (a) η, (b) p_T, (c) n_{ch}, and (d) $\langle p_T \rangle$ dependence on multiplicity. The dots represent the data and the curves the predictions from different MC models. The bottom panels show the ratio of the MC over the data. Taken from Ref. [3].

4 Underlying events at 7 and 13 TeV

A detector level measurement of track distributions sensitive to the properties of the underlying event is given in Ref. [4]. It is based on 9 million events collected using the ATLAS detector in pp collisions at 13 TeV, corresponding to an integrated luminosity of $\approx 170\ \mu b^{-1}$. The underlying event (UE) is

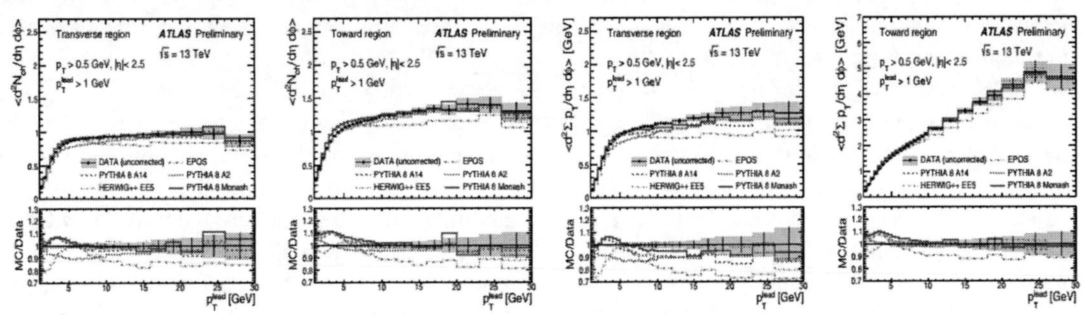

(a) (b) (c) (d)

Figure 3. Comparison of detector level data and MC predictions for (a, b) $\langle d^2N_{ch}/d\eta d\phi \rangle$ and (c, d) $\langle d^2 \sum p_T/d\eta d\phi \rangle$ as a function of p_T^{lead} in the transverse (a, c) and toward (b, d) regions. The bottom panels in each plot show the ratio of MC predictions to data. Taken from Ref. [4].

defined as the activity accompanying any hard scattering in a collision event. This includes partons

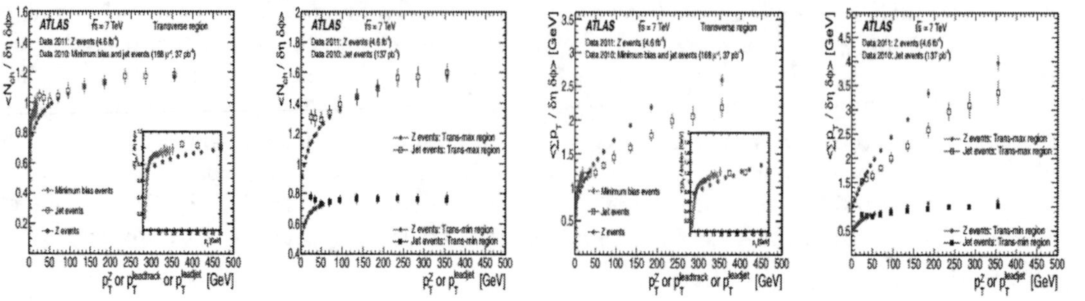

(a) (b) (c) (d)

Figure 4. Charged particle multiplicity average values, $N_{\text{ch}}/\delta\eta\delta\phi$ (a, b), and scalar $\sum p_{\text{T}}$ density average values, $\sum p_{\text{T}}/\delta\eta\delta\phi$ (c, d), compared between leading charged particle MB, leading jet and Z-boson events, respectively as functions of leading track $p_{\text{T}}^{\text{lead}}$, leading jet $p_{\text{T}}^{\text{leadjet}}$ and Z-boson p_{T}^{Z}, in the transverse region, trans-max and trans-min-regions, respectively. Taken from Ref. [5].

not participating in the hard-scattering process (beam remnants), and additional scatters in the same pp collision, termed multiple parton interactions (MPI). Initial and final state gluon radiation (ISR, FSR) also contribute to the UE activity. The direction of the leading track is used to define regions in the azimuthal plane that have different sensitivity to the UE. In Fig. 3 the average densities of track multiplicity and scalar $\sum p_{\text{T}}$ are shown. In the transverse region, both show a gradual increase, rising to an approximately constant plateau for $p_{\text{T}}^{\text{lead}} > 6$ GeV. For higher values of $p_{\text{T}}^{\text{lead}}$, the toward and away regions include contributions from jet-like activity, yielding gradually rising average $\sum p_{\text{T}}$ density. Among the MC models, Pythia 8 A14, Monash [12] and Herwig++ [13] predictions are closest to data in the plateau part of the transverse region, but none of the models describe the initial rise well. The EPOS [11] generator predicts significantly less activity at higher $p_{\text{T}}^{\text{lead}}$, indicating the absence of semi-hard MB events.

A measurement of charged particle distributions sensitive to the properties of the underlying event is presented for an inclusive sample of events containing a Z-boson [5] and jets [6]. The measurement is based on data collected using the ATLAS detector in pp collisions at 7 TeV with an integrated luminosity of 4.6 fb^{-1} for Z-boson and 37 pb^{-1} for jets. The certain common qualitative features can be observed by comparing $\sum p_{\text{T}}/\delta\eta\delta\phi$ and $N_{\text{ch}}/\delta\eta\delta\phi$ as functions of the leading object p_{T} in the transverse region, and also separated into the trans-max/min regions as shown in Fig. 4. The measurements with a leading jet are complementary to the measurements with a leading track, and a smooth continuation at 20 GeV is observed (in Fig. 4 (a), (c)), corresponding to the lowest jet p_{T} for which the jet measurement could be performed and the highest leading track momentum included in the leading track analysis. Where the p_{T} of the leading object is less than 50 GeV, a large difference is observed both for the N_{ch} and $\sum p_{\text{T}}$ average values between the jet and Z-boson measurements in Fig. 4 (a), (c); the increase of the associated activity as a function of the hard scale p_{T} is very different in track/jets events from the Z-boson events. The N_{ch} density is similar in the underlying event associated with a jet to that with a Z-boson for higher values of the hard scale ($\geq$ 50 GeV), there are residual differences in the average $\sum p_{\text{T}}$ densities. The activity in events with a Z-boson is systematically higher than that in events with jets. From the behaviour of the underlying event properties in the trans-max/min regions in Fig. 4 (b), (d) this difference originates mostly from the trans-max region, due to selection bias. The trans-min region is very similar between the two measurements, despite the different hard scales, indicating again that this region is least sensitive to the hard interaction and

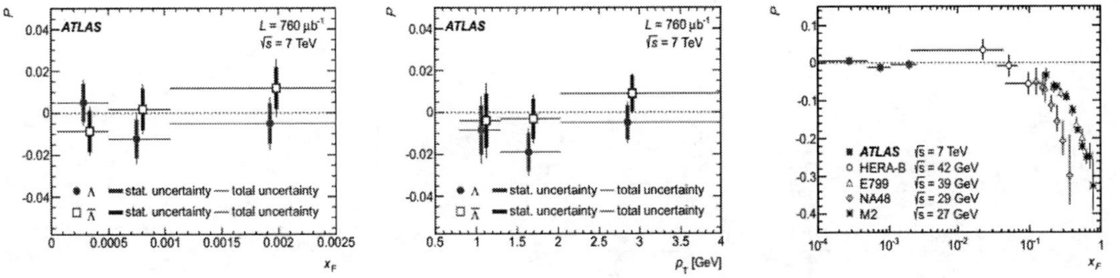

(a) (b) (c)

Figure 5. Transverse polarization of Λ and $\bar{\Lambda}$ hyperons as a function of (a) x_F and (b) p_T (taken from Ref. [7]). The Λ transverse polarization measured by ATLAS compared to the measurements from lower center-of-mass energy experiments HERA-B data taken from Ref. [15], NA48 from Ref. [16], E799 data from Ref. [17], and M2 from Ref. [18].

most sensitive to the MPI component. The measured distributions are compared to the predictions of various Monte Carlo generators implementing different underlying event models.

5 Particleproduction at 7 TeV

The transverse polarization of Λ and $\bar{\Lambda}$ hyperons produced in pp collisions at a center-of-mass energy of 7 TeV is measured in Ref. [7]. The analysis uses 760 μb^{-1} of MB data collected by the ATLAS detector. The measured transverse polarization averaged over Feynman x_F from 5×10^5 to 0.01 and transverse momentum p_T from 0.8 to 15 GeV is 0.010 ± 0.005 (*stat*) $\pm$ 0.004 (*syst*) for Λ and 0.002 ± 0.006 (*stat*) $\pm$ 0.004 (*syst*) for $\bar{\Lambda}$. It is also measured as a function of x_F (Fig. 5 (a)) and p_T (Fig. 5 (b)), but no significant dependence on these variables is observed. Prior to this measurement, the polarization was measured at fixed target experiments with center-of-mass energies up to about 40 GeV. The ATLAS results are compatible with the extrapolation of a fit from previous measurements to the x_F range covered by this measurement (Fig. 5 (c)).

A measurement of the $\phi(1020) \times BR(\phi \to K^+K)$ production cross section at 7 TeV is presented using pp collision data corresponding to an integrated luminosity of 383 μb^{-1}, collected with the ATLAS experiment [8]. Selection of $\phi(1020)$-mesons is based on the identification of $K^\pm$ by their energy loss in the pixel detector. The differential cross section is measured as a function of the transverse momentum, $p_{T,\phi}$, (Fig. 6 (a)) and rapidity, y_ϕ, (Fig. 6 (b)) of the $\phi(1020)$-meson in the fiducial region $500 < p_{T,\phi} < 1200$ MeV, $|y_\phi| < 0.8$, kaon $p_{T,K} > 230$ MeV, and kaon momentum $p_K <$ 800 MeV. The integrated $\phi(1020)$-meson production cross section in this fiducial range is measured to be $\phi(1020) \times BR(\phi \to K^+K) = 570 \pm 8\,(stat) \pm 66\,(syst) \pm 20\,(lumi)\,\mu$b. The extrapolated cross section is compared to the measurement by the ALICE Collaboration of the $\phi(1020)$ production cross section (see Ref. [19]). A comparison between the cross section measurements is shown in Fig. 6 (c). The measurements as a function of $p_{T,\phi}$ are in agreement to within 10% in the first two bins and to within 3% in the other bins, which is well within the systematic uncertainties. The $\phi(1020)$ production cross section is in agreement with the predictions of the MC generators EPOS [11] and Pythia 6 DW [12].

6 Summary

The Bose-Einstein correlations at 0.9 and 7 TeV were studied. For the first time a saturation effect in the multiplicity dependence of the extracted BEC radius parameter is observed. The k_T dependence

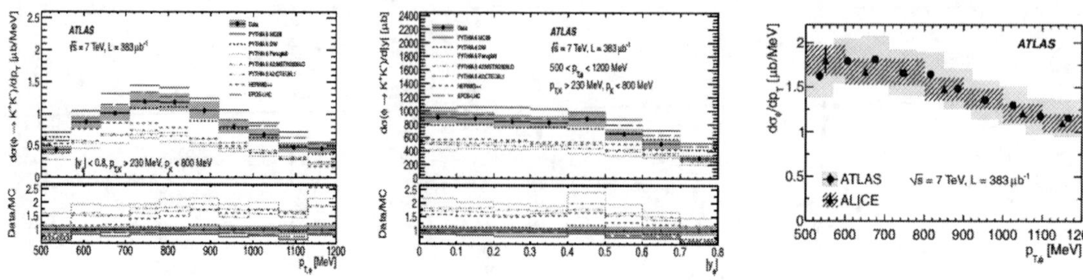

(a) (b) (c)

Figure 6. The $\phi(1020) \times BR(\phi \to K^+K)$ cross section as a function of (a) $p_{T,\phi}$ and (b) $|y_\phi|$ (taken from Ref. [8]). The data are compared to various MC expectations as described in the legends. The $\phi(1020)$-meson cross section as a function of (c) $p_{T,\phi}$ is compared to the measurement by the ALICE Collaboration [19].

on R is observed to increase with multiplicity. Charged particle multiplicity measurements at 13 TeV using pp collisions are presented. Among the models considered, EPOS reproduces the data the best way, Pythia 8 A2 and Monash give reasonable descriptions. Underlying event analysis with 13 TeV data are shown: one observes a reasonable agreement of tunes used in Atlas MC with the new data. Further UE studies were done at 7 TeV for Z, leading jet and leading track. These are required for tuning of the soft part of Monte Carlo simulation. Transverse polarisation of Λ and $\bar{\Lambda}$ hyperons at 7 TeV were found to be consistent with 0 at low x_F, confirming the behaviour of previous experiments showing a decrease of polarization as function of x_F. The measurement of $\phi(1020)$ differential cross section at 7 TeV can provide useful input for turning and development of phenomenological models.

References

[1] ATLAS Collaboration, JINST **3** S08003 (2008)
[2] ATLAS Collaboration, Eur. Phys. J. C **75**, 10, 466 (2015) [arXiv:1502.07947]
[3] ATLAS Collaboration, ATLAS-CONF-2015-028, https://cds.cern.ch/record/2037701
[4] ATLAS Collaboration, ATL-PHYS-PUB-2015-019, https://cds.cern.ch/record/2037684
[5] ATLAS Collaboration, Eur. Phys. J. C **74**, 12, 3195 (2014) [arXiv:1409.3433]
[6] ATLAS Collaboration, Eur. Phys. J. C **74**, 8, 2965 (2014) [arXiv:1406.0392]
[7] ATLAS Collaboration, Phys. Rev. D **91**, 3, 032004 (2015) [arXiv:1412.1692]
[8] ATLAS Collaboration, Eur. Phys. J. C **74**, 7, 2895 (2014) [arXiv:1402.6162]
[9] T. Alexopoulos *et al.*, Phys. Rev. D **48**, 1931 (1993)
[10] M.M. Aggarwal *et al.*, STAR Collaboration, Phys. Rev. C **83**, 064905 (2011)
[11] S. Porteboeuf, T. Pierog and K. Werner [arXiv:1006.2967]
[12] T. Sjöstrand, S. Mrenna and P. Z. Skands, Comput. Phys. Commun. **178**, 852867 (2008)
[13] M. Bahr *et al.*, Eur. Phys. J. C **58** 639707 (2008)
[14] S. Ostapchenko, Phys. Rev. D **83** 014018 (2011)
[15] I. Abt *et al.*, HERA-B Collaboration, Phys. Lett. B **638**, 415 (2006)
[16] V. Fanti *et al.*, NA48 Collaboration, Eur. Phys. J. C **6**, 265 (1999)
[17] E. Ramberg *et al.*, E799 Collaboration, Phys. Lett. B **338**, 403 (1994)
[18] B. Lundberg *et al.*, Phys. Rev. D **40**, 3557 (1989)
[19] ALICE Collaboration, Eur. Phys. J. C **72**, 2183 (2012)

EPJ Web of Conferences **120**, 01002 (2016) DOI: 10.1051/epjconf/201612001002
ISMD 2015

Searching for a hidden sector in multiparticle production at LHC

Miguel-Angel Sanchis-Lozano[1,a], Edward Sarkisyan-Grinbaum[2,3,b], Salvador Moreno-Picot[4,c]

[1]*Departamento de Física Teórica, IFIC, CSIC-University of Valencia, 46100 Burjassot, Spain*
[2]*Department of Physics, The University of Texas at Arlington, Arlington, TX 76019, USA*
[3]*Department of Physics, CERN, 1211 Geneva 23, Switzerland*
[4]*Departamento de Informática, ETSE, University of Valencia, 46100 Burjassot, Spain*

Abstract. A hidden sector beyond the Standard Model can show up in multiparticle production altering inclusive correlations and factorial cumulants of multiplicity distributions. In this report such a study is advocated with a special emphasis on the searches at LHC.

1 Introduction

Most signatures of New Physics in colliders are expected to be found in hard events, on the transverse plane with respect to the beams' direction (*i.e.* emitting particles with high transverse momentum $p_\perp$), where background is much reduced. In this report, however, we focus on rather diffuse soft signals in pp inelastic interactions, though expectedly tagged by hard decay products and appropriate cuts on events. For example, a non-standard state of matter from a Hidden Sector (HS) [1] might alter particle correlations [2–4] which can be measured to a large accuracy at the LHC.

Hadron interactions at high energy are usually considered as resulting from collisions of their constituent partons, likely dominated by pairwise parton interactions. In this work we study the effects of a new physics contribution on the conventional parton cascade. To this end, we extend the phenonemological approach of the Independent Pair Parton Interaction (IPPI) model [5, 6] including an extra step of a hypothetical new stage of matter associated with HS.

1.1 Factorial moments of multiplicity correlations

The study of inclusive particle correlations in multiparticle production can be performed by analyzing n-particle correlation functions and/or normalised factorial moments of multiplicity distributions [7–9]. Here we focus on the latter.

The normalized factorial moments of rank $q = 2, 3, \ldots,$ are defined as

$$F_q = \frac{\sum_n P(n)\, n(n-1)\cdots(n-q+1)}{(\sum_n P(n)\, n)^2}\,, \qquad (1)$$

[a]e-mail: Miguel.Angel.Sanchis@ific.uv.es
[b]e-mail: sedward@mail.cern.ch
[c]e-mail: Salvador.Moreno@uv.es

Table 1. Probability distribution of the number of active pairs in proton-proton collisions for different TeV energies according to the IPPI model.

$\sqrt{s}$	w_1	w_2	w_3	w_4	w_5	w_6	w_7
1.8 TeV	0.519	0.269	0.140	0.072	0.0	0.0	0.0
7.0 TeV	0.504	0.254	0.128	0.065	0.033	0.016	0.0
13 TeV	0.5020	0.2520	0.1265	0.0635	0.0319	0.0160	0.0080

where $P(n)$ denotes the probability for n final-state particles (charged hadrons).

The factorial moments represent any correlation between the emitted particles in events. To extract the genuine q-particle correlations, not reducible to the product of the lower-order correlations, one uses the normalised cumulant functions, or cumulants, defined as

$$K_q = F_q - \sum_{r=1}^{q-1} \frac{(q-1)!}{r!(q-r-1)!} K_{q-r} F_r \,. \tag{2}$$

Since F_q and $|K_q|$ grow rapidly as the rank q increases, it is convenient to consider the ratio

$$H_q = \frac{K_q}{F_q} \,, \tag{3}$$

Factorial moments and cumulants have been extensively applied to the analysis of multihadron dynamics in different types of collisions, from e^+e^- to nucleus-nucleus interactions, in a broad range of energies [7–9]. In particular, normalized H_q moments are extremely sensitive to the details of multiplicity distributions [10] and can be used to distinguish between different multiparticle production models and eventually the contribution of a HS as advocated in this paper.

2 Multiparticle production as a multi-step cascade

The IPPI model [5, 6] was proposed in order reproduce the moments of multiplicity distributions in pp collisions at high energy with minimum adjustable parameters. The IPPI picture corresponds to a simplified 2-step scenario: parton binary collisions become seeds of independent cascades which hadronize (*e.g.* via string fragmentation) to the final-state multiparticle state.

Moreover, it is assumed that each pair parton interaction gives rise to a negative binomial distribution (NBD), while the total distribution is ultimately described by means of the weighted sum:

$$P^{(2)}(n) = \sum_{j=1}^{j_{max}} w_j \sum_{n_i} \prod_{i=1}^{j} P_{\rm NBD}(n_i, m^{(1)}, k^{(1)}) = \sum_{j=1}^{j_{max}} w_j \, P_{\rm NBD}(n; jm^{(1)}, jk^{(1)}) \,, \tag{4}$$

where w_j denotes the probability for a j-pair interaction, $m^{(1)}$ and $k^{(1)}$ correspond to the mean multiplicity and dispersion for a single pair interaction, respectively. Note that no new adjustable parameters appear in Eq.(4) besides the distribution for j binary parton interactions which can be evaluated if some model is adopted [11, 12].

In the IPPI, the probability for j binary parton interactions per event is simply estimated as $w_j = w_1^j$, where w_1 refers to a single pair, with the normalization condition $\sum_{j=1}^{j_{max}} w_j = 1$. In Table 1 we show the values of w_j up to $w_{max} = 7$, corresponding to pp collisions at the c.m. energy $\sqrt{s}$ of 13 TeV taken from [6].

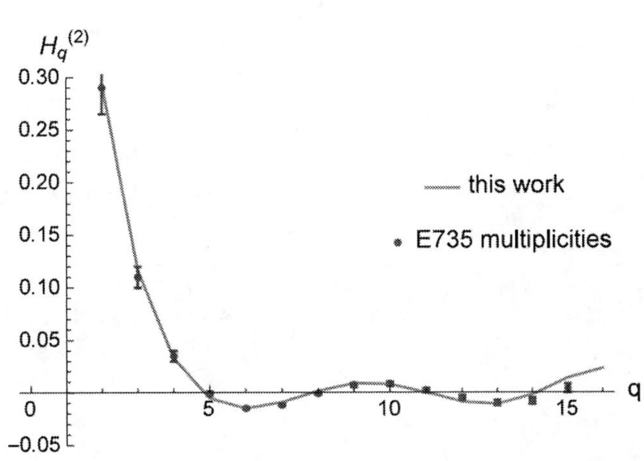

Figure 1. $H_q^{(2)}$ moments up to $q = 16$ in $p\bar{p}$ collisions at 1.8 TeV obtained in this work using expressions for a 2-step cascade and the same parameters as in [5, 6]. Very good agreement is found with their results and experimental data [17], shown by circles with error bars.

2.1 Two-step cascade

Let us rewrite Eq.(4) for arbitrary particle production distributions and sources:

$$P^{(2)}(n) = \sum_{N_s} P(N_s) \sum_{n_i} \prod_{i=1}^{N_s} P^{(1)}(n_i). \tag{5}$$

Here n and N_s denote the number of (charged) particles and sources, respectively. Note that the number of sources N_s in Eq.(5) corresponds to the number of parton pair collisions j in Eq.(4). In the notation used here, $P(N_s)$ stands now for the distribution of (fragmenting string) sources, equivalent to the parton pair interaction distribution w_j. Correspondingly, the average multiplicity can be written as $\langle n \rangle = \langle N_s \rangle\, m^{(1)}$ according to a 2-step description of multiparticle production.

On the other hand, the authors of [5, 6] benefit from a dramatic reduction of free parameters when assuming a weighted superposition of NBDs with shifted parameters, as can be seen in Eq.(4). In addition, since $m^{(1)}$ should be the same for any value of the rank q, only $k^{(1)}$ remains a free parameter (w_{max} was determined using a particular model).

In order to make a comparison of the results of the current study and those from [5, 6], below we assume that all $P^{(1)}(n_i)$ are NBDs. Moreover, $P(N_s)$ and w_j distributions can be formally identified.

The computation of high rank $F_q^{(2)}$ moments becomes extremely involved at large q. Therefore, we have written a Prolog code [13] which provides the expressions $F_q^{(p)}$ for any value of the rank q and any number of steps p in the cascade $F_q^{(p)}$, depending on the computer capacity available.

As shown below, we are able to reproduce (up to the percent level) the $H_q^{(2)}$ moments [1] using the same values and assumptions as in Refs. [5, 6]. This accordance suggests to proceed further in the approach given here by incorporating a new step in the parton cascade following the mIPPI scheme.

[1]The superindex p in $H_q^{(p)}$ indicates the number of steps in the cascade: a two-step conventional cacade with $p = 2$, and the three-step cascade with $p = 3$ once a HS is included.

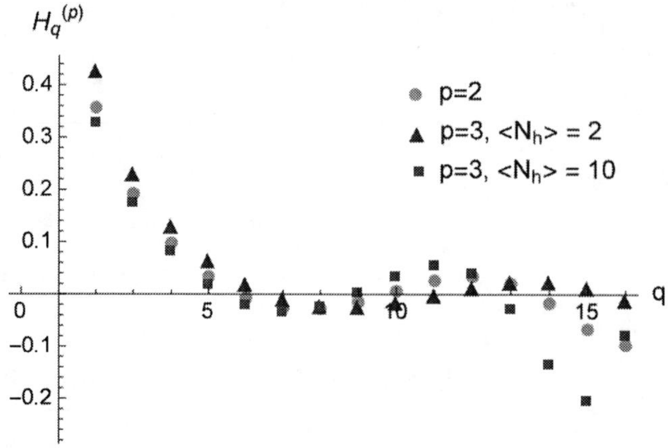

Figure 2. Predictions for $H_q^{(p)}$ moments as a function of the rank q for pp collisions at $\sqrt{s} = 13$ TeV. The circles correspond to a conventional 2-step cascade ($p = 2$) from extrapolation at lower energies using the IPPI model. The triangles and the squares correspond to a 3-step cascade ($p = 3$) using the mIPPI model (this work) with $\langle N_h \rangle = 2$ and $\langle N_h \rangle = 10$, respectively. A different pattern in the amplitude of the oscillations at high q values can be clearly observed.

2.2 Three-step cascade

Let us now include an extra step in the cascade to simulate a hypothetical new stage of matter associated to a HS. The resulting multiplicity in a 3-step process should obey the following distribution:

$$P^{(3)}(n) = \sum_{N_s} P(N_s) \sum_{n_j} \prod_{j=1}^{N_s} P^{(2)}(n_j) \ ; \ P^{(2)}(n) = \sum_{N_h} P(N_h) \sum_{n_i} \prod_{i=1}^{N_h} P^{(1)}(n_i) , \tag{6}$$

with N_h denoting the number of active hidden sources in a collision. In what follows, for the sake of simplicity we assume that $P(N_h)$ follows a Poisson distribution, *i.e.* independent production of hidden sources resulting from binary parton interactions.

In other words, the probability distribution of parton interactions remains the same as in the conventional cascade (being already adjusted to reproduce experimental data in pp collisions) while one adds another step subsequent to the initial binary parton interaction.

3 H_q-moment oscillations as a function of the rank q

QCD next-to-leading order calculations [14, 15] predict that the ratios H_q defined in Eq.(3) oscillate as a function of the rank q, crossing the q-axis and becoming negative with $q_{min} \approx 5$ at LHC energies, shifting to larger values at higher energies. This prediction has been tested against experimental data and found to be observed not only in e^+e^- collisions but also in a variety of colliding particles and energies, including pp, pA and AA collisions [16].

In Fig. 1 we plot the values of the $H_q^{(2)}$ moments ($q = 2$ to 16) for $\sqrt{s} = 1.8$ TeV multiplicity data, obtained through Eqs.(2) and (3) from the expressions of $F_q^{(2)}$ (a 2-step cascade). We fix the parameters for the plot alike it is done in Ref. [5], *i.e.*, assuming NBDs for all binary parton collisions with $k^{(1)} = 4.4$, and $P(N_s)$ (equivalent to the w distribution) from Table 1.

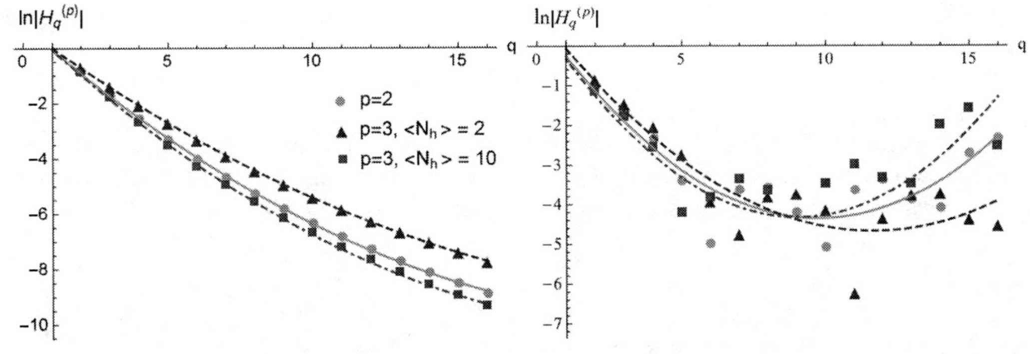

Figure 3. Values of $\ln|H_q^{(p)}|$ at $\sqrt{s} = 13$ TeV versus q for a 2-step scenario ($p = 2$, circles) and a 3-step scenario ($p = 3$) with $\langle N_h \rangle = 2$ (triangles) and $\langle N_h \rangle = 10$ (squares). Solid, dashed and dot-dashed lines correspond to parabolic fits, respectively. Left panel: superposition of NBDs with different reference values for their k parameter (see [4]). Right panel: $P(N_s)$ incorporates the values of Table 1 (not the NBD case). Notice that there are no oscillations when *all* the distributions of the convolution are of the NBD type.

One can see the two minima in the Fig. 1. This oscillatory pattern is due to the fact that the probability distribution for the number of sources $P(N_s)$ (equivalently, the distribution for the number of parton pair collisions) does **not** follow a NBD. In case all the distributions are NBD, the resulting distribution turns out to be of the NBD type too and no oscillation pattern for $H_q^{(2)}$ shows up.

The overall good agreement with the results of Refs. [5, 6] and experimental data [17] suggests the further introduction of a new step in the cascade to be interpreted as a HS, thereby studying the eventual variation of the crossing points/minima and the amplitude of the H_q oscillations.

4 HS-cascade versus a conventional cascade

In Fig. 2 three sets of points corresponding to different scenarios at pp collisions at $\sqrt{s} = 13$ TeV are shown. The circles correspond to a conventional cascade, while the triangles and squares correspond to an extra step in the mIPPI model setting $\langle N_h \rangle = 2$ and $\langle N_h \rangle = 10$, respectively. One can see that the crossing point (and minimum) moves by about one unit to the left for the $\langle N_h \rangle = 10$, and by the same amount to the right for $\langle N_h \rangle = 2$ compared to the case of a conventional cascade. Such an altered behaviour could become a hint of a HS affecting the parton evolution in multiparticle production, deserving a more detailed study.

Next we examine the amplitude of the H_q oscillations as a function of the rank q, and its dependence on the parameters used in the mIPPI model as can be seen from in Fig. 2.

Depending on the number of hidden sources two different behaviours of the oscillation pattern of $H_q^{(3)}$ moments can be distinguished:

- For a small number of hidden sources, the oscillation amplitude becomes appreciably dumped for high q values as compared to a conventional (2-step) cascade.
- For a large number of hidden sources, the oscillation amplitude is considerably larger for high q values as compared to a conventional (2-step) cascade.

These conclusions are indeed confirmed in Fig. 3 where the values of $\ln|H_q|$ are plotted against q for different scenarios depending on the type of the distributions used. The calculated points are shown together with the parabolic fits to them. One can see that the behaviour of the fitted curves is very different for different scenarios especially at large q values.

The fitted curves pass through the points in the left panel, whereas the points scatter around the curves in the right panel. Hence no oscillations appear whenever all distributions in the superposition of Eqs.(6), including $P(N_s)$, are of the NBD type. This behaviour can be easily understood in the mIPPI model inasmuch the convolution of NBDs in Eq.(6) leads again to a NBD. Conversely, the oscillation pattern in the right panel emerges as a consequence of $P(N_s)$ not being a NBD.

5 Summary

Using the modified Independent Parton Pair Interaction we have studied the effect of a HS on the cumulant-to-factorial moment ratio H_q of the multiplicity distributions of final-state particles in pp collisions at the LHC. We conclude that a large (small) number of hidden sources would lead to an enhancement (softening) of the oscillation amplitude at high q values. Moreover, the crossing of the q-axis and the minimum of the H_q-moments interpolating curve shifts to smaller (larger) q values for a large (few) number of hidden sources.

Acknowledgments

Many thanks to the organizers of ISMD 2015 for a very pleasant working atmosphere and excellent organization. This work has been partially supported by MINECO under grants FPA2011-23596 and FPA2014-54459-P, and Generalitat Valenciana under grant PROMETEOII/2014/049.

References

[1] M. J. Strassler and K. M. Zurek, Phys. Lett. B **651**, 374 (2007) [hep-ph/0604261].
[2] M.-A. Sanchis-Lozano, Int. J. Mod. Phys. A **24** 4529 (2009) [arXiv:0812.2397 [hep-ph]].
[3] M.-A. Sanchis-Lozano and E. Sarkisyan-Grinbaum, arXiv:1409.5262 [hep-ph].
[4] M.-A. Sanchis-Lozano, E. K. Sarkisyan-Grinbaum, S. Moreno-Picot, arXiv:1510.08738 [hep-ph].
[5] I. M. Dremin and V. A. Nechitailo, Phys. Rev. D **70** 034005 (2004) [hep-ph/0402286].
[6] I. M. Dremin and V. A. Nechitailo, Phys. Rev. D **84** 034026 (2011) [arXiv:1106.4959 [hep-ph]].
[7] E. A. De Wolf, I. M. Dremin and W. Kittel, Phys. Rept. **270** 1 (1996) [hep-ph/9508325].
[8] I. M. Dremin and J. W. Gary, Phys. Rept. **349**, 301 (2001) [hep-ph/0004215].
[9] W. Kittel, E. A. De Wolf, *Soft Multihadron Dynamics* (World Scientific, Singapore, 2005).
[10] R. Ugoccioni, A. Giovannini and S. Lupia, Phys. Lett. B **342** 387 (1995) [hep-ph/9410340].
[11] A. B. Kaidalov and K. A. Ter-Martirosyan, Phys. Lett. B **117** 247 (1982)
[12] S. G. Matinyan and W. D. Walker, Phys. Rev. D **59** 034022 (1999) [hep-ph/9801219].
[13] W.F. Clocksin and C.S. Mellish, *Programming in Prolog*, 5th edition. 2003. Springer-Verlag.
[14] I. M. Dremin, Phys. Lett. B **341** 95 (1994), [Phys. Lett. B **348** 711 (1995) (E)] [hep-ph/9408300].
[15] M. A. Buican, C. Forster and W. Ochs, Eur. Phys. J. C **31** (2003) 57 [hep-ph/0307234].
[16] I. M. Dremin, V. A. Nechitailo, M. Biyajima and N. Suzuki, Phys. Lett. B **403**, 149 (1997) [hep-ph/9704318].
[17] E735 Collaboration, F. Turkot *et al.*, Nucl. Phys A **525** 165 (1991)

EPJ Web of Conferences **120**, 01003 (2016) DOI: 10.1051/epjconf/201612001003
ISMD 2015

The τ-model of Bose-Einstein Correlations: Some recent results

Wesley J. Metzger[1,a]

[1] *IMAPP, Radboud University, 6525 AJ Nijmegen, The Netherlands*

Abstract. Bose-Einstein correlations of pairs of identical charged pions produced in hadronic Z decays and in 7 TeV pp minimum bias interactions are investigated within the framework of the τ-model.

1 Introduction

After a brief review of relevant previous results, new *preliminary* results are presented on the dependence of the Bose-Einstein correlation function on track and jet multiplicity, pair transverse momentum and pair rapidity, using a parametrization which has been found [1] to describe well Bose-Einstein correlations in hadronic Z decay, namely that of the τ-model [2, 3].

1.1 Parametrization

The Bose-Einstein correlation function, R_2, is often parametrized as

$$R_2 = \gamma\left[1 + \lambda \exp\left(-(rQ)^2\right)\right](1 + \epsilon Q) , \tag{1}$$

and is measured by $R_2(Q) = \rho(Q)/\rho_0(Q)$, where $\rho(Q)$ is the density of identical boson pairs with invariant four-momentum difference $Q = \sqrt{-(p_1 - p_2)^2}$ and $\rho_0(Q)$ is the similar density in an artificially constructed reference sample, which should differ from the data only in that it does not contain the effects of Bose symmetrization of identical bosons.

However, the "classic" parametrization of Eq. (1) is found to be inadequate, even when it is generalized to allow for a Lévy distribution of the source:

$$R_2 = \gamma\left[1 + \lambda \exp\left(-(rQ)^\alpha\right)\right](1 + \epsilon Q) , \quad 0 < \alpha \leq 2 \tag{2}$$

This was not realized for a long time because the correlation function was only plotted up to $Q =$ 2 GeV or less. In Ref. [1] Q was plotted to 4 GeV, and it became apparent that there is a region of anti-correlation ($R_2 < 1$) extending from about $Q = 0.5$ to 1.5 GeV. This anti-correlation, which one might term Bose-Einstein Anti-Correlations (BEAC), as well as the Bose-Einstein correlations (BEC) are well described by the τ-model.

In the τ-model R_2 is found to depend not only on Q, but also on quantities a_1 and a_2. For two-jet events $a = 1/m_t$, where $m_t = \sqrt{m^2 + p_t^2}$ is the transverse mass of a particle). Parameters of the model

[a] e-mail: W.Metzger@science.ru.nl

are the parameters of the Lévy distribution which describes the proper time of particle emission: α, the index of stability of the Lévy distribution; a width parameter $\Delta\tau$; and the proper time τ_0 at which particle production begins.

We shall use a simplified parametrization [1] where τ_0 is assumed to be zero and a_1 and a_2 are combined with $\Delta\tau$ to form an effective radius R:

$$R_2(Q) = \gamma\left[1 + \lambda\cos\left((R_a Q)^{2\alpha}\right)\exp\left(-(RQ)^{2\alpha}\right)\right](1+\epsilon Q)\,, \tag{3a}$$

$$R_a^{2\alpha} = \tan\left(\frac{\alpha\pi}{2}\right)R^{2\alpha}\,. \tag{3b}$$

Note that the difference between the parametrizations of Eqs. (2) and (3) is the presence of the cosine term, which provides the description of the anti-correlation. The parameter R describes the BEC peak, and R_a describes the anti-correlation region. While one might have had the insight to add, *ad hoc*, a cos term to Eq. (2), it is the τ-model which provides a physical reason for it and which predicts a relationship, Eq. (3b), between R and R_a.

A fit of Eq. (3) to L3 two-jet events is shown in Fig. 1, from which it is seen that the τ-model describes both the BEC peak and the BEAC dip quite well. Also the three-jet data is well described [1], which is perhaps surprizing since the τ-model is inspired by a picture of fragmentation of a single string.

It must also be pointed out that the τ-model has its shortcomings: The τ-model predicts that R_2 depends on the two-particle momentum difference only through Q, not through components of Q. However, this is found not to be the case [1]. Nevertheless, regardless of the validity of the τ-model, Eq. (3) provides a good description of the data. Accordingly, we shall use it in most of the following.

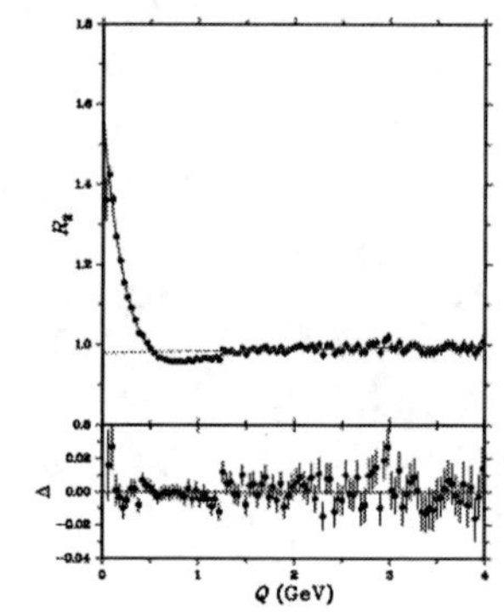

Figure 1. The Bose-Einstein correlation function R_2 for two-jet events. The curve corresponds to the fit of Eq. (3). Also plotted is Δ, the difference between the fit and the data. The dashed line represents the long-range part of the fit, *i.e.*, $\gamma(1+\epsilon Q)$. The figure is taken from Ref. [1].

1.2 Data

The e^+e^- data were collected by the L3 detector at a center-of-mass energy of $\sqrt{s} \simeq 91.2\,\text{GeV}$. Approximately 36 million like-sign pairs of well-measured charged tracks from about 0.8 million hadronic Z decays are used. This data sample is identical to that of Ref. [1]. The same event mixing technique is used to construct ρ_0 as in Ref. [1].

The minimum-bias pp data were collected by the ATLAS detector at a center-of-mass energy of $\sqrt{s} = 7\,\text{TeV}$. The sample contains approximately 1.8×10^9 like-sign charged track pairs from approximately 10^7 events. The acceptance in pseudorapidity is $|\eta| < 2.5$. This data sample is identical to the 7 TeV sample used in the recent ATLAS paper on BEC [4]. However the new results reported here, which are taken from a Ph.D. thesis [5], use, unless otherwise stated, a reference sample constructed by the opposite-hemisphere (OHP) method where the three-momentum, $\vec{p}$ of one of the particles of the pair is replaced by $-\vec{p}$. The recent ATLAS paper [4] used unlike-sign pairs to construct ρ_0, which is unsuitable for studying the region of anti-correlation, since the ρ^0 meson is a dominant feature there.

The Durham and JADE algorithms are used to classify e^+e^- events according to the number of jets. The number of jets in a particular event depends on the jet resolution parameter of the algorithm, y_{cut}. We define y_{23}^{D} (y_{23}^{J}) as the value of y_{cut} at which the number of jets in the event changes from two to three using the respective algorithms. Small y_{23} corresponds to narrow two-jet events, large y_{23} to events with three (or more) well-separated jets. The two algorithms lead to similar results.

DOI: 10.1051/epjconf/201612001003

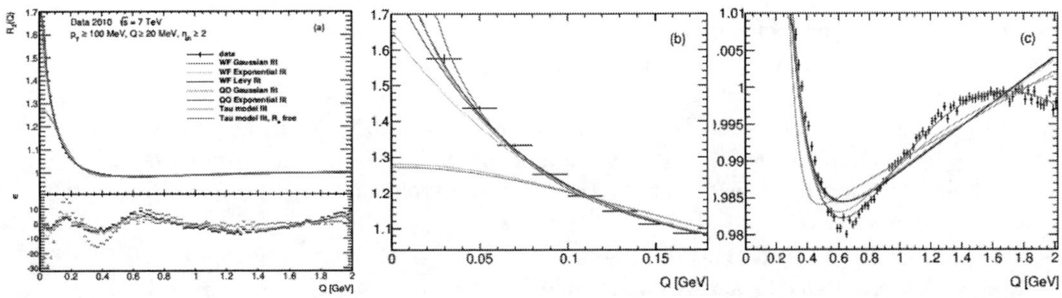

Figure 2. R_2 for 7 TeV pp minimum bias interactions with the results of fits of several parametrizations [5].

2 New *Preliminary* Results

As in e^+e^- annihilation [1] , the pp minimum bias data are found [5] to be best described by Eq. (3a). While other parametrizations can describe the BEC peak, only this parametrization (with R_a a free parameter) comes close to describing not only the BEC peak but also the BEAC region, as is seen in Fig. 2. Previously, CMS had also observed the BEAC and fit it with the τ-model parametrization [6].

As has frequently been pointed out (see, *e.g.*, Ref. [7]) the values of the parameters of BEC depend strongly on the choice of reference sample. This is seen again in Fig. 3, where the dependence of R on the charged track multiplicity, N, and on the average pair transverse momentum, $k_t = (\vec{p_{t1}} + \vec{p_{t2}})/2$, is shown for the exponential parametrization of R_2, *i.e.*, Eq. 2 with $\alpha = 1$, which is the parametrization used in the ATLAS paper [4]. The dependences with unlike-sign (ULS) reference sample are markedly different from those with the other samples. Whereas R increases somewhat less than linearly with N for the other reference samples, it appears to saturate for the ULS sample. And, whereas R decreases approximately linearly with k_t for the ULS sample, for the rotated reference sample it decreases only up to $k_t \approx 450$ MeV and then increases, and for the other reference samples it is first roughly constant with k_t and then increases.

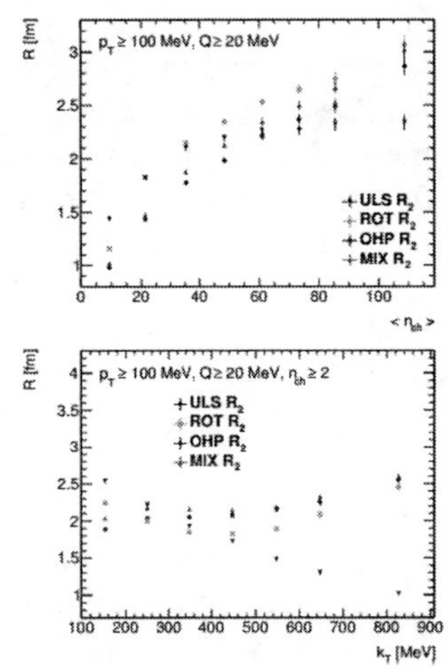

Figure 3. The dependence of R (top) on N and (bottom) on k_t with respect to the beam axis for different reference samples in 7 TeV pp minimum bias events. The exponential parametrization is used [5].

An unfortunate property of the τ-model parameterization is that the estimates of α, R, and R_a from the fits tend to be highly correlated. Therefore, when studying the dependence of R on quantities such as N or k_t, in order to stabilize the fits α is fixed to the value obtained in a fit to all events. In examining the de-

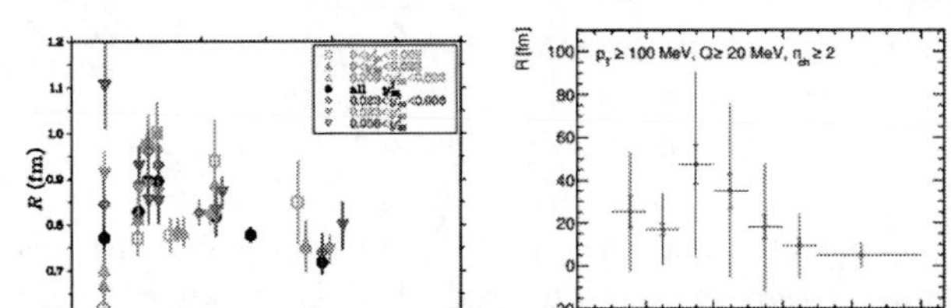

Figure 4. The dependence of R on k_t for e^+e^- (top) and pp (bottom). The e^+e^- results are shown for various selections on y_{23}^J. The τ-model parametrization is used with R_a a free parameter.

pendence of R on charged track multiplicity and on k_t, the parametrization of Eq. (3a) is used (with R_a given by Eq. (3b) for e^+e^- and a free parameter for pp).

The dependence of R on k_t with respect to the thrust axis for e^+e^- and with respect to the beam axis for pp is shown in Fig. 4 for the τ-model parametrization with R_a a free parameter is shown in Fig. 4. The behavior of two-jet e^+e^- seems similar to pp, although the error bars are large.

For both e^+e^- and pp, R is found [5, 7] to increase with multiplicity (not shown). Further, for e^+e^- R is found [7] to increase with 'jettiness', as measured by y_{23}.

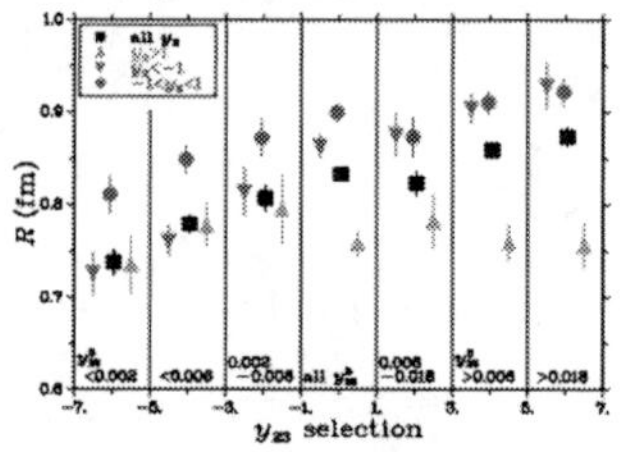

Figure 5. R obtained in fits of Eq. (3) for various rapidity and y_{23}^{D} intervals.

In e^+e^-, R is also found to depend on the rapidity of the pair, as shown in Fig. 5. Here the rapidity, y_E, is defined with respect to the thrust axis with the positive thrust axis chosen to be in the same hemisphere as the most energetic jet. Then $y_E > 1$ selects almost always tracks from the most energetic quark jet, and $y_E < -1$ selects mostly tracks from the other quark jet with the contribution of tracks from the gluon jet increasing as y_{23} increases, *i.e.*, as the events become more three-jetlike. The intermediate y_E region contains tracks from the gluon jet and low-energy tracks from both quark jets. One observes that R is roughly independent of y_{23} for $y_E > 1$. This value of R is also found for $y_E < -1$ in the case of two-jet events. These are the situations of 'pure' quark jets. As y_{23} increases R for $y_E < -1$ also increases, reflecting the increasing contribution of tracks from the gluon. The region $-1 < y_E < 1$ has a larger value of R for two-jet events, and this value increases with y_{23}. For three-jet events, where the gluon contributes to both $-1 < y_E < 1$ and $y_E < -1$, the values of R are approximately equal.

As mentioned above, the τ-model is known to break down. In the τ-model R_2 depends on Q, and not on components of Q. This was tested in Ref. [1]. In Eq. (3) R^2Q^2 is replaced by $R_L^2Q_L^2 + R_{side}^2Q_{side}^2 + R_{out}^2Q_{out}^2$ and fits for two-jet events performed in the LCMS. Different values were found for R_L, R_{out}, and R_{side}. Here results are shown in Fig. 6 for R_L and R_{side} for various selections of y_{23}^{D}. One sees that R_L is independent of y_{23}, while R_{side} increases with 'jettiness'. Thus the increase in R occurs mainly in the transverse plane. This is consistent with the increase in R being due to the increasing hardness of the gluon, which was seen in the rapidity dependence above.

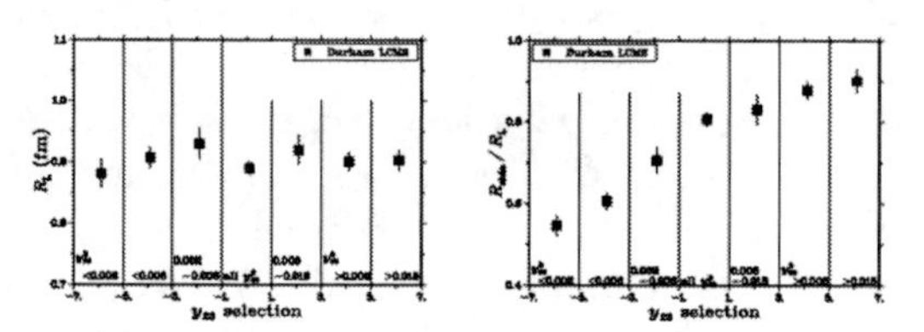

Figure 6. R_L and R_{side}/R_L obtained in fits of Eq. (3) for various rapidity and y_{23}^{D} intervals.

A large source of systematic uncertainty on the parameter values is the range of Q over which the fit is performed. A higher upper limit, Q_U, of the fit range gives more data to determine the baseline, which in turn affects the values of the parameters describing BEC and BEAC. Clearly, Q_U should be chosen well beyond the anti-correlation region. Fig. 7 and Table 1 show this effect. The values of the parameters change drastically when the upper limit, Q_U, is increased from 2 to 3 GeV. As Q_U is further increased the change is less, but still large. Clearly the parametrization does not fully describe the data. Nevertheless, it is much better than all the other parametrizations that were tried. It is also worth noting that the value of α is quite different from the value $\alpha = 0.41 \pm 0.02^{+0.04}_{-0.06}$ found for two-jet e^+e^- events [1].

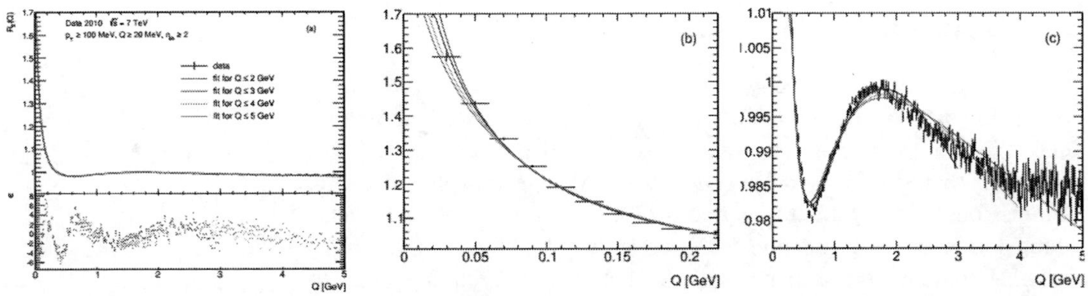

Figure 7. Fits of Eq. (3a), with R_a a free parameter, to 7 TeV minimum bias data for various choices of Q_U [5].

Table 1. Results of fits of Eq. (3a), with R_a a free parameter, for various choices of Q_U [5].

Q_U	2 GeV	3 GeV	4 GeV	5 GeV
α	0.108 ± 0.001	0.186 ± 0.005	0.235 ± 0.003	0.261 ± 0.003
R (fm)	17.8 ± 0.7	6.7 ± 0.5	4.1 ± 0.2	3.3 ± 0.1
R_a (fm)	43.4 ± 1.2	3.0 ± 0.2	1.80 ± 0.04	1.52 ± 0.02
λ	3.08 ± 0.05	1.91 ± 0.10	1.36 ± 0.05	1.15 ± 0.03

3 BEAC in more detail

In the τ-model BEAC arises through the correlation of coordinate space and momentum space. Recently, another explanation has been proposed, namely the non-zero size of the pion [8, 9]. A detailed investigation of the BEAC region seems therefore warranted, in the hope of distinguishing between these explanations.

Since fits of Eq. (3a), with R_a a free parameter, provide a reasonable description of the anti-correlation region, they are used to show how BEAC depend on 'jettiness' in e^+e^- and to compare the dependence of BEAC on multiplicity and k_t in two-jet e^+e^- and pp minimum bias events.

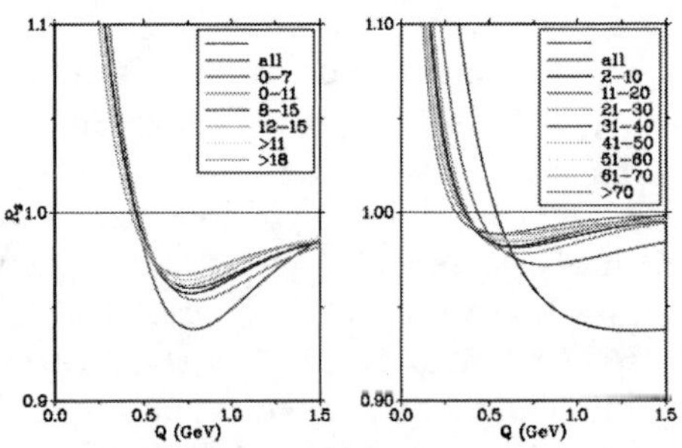

Figure 8. The result of fits of Eq. (3a) to (left) e^+e^- two-jet ($y_{23}^J < 0.023$) and (right) pp minimum bias data for various intervals of multiplicity, observed in the case of e^+e^- and corrected in the case of pp.

Fig. 8 shows the dependence of the BEAC on track multiplicity. The anti-correlation dip is deeper and at somewhat higher Q for two-jet e^+e^- than for pp. With increasing N the minimum moves to lower Q; this effect is larger in pp than in e^+e^-. The dip also becomes less deep as N increases, an effect also noticed by CMS [6].

Fig. 9 shows the dependence of the BEAC on y_{23}^J. It is seen that the anti-correlation dip becomes deeper and its minimum moves slighlty lower in Q as the 'jettiness' increases, i.e., as one moves from two- to three-jet events.

Fig. 10 shows the dependence of the BEAC on k_t. Little dependence is seen for two-jet e^+e^- data. For pp minimum bias data, the depth of the dip decreases with k_t, but the position of the minimum is approximately constant.

4 Conclusions

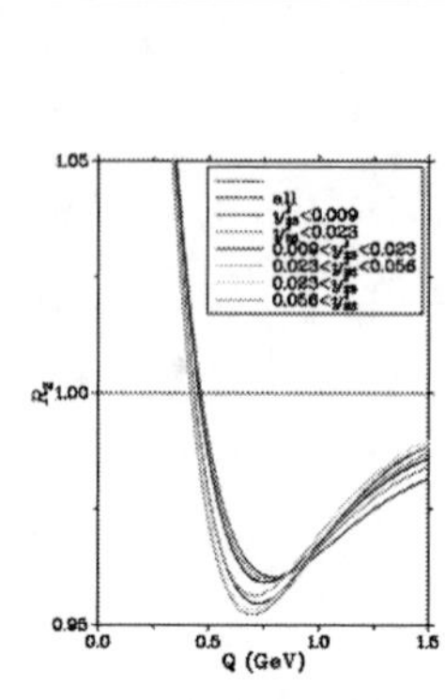

Figure 9. The result of fits of Eq. (3) to e^+e^- for various y_{23}^J selections.

The τ-model, which is closely related to the string picture, provides a reasonable explanation of both the BEC peak and the BEAC dip in both e^+e^- and minimum bias pp interactions. Of all the parametrizations tried, only that of the τ-model survives as a candidate to explain the data. Another possible explanation is that the anti-correlation arises because of the non-zero size of the pion. To discover which of these explanations (or what combination of them) is the best explanation requires detailed investigation of both BEC and BEAC.

R and R_a are found to depend on track multiplicity and transverse momentum in both pp and e^+e^- and on jets and rapidity in e^+e^-. What more does it depend on? If the BEAC are due to the correlation between coordinate space and momentum space, as in the τ-model, it is reasonable that the dependence on jet structure, as seen in e^+e^-, occurs. Then one would also expect similar dependences in pp. This possibility makes the study of BEC (and BEAC) in jet events interesting. One might also expect a rapidity dependence in minimum bias events, as the number of strings involved or the presence of color reconnection may vary with rapidity. On the other hand, it is not clear (at least to me) why the effect of the pion size should depend on either rapidity or the presence of a jet.

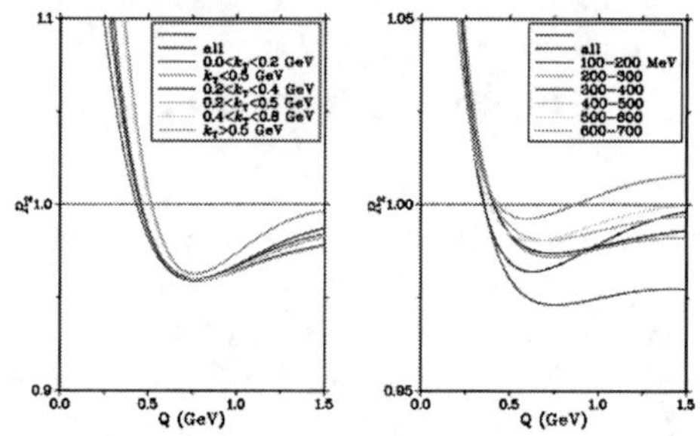

Figure 10. The result of fits of Eq. (3a) to (left) e^+e^- two-jet ($y_{23}^J < 0.023$) and (right) pp minimum bias data for various intervals of k_t.

In studying these correlations, the reference sample used plays a crucial role. The use of an unlike-sign reference sample is not possible in studying BEAC. Further, the upper limit of the fit range must be well above the BEAC region. It is very difficult, if not impossible, to compare quantitatively the results using different reference samples. It would therefore be useful if the LHC experiments could agree on a standard method of constructing the reference sample, and give results using this standard method as well as any other method deemed superior. This is especially important to investigate, *e.g.*, dependence on rapidity, where the acceptances of the experiments are very different, *e.g.*, LHCb and ATLAS, CMS, ALICE.

References

[1] L3 Collab., P. Achard, *et al.*, Eur. Phys. J. **C71**, 1648 (2011)
[2] T. Csörgő, J. Zimányi, Nucl. Phys. **A517**, 588 (1990)
[3] T. Csörgő, W. Kittel, W. Metzger, T. Novák, Phys. Lett. **B663**, 214 (2008)
[4] ATLAS Collab., G. Aad, *et al.*, Eur. Phys. J. **C75**, 466 (2015), `1502.07947`
[5] R. Astaloš, Ph.D. thesis, Radboud University (2015), `http://hdl.handle.net/2066/143448`
[6] CMS Collab., V. Khachatryan, *et al.*, JHEP **5**, 29 (2011), `1101.3518`
[7] Wesley J. Metzger, EPJ Web of Conferences **90**, 01006 (2015), `arXiv:1411.3898`
[8] A. Białas, K. Zalewski, Phys. Lett. **B727**, 182 (2013), `1309.6169`
[9] A. Białas, W. Florkowski, K. Zalewski, Phys. Lett. **B748**, 9 (2015), `1503.02807`

EPJ Web of Conferences **120**, 01004 (2016) DOI: 10.1051/epjconf/201612001004
ISMD 2015

Composite nature of hadrons and Bose-Einstein correlations

A.Bialas[1,a]

[1] *M.Smoluchowski Institute of Physics, Jagellonian University, , ul.Lojasiewicza 11, 30-348 Krakow, Poland*

Abstract. I am reporting results of two papers, written together with W.Florkowski and K.Zalewski [1, 2], discussing the consequences of the observation [3] that, due to their composite nature and thus finite size, hadrons observed in the HBT measurements must be correlated in space-time. Using the blast-wave model [4] adjusted [1] to ALICE data on the measured HBT radii in *pp* collisions at 7 TeV [5], the full Bose-Einstein correlation functions in three direction (*out, side, long*) are evaluated. The results are presented together with some additional comments.

1 Introduction

Interpretation of the measurements of Bose-Einstein correlation function is usually discussed using the formula

$$C(P_{12},Q) \equiv \frac{N(p_1,p_2)}{N(p_1)N(p_2)} = 1 + \frac{\left|\int dx_1 dx_2 e^{iQx} w(P_{12},x)\right|^2}{\int dx_1 w(p_1,x_1)\int dx_2 w(p_2,x_2)} \equiv C_u(P_{12},Q) \geq 1 \qquad (1)$$

where p_1, x_1 and p_2, x_2 are momenta and positions of particles, $w(p,x)$ is their corresponding distribution (Wigner function). $N(p_1,p_2)$ is the measured number of particle pairs, $N(p)$ is the measured single particle distribution, and

$$P_{12} = (p_1+p_2)/2; \quad Q = p_1 - p_2, \qquad (2)$$

Several experiments have shown, however, that this formula does not describe correctly the existing data [6–8] because the inequality on the R.H.S. of (1) is not satisfied.

The explanation of this problem can be found by observing that (1) is only an approximation to the general formula [9, 10]

$$C(P_{12},Q) = 1 + \frac{\int dx_1 dx_2 e^{iQ(x_1-x_2)} W(P_{12},P_{12};x_1,x_2)}{\int dx_1 w(p_1,x_1)\int dx_2 w(p_2,x_2)} \qquad (3)$$

where $W(P_{12},P_{12};x_1,x_2)$ is the distribution in momentum and in space-time (Wigner function) of the two particles. Eq. (1) is obtained from (3) under the assumption that the produced particles are uncorrelated, i.e.

$$W(p_1,p_2;x_1,x_2) = w(p_1,x_1)w(p_2,x_2) \qquad (4)$$

[a] e-mail: bialas@th.if.uj.edu.pl

Figure 1. Schematic illustration of the mixing of the constituents of two hadrons

Thus violation of the inequality (1) implies that the produced particles must be correlated.

It was suggested in [3] that such correlations in space-time must be necessarily present because observed hadrons are not point-like particles but rather systems composed of a certain number of more elementary constituents. Consequently, when two hadrons happen to occur at a distance smaller than their actual size, their constituents mix, their wave functions are distorted and the quantum interference cannot take place. In other words, the HBT experiment cannot see hadrons which happen to be too close to each other. The idea is illustrated schematically in Fig 1.

To take this effect into account, we propose to replace (4) by

$$W(p_1, p_2; x_1, x_2) = w(p_1, x_1)w(p_2, x_2)[1 - D(x_1 - x_2)] \tag{5}$$

where the "cut-off" function $D(x_1 - x_2)$ equals 1 at $x_1 = x_2$ and vanishes at distances larger than, say, 1 fm.

Introducing (5) into (3) one obtains for the correlation function

$$C(P_{12}, Q) = C_u(P_{12}, Q) - C_D(P_{12}, Q) \tag{6}$$

where $C_u(P_{12}, Q)$ is given by (1) and

$$C_D(P_{12}, Q) = \frac{\int dx_1 dx_2 e^{iQ(x_1-x_2)} w(P_{12}, x_1)w(P_{12}, x_2)D(x_1 - x_2)}{\int dx_1 w(p_1, x_1) \int dx_2 w(p_2, x_2)} \tag{7}$$

One sees from (6) that the inequality (1) is no longer compulsory.

2 Qualitative picture

It is essential to realise that the two terms on the R.H.S. of (6) differ drastically in their Q-dependence. Indeed, in the first one the integral takes contributions from the full space-time volume where hadrons are produced and, consequently, the dependence on Q is rather sharp, reflecting the size of the system. On the other hand, in the second term the integration volume is strongly restricted by the cut-off function $D(x_1 - x_2)$ and thus the dependence on Q is much weaker. Therefore, although the second term gives only a small correction at $Q \approx 0$, it may dominate at large Q, where the data show that the inequality (1) is violated.

This observation allows to formulate the approximate procedure to estimate the magnitude of the effects we are discussing:

STEP ONE: Choose a model describing the considered process and fix its parameters by adjusting to data at small Q, *ignoring the correction from the space-time correlation.*

STEP TWO: Introduce the cut-off function and evaluate the full correlation function in the wide region of Q.

STEP THREE: verify that the description at small Q is not destroyed.

Then the results of STEP TWO represent a prediction of the model which may be compared with data at large Q in order to find a proper form of the cut-off function $D(x_1 - x_2)$

3 Blast-wave model

To realise this program, we employed the blast-wave model [4, 9, 11]. (its details are described in [1]). The parameters were adjusted to account for the HBT radii measured by the ALICE collaboration in pp collisions at 7 TeV [5] (STEP ONE). Then the HBT correlation functions were evaluated in the region up to $Q = 1$ GeV (STEP TWO), using the cut-off function in the Gaussian form

$$D(x_1 - x_2) = e^{-l^2/\Delta^2} \tag{8}$$

where l is the space-time distance (along the hyperbola $\tau = \sqrt{t^2 - z^2} = \text{const}$), and $\Delta = 1$ fm. It was checked that introducing the cut-off function does not change the HBT radii (STEP THREE).

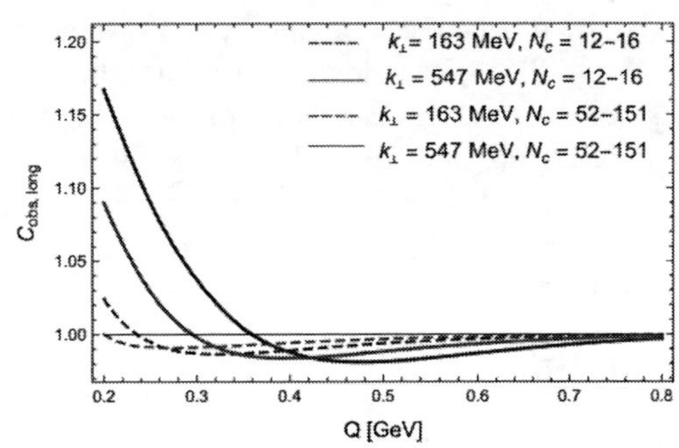

Figure 2. (Color online) Correlation function C_{obs} for the *long* direction in the interval 0.2 GeV $\leq Q \leq$ 0.8 GeV (normalised to 1 at Q= 1GeV). The dashed lines describe the results for $k_\perp$ = 163 MeV and the two multiplicity classes: N_c = 12–16 and N_c = 52–151. The solid lines describe the results for $k_\perp$ = 547 MeV and the same two multiplicity classes.

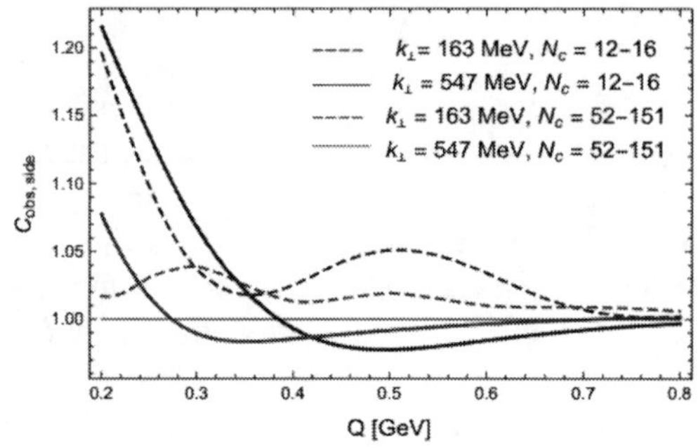

Figure 3. The same as Fig. 2 but for the side direction.

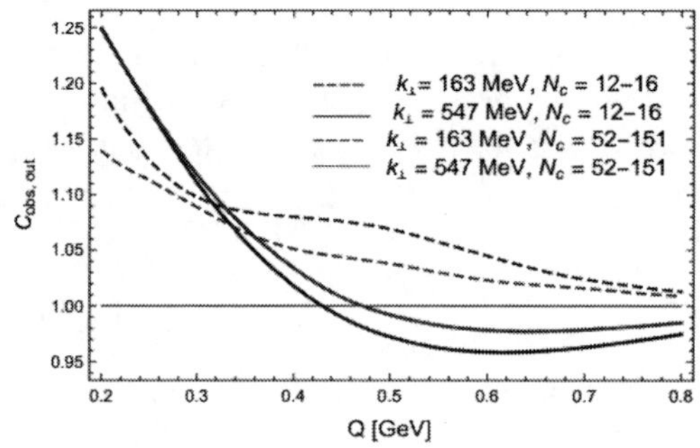

Figure 4. The same as Fig. 2 but for the out direction.

It turns out that the results (published in [2]) depend substantially on (i) the direction in which the measurement is performed, (ii) the transverse momentum of the pair and (iii) the multiplicity of the collision. They are shown in Figs. 2- 4 (for a detailed description, see [2]). We also found that the magnitude of the corrections decreases with decreasing ratio of the cut-off parameter Δ to the total size of the system. Thus one expects that the effects shall be less visible in collisions involving nuclei.

Unfortunately, there is no substantial sensitivity to the shape of the cut-off function: Calculations with the sharp cut-off at the distance $\Delta \approx 0.75$ fm gave very similar results.

4 Summary and comments

To summarize, we estimated the effects of the space-time correlations, implied by the composite nature of the produced hadrons, on the HBT measurements. Our calculations indicate that the precise measurements of the HBT correlation functions in the region of Q from 200 MeV up to 1 GeV may reveal a rich structure, depending nontrivially on the direction of the measurement and on other characteristics of the produced hadronic system. Detailed comparison of data with the model calculations may, eventually, pin down the value of the cut-off parameter (and thus also the density) at which the produced hadrons melt into the quark-gluon plasma. Although rather difficult, it seems that such measurements are worth to undertake.

Acknowledgements

I would like to thank my collaborators for help in preparing this report. Thanks are also due to Adam Kisiel for introduction into data from the ALICE experiment. This investigation was supported by the grant from the National Science Centre (Narodowe Centrum Nauki) DEC-2013/09/B/ST2/00497.

References

[1] A. Bialas, W. Florkowski and K. Zalewski, J. Phys. G: Nucl.Part.Phys. **42** (2015) 045001.
[2] A. Bialas, W. Florkowski and K. Zalewski, : Phys. Letters **B748**, 9 (2015)
[3] A. Bialas and K. Zalewski, Phys. Lett. **B727**, 182 (2013)
[4] F. Retiere and M. A. Lisa, Phys.Rev. **C70**, 044907 (2004).
[5] K. Aamodt *et al.* [ALICE Collaboration], Phys. Rev. D **84**, 112004 (2011).
[6] M. Acchiari et al. [L3 coll.] Phys. Lett. **B458** (1999) 517; P. Achard et al. [L3 coll.], Eur. Phys.J. **C71** (2011) 1648.
[7] G. Abbiendi et al. [OPAL coll.], Eur. Phys. J. **C16** (2000) 423; P. Abreu et al. [DELPHI coll.] Phys. Lett. **B471** (2000) 460; A. Heister et al. [ALEPH coll.] Eur. Phys. J. **C36** (2004) 460.
[8] V. Khachatryan et al. [CMS coll.], JEHP **1105** (2011) 029; L. Perroni, PoS (WPCF2011) 015.
[9] For a review, see W. Florkowski, "Phenomenology of Ultra-Relativistic Heavy-Ion Collisions", (Singapore, World Scientific, 2010); M. A. Lisa, S. Pratt, R. Soltz and U. Wiedemann, Ann. Rev. Nucl. Part. Sci. **55**, 357 (2005).
[10] For an elementary derivation, see, e.g. A. Bialas and A. Krzywicki, Phys. Lett. B **354**, 134 (1995).
[11] A. Bialas, W. Florkowski and K. Zalewski, Acta Phys. Pol. **B45** (2014) 1883.

EPJ Web of Conferences **120**, 01005 (2016)
ISMD 2015 DOI: 10.1051/epjconf/201612001005

Quantum properties of QCD string fragmentation

Šárka Todorova-Nová[1,a]

[1] *Institute of Particle and Nuclear Physics of Charles University, V Holešovičkách 2, 180 00 Praha 8, Czech Republic*

Abstract. A simple quantization concept for a 3-dim QCD string is used to derive properties of QCD flux tube from the mass spectrum of light mesons and to predict observable quantum effects in correlations between adjacent hadrons. The quantized fragmentation model is presented and compared with experimental observations.

1 Introduction

The Lund string fragmentation[1], which is using a 1-dimensional string to model the QCD confinement, imposes a space-like distance between the string breakup vertices forming a hadron. The model has to rely on the concept of quantum tunneling to generate intrinsic transverse momenta of hadrons. The local charge and momentum conservation holds in the break-up vertex but there are no correlations, in the string transverse plane, between non-adjacent hadrons.

The situation changes substantially if the 1-dimensional string is replaced by a 3-dimensional string and the quantum tunneling turned into gluon splitting into quark-antiquark pair with a negligible momentum in the rest frame of the string. Fragmentation of such a string generates intrinsic transverse momenta which depends on the folding of the string and implies azimuthal correlations between hadrons. On the basis of angular properties of gluon emission under helicity conservation, it has been shown that the shape of the QCD string should be helix-like [2].

The helix-like shaped QCD string (or any 3-dimensional string) allows to develop fragmentation model where cross-talk between breakup vertices is possible (their distance is time-like). When the cross-talk (i.e. causal relation) between breakup vertices is imposed, the transverse shape of the string generates both transverse momentum and mass of the hadron. Quantization of the model then opens the way to build-up of hadron mass spectrum. It turns out that the causal constraint applied to the helical QCD field reveals a rather simple quantization pattern for particles with mass below 1 GeV. In particular, the pseudoscalar mesons (π, η, η') can be regarded as string pieces fragmenting into (n=1,3,5) ground state hadrons (pions), with transverse mass (E_T) and momentum (p_T) of mesons defined by helical string properties [3]

$$E_T = \sqrt{m_n^2 + p_T^2} = n\,\kappa\,R\,\Delta\Phi, \qquad (1)$$

$$p_T = 2\,\kappa\,R\,|\sin(n\Delta\Phi/2)\,|, \qquad (2)$$

[a] e-mail: sarka.todorova@cern.ch

where R stands for the radius of the helix, $\kappa \sim 1$GeV/fm is the string tension, $\Delta\Phi$ is the quantized helix phase difference describing the minimal piece of string which can form a hadron, and m_n is the (quantized) meson mass spectrum.

The mass spectrum of pseudoscalar mesons is used to extract the parameters of the helical QCD field and its quantum properties; the fit of the spectrum indicates a rather narrow radius of the helical string (κR= 68 ±2 MeV) and the quantized phase difference $\Delta\Phi$ =2.82 ± 0.06 rad which translates into a quantized ground state transverse energy ($E_T|_{n=1} \sim 0.193$ GeV). It is the numerical value of $\Delta\Phi$ which suggests there should be a significant charge-combination asymmetry in the production of chains of ground state pions: while adjacent (unlike-sign) pairs of pions have to go apart in the transverse plane, the like-sign pion pairs with rank 2[1] should have a relatively small opening angle of $2.(\pi - 2.8) \sim 0.7$ rad. Charge–combination asymmetry in the production of pairs of charged pions thus may be understood as a consequence of local charge conservation in (coherent) QCD string fragmentation into a quantized chain of ground state pions.

It is an interesting observation that the κR obtained from the fit of the mass spectrum of pseudoscalar mesons agrees with the estimate of the size of the QCD flux tube derived from the fit of mass spectrum of glueball states using topological constants of knotted flux tubes [4].

Neglecting the longitudinal momentum differences between adjacent hadrons (i.e. assuming local homogenity of the fragmenting QCD field), the momentum difference between ground state pions can be written as function of their rank r:

$$Q(\mathrm{r}) = \sqrt{-(p_i - p_{i+r})^2} \cong 2\, p_T^{thr} |\sin(\mathrm{r}\, \Delta\Phi/2)|, \quad (3)$$

where p_T^{thr} (~ 134 MeV) stands for the intrinsic p_T of ground state pions.

The numerical values of the predicted momentum difference separating pairs of ground state hadrons with rank up to 5 are given in Table 1.

The chain of n adjacent ground state pions has the mass

$$M_{nh} = \sqrt{n^2 m_\pi^2 + \sum_{i \neq j} Q_{ij}^2}, \quad (4)$$

where m_π is the pion mass, and Q_{ij} the momentum difference between pairs of hadrons forming the chain.

Pair rank difference	1	2	3	4	5
Q expected [MeV]	266 ± 8	91 ± 3	236 ± 7	171 ± 5	178 ± 5

Table 1. The expected momentum difference between ground state direct hadrons, in the quantized helix string model. The uncertainty (3%) is derived from the precision of the fit of the mass spectrum of light pseudoscalar mesons[3].

2 Coherent particle production

2-particle correlations are usually studied as collective, incoherent effects. Experimentally, one tries to compare the measured inclusive 2-particle spectra with an 'uncorrelated' reference inclusive spectra,

[1]The rank of direct particles is equivalent to their ordering along the string : rank 0 is attributed to the decay products of a direct hadron, rank 1 designs adjacent hadrons, rank 2 hadrons separated by a common adjacent hadron et cetera; rank -1 is used to classify pairs of hadrons originating from different string/colour singlets. The decay products inherit the rank from their direct mother.

where 'uncorrelated' means containing all dynamical correlations except those being measured. A well known example of this type of measurements is the study of Bose-Einstein effect, which is supposed to describe the excess of like-sign pairs of charged hadrons observed in all hadronic data samples. Since the Bose-Einstein symmetrization does not concern the pairs of hadrons with opposite charge, the inclusive spectrum of unlike-sign pairs can be used as a reference sample. The notion of the collective effect influences strongly the choice of the correlation function as ratio of the observed and the reference distribution. Assuming particle correlations arise in the hadronization process and

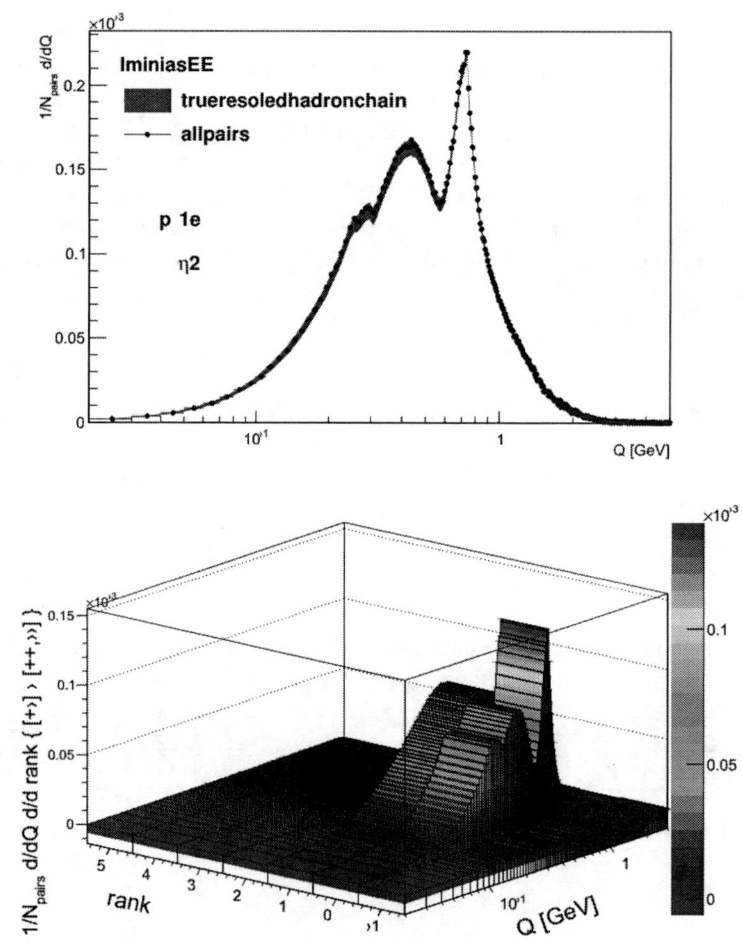

Figure 1. Upper plot: Comparison of the Q spectrum obtained by the means of subtraction of like-sign pair distribution from that of unlike-sign pairs, with the true Q spectrum for adjacent charged hadrons, obtained from MC truth (the uncertainty marked by red band reflects the uncertainty in the resolving the ordering within a decay of a direct hadron). Bottom plot: Detailed view of the rank dependence of the subtraction between unlike-sign and like-sign pairs in MC; pairs of hadrons originating from different strings (color-disconnected chains) are assigned rank=-1. Rank 0 designs pairs of hadrons from the decay of a direct common resonance. PYTHIA8[5] minimum bias sample for pp collisions at $\sqrt{s}$=7 TeV c.m.s.energy.

 DOI: 10.1051/epjconf/201612001005

reflect dynamical properties of the QCD field, the charge-combination asymmetry in the production of particles should be limited to adjacent hadrons (bound by local charge conservation in the breakup of QCD field), which can never form a like-sign pair. However, due to the (presumably random) presence of neutral hadrons in the QCD string fragmentation, it is unlikely that the charge-combination asymmetry holds for pairs with rank above 2. If this is the case, the number of correlated hadron pairs should grow linearly with the number of charged hadrons (i.e. with the string length). Since the combinatorial background rises quadratically with the charged multiplicity, it represents a huge challenge, unless it is recognized that it can be completely eliminated by subtraction of like-sign and unlike-sign inclusive spectra. Thus it seems reasonable, for the study of adjacent hadron spectra, to replace the ratio of Q distributions by their difference. Indeed, the Monte-Carlo studies confirm the extraordinary power of the subtraction technique as illustrated by Fig. 1 - the level of precision with which the true shape of momentum difference for adjacent charged hadrons can be reproduced is within limits of the uncertainty in the resolving the order of the daughter hadrons in a multibody decay of a heavy resonance.

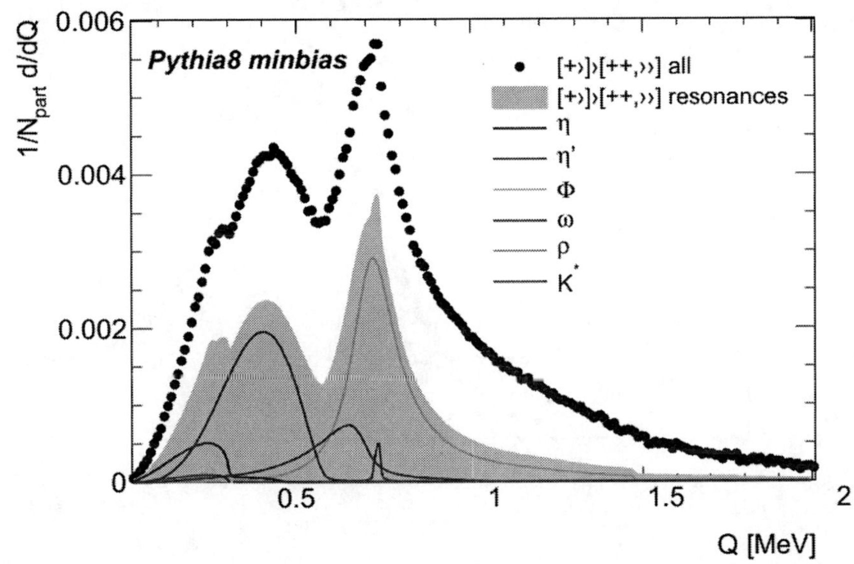

Figure 2. The Q spectrum generated by Pythia8 and the decomposition of its resonant part into leading contributions. The peak with lowest mass is formed by decay products of pseudoscalar mesons (η, η') and the 2-body decay of the Φ resonance.

The shape of the subtracted distribution is expected to be driven by the resonance decays (Fig. 2). The dominant contribution comes from [η, η',Φ], [ω] and [ρ, K^{*0}], which combine in 3 broad peaks. Please note that definition of primary particles[2] does not include decay products of K mesons which therefore do not appear on the plot (K mesons are treated as stable particles).

[2]Primary particles are defined as all particles with lifetime longer than 0.3x10^{-10}s originating from the primary interaction or from subsequent decay of particles with shorter lifetime.

3 Experimental data

The predictions of the quantized fragmentation model can be verified using the Q distributions published by ATLAS[6]. The subtracted quantity

$$\Delta Q(= N^{+-}(Q) - N^{++,--}(Q)) \tag{5}$$

is fitted - in the region up to Q=0.4 GeV - by 3 gaussians describing the enhanced production of pairs of like-sign hadrons (seen as depletion at low Q), and the first 2 (positive) peaks for adjacent unlike-sign pairs of hadrons. The shape of the enhanced production of like-sign pairs is in good agreement with the model prediction for the size of momentum difference between rank 2 hadrons in the chain of ground state hadrons (pions), with the maximum (turned minimum) at Q=0.09 GeV. The position of the lowest peak for unlike-sign pairs agrees with the model prediction for the minimal momentum difference between a pair of adjacent ground state hadrons (0.26 GeV).

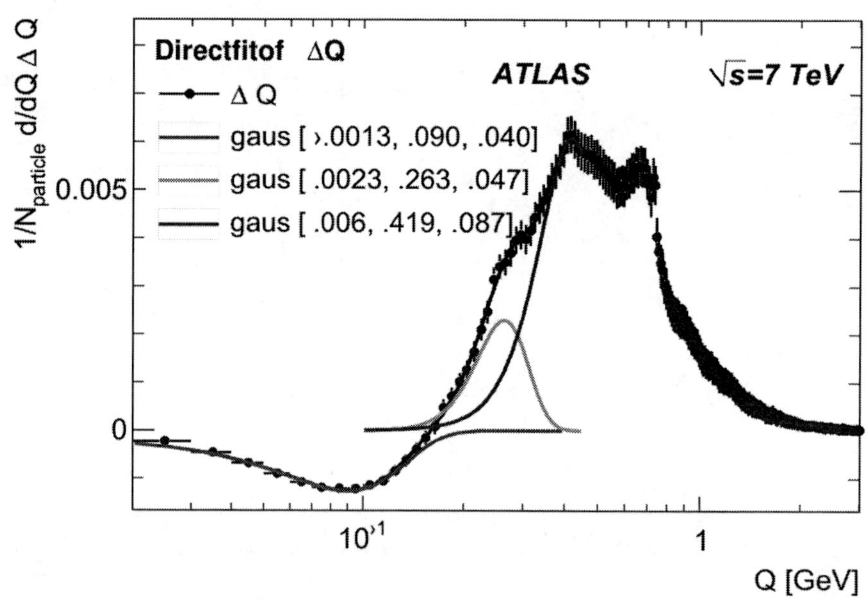

Figure 3. Direct fit of the triple gaussian shape at low Q in the inclusive ΔQ distribution. The onset of adjacent pair production coincides with the position of (the first) Q^{+-} peak at Q=0.26 GeV. The ATLAS data are taken from Ref [6].

4 Conclusion

The combination of causal and quantum constraints allows to develop model of QCD string fragmentation where the shape of QCD field translates into observable properties (mass, transverse momentum) of hadrons. Parameters of the QCD string are fitted using mass spectrum of pseudoscalar mesons. Due to the reduction of the number of free parameters, the model acquires considerable predictive power. The measured shape of 2-particle correlations (commonly attributed to Bose-Einstein effect) agrees with model predictions, which means the effect (i.e.enhanced production of pairs of like-signed charged hadrons) can be attributed to the coherent particle production.

References

[1] B. Andersson, et al. *Parton Fragmentation and String Dynamics*, Phys. Rept. **97** (1983) 31.
[2] B. Andersson, G. Gustafson, J. Hakkinen, M. Ringner and P. Sutton, *Is there screwiness at the end of the QCD cascades?*, JHEP **9809** (1998) 014.
[3] Š. Todorova, *Quantization of the QCD string with a helical structure*, Phys.Rev.D**89**, 015002 (2014).
[4] R.V.Buniy, et al.,*The tight knot spectrum in QCD*, Phys.Rev.D**89**, 054513 (2014).
[5] T. Sjostrand, S. Mrenna and P. Z. Skands, Comput. Phys. Comm. 178 (2008)
[6] ATLAS Collaboration, Eur.Phys.J.C75 (2015) 466.

EPJ Web of Conferences **120**, 01006 (2016) DOI: 10.1051/epjconf/201612001006
ISMD 2015

Bayesian model comparison for one-dimensional azimuthal correlations in 200GeV AuAu collisions

Hans C. Eggers[1,2] Michiel B. de Kock[1] and Thomas A. Trainor[3]

[1] *Department of Physics, Stellenbosch University, 7600 Stellenbosch, South Africa*
[2] *National Institute for Theoretical Physics (NITheP), 7600 Stellenbosch, South Africa*
[3] *CENPA 354290, University of Washington, Seattle, Washington 98195, United States*

Abstract. In the context of data modeling and comparisons between different fit models, Bayesian analysis calls that model best which has the largest evidence, the prior-weighted integral over model parameters of the likelihood function. Evidence calculations automatically take into account both the usual chi-squared measure and an Occam factor which quantifies the price for adding extra parameters. Applying Bayesian analysis to projections onto azimuth of 2D angular correlations from 200 GeV AuAu collisions, we consider typical model choices including Fourier series and a Gaussian plus combinations of individual cosine components. We find that models including a Gaussian component are consistently preferred over pure Fourier-series parametrizations, sometimes strongly so. For 0–5% central collisions the Gaussian-plus-dipole model performs better than Fourier Series models or any other combination of Gaussian-plus-multipoles.

1 Introduction

A significant problem has emerged in the interpretation of high-energy collision data from the relativistic heavy ion collider (RHIC) and the large hadron collider (LHC). Two classes of data models with divergent physics implications are invoked to support competing narratives. One narrative interprets results with concepts from high-energy physics in which the essential phenomenon is dijet production; the other uses quark-gluon plasma and collective flow concepts in which the essential phenomenon is a flowing, dense medium while dijets play only a subordinate role.

To place evaluation and comparison of competing models on a scientific footing we require a formally consistent and statistically sound framework. We suggest that Bayesian inference provides the formal consistency and the necessary mathematical framework for representing prior knowledge, evaluating candidate models for different parameter values and thereby establishing value judgments on models as a whole. In this contribution, we show by example how the Bayesian framework successfully evaluates and compares models for that type of one-dimensional correlation data which has played a large role in claimed support for these narratives. The exercise also serves to pinpoint the weaknesses of the conventional minimum-χ^2 and maximum-likelihood approaches.

2 Data and parametrization

The data analyzed below were published by the STAR Collaboration [1] in the form of two-dimensional binned histograms derived from 1.2 million 200GeV AuAu collision events in eleven centrality classes. The relevant autocorrelations

$$A(\phi_\Delta, \eta_\Delta) = \Delta\rho / \sqrt{\rho_{\text{ref}}} = \rho_0 \left[(\rho/\rho_{\text{ref}}) - 1 \right] \tag{1}$$

are measured on azimuthal and pseudorapidity difference variables $\phi_\Delta = \phi_1 - \phi_2$ and $\eta_\Delta = \eta_1 - \eta_2$, pair densities $\rho(\phi_\Delta, \eta_\Delta)$ and $\rho_{\text{ref}}(\phi_\Delta, \eta_\Delta)$ and the one-particle differential cross section, $\rho_0 = d^2 n_{ch}/d\eta d\phi$.

Here, we consider one-dimensional projections of the 2D autocorrelations onto azimuth difference ϕ_Δ with 24 bins of size $\pi/12$. As the data are symmetrized on both η_Δ and ϕ_Δ, there are only 13 unique points. These are reduced to $N = 11$ data points by excluding from fits the $\phi_\Delta = 0$ bin containing the Bose-Einstein peak and constraining the 1D projections to sum to zero.

Since the 11-parameter 2D parametrization (or data model) described in Ref. [1] describes the corresponding 2D data rather well, we adapted it to the present 1D projection. Taking into account the zero-sum constraint, this results in what we call the Basic Model; abbreviating $\phi \equiv \phi_\Delta$,

$$f(\phi, \boldsymbol{w} \,|\, \mathcal{H}_{\text{BM}}) = A_{\text{1D}} \left\{ \exp\left[-\frac{1}{2} \left(\frac{\phi}{\sigma_\phi} \right)^2 \right] - \frac{\sigma_\phi}{\sqrt{2\pi}} \right\} - A'_{\text{D}}\, 2\cos(\phi) \tag{2}$$

with three parameters $\boldsymbol{w} = (A_{1D}, \sigma_\phi, A'_{\text{D}})$ reflecting the basic same-side (SS) peak and away-side (AS) dipole structure. The Basic Model $\mathcal{H}_{\text{BM}}$ is supplemented in Augmented Models $\mathcal{H}_{\text{AM}}$ by a quadrupole term $-A'_{\text{Q}}\, 2\cos(2\phi)$, a sextupole $-A'_{\text{S}}\, 2\cos(3\phi)$, and an octupole $-A'_{\text{O}}\, 2\cos(4\phi)$ in various combinations. The factor 2 entering the cosine terms is designed to reflect a corresponding Fourier Series (FS) parametrization popular within the collective-flow paradigm [2],

$$f(\phi, \boldsymbol{w} \,|\, \mathcal{H}_{\text{FS}}) = 2 \sum_{j=1}^{K} w_j \cos(j\phi). \tag{3}$$

In our analysis, the number of free parameters K varies from 1 to 11 in the FS case, while in the Basic and Augmented Models K varies from 3 to 6.

3 Model comparison

The two goals of fitting a parametrization to data are model comparison and, given a particular model $\mathcal{H}$, finding the best-fit values of its parameters. In the present case, the data $\boldsymbol{D} = (y_1, \ldots, y_N)$ are given by binwise autocorrelations $A(\phi)$ at mid-bin sample points ϕ_n while the competing models include $\mathcal{H}_{\text{BM}}$, $\mathcal{H}_{\text{AM}}$ and $\mathcal{H}_{\text{FS}}$ with various values of K.

In the usual **frequentist approach**, model comparison is effected by searching the space of parameters $\boldsymbol{w}$ to find the global minimum χ_0^2 of

$$\chi^2 = \sum_{n=1}^{N} \left(\frac{y_n - f(\phi_n, \boldsymbol{w} \,|\, \mathcal{H})}{\sigma_n} \right)^2 \tag{4}$$

with σ_n the experimental standard errors. Model $\mathcal{H}_1$ with K_1 parameters is deemed to be better than Model $\mathcal{H}_2$ with K_2 parameters if $\min(\chi_1^2)/(N-K_1) < \min(\chi_2^2)/(N-K_2)$. Underlying the χ^2 criterion is the assumption of independent Gaussian fluctuations of deviations between data and parametrization,

$$y_n - f(\phi_n, \boldsymbol{w} \,|\, \mathcal{H}) = \varepsilon_n \qquad \text{with} \qquad p(\varepsilon_n) = \text{Gaussian with standard deviation } \sigma_n \tag{5}$$

which results in a likelihood

$$p(\boldsymbol{D}\,|\,\boldsymbol{w},\mathcal{H}) = \frac{e^{-\chi^2/2}}{(2\pi)^{N/2}\prod_n \sigma_n}. \tag{6}$$

When parametrizations $f(\phi,\boldsymbol{w}\,|\,\mathcal{H})$ are linear in $\boldsymbol{w}$, the Hessian

$$\mathbb{H}_{k\ell} = \frac{1}{2}\frac{\partial^2\chi^2}{\partial w_k\,\partial w_\ell} \tag{7}$$

is $\boldsymbol{w}$-independent and the likelihood takes on a Gaussian form in parameter space around the mode $\tilde{\boldsymbol{w}}$ in terms of $\mathbb{H}$ or equivalently the covariance matrix $\mathbb{C} = \mathbb{H}^{-1}$.

Bayesian model comparison utilizes the same likelihood (6) and hence also χ^2 and the covariance matrix. Crucially, however, Bayesian model comparison is based not on the maximum likelihood or χ_0^2 but on the so-called *evidence* for data $\boldsymbol{D}$ within model $\mathcal{H}$,

$$p(\boldsymbol{D}\,|\,\mathcal{H}) = \int d\boldsymbol{w}\; p(\boldsymbol{D}\,|\,\boldsymbol{w},\mathcal{H})\;p(\boldsymbol{w}\,|\,\mathcal{H}), \tag{8}$$

which averages the likelihood over all values of $\boldsymbol{w}$, weighted by the *prior* $p(\boldsymbol{w}\,|\,\mathcal{H})$ for the parameters.

The reason why evidence is a good measure for model comparison lies within Bayes' Theorem itself. Starting with the joint data-parameter probability $p(\boldsymbol{D},\boldsymbol{w}\,|\,\mathcal{H}) = p(\boldsymbol{D}\,|\,\boldsymbol{w},\mathcal{H})\,p(\boldsymbol{w}\,|\,\mathcal{H}) = p(\boldsymbol{w}\,|\,\boldsymbol{D},\mathcal{H})\,p(\boldsymbol{D}\,|\,\mathcal{H})$, Bayes' Theorem states that, once data $\boldsymbol{D}$ is known, the posterior probability for parameters $\boldsymbol{w}$ can be found from the likelihood, prior and evidence by

$$p(\boldsymbol{w}\,|\,\boldsymbol{D},\mathcal{H}) = \frac{p(\boldsymbol{D}\,|\,\boldsymbol{w},\mathcal{H})\,p(\boldsymbol{w}\,|\,\mathcal{H})}{p(\boldsymbol{D}\,|\,\mathcal{H})} \qquad (\text{posterior}) = \frac{(\text{likelihood})\,(\text{prior})}{(\text{evidence})}. \tag{9}$$

Similarly on a higher level $p(\boldsymbol{D},\mathcal{H}) = p(\boldsymbol{D}\,|\,\mathcal{H})\,p(\mathcal{H}) = p(\mathcal{H}\,|\,\boldsymbol{D})\,p(\boldsymbol{D})$ implies that, given data $\boldsymbol{D}$, the probability for the entire model $\mathcal{H}$ including all possible parameter values is

$$p(\mathcal{H}\,|\,\boldsymbol{D}) = \frac{p(\boldsymbol{D}\,|\,\mathcal{H})\,p(\mathcal{H})}{p(\boldsymbol{D})} \tag{10}$$

which, used twice for competing models $\mathcal{H}_1$ and $\mathcal{H}_2$, says that model comparison given data $\boldsymbol{D}$ can be expressed as a ratio of evidences,

$$\frac{p(\mathcal{H}_1|\boldsymbol{D})}{p(\mathcal{H}_2|\boldsymbol{D})} = \frac{p(\boldsymbol{D}|\mathcal{H}_1)}{p(\boldsymbol{D}|\mathcal{H}_2)}\,\frac{p(\mathcal{H}_1)}{p(\mathcal{H}_2)} = \frac{p(\boldsymbol{D}|\mathcal{H}_1)}{p(\boldsymbol{D}|\mathcal{H}_2)} = \frac{\text{evidence for } \mathcal{H}_1}{\text{evidence for } \mathcal{H}_2} \tag{11}$$

on setting equal *a priori* probabilities for models, $p(\mathcal{H}_1) = p(\mathcal{H}_2) = 1/2$.

The choice of **parameter prior** $p(\boldsymbol{w}\,|\,\mathcal{H})$ is governed by available information or lack thereof. We use uniform priors, implying indifference for any value of $\boldsymbol{w}$ within the bounds set by cutoffs or prior widths Δ_k,

$$p(\boldsymbol{w}|\mathcal{H}) = \textstyle\prod_{k=1}^{K}\Delta_k^{-1}. \tag{12}$$

We assign the same widths across different models for corresponding quantities. In particular, we set $\Delta_k = 1/3$ for all $k = 1,\ldots,K$ cosine coefficients, based on knowledge of typical coefficient magnitude. The interval for σ_ϕ is set to $[0,\pi/2]$ in accordance with its definition as a same-side peak width. To the prior for the Gaussian amplitude parameter $A'_{\rm D}$ we assign width $\Delta = 1$ since correlation-structure amplitudes such as peak-to-peak excursions are generally $< O(1)$.

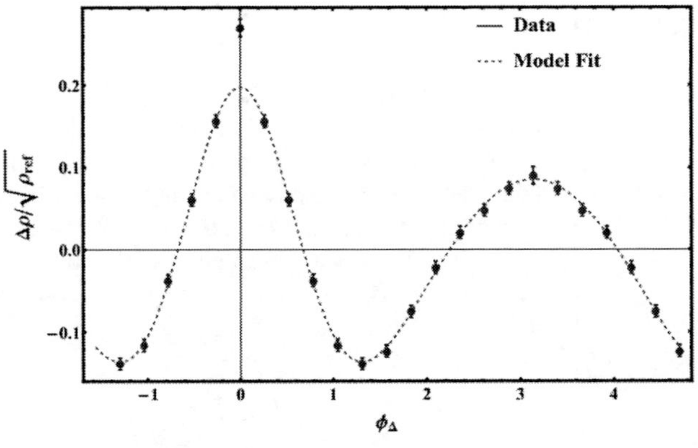

Figure 1. (Color online) Autocorrelation data projected onto one dimension for the most central events in 200 GeV AuAu collisions measured by STAR [1] along with the Basic Model best fit. Corresponding Fourier-Series fits with four or more terms appear identical on the scale of the figure. Eyeball viewing is not a reliable basis for model comparison.

For uniform priors, the evidence integrals (8) can be carried out analytically in terms of the **Laplace approximation** which neglects third- and higher-order terms in the Taylor expansion of the log likelihood and is exact for linear systems. In our case, the log likelihood is nonlinear only in the parameter σ_ϕ. The Laplace Approximation also requires that the effect of finite integration limits set by the Δ_k be negligible. Within these constraints, the evidence takes on a simple form in terms of the maximum likelihood, prior and covariance matrix,

$$p(\boldsymbol{D}|\mathcal{H}) = p(\boldsymbol{D}|\tilde{\varpi}, \mathcal{H})\; p(\tilde{\varpi}\,|\,\mathcal{H})\,(2\pi)^{K/2}\,\sqrt{\det\mathbb{C}}\,. \tag{13}$$

Since χ_0^2 is related to the maximum likelihood by $\chi_0^2 = -2\log p(\boldsymbol{D}\,|\,\tilde{\varpi}, \mathcal{H})$, we can write the corresponding log evidence in terms of χ_0^2 and an *information* $\mathcal{I}$,

$$-2\text{LE} = \chi_0^2 + 2\mathcal{I}, \tag{14}$$

$$\mathcal{I} = -\log\Big[p(\tilde{\varpi}|\mathcal{H})\,(2\pi)^{K/2}\,\sqrt{\det\mathbb{C}}\,\Big]. \tag{15}$$

The smaller −2LE for a given model, the larger its evidence. As shown in Eq. (11), −2LE for different models have no absolute meaning but must always be interpreted in terms of evidence ratios (also called *odds* in the literature) or of log differences.

4 Results

In this conference summary, we can show results only for the STAR "Bin-10" data covering the most-central 0–5% collision events; further results for Bin-0 peripheral collisions (83–94%) and Bin-8 intermediate collisions (9–18%) are presented and discussed in Ref. [3].

In Figure 1, the projected 1D data along with the best-evidence fit for the Basic Model (2) is shown. Experimental error bars were scaled up by a factor 2 to make them visible in the plot. As stated, the point at $\phi = 0$ was excluded since it contains the Bose-Einstein peak and charge-neutral electron pairs from photoconversions. The best-fit parameter values are $A_{1D} = 0.57 \pm 0.007$, $A'_D = 0.115 \pm 0.002$ and $\sigma_\phi = 0.635 \pm 0.007$ while $\chi_0^2 = 12.5$ for $11 - 3$ degrees of freedom. A fit of the Fourier Series model (3) with $K \geq 4$ is indistinguishable to the eye on the scale of the plot. That does not mean that the BM and FS models have the same −2LE.

Figure 2 illustrates for the example of Fourier-Series models why model comparison based on minimum χ^2 alone is misleading when the number of data points is small, as is the case here since $N = 11$. Plotted are various quantities as a function of the number of free parameters K in Eq. (3).

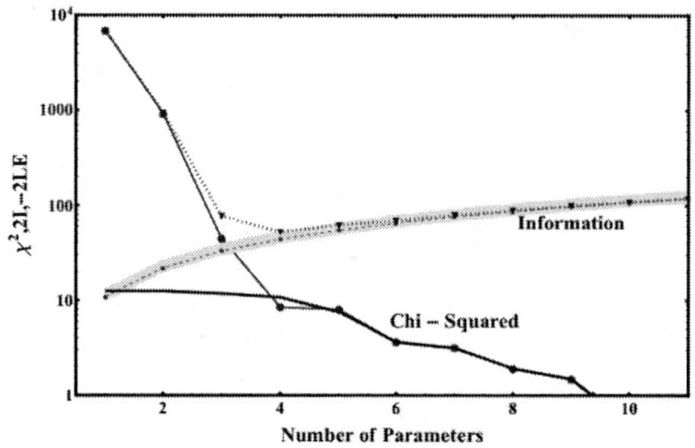

Figure 2. (Color online) Soid blue line: the usual minimum χ_0^2, which becomes identical with the χ_0^2 of the noise for $K \geq 5$ shown as the black solid line. Red dashed line and band: Information $2\mathcal{I}$ plus possible uncertainty due to the choice of prior. The negative log-evidence −2LE, the sum of χ_0^2 and $2\mathcal{I}$, is shown as the dashed black line. It has a clear minimum at $K = 4$.

The solid blue line traces the evolution of χ_0^2 as more cosine terms are added. As expected, there is a dramatic decrease in χ_0^2 up to $K = 4$ but little benefit for adding further terms beyond that as $K \geq 5$ Fourier Series models increasingly fit the noise represented by the lower solid curve showing fits to the residuals (data minus Basic Model). The naive criterion of minimum χ_0^2 is oblivous to this effect: χ_0^2 keeps decreasing and purely χ_0^2-based model comparison would hence select $K = 9$ or higher. Note that the equivalent plot for $\chi_0^2/(N - K)$ (not shown) barely changes the general decrease with K.

The contribution of information $2\mathcal{I}$ of Eq. (15) is shown as the red dashed curve. The shaded band shows the sensitivity to the $\Delta_k = 1/3$ prior width, using $\Delta_k = 1$ and $\Delta_k = 1/5$ as extremes. The resulting log-evidence −2LE of Eq. (14), shown as the black dashed curve, has a clear minimum at $K = 4$, confirming the above conclusion that FS models with $K \geq 5$ are not preferred.

In Figure 3(a), the scope of comparison between pure χ_0^2 and evidence-based model comparison is widened to include FS, Basic Model and Augmented Models with various combinations of higher-order multipoles. Consideration of χ_0^2 alone as in Fig. 3(a) would make the $K \geq 6$ FS models the winner. By contrast, the Occam penalty of $2\mathcal{I}$ forming part of −2LE in Figure 3(b) effects a dramatic re-ordering: the 3-parameter Basic Model has substantially smaller −2LE (larger evidence) than all other combinations. In addition, FS models are universally rejected in favour of Gaussian-plus-multipole models with the same number of parameters.

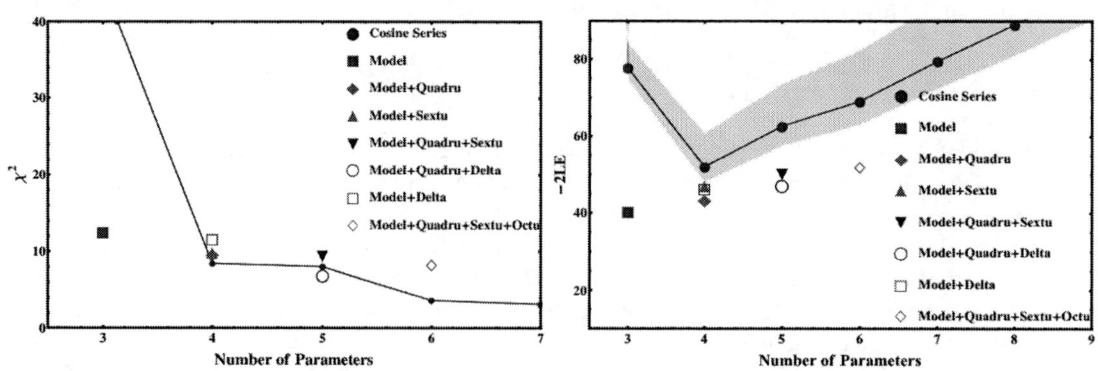

Figure 3. (Color online) (a) Minimum-χ^2 comparison of FS models (solid blue line) with Gaussian plus various combinations of multipoles. (b) The same models assessed with the Bayesian log-evidence criterion. The 3-parameter Basic Model of Eq. (2) (filled square) is the clear winner. Note that −2LE is a logarithmic scale.

5 Conclusions

Physics interpretation: As shown, the evidence for a simple same-side Gaussian plus away-side dipole is larger than for any Fourier Series-based model. While our approach and results must naturally be tested and confirmed with other data, measurement quantities and models, our results do cast serious doubt on the interpretation of azimuthal correlations as indicators of collective flow.

Lessons for data analysis: Physics conclusions must rely on the integrity of the underlying data analysis methods and assumptions. Below, we provide in tabular form a motivation why Bayesian evidence-based model comparison is superior to simple χ^2 fitting. The sometimes very different conclusions shown above suggest that any physical interpretation based on simple χ^2 fitting of one-dimensional correlation data may be debatable or outright wrong. We secondly emphasize that no model or "goodness of fit" has any meaning unless it is compared to another competing model.

Traditional use of χ^2	**Bayesian Inference**
Parameter estimation: Find maximum likelihoods max $p(\boldsymbol{D}\,|\,\boldsymbol{w}_1,\mathcal{H})$ and max $p(\boldsymbol{D}\,|\,\boldsymbol{w}_2,\mathcal{H}_2)$ which determine **single best fit parameters** $\tilde{\boldsymbol{w}}_1,\tilde{\boldsymbol{w}}_2$. This is equivalent to $\min\chi_1^2 = -2\ln p(\boldsymbol{D}\,|\,\tilde{\boldsymbol{w}}_1,\mathcal{H}_1)$ $\min\chi_2^2 = -2\ln p(\boldsymbol{D}\,|\,\tilde{\boldsymbol{w}}_2,\mathcal{H}_2)$	**Parameter estimation:** Find evidences $p(\boldsymbol{D}\,|\,\mathcal{H}_1)$ and $p(\boldsymbol{D}\,|\,\mathcal{H}_2)$ which consider **all possible parameter values**. This is equivalent to $-2\,\mathrm{LE}_1 = -2\ln p(\boldsymbol{D}\,|\,\mathcal{H}_1)$ $-2\,\mathrm{LE}_2 = -2\ln p(\boldsymbol{D}\,|\,\mathcal{H}_2)$
Estimating uncertainties: $\sigma(\boldsymbol{w})$ is estimated from parameter covariance matrix $\mathbb{C}$.	**Estimating uncertainties:** Posterior $p(\boldsymbol{w}\,|\,\boldsymbol{D},\mathcal{H})$ provides comprehensive information on parameters, including $\sigma(\boldsymbol{w})$.
Model comparison: Comparison of $\chi_1^2/(N-K_1)$ to $\chi_2^2(N-K_2)$ uses only maximum likelihood $p(\boldsymbol{D}\,|\,\tilde{\boldsymbol{w}},\mathcal{H})$ and ignores covariances of $\boldsymbol{w}$.	**Model comparison:** Comparison of $-2\,\mathrm{LE}_1$ to $-2\,\mathrm{LE}_2$ uses covariances of $\boldsymbol{w}$ through $\mathbb{C}$ since $p(\boldsymbol{D}\,|\,\mathcal{H}) = p(\boldsymbol{D}\,|\,\tilde{\boldsymbol{w}},\mathcal{H})\;p(\tilde{\boldsymbol{w}}\,|\,\mathcal{H})\,(2\pi)^{K/2}\sqrt{\det\mathbb{C}}$.
Model construction: Explore different likelihoods	**Model construction:** Explore different priors and likelihoods
Accurate only for large N	Accurate for all N

Acknowledgements

HCE thanks the Alexander von Humboldt Foundation and the *Excellence Cluster Universe* for support. This work was supported in part by a Stellenbosch Consolidoc fellowship, the National Institute for Theoretical Physics and the National Research Foundation of South Africa.

References

[1] G. Agakishiev, *et al.* (STAR Collaboration), Phys. Rev. C **86**, 064902 (2012).
[2] J. Adams *et al.* (STAR Collaboration), Phys. Rev. C **72**, 014904 (2005).
[3] M.B. de Kock, H.C. Eggers and T.A. Trainor, Phys. Rev. C **92**, 034908 (2015).

EPJ Web of Conferences **120**, 01007 (2016) DOI: 10.1051/epjconf/201612001007
ISMD 2015

Universality of particle production and energy balance in hadronic and nuclear collisions

Aditya Nath Mishra[1,a], Edward K. G. Sarkisyan[2,3,b], Raghunath Sahoo[1,c], and Alexander S. Sakharov[2,4,5,d]

[1] *Discipline of Physics, School of Basic Science, Indian Institute of Technology Indore, Indore-452017, India*
[2] *Department of Physics, CERN, 1211 Geneva 23, Switzerland*
[3] *Department of Physics, The University of Texas at Arlington, Arlington, TX 76019, USA*
[4] *Department of Physics, New York University, New York, NY 10003, USA*
[5] *Physics Department, Manhattan College, New York, NY 10471, USA*

Abstract. The multihadron production in nucleus-nucleus and (anti)proton-proton collisions is studied by exploring the collision-energy and centrality dependencies of the mean multiplicity in the existing data. The study is performed in the framework of the recently proposed effective-energy approach which combines the constituent quark picture and Landau hydrodynamics counting for the centrality-defined effective energy of participants. Within this approach, the multiplicity energy dependence and the pseudorapidity spectra from the most central nuclear collisions are well reproduced. The study of the multiplicity centrality dependence reveals a new scaling between the measured pseudorapidity spectra and the calculations. Using this scaling, called the energy balanced limiting fragmentation scaling, the pseudorapidity spectra are well reproduced for all centralities. The scaling clarifies some differences in the multiplicity and midrapidity density centrality dependence from RHIC and LHC. A similarity in the multiplicity energy dependence in the most central collisions and centrality data is shown. Predictions are drawn for the mean multiplicities to be measured in hadronic and heavy-ion collisions at the LHC.

1. In this report, we discuss our recent results on the universality of multiparticle production in nucleus-nucleus (AA) and hadronic interactions in view of a new scaling obtained [1]. The study exploites concept of effective energy [2] employed for the data interpreted in terms of the approach of the dissipating energy of quark participants [3, 4], or, for brevity, the participant dissipating energy (PDE) approach. This approach combines the constituent quark picture together with Landau relativistic hydrodynamics and interrelates different types of collisions. The earlier observations [2] are made by studying the dependencies of the pseudorapidity density and transverse energy pseudorapidity density at midrapidity on the collision center-of-mass (c.m.) energy in hadronic and the most central (head-on) AA collisions and on the number of nucleon participants, or centrality, in AA collisions in the entire available energy range of the existing data. The complementarity of the measurements in

[a] e-mail: Aditya.Nath.Mishra@cern.ch
[b] e-mail: sedward@cern.ch
[c] e-mail: Raghunath.Sahoo@cern.ch
[d] e-mail: Alexandre.Sakharov@cern.ch

non-central and head-on AA collisions is shown. Here, in the framework of the PDE approach, we extend the previous studies of the charged particle mean multiplicity [3, 4] to LHC energies.

This approach quantifies the process of particle production in terms of the amount of energy deposited by interacting constituent quark participants inside the small Lorentz-contracted volume. The whole process of a collision is then represented as the expansion of an initial state and the subsequent break-up into particles. This approach resembles the Landau phenomenological hydrodynamic approach of multiparticle production in relativistic particle interactions [5]. In the picture considered here, the Landau hydrodynamics is employed in the framework of constituent (or dressed) quarks, in accordance with the additive quark model [6]. This makes the secondary particle production to be basically driven by the amount of the initial *effective* energy deposited by constituent quarks. In $pp/\bar{p}p$ collisions, a single constituent quark from each nucleon is considered to take part in a collision and the remaining quarks are treated as spectators. Thus, the effective energy for the production of secondary particles is 1/3 of the entire nucleon energy. On the contrary, in the head-on AA collisions, the participating nucleons are considered colliding with all three constituent quarks from each nucleon. This makes the whole energy of the colliding nucleons (participants) available for the secondary particle production. Within this picture, one expects the results for bulk observables from head-on AA collisions at the c.m. energy per nucleon, $\sqrt{s_{NN}}$, to be similar to those from the $pp/\bar{p}p$ collisions at $\sqrt{s_{pp}} \simeq 3\sqrt{s_{NN}}$. Such an universality is found to correctly predict [3] the value of the midrapidity density in pp interactions measured at the TeV LHC energies [7].

Combining the above-discussed ingredients, one obtains the relationship between the rapidity density per participant pair, $\rho(\eta) = (2/N_{\rm part})dN_{\rm ch}/d\eta$ at midrapidity ($\eta \approx 0$), in AA and $pp/\bar{p}p$ collisions:

$$\rho(0)/\rho_{pp}(0) = 2N_{\rm ch}/(N_{\rm part}\, N_{\rm ch}^{pp})\sqrt{L_{pp}/L_{NN}}\,, \qquad \sqrt{s_{pp}} = 3\sqrt{s_{NN}}\,. \tag{1}$$

In Eq.(1), the relation of the pseudorapidity density and the mean multiplicity is applied in its Gaussian form as obtained in Landau hydrodynamics. The factor L is defined as $L = \ln(\sqrt{s}/2m)$. According to the approach considered, m is the proton mass, m_p, in AA collisions and the constituent quark mass in $pp/\bar{p}p$ collisions set to $\frac{1}{3}m_p$. $N_{\rm ch}$ and $N_{\rm ch}^{pp}$ are the mean multiplicities in AA and nucleon-nucleon collisions, respectively, and $N_{\rm part}$ is the number of participants.

Solving Eq. (1) for the multiplicity $N_{\rm ch}$ at a given rapidity density $\rho(0)$ at $\sqrt{s_{NN}}$, and for the rapidity density $\rho_{pp}(0)$ and the multiplicity $N_{\rm ch}^{pp}$ at $3\sqrt{s_{NN}}$, one finds:

$$2N_{\rm ch}/N_{\rm part} = N_{\rm ch}^{pp}\,\rho(0)/\rho_{pp}(0)\,\sqrt{1 - 2\ln 3/\ln(4.5\sqrt{s_{NN}}/m_p)}\,, \qquad \sqrt{s_{NN}} = \sqrt{s_{pp}}/3\,. \tag{2}$$

Further development [2] treats this dependence in terms of centrality. The most central collisions correspond to the lowest centrality while the larger centrality to the more peripheral collisions. The centrality is related to the number of participants and, then, to the amount of the energy released in the collisions, *i.e.*, to the effective energy, ε_{NN}. In the framework of the PDE approach, ε_{NN} can be defined as a fraction of the c.m. energy available in a collision according to the centrality, α:[1]

$$\varepsilon_{NN} = \sqrt{s_{NN}}(1 - \alpha). \tag{3}$$

Then, for the effective c.m. energy ε_{NN}, Eq. (2) reads:

$$2\,N_{\rm ch}/N_{\rm part} = N_{\rm ch}^{pp}\,\rho(0)/\rho_{pp}(0)\,\sqrt{1 - 2\ln 3/\ln(4.5\,\varepsilon_{NN}/m_p)}\,, \qquad \varepsilon_{NN} = \sqrt{s_{pp}}/3\,, \tag{4}$$

[1]Conventionally, the data are divided into centrality intervals, so that α is the average centrality per centrality interval, *e.g.* $\alpha = 0.25$ for the centrality interval of 20–30% centrality.

where $\rho(0)$ is the midrapidity density in central AA collisions measured at $\sqrt{s_{NN}} = \varepsilon_{NN}$.

2. Figure 1(a) shows the c.m. energy dependence of the multiplicity measured in head-on AA collisions [8–13] and the fits made in the energy range of $\sqrt{s_{NN}} = 2$ GeV to 2.76 TeV. Given the fact that the measurements support the 2nd-order logarithmic dependence on $\sqrt{s_{NN}}$ up to the top RHIC energy [3, 10] while the power-law dependence is obtained for the LHC data [8], we fit the head-on data by the "hybrid" fit function. In addition to the hybrid fit, we show the $\log^2(s_{NN})$-fit [3, 4] up to the top RHIC energy and the power-law fit. One can see that the power-law fit well describes the data and is almost indistinguishable from the hybrid fit up to the LHC data. However, the 2nd-order log-polynomial lies below the data for $\sqrt{s_{NN}} > 200$ GeV. This observation supports a possible transition to a new regime in AA collisions at $\sqrt{s_{NN}}$ of about 1 TeV, as indicated earlier [2].

Addressing now Eq. (2), one calculates $N_{\rm ch}/(N_{\rm part}/2)$ for AA interactions using the $pp/\bar{p}p$ data. The $\rho_{pp}(0)$ and $N_{\rm ch}^{pp}$ values are taken from the existing data [15] or, where not available, calculated using the corresponding experimental $\sqrt{s_{pp}}$ fits [2] at $\sqrt{s_{pp}} = 3\sqrt{s_{NN}}$. The $\rho(0)$ values are also from the measurements in central AA collisions where available, otherwise the hybrid fit [2] is used.

One can see that the calculated $N_{\rm ch}/(0.5N_{\rm part})$ values follow the measurements from AA collisions at $\sqrt{s_{NN}}$ from a few GeV up to the TeV LHC energy. This points to the universality of the multiparticle production in the different types of collisions.

Solving Eq. (2) for $N_{\rm ch}^{pp}$, one estimates [1] its values to be about 48 at $\sqrt{s_{pp}} = 2.36$ TeV, 69 at 7 TeV, and 79 at 13 TeV with 5 to 10% uncertainties. The study [1] shows that the $\log^2(s)$-polynomial fit function is very close to the power-law function in the c.m. energy range from a few GeV up to about a few TeV, similar to the earlier observations for $\sqrt{s_{pp}} > 53$ GeV [14]. The $\log^2(s)$-polynomial function is not far from the power-law fit even for $\sqrt{s_{pp}} > 2$ TeV. This may point to apparently no change in the multihadron production in pp interactions up to the highest LHC energy, in contrast to a new regime possibly emerging at $\sqrt{s_{NN}} \approx 1$ TeV in AA collisions.

In Fig. 1(b), we show the $N_{\rm part}$-dependence of $N_{\rm ch}/(N_{\rm part}/2)$. One can see that the ε_{NN}-calculations well reproduce the LHC data except slightly underestimating a couple of the most peripheral measurements. For the RHIC data, however, the difference between the calculations and the measurements is visible already for medium centralities, *i.e.*, for more central collisions. These observations are also interrelated with the difference observed in the measurements at RHIC *vs.* those from LHC. This becomes even clearer when the 200 GeV PHOBOS data are multiplied by a factor of 2.87 which allows matching the ALICE data from the highly central collisions.

To clarify the observed differences, in the following sections the distributions of the pseudorapidity density are investigated in the context of the PDE picture.

3. Fig. 2 shows the distributions $\rho(\eta)$ of charged particles measured in head-on and very central AA collisions and $\rho(\eta)_{pp}$ in $pp/\bar{p}p$ interactions at $\sqrt{s_{pp}} \approx 3\sqrt{s_{NN}}$ or $3\,\varepsilon_{NN}$.

Within the considered model of constituent quarks and the Gaussian form of the pseudorapidity distribution in Landau hydrodynamics, the relationship between $\rho(\eta)$ and $\rho_{pp}(\eta)$ reads

$$\rho(\eta)/\rho_{pp}(\eta) = 2N_{\rm ch}/(N_{\rm part}\,N_{\rm ch}^{pp})\,\sqrt{1+2\ln 3/L_{NN}}\,\exp\!\left[-\eta^2/(L_{NN}\,(2+L_{NN}/\ln 3))\right]. \quad (5)$$

The calculations are shown in Fig. 2 and demonstrate very good agreement with the measurements. Minor deviations are due to some mismatch between $\sqrt{s_{pp}}$ and $3\sqrt{s_{NN}}$ (or $3\,\varepsilon_{NN}$) and, as expected, due to a slight non-centrality. It is noticeable how well the PDE picture allows one to reproduce the pseudorapidity density distributions in AA interactions in the full-η range, from central-η to forward-η

[2]The hybrid fit [1] is used for $N_{\rm ch}^{pp}$, while $\rho_{pp}(0)$ is calculated using the linear-log fit $\rho_{pp}(0) = -0.308 + 0.276\,\ln(s_{pp})$ [14] and the power-law fit by CMS [16], $\rho_{pp}(0) = -0.402 + s_{pp}^{0.101}$, at $\sqrt{s_{pp}} \leq 53$ GeV and at $\sqrt{s_{pp}} > 53$ GeV, respectively.

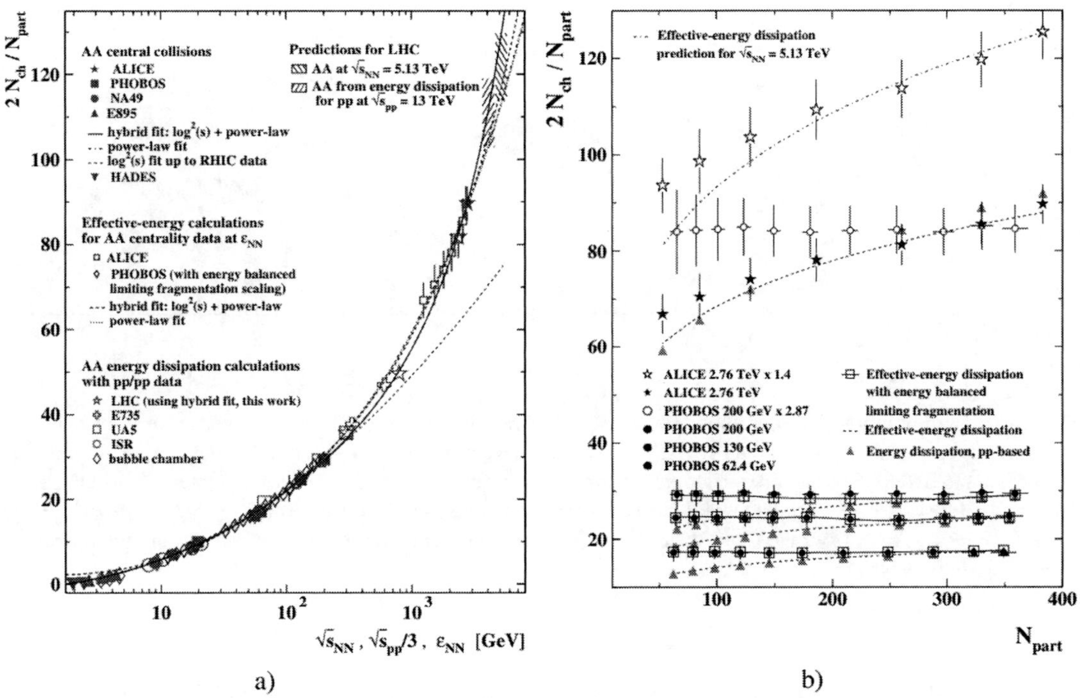

Figure 1. **(a)** The energy dependence of the charged particle mean multiplicity per participant pair. The large solid symbols show the measurements from the most central nucleus-nucleus (AA) collisions [8–13] given as a function of the nucleon-nucleon c.m. energy, $\sqrt{s_{NN}}$. The calculations by Eq. (2) based on $pp/\bar{p}p$ data [14, 15] at the c.m. energy $\sqrt{s_{pp}} = 3\,\sqrt{s_{NN}}$ are shown *vs.* $\sqrt{s_{pp}}/3$ by large open symbols. The small open symbols show the AA data at different centralities by using Eq. (4) at the effective energy ε_{NN} (Eq. (3)). The RHIC centrality data are shown after removing the energy balanced limiting fragmentation scaling ingredient from the calculations of Eq. (4), while the calculations do not take into account this ingredient for the LHC centrality data (see text). The LHC multiplicities in pp interactions are calculated using the hybrid fit from [1]. The solid and the dashed-dotted lines show, correspondingly, the hybrid fit, $-0.577 + 0.394\ln(s_{NN}) + 0.213\ln^2(s_{NN}) + 0.005 s_{NN}^{0.551}$, and the power-law fit, $-6.72 + 5.42\, s_{NN}^{0.181}$, to the most central AA data. The thin dashed line shows the 2nd-order log-fit $-0.35 + 0.24\ln(s_{NN}) + 0.24\ln^2(s_{NN})$ to the most central AA data up to the top RHIC energy [3, 4]. The dashed and the dotted lines show, correspondingly, the hybrid fit, $2.45 - 1.06\ln(\varepsilon_{NN}) + 1.04\ln^2(\varepsilon_{NN}) + 0.082\,\varepsilon_{NN}^{0.744}$, and the power-law fit, $-6.55 + 5.39\,\varepsilon_{NN}^{0.362}$, to the centrality AA data. The right-inclined hatched area shows the prediction for AA collisions at $\sqrt{s_{NN}}$ = 5.13 TeV and the left-inclined hatched area gives the prediction expected from pp collisions at $\sqrt{s_{pp}}$ = 13 TeV. **(b)** The charged particle mean multiplicity per participant pair as a function of the number of participants, N_{part}. The solid circles show the dependence measured in AuAu collisions at RHIC by the PHOBOS experiment at $\sqrt{s_{NN}}$ = 62.4, 130 and 200 GeV [10] (bottom to top). The solid stars show the measurements from PbPb collisions at the LHC by the ALICE experiment at $\sqrt{s_{NN}}$ = 2.76 TeV [8, 17]. The triangles show the calculations by Eq. (4) using $pp/\bar{p}p$ data. The lines represent the calculations within the effective-energy approach based on the hybrid fit obtained for the c.m. energy dependence of the mean multiplicity in the most central AA collisions shown in (a). The open squares show the effective-energy calculations which include the energy balanced limiting fragmentation scaling (see text); the solid lines connect the calculations to guide the eye. The open circles show the PHOBOS measurements at $\sqrt{s_{NN}}$ = 200 GeV multiplied by 2.87. The open stars show the ALICE measurements at $\sqrt{s_{NN}}$ = 2.76 TeV multiplied by 1.4.

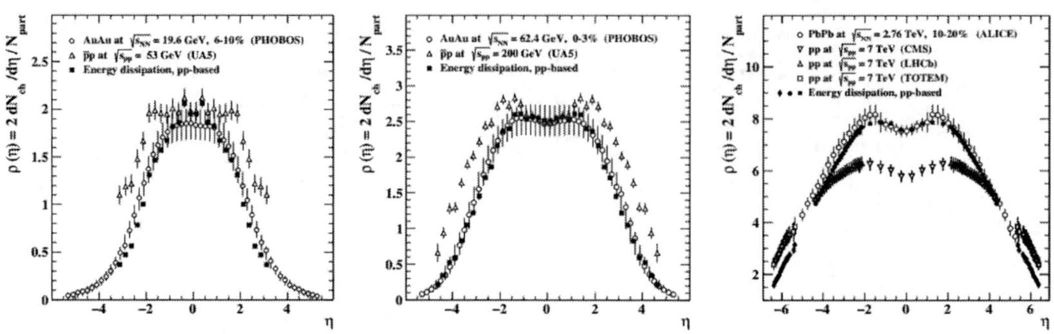

a) b) c)

Figure 2. The pseudorapidity distributions of charged particle pseudorapidity density per participant pair. The open circles show the measurements at RHIC at **(a)** $\sqrt{s_{NN}}$ = 19.6 GeV, **(b)** 62.4 GeV [10], and **(c)** at the LHC at $\sqrt{s_{NN}}$ = 2.76 TeV [8]. The open triangles show the distributions measured in $pp/\bar{p}p$ interactions at $\sqrt{s_{pp}}$ = 53 GeV to 7 TeV [15, 18, 19]. The solid markers show the calculations by Eq. (5) using $pp/\bar{p}p$ data at $\sqrt{s_{pp}} \approx 3\sqrt{s_{NN}}$ or $3\,\varepsilon_{NN}$. Apart from the CMS data, the negative-η data points for $pp/\bar{p}p$ interactions are the reflections of the measurements taken in the positive-η region.

regions, in the $\sqrt{s_{NN}}$ range spanning over more than two orders of magnitude and for different relative heights of AA *vs.* pp spectra.

Let us now address peripheral collisions to clarify the deviation between the data and the calculations in Fig 1(b).

In Fig. 3(a) the distributions $\rho(\eta)$ measured [10] in AuAu collisions at $\sqrt{s_{NN}}$ = 130 GeV at 45-50% centrality, $\alpha = 0.475$, and in $\bar{p}p$ collisions at $\sqrt{s_{pp}}$ = 200 GeV [15], *i.e.*, at $\sqrt{s_{pp}} \approx 3\,\varepsilon_{NN}$. The $\rho(\eta)$ spectrum, calculated by Eq. (5), is also shown in Fig. 3(a) and provides a good agreemnet with the measurements in the central-η region while fall below the data outside this region. This finding shows that in non-central collisions, the calculations within the PDE approach reproduces well the pseudorapidity density around the midrapidity while underestimate the mean multiplicity. The former conclusion is well confirmed by our recent studies reported in [2] for the midrapidity observables, and the latter one is demonstrated by Fig. 1(b).

The following comments are in order.

In the PDE picture, the global observables are defined by the energy of participants, deriving the bulk production by the initial energy deposited at zero time at rapidity $\eta = 0$, similar to the Landau hydrodynamics. Then, as it is expected and just commented, the pseudorapidity density *at midrapidity* is well reproduced for *all* types of AA collisions, from the most central to peripheral ones.

From Fig. 3(a), one can see that the calculated distribution $\rho(\eta)$ is narrower than that of the data. This is explained by a smaller value of ε_{NN} compared to the value of the actual collision energy $\sqrt{s_{NN}}$ while the calculations in Eq. (5) are made with the multiplicity N_{ch} taken from the most central AA collisions at the c.m. energy equal to ε_{NN}. In other words, in the approach applied here, similar to the Landau hydrodynamics, AA collisions are treated head-on-like.

4. It is established that at high enough energies, in different types of interactions the pseudorapidity density spectra, measured at different c.m. energies, become similar in the fragmentation region. It means that they are independent of a projectile state for the same type of colliding objects, *i.e.* being

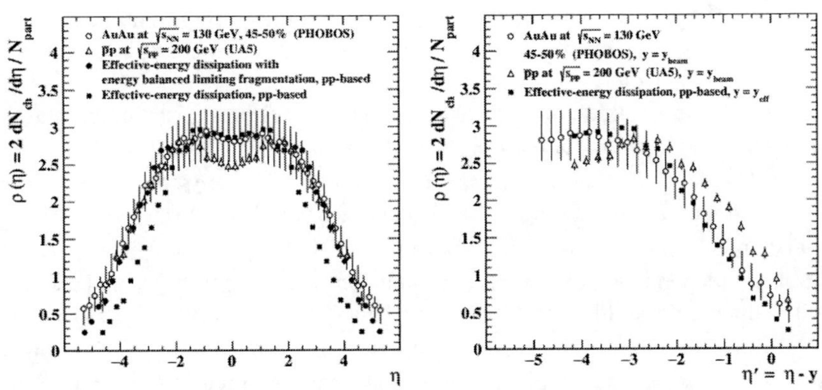

a) b)

Figure 3. **(a)** The charged particle pseudorapidity density distributions per participant pair. The open markers show the data from AuAu collisions [10] and $\bar{p}p$ interactions [15]. The solid squares show the distribution calculated from Eq. (5). $\sqrt{s_{pp}} \approx 3\,\varepsilon_{NN}$ (see Eq. (3) for the definition of ε_{NN}). The solid circles show the beyond-midrapidity part obtained from the calculations using the energy balanced limiting fragmentation scaling, i.e. under the shift $\eta \to \eta - \ln(\varepsilon_{NN}/\sqrt{s_{NN}})$. The negative-$\eta$ data points for $\bar{p}p$ interactions are the reflections of the measurements taken in the positive-η region. **(b)** Same as (a) but the measured distributions of AuAu and $\bar{p}p$ collisions are shifted by the beam rapidity, $\eta' = \eta - y_{\rm beam}$, with $y_{\rm beam} = \ln(\sqrt{s}/m_p)$, where $s = s_{NN}$ and s_{pp}, correspondingly, and the calculated distribution is shifted to $\eta' = \eta - y_{\rm eff}$ with $y_{\rm eff} = \ln(\varepsilon_{NN}/m_p)$. The shifted distributions – the one measured in AuAu collisions and the calculated one – agree well in the fragmentation region that represents the energy balanced limiting fragmentation scaling.

considered as a function of $\eta' = \eta - y_{\rm beam}$, where $y_{\rm beam} = \ln(\sqrt{s_{NN}}/m_p)$ is the beam rapidity [14, 20]. This observation obeys a hypothesis of the limiting fragmentation scaling [21].

Considering the limiting fragmentation hypothesis within the effective-energy approach, one expects the limiting fragmentation scaling of the distribution $\rho(\eta)$, which is measured at $\sqrt{s_{NN}}$, to be similar to that of the calculated distribution but taken at the effective energy ε_{NN}.

In Fig. 3(b), the limiting fragmentation hypothesis is applied to both the measured and the calculated pseudorapidity density distributions $\rho(\eta)$ from Fig. 3(a) using the c.m. energy and the effective energy, respectively. Therefore the measured distribution $\rho(\eta)$ is shifted by the beam rapidity, $y_{\rm beam}$, while the calculated distribution from Eq. (5) is shifted by $y_{\rm eff} = \ln(\varepsilon_{NN}/m_p)$ and becomes a function of $\eta' = \eta - y_{\rm eff}$. One can see that the calculated $\rho(\eta')$ distribution of non-central AA collisions agrees well with the measured distribution $\rho(\eta')$. This finding points to a new energy scaling as soon as the effective-energy approach is applied. In analogy with the limiting fragmentation scaling, we call the observed scaling the *energy balanced* limiting fragmentation scaling. Due to this scaling, the calculated pseudorapidity density is getting corrected outside the central-η region accordingly.

To this end, in Fig. 3(a), the calculated distribution $\rho(\eta)$ is shifted by the difference $(y_{\rm eff} - y_{\rm beam})$ in this region: $\eta \to \eta - \ln(\varepsilon_{NN}/\sqrt{s_{NN}})$, or, using the effective energy definition, Eq. (3), $\eta \to \eta - \ln(1-\alpha)$. The shift *balances* the energy and this brings the calculations to the measured pseudorapidity density distribution in the full-η range in non-central AA collision.

This finding allows obtaining $N_{\rm ch}$ within the PDE approach. Namely, the difference between the two $N_{\rm ch}$ values, one obtained by integrating the calculated pseudorapidity density distribution from Eq. (5), and another one of the same distribution but being shifted to the left by $\ln(1-\alpha)$, is added to the $N_{\rm ch}$ value obtained from Eq. (4). Where no pseudorapidity density distributions are available in

$pp/\bar{p}p$ measurements at $\sqrt{s_{pp}} = 3\,\varepsilon_{NN}$, the energy balanced limiting fragmentation scaling is applied to reproduce the calculated $\rho(\eta)$: the measured distribution from a non-central AA collision is shifted by $(y_{\rm beam} - y_{\rm eff})$, *i.e.* $\eta \to \eta + \ln(1-\alpha)$. Then $N_{\rm ch}$ is calculated as above, by adding to the calculation of Eq. (4) the difference between the integral from the obtained shifted distribution and the measured multiplicity in this non-central AA collision.

Using this ansatz, one finds that the calculated values of $N_{\rm ch}$ well reproduce the measurements from RHIC, with no deficit in non-central collisions, as shown in Fig. 1(b).

The energy balanced limiting fragmentation scaling provides an explanation of the "puzzle" between the centrality independence of the $N_{\rm part}$-normalized mean multiplicity and the monotonic decrease of the normalized midrapidity pseudorapidity density with the centrality, as observed at RHIC.

The multiplicity is measured in the full η-region, so it gets additional contribution from beyond the midrapidity. In the context of the picture proposed here, this contribution is due to the balance between the collision c.m. energy shared by all nucleons of colliding nuclei and the centrality-defined effective energy of the interacting participants. From Fig. 1(b) one can conclude that, in contrast to the RHIC measurements, almost no additional contribution is needed for the PDE calculations of Eq. (4) to describe the LHC data. From this one concludes that in AA collisions at the LHC at TeV energies the multihadron production obeys a head-on collision regime, at least for the centrality intervals measured so far. This points to apparently different regimes of hadroproduction occurring in AA collisions with $\sqrt{s_{NN}}$ between a few hundred GeV and TeV energies. This supports the conclusion made above, which is suggested from the observation of a change of the fit type needed to describe the energy behaviour as soon as the LHC data are included, see Fig. 1(a).

5. Given the obtained agreement between the data and the calculations, and considering the similarity put forward for ε_{NN} and $\sqrt{s_{NN}}$, one would expect the measured centrality data at ε_{NN} to follow the $\sqrt{s_{NN}}$ dependence of the mean multiplicity in the most central AA collisions. In Fig. 1(a), the measurements of the charged particle mean multiplicity of *head-on* AA collisions are added by the *centrality* measurements by the PHOBOS [10] and the ALICE [8, 17] experiments (Fig. 1(b)) where the centrality data are plotted as a function of ε_{NN}. Due to the above finding of the energy balanced limiting fragmentation scaling these data are plotted by subtracting the energy balanced contribution. From Fig. 1(a), one concludes that effective-energy dependence of the centrality data complements the c.m. energy behaviour of the head-on collision data.

To better trace the similarity between the head-on collision and centrality data, we fit the ε_{NN}-dependence of the centrality data by the hybrid and the power-law functions, similarly to the head-on collisions. The fits agree well with the same type of fits to the head-on collision data in the entire available energy range.

As soon as the hybrid fit for the head-on collision data and the fit to the centrality data show slightly different increase with c.m. energy, the predictions of the two fits are averaged. Hence, the mean multiplicity $2N_{\rm ch}/N_{\rm part}$ value is predicted to be about 128 with 5% uncertainty in the most central AA collisions at $\sqrt{s_{NN}} = 5.13$ TeV. The prediction is shown by the right-inclined hatched area in Fig. 1(a). In addition, the fit-averaged prediction based on pp collisions at $\sqrt{s_{pp}} = 13$ TeV, recalculated within the PDE approach, is shown in Fig. 1(a) as the left-inclined hatched area.

Similarly to the existing data on the mean multiplicity $N_{\rm part}$-dependence, the head-on data hybrid fit is used to make the predictions for the centrality dependence at $\sqrt{s_{NN}} = 5.13$ TeV. The predictions are shown by the dashed-dotted line in Fig. 1(b), where the centrality and $N_{\rm part}$ values are taken from the 2.76 TeV data. The expectations show an increase of the mean multiplicity with $N_{\rm part}$ (decrease with centrality) from about 82 to about 128. The increase looks to be slightly faster than at $\sqrt{s_{NN}} = 2.76$ TeV, especially for the peripheral region. One can see that, except a couple of points from peripheral collisions, the predictions are well reproduced by the LHC data, when the latter are scaled by a factor of 1.4.

6. In this report, we present our recent results of the study [1] of the energy and centrality dependences of the mean multiplicity by extending the earlier energy-dependence analysis [3, 4] above the RHIC energies and adding to that the centrality dependence study. In the entire available $\sqrt{s_{NN}}$ range of about a few TeV, the energy dependence of the multiplicity in head-on collisions is found to be well described by the calculations performed within the the approach of the dissipating energy of quark participants. Meanwhile, depending on the data sample, the calculations are found either to describe the measured centrality dependence or to show some deviation between the calculations and the data. For the RHIC data, the deficit in the predictions is observed for non-central collisions so that the predictions do not follow a constancy with the centrality as it is observed at RHIC. The LHC mean multiplicity centrality dependence, however, is found to be well described by the calculations including the increase towards the most central collisions.

To clarify the observations, the pseudorapidity density distribution measured in AA collisions are calculated in the framework of the approach considered here. The energy balanced limiting fragmentation scaling is introduced based on assumption of the similarity of the fragmentation region of the measured distribution in the beam rest frame and that determined from the calculations by using the effective energy. The revealed scaling allows us to reproduce the pseudorapidity distributions independently of the centrality of collisions and then to correctly describe the centrality independence of the mean multiplicity measured at RHIC. Moreover, this finding provides a solution to the RHIC "puzzle" of the difference between the centrality independence of the mean multiplicity *vs.* the monotonic decrease of the midrapidity pseudorapidity density with the increase of centrality. The mean multiplicity is shown to get a fraction of additional contribution to account for the balance between the collision c.m. energy shared by all nucleons and the effective energy of the participants. However, the midrapidity pseudorapidity density is fully defined by the effective energy of colliding participants.

Given the calculations made in the context of the proposed approach are considering central collisions of nuclei, an agreement between the calculations and the LHC data indicates that at TeV energies the collisions seem to present head-on collisions of the participants at the c.m. energy of the scale of the effective energy. Then, no energy-balanced additional contribution is required even for relatively small number of participants at TeV energies.

Based on the above findings, the complementarity of the head-on collisions and the centrality data is shown resulting in the similar energy behaviour of the mean multiplicity measurements as soon as the data are considered in terms of the effective energy. A departure of the c.m. energy dependence of the data from the logarithmic behaviour to the power-law one observes at the region of 1 TeV suggests a transition to a new regime in AA collisions at TeV energies. This is in accordance with the change of the multiplicity dependence on centrality which also indicates a possible change of the regime of multihadron production as one moves from RHIC to LHC. In contrast to heavy ions, no change in multihadron production in *pp* collisions is expected up to the foreseen LHC energy.

Based on the fits obtained, the predictions are made for the charged particle mean multiplicity in head-on AA and *pp* collisions at the LHC. Within the obtained complementarity of head-on collisions and centrality data, the predictions are made as well for the mean multiplicity centrality dependence.

Acknowledgements

The authors are grateful to the XLV International Symposium on Multiparticle Dynamics (ISMD-2015) organizers for kind invitation, warm hospitality and financial support. The work of Alexander Sakharov is partially supported by the US National Science Foundation under Grants No.PHY-1205376 and No.PHY-1402964.

References

[1] E.K.G. Sarkisyan, A.N. Mishra, R. Sahoo, A.S. Sakharov, arXiv:1506.09080 [hep-ph]
[2] A.N. Mishra, R. Sahoo, E.K.G. Sarkisyan, A.S. Sakharov, *Eur. Phys. J. C* **74** (2014) 3147
[3] E.K.G. Sarkisyan, A.S. Sakharov, *Eur. Phys. J. C* **70** (2010) 533
[4] E.K.G. Sarkisyan, A.S. Sakharov, *AIP Conf. Proc.* **828** (2006) 35
[5] L.D. Landau, *Izv. Akad. Nauk: Ser. Fiz.* **17** (1953) 51. English transl.: *Collected Papers of L. D. Landau*, ed. by D. Ter-Haarp (Pergamon, Oxford, 1965), p. 569
[6] For review, see: V.V. Anisovich, N.M. Kobrinsky, J. Nyiri, Yu.M. Shabelsky, *Quark Model and High Energy Collisions* (World Scientific, Singapore, 2004)
[7] See *e.g.* R. Rougny (for the CMS Collab.), *Nucl. Phys. B (Proc. Suppl.)* **207-208** (2010) 29
[8] ALICE Collab., E. Abbas *et al.*, *Phys. Lett. B* **726** (2013) 610
[9] PHOBOS Collab., B.B. Back *et al.*, *Nucl. Phys. A* **757** (2005) 28
[10] B. Alver *et al.*, *Phys. Rev. C* **83** (2011) 024913
[11] NA49 Collab., S.V. Afanasiev *et al.*, *Phys. Rev. C* **66** (2002) 054902
[12] E895 Collab.: J.L. Klay, PhD Thesis (U.C. Davis, 2001), see [9]
[13] HADES Collab., G. Agakishev *et al.*, *Eur. Phys. J. A* **40** (2009) 45
[14] J.F. Grosse-Oetringhaus, K. Reygers, *J. Phys. G* **37** (2010) 083001
[15] The Review of Particle Physics, K.A. Olive *et al.* (Particle Data Group), Chin. Phys. C **38**, 090001 (2014).
[16] CMS Collab., S. Chatrchyan *et al.*, *J. High Energy Phys.* **08** (2011) 141
[17] A. Toia (for the ALICE Collab.), *J. Phys. G* **38** (2011) 124007
[18] LHCb Collab., R. Aaij *et al.*, *Eur. Phys. J. C* **72** (2012) 1947
[19] TOTEM Collab.., G. Antchev *et al.*, *Europhys. Lett.* **98** (2012) 31002
[20] W. Kittel, E.A. De Wolf, *Soft Multihadron Dynamics* (World Scientific, Singapore, 2005)
[21] J. Benecke, T.T. Chou, C.N. Yang, E. Yen, *Phys. Rev.* **188** (1969) 2159

EPJ Web of Conferences **120**, 02001 (2016)   DOI: 10.1051/epjconf/201612002001
ISMD 2015

Measurement of diffractive and exclusive processes with the ATLAS detector

Grzegorz Gach[1,a] on behalf of the ATLAS Collaboration.

[1] *AGH University of Science and Technology*

Abstract. The ATLAS Collaboration has carried out a study of diffractive dijet production in proton–proton collisions at a centre-of-mass energy of $\sqrt{s}$ = 7 TeV at the LHC. The data distributions are compared with Monte Carlo models and the rapidity gap survival probability has been estimated in the kinematic region with high diffractive contribution. Prospects for exclusive jet production studies with the forward proton tagging capability of the AFP sub-detector of ATLAS are also discussed. First results based on data taken jointly with the ATLAS and the LHCf detectors in a p+Pb run will also be shown. In addition, the measurement of the cross-section for the exclusive production of di-lepton pairs in pp collisions at $\sqrt{s}$ = 7 TeV is discussed.

1 Introduction

Diffractive interaction can be defined as a process in which an object carrying vacuum quantum numbers is exchanged. In case of proton–proton collisions such interaction can be divided into three main classes:

- single diffractive dissociation — when one of the incoming protons emerges from the interaction and the other one dissociates into a system X which quantum numbers are the same as of incoming proton,
- double diffractive dissociation — when each of the incoming protons dissociate into systems X and Y, respectively, each of the systems has proton quantum numbers,
- central production — when a system Z is produced in the final state and the incoming protons can dissociate into separate hadronic systems with proton quantum numbers or stay intact.

In the non-diffractive interactions the produced systems and proton remnants are not clearly separated, because the mediating particle carries a charge and thus radiates particles and fills the space between the systems. This does not happen in the diffractive interactions and thus the probability of the dissociated systems to be well separated in pseudorapidity is relatively large. In cases when it is not possible to measure the scattered proton, this property is utilised to experimentally identify the diffractive interactions.

[a]e-mail: grzegorz.gach@cern.ch

2 Dijet Production with Large Rapidity Gaps at 7 TeV

The ATLAS [1] Collaboration published first studies of the single diffractive dissociation at $\sqrt{s} = 7$ TeV already in 2012 [2]. It was an inclusive measurement without the requirement of a hard scale. Since there is no information about the scattered proton, a rapidity gap i.e. a region in η[1] devoid of any hadronic activity, was required between the edge of the calorimeter (which covers $|\eta| < 4.9$) and the nearest energy deposit which is not identified as noise. This procedure provides two rapidity gaps: one in the forward and one in the backward direction. In these studies the larger one is used as the rapidity gap, because single diffractive dissociation events are studied and background contributions are subtracted.

The measurement of dijet production in events with large rapidity gaps [3] is an extension of the study performed in ref. [2]. The presence of jets provides a hard scale in the interaction and allows comparisons with perturbative calculations. This process gives an insight into the underlying parton dynamics and colour singlet exchange. It is also sensitive to the soft survival probability S^2 [4].

The measurement is based on the low pile-up data with $\langle\mu\rangle \sim 0.04 - 0.14$. The jets were reconstructed using the anti-k_t algorithm with two radii: $R = 0.6$ and $R = 0.4$. The latter results are not presented in the paper but are available at HepData [5]. Each jet was required to have transverse momentum greater than 20 GeV and pseudorapidity between -4.4 and 4.4.

The measured differential cross-section for dijet production as a function of rapidity gap ($\Delta\eta^F$) is presented in Fig. 1. In contrast to the soft rapidity gap result no clear diffractive plateau can be observed. This is caused by the reduction of phase space for the dijet system to develop with a growing rapidity gap. The measured cross-section is compared with Pythia 8 Monte Carlo with the ATLAS tune AU2-CT10 [6]. The non-diffractive contribution was scaled by 1/1.4 so that the predicted cross-section for dijet production with a rapidity gap between 0 and 0.5 is the same as the measured one. This region was chosen, because most of the events with a small rapidity gap are expected to have a non-diffractive origin. The Monte Carlo describes the data well in the whole measured rapidity gap range. According to the Pythia 8 predictions the diffractive and non-diffractive cross-sections are equal for rapidity gaps close to 3. Larger rapidity gaps are dominated by the contribution from single diffractive dissociation.

Diffractive and non-diffractive distributions can also be separated based on a variable

$$\tilde{\xi} \simeq M_X^2/s = \sum_{\text{particles}} \left(p_T e^{-|\eta|}\right) / \sqrt{s},$$

where M_X is the mass of the proton dissociated system, p_T is the particle transverse momentum and η is the particle pseudorapidity. The sum runs over all particles in the hadronic final state. In Fig. 1 the distribution of $\log_{10}(\tilde{\xi})$ is presented and it can be observed that the smaller the logarithm the larger the diffractive contribution. Both of the methods introduced above to improve purity of diffractive sample were used in further studies i.e. rapidity gap $\Delta\eta^F > 2$ and $\log_{10}(\tilde{\xi}) < -2$ were required.

The comparison of POMWIG [7] prediction with the data is presented in Fig. 2. The predicted cross-section is about three times larger than the measured one. This discrepancy can be fixed by modifying the rapidity gap survival probability, which in the basic POMWIG is equal to one. The extracted probability is model dependent and was measured in the region of $-3.2 < \log_{10}(\tilde{\xi}) < -2.5$, where the contribution from single diffractive dissociation is expected to dominate by far. The proba-

[1]ATLAS uses a right-handed coordinate system with its origin at the nominal interaction point (IP) in the centre of the detector and the z-axis along the beam pipe. The x-axis points from the IP to the cetnre of the LHC ring, and the y-axis points upward. Cylindrical coordinates (r, ϕ) are used in the transverse plane, ϕ being the azimuthal angle around the beam pipe. The pseudorapidity is defined in terms of the polar angle θ as $\eta = -\ln\tan(\theta/2)$.

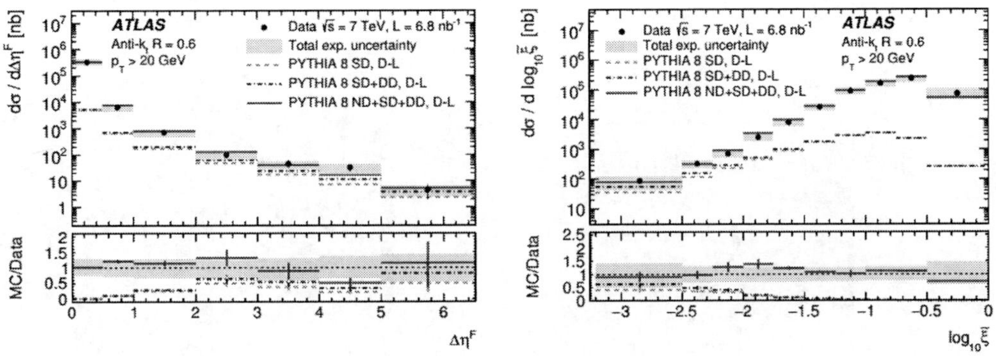

Figure 1. Differential cross-sections for dijet production [3] as functions of rapidity gap (left) and $\log_{10}(\tilde{\xi})$ (right), where $\tilde{\xi} \simeq M_X^2/s$. The rapidity gap is defined as a distance between the calorimeter edge and the nearest energy deposit not identified as noise.

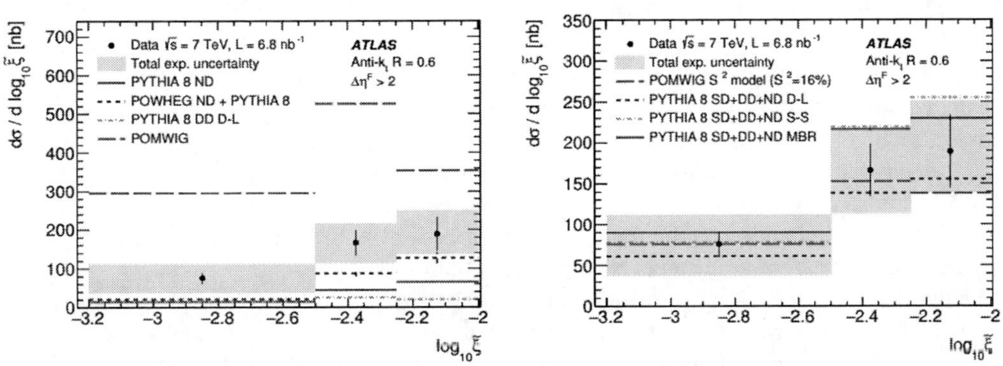

Figure 2. Differential cross-sections for dijet production [3] with rapidity gap greater than 2 as a function of $\log_{10}(\tilde{\xi})$, where $\tilde{\xi} \simeq M_X^2/s$. The left plot shows comparison between the measured cross-sections and POMWIG MC that was used to extract rapidity gap survival probability. The right plot shows the comparison between the data and predictions obtained with three Pomeron flux models.

bility was measured to be $S^2 = 0.16 \pm 0.4(\text{stat.}) \pm 0.8(\text{syst.})$. It is worth noting that Pythia 8 MC does not need any additional rapidity gap survival probability in order to describe the data well.

The measured dijet production differential cross-section as a function of $\log_{10}(\tilde{\xi})$ was used to test Pomeron flux models. Pythia 8 MC with: Donnachie and Landshoff (D–L) [8], Schuler and Sjöstrand (S–S) [9] and Minimum Bias Rockefeller (MBR) [10] is compared with the data in Fig. 2. All predictions agree with the measurements within the experimental uncertainties.

3 Exclusive Dijet Production

One of the possible follow-ups of the dijet production with a large rapidity gap is the measurement of exclusive dijet production. In this process both of the incoming protons stay intact and a gluon pair, which hadronises into a dijet system, is produced. There are no particles that would not be part

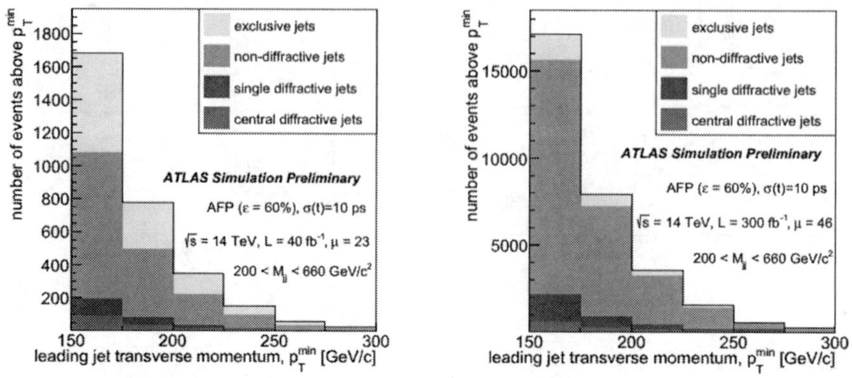

Figure 3. Results of exclusive dijet production feasibility studies [11]. The number of events, separated into signal and background categories, as a function of the leading jet transverse momentum is presented. The left plot shows studies with medium pile-up conditions of $\langle\mu\rangle = 23$. The right plot presents results for high pile-up conditions $\langle\mu\rangle = 46$.

of the dijet system. Only gluons take part in this process, thus this measurement provides a way to probe their distributions in proton. The measurement of exclusive dijet production is very challenging, because of relatively low cross-section ($\sigma = 0.5$ pb) and large backgrounds [11]. The scattered proton tagging is a must in this measurement. In the future it will be provided by the ATLAS Forward Proton (AFP) detectors [12], which were approved by the Research Board in 2015 and will be installed 220 m away from the nominal interaction point. The AFP stations will be equipped with timing detectors which are crucial for measurements in high pile-up conditions.

The feasibility studies of this measurement are published in an ATLAS note [11]. They reveal that the main source of uncertainty is the background estimation. It was assessed that in case of medium pile-up conditions ($\langle\mu\rangle = 23$) and measurement of both scattered protons the signal to background ratio is 0.57. The number of events, separated into signal and background categories, as a function of the leading jet transverse momentum is presented in the left plot in Fig. 3. This proves that although the measurement will be difficult, it is nonetheless feasible. The same studies were repeated for high pile-up conditions of $\langle\mu\rangle = 46$. The results are presented in the right plot in Fig. 3. The estimated signal to background ratio is 0.16, which makes the measurement virtually impossible with this technique.

4 Exclusive Dilepton Production

In the previous section it was shown that measurements of exclusive processes are very challenging. Yet, the ATLAS Collaboration successfully measured the two-photon exclusive production of lepton pairs [13]. The studies were done using proton–proton collisions at $\sqrt{s} = 7$ TeV and mean pile-up between 6.3 and 11.6. The luminosity of the data amounts to 4.6 fb^{-1}. A clean sample of signal events was selected by requiring exactly two opposite charge tracks with transverse momentum greater than 400 MeV. The net transverse momentum of the tracks pair had to be smaller than 1.5 GeV and the centre-of-mass energy of the system could not be in the Z-boson mass region i.e. $m_{ll}/\mathrm{GeV} \notin (70, 105)$.

The measurement was performed in electron and muon channels. Electrons with transverse momentum greater than 12 GeV and pseudorapidity within the range of -2.4 and 2.4 were used in this analysis. The exclusive electron–positron production cross-section was measured to be

 DOI: 10.1051/epjconf/201612002001

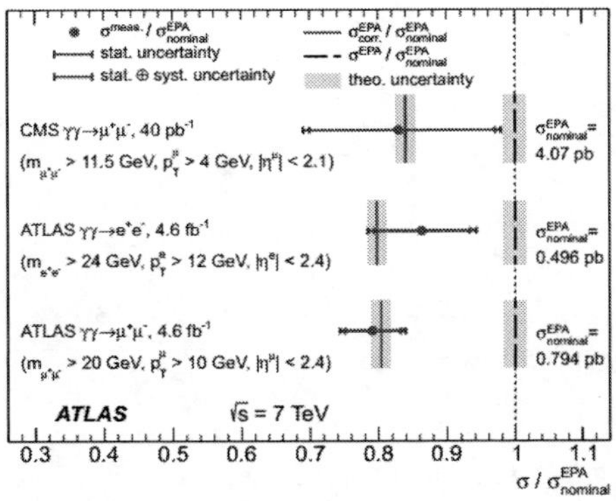

Figure 4. A summary of exclusive dilepton production measurements done by the ATLAS and CMS collaborations [13]. The measured and predicted cross-sections are normalised to the results of calculations based on the uncorrected Equivalent Photon Approximation model.

$\sigma_{\gamma\gamma\to e^+e^-} = 0.428 \pm 0.035(\text{stat.}) \pm 0.018(\text{syst.})$ pb. This result is in agreement with the prediction of the Equivalent Photon Approximation (EPA) model corrected for the proton absorption effects $\sigma^{\text{EPA}}_{\gamma\gamma\to e^+e^-} = 0.398 \pm 0.005(\text{theor.})$ pb.

The muon channel measurement was based on the particles identified as muons and with transverse momentum greater than 10 GeV. The tracks had to be reconstructed with $|\eta| < 2.4$. The measured cross-section $\sigma_{\gamma\gamma\to\mu^+\mu^-} = 0.628 \pm 0.032(\text{stat.}) \pm 0.021(\text{syst.})$ pb agrees with the (EPA) model corrected for the proton absorption effects $\sigma^{\text{EPA}}_{\gamma\gamma\to\mu^+\mu^-} = 0.638 \pm 0.011(\text{theor.})$ pb.

In Fig. 4 a summary of exclusive dilepton production measurements is presented. The CMS measurement in the muon channel suggests that absorption corrections may play an important role in this process, but it is in agreement with uncorrected EPA prediction within a little bit more than one standard deviation. The ATLAS measurement in the same channel proves that the Equivalent Photon Approximation model alone is not sufficient to describe the data and the absorption corrections are crucial. The measured suppression with respect to the uncorrected EPA is of the order of 20%.

5 First Results from ATLAS and LHCf Experiments Combination

The LHCf experiment [14] consists of detectors placed about 140 metres away from the nominal interaction point that measure only neutral particles with pseudorapidity smaller than −8.4. Its main purpose was to measure particle showers. The first successful attempt of combining the data from the ATLAS and LHCf detectors is published in an ATLAS note [15].

The data from proton–lead runs at centre-of-mass energy of 5.02 TeV were used. Approximately 127000 common events were recorded. Thanks to the combination of information from both detectors it was possible to measure the energy spectra of neutral particles produced at very small angles with and without activity in the central rapidity region. These results are presented in the left plot in Fig. 5. Additionally, it was possible to measure the scattering angle of these particles (the right plot in Fig. 5). In the majority of measured events no activity in the central rapidity region is observed which suggests ultra-peripheral-collisions. In such events neutral hadron-like particles energy distribution peaks at 3.5 TeV and scattering angle of these particles is close to zero. This indicates that the dominating process is the production of the Δ^+ baryon: $p + \gamma \to \Delta^+ \to n + \pi^+$.

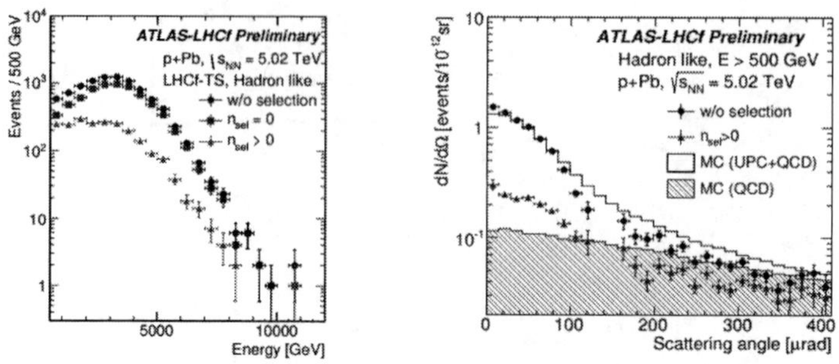

Figure 5. Results of the first ATLAS and LHCf data combination [15]. The left plot shows the energy spectrum of hadron-like neutral particles with and without activity in the central rapidity region. The right plot presents the scattering angle of these particles.

6 Summary

ATLAS diffractive and exclusive physics program is rich and consistent. The first diffractive measurement of soft rapidity gaps was followed by an analysis of the dijet production with a large rapidity gap. It turned out that a tuned Pythia 8 is able to describe the data. The diffractive dijet measurement is planned to be extended by the exclusive dijet production studies with AFP detectors. From the published exclusive measurement of dilepton production it can be concluded that absorption corrections are crucial. The first combined measurement with the ATLAS and the LHCf detectors has been published.

Acknowledgements

This work was supported in part by the National Science Center grants DEC-2011/03/B/ST/02631.

References

[1] ATLAS Collaboration, JINST **3**, S08003 (2008)
[2] ATLAS Collaboration, Eur. Phys. J. **C72**, 1926 (2012), 1201.2808 [hep-ex]
[3] ATLAS Collaboration, CERN-PH-EP-2015-227 (2015), 1511.00502 [hep-ex]
[4] J.D. Bjorken, Phys. Rev. D **47**, 101 (1993)
[5] *The Durham HepData Project*, http://hepdata.cedar.ac.uk/
[6] ATLAS Collaboration, ATL-PHYS-PUB-2012-003 (2012), https://cds.cern.ch/record/1474107
[7] B.E. Cox, J.R. Forshaw, Comput. Phys. Commun. **144**, 104 (2002), hep-ph/0010303
[8] A. Donnachie, P. Landshoff, Nuclear Physics B **244**, 322 (1984)
[9] G.A. Schuler, T. Sjöstrand, Phys. Rev. D **49**, 2257 (1994)
[10] R. Ciesielski, K. Goulianos, PoS **ICHEP2012**, 301 (2013), 1205.1446 [hep-ex]
[11] ATLAS Collaboration, ATL-PHYS-PUB-2015-003 (2015), https://cds.cern.ch/record/1993686
[12] L. Adamczyk et al., Tech. Rep. ATL-COM-LUM-2011-006, CERN, Geneva (2011), AFP technical proposal, https://cds.cern.ch/record/1331213
[13] ATLAS Collaboration, Phys. Lett. **B749**, 242 (2015), 1506.07098 [hep-ex]
[14] O. Adriani et al. (LHCf), JINST **3**, S08006 (2008)
[15] ATLAS Collaboration, ATL-PHYS-PUB-2015-038 (2015), https://cds.cern.ch/record/2047832

EPJ Web of Conferences **120**, 02002 (2016) DOI: 10.1051/epjconf/201612002002
ISMD 2015

Hard diffraction in Pythia 8

Christine Overgaard Rasmussen[1,a]

[1] *Theoretical Particle Physics,*
Department of Astronomy and Theoretical Physics,
Lund University,
Sölvegatan 14A,
SE-22362 Lund,
Sweden.

Abstract. We present an overview of the options for diffraction implemented in the general-purpose event generator Pythia 8 [1]. We review the existing model for soft diffraction and present a new model for hard diffraction. Both models use the Pomeron approach pioneered by Ingelman and Schlein, factorising the diffractive cross section into a Pomeron flux and a Pomeron PDF, with several choices for both implemented in Pythia 8. The model of hard diffraction is implemented as a part of the multiparton interactions (MPI) framework, thus introducing a dynamical gap survival probability that explicitly breaks factorisation.

1 Introduction

Despite the great success of QCD many soft phenomena still remain as mysteries because of their nonperturbative origin. Amongst these lies soft hadron collisions, in which one or both of the hadrons survive the collision. In this regime we cannot predict the collisions or their rate with our usual perturbative framework of Feynman diagrams and factorisation theorems, as these require a breakup of the hadrons into their constituents, which then collide. We must therefore base our work on phenomenological models, many of which has sprung out of theories invented before QCD, such as S-matrix theory and the related Regge theory.

In Regge theory, poles in the plane of complex spin, α, can be seen as hadronic resonances. These appear to lie on linear trajectories, $\alpha(t) = \alpha(0) + \alpha' t$, of which the most important for high-energy applications is the Pomeron ($\mathbb{P}$) trajectory, with its $\alpha(0) > 1$ explaining the rise of the total cross section. The Pomeron must be a colour singlet carrying the quantum numbers of the vacuum, as it should be able to describe elastic scattering of the hadrons. From a more modern viewpoint it would mainly consist of gluons, but a quark content could also be envisioned – as long as this content come in quark-antiquark pairs, in order to preserve the singlet nature of the $\mathbb{P}$. Having defined the state and its possible partonic content, one can perform a topological expansion similar to that of perturbative QCD, where the simplest possible exchange is a single-$\mathbb{P}$ one, describing elastic scattering. More complex processes can be constructed through the triple-$\mathbb{P}$ vertex, resulting in the different

[a] e-mail: christine.rasmussen@thep.lu.se

diffractive topologies. We generally work with three different diffractive processes, all involving the triple-$\mathbb{P}$ vertex; the single-diffractive (SD) processes (one surviving hadron and one triple-$\mathbb{P}$ vertex), double-diffractive (DD) processes (no surviving hadrons, two triple-$\mathbb{P}$ vertices) and central-diffractive (CD) processes (two surviving hadrons and two triple-$\mathbb{P}$ vertices). In this paper we focus on the SD processes, being the simplest possible configuration and the one with the largest cross section.

Ingelman and Schlein proposed a model [2] in which the $\mathbb{P}$ can be viewed as a hadronic state with partonic content. This opened up the possibility for $\mathbb{P}$ parton distribution functions (PDFs) to be combined with the probability of extracting a $\mathbb{P}$ from the colliding hadron, denoted the $\mathbb{P}$ flux. Hence the diffractive system could be envisioned as a hadron-hadron collision at reduced energy, which opened up the possibility for modelling the diffractive processes with existing hadron-hadron event generators.

A model for diffractive systems should be able to describe all aspects of these collisions, such as differential cross sections, one-particle distributions and global event characteristics. The framework used in Pythia 8 describes all of these aspects in detail. As rapidity gaps are crucial for diffraction, this framework does not permit MPIs in the pp system, as these would fill up the would-have-been rapidity gap. Thus we introduce the concept of rapidity gap survival probability (RGSP) (cite Bjorken), a concept unique to hadron-hadron collisions, given credibility by the lower observed rate for diffraction at the Tevatron as expected from HERA flux and PDF determinations.

2 Soft diffraction in Pythia 8

The soft diffraction framework available in Pythia 8 was originally developed for its predeccesor, Pythia 6 [3], but rewritten and expanded for the new version [4]. The total hadronic cross section is calculated using the Donnachie-Landshoff [5] parametrisation, including both a Pomeron and Reggeon term. The elastic and diffractive cross sections are calculated using the Schuler-Sjöstrand model [6], with the nondiffractive cross section inferred from the above.

The Schuler-Sjöstrand model is based on Regge theory, giving an exponential t-dependence and an approximate dM^2/M^2 mass dependence, with fudge factors introduced in order to dampen the cross sections close to the kinematical limits, as well as dampen the DD cross section in regions where the two diffractive systems overlap. Other $\mathbb{P}$-flux models have also been implemented, all of which have been expanded to include both DD and CD where needed [7]. The subsequent evolution and hadronisation of the $\mathbb{P}$p system is separated from the $\mathbb{P}$-parametrisation, hence being the same for almost all models.

2.1 Low-mass soft diffraction

In the low-mass regime, $M \leq 10$ GeV, the energies are not sufficiently high to apply a perturbative framework to the $\mathbb{P}$p subsystem. The framework is setup as an interaction, where a $\mathbb{P}$ has "kicked out" a quark or a gluon from the incoming hadron. In the former case a colour string is stretched between the "free" quark and the diquark remnant inside the proton. In the latter case a colour string is stretched between one valence quark inside the hadron through the "free" gluon and back to the diquark in the remnant, resulting in a hairpin-like structure. The probability for the $\mathbb{P}$ to interact in either of the two ways is mass-dependent and tunable, $\mathcal{P}(q)/\mathcal{P}(g) = N/M^p$, set up such that the gluons dominate at higher masses. Since this is a nonperturbative framework, no additional evolution of the colour strings is applied. Instead the system is hadronised directly using the Lund string fragmentation model [8], thus giving rise only to low-$p_\perp$ activity in the diffractive system.

DOI: 10.1051/epjconf/201612002002

2.2 High-mass soft diffraction

In the high-mass regime, $M > 10$ GeV, energies are sufficiently high to use a perturbative framework. The high-mass regime is smoothly connected to the low-mass one, with the fraction of perturbative events gradually increasing from the 10 GeV starting point and dominating at diffractive masses above 20 GeV. In this regime the $\mathbb{P}$ is viewed as having a partonic content. Thus once M and t have been selected according to the model of the $\mathbb{P}$ flux, the $\mathbb{P}$p subsystem is set up and a semi-hard perturbative $2 \to 2$ QCD process is selected by the MPI framework as the hardest collision. The subsystem is then further evolved using the interleaved parton evolution of Pythia 8, including both MPI as well as initial and final state showers (ISR, FSR), using the $\mathbb{P}$ PDFs on the diffractive side.

The MPI acitivity in the $\mathbb{P}$p system has been tuned to give approximately the same amount of activity as nondiffractive events of the same mass by introducing an effective total $\mathbb{P}$p cross section of 10 mb. This is a tunable parameter, and can be made to depend on the mass of the diffractive system, just as the total pp cross section does. The evolution results in a larger number of colour strings in the event, all hadronised with the Lund string fragmentation model, giving rise to low- to high-$p_\perp$ activity in the event.

Although largely successful, the model for soft diffraction in Pythia 8 has minor issues. Not all aspects of data are described by the default model and settings, including both differential cross sections and particle spectra. A full retune could improve some of these discrepancies, but a more flexible shape for the $\mathbb{P}$ flux may be needed in order to describe data better. Such studies are planned for the near future.

3 Hard diffraction in Pythia 8

The model described in the previous section allows for $2 \to 2$ QCD processes at all $p_\perp$ scales, but is primarily intended for lower values of $p_\perp$. For truly hard diffractive processes, the new model for hard diffraction [9] was developed, not only for high-$p_\perp$ jets, but also allowing for $W^\pm, Z^0, H$ etc.

The model is based on the assumption that the proton PDF can be split into a diffractive and a nondiffractive part,

$$f_{i/\mathrm{p}}(x, Q^2) = f_{i/\mathrm{p}}^{\mathrm{ND}}(x, Q^2) + f_{i/\mathrm{p}}^{\mathrm{D}}(x, Q^2), \tag{1}$$

with the diffractive part being described using the factorisation approach of Ingelman and Schlein,

$$\begin{aligned} f_{i/\mathrm{p}}^{\mathrm{D}}(x, Q^2) &= \int_0^1 \mathrm{d}x_{\mathbb{P}}\, f_{\mathbb{P}/\mathrm{p}}(x_{\mathbb{P}}) \int_0^1 \mathrm{d}x'\, f_{i/\mathbb{P}}(x', Q^2)\, \delta(x - x_{\mathbb{P}}x') \\ &= \int_x^1 \frac{\mathrm{d}x_{\mathbb{P}}}{x_{\mathbb{P}}}\, f_{\mathbb{P}/\mathrm{p}}(x_{\mathbb{P}})\, f_{i/\mathbb{P}}\left(\frac{x}{x_{\mathbb{P}}}, Q^2\right), \end{aligned} \tag{2}$$

where $f_{\mathbb{P}/\mathrm{p}}(x_{\mathbb{P}}) = \int_{t_{\mathrm{min}}}^{t_{\mathrm{max}}} \mathrm{d}t\, f_{\mathbb{P}/\mathrm{p}}(x_{\mathbb{P}}, t)$, as t for the most part is not needed.

The probability for side A, B, to be diffractive is then given by the ratio of diffractive to inclusive PDFs,

$$\begin{aligned} \mathcal{P}_A^{\mathrm{D}} &= \frac{f_{i/B}^{\mathrm{D}}(x_B, Q^2)}{f_{i/B}(x_B, Q^2)} \quad \text{for } AB \to XB\,, \\ \mathcal{P}_B^{\mathrm{D}} &= \frac{f_{i/A}^{\mathrm{D}}(x_A, Q^2)}{f_{i/A}(x_A, Q^2)} \quad \text{for } AB \to AX\,. \end{aligned} \tag{3}$$

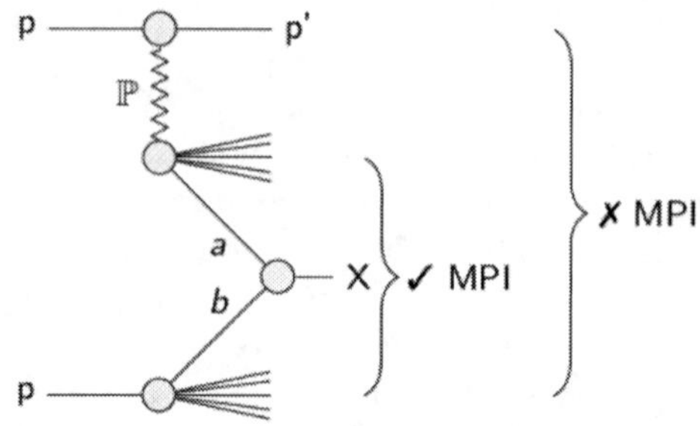

Figure 1. The dynamical gap survival implemented in Pythia 8. The model do not allow for MPIs in the pp system, but allows for additional MPIs in the ℙp system.

Table 1. Diffractive fractions for the $2 \to 2$ QCD processes with $p_{\perp} > 20$ GeV in pp collisions at $\sqrt{s} = 8$ TeV obtained with Pythia 8. The samples have been produced without any phase-space cuts. ℙ parametrizations not cited in the text are: H1 Fit A NLO PDF [11] and H1 Fit A flux [11, 13]

ℙ PDF ℙ flux	PDF selection	MPI selection
H1 Fit B LO Schuler-Sjöstrand	(14.33 ± 0.11) %	(0.98 ± 0.03) %
H1 Fit B LO MBR	(14.79 ± 0.11) %	(0.96 ± 0.03) %
H1 Jets Schuler-Sjöstrand	(13.70 ± 0.11) %	(0.92 ± 0.03) %
H1 Fit A NLO H1 Fit A	(20.55 ± 0.13) %	(1.35 ± 0.04) %
H1 Fit B LO H1 Fit A	(18.49 ± 0.12) %	(1.32 ± 0.04) %

The model also implements a dynamical gap survival, cf. Fig. 1. On an event-by-event basis we evaluate the possibility for additional MPIs in the pp system, and if no further MPIs are found, then the event is diffractive. Thus initially eq. (3) has no consequence, with all events still handled as being nondiffractive. Only if no additional MPIs are found is the event classified as diffractive, and the ℙp system set up. A full evolution is then performed in this subsystem, along with the hadronisation of the colour strings in the event. At this point, the nondiffractive events can be discarded if an exclusively diffractive sample is wanted, otherwise they can be kept for an inclusive sample consisting of both non- and single-diffractive events.

The dynamical gap survival introduces an additions suppression of the diffractive events, such that the total probability for hard diffraction drops from $\sim 10\%$ to $\sim 1\%$. In Table 1 we show the number of events passing either the probabilistic criterion of eq. (3) (PDF selection) or both the probabilistic and the dynamical gap survival (MPI selection) for a variety of the ℙ fluxes and PDFs available in Pythia 8. These fractions are very model dependent, of course. They depend both on the ℙ parametrization and on the free parameters of the MPI framework, although many of the distributions tend to be mainly driven by only one of them.

Table 2. Diffractive fractions for the $2 \to 2$ QCD processes with $p_\perp > 20$ GeV in $\sqrt{s} = 8$ TeV pp collisions. Pythia is run with the Schuler-Sjöstrand flux and the H1 Fit B LO PDF.

	PDF selection	MPI selection
$p_{\perp 0}^{\mathrm{ref}} = 1.78$	(14.50 ± 0.11) %	(0.39 ± 0.02) %
$p_{\perp 0}^{\mathrm{ref}} = 2.28$	(14.33 ± 0.11) %	(0.98 ± 0.03) %
$p_{\perp 0}^{\mathrm{ref}} = 2.78$	(14.19 ± 0.11) %	(2.00 ± 0.04) %

Distributions affected by the modelling of the $\mathbb{P}$ flux and PDF are e.g. the chosen $x_{\mathbb{P}}$ value, hence the mass of the diffractive system, as well as the value of the squared momentum transfer, t and through that the angle at which the proton is scattered. The mass of the diffractive system is shown in Fig. 2a, using the following parametrizations: The Schuler-Sjöstrand flux (SaS) [6], the Minimum Bias Rockefeller lux (MBR) [10], the H1 Fit B LO PDF [11] and the H1 Jets PDF [12]. It is observed that the variation of the $\mathbb{P}$ flux affects the distribution more than the variation of the $\mathbb{P}$ PDF.

Those affected by the MPI parameters are several particle distributions such as multiplicities and $p_\perp$ spectra. One of these is shown in Fig. 2b, where the transverse matter profile of the $\mathbb{P}$ is varied. In the end, what really matters is the convolution of the matter profile of the $\mathbb{P}$ and the p, resulting in an overlap function, from where the impact parameter b, describing the amount of overlap between the $\mathbb{P}$ and p, is chosen. We have implemented three options:

- $b_{\mathbb{P}\mathrm{p}} = b_{\mathrm{pp}}$, implying that the $\mathbb{P}$ and p are equally big and have equal matter profiles.
- $b_{\mathbb{P}\mathrm{p}} = \sqrt{b_{\mathrm{pp}}}$, motivated if the $\mathbb{P}$ is a point particle. The variation of the overlap function with b is then that of one proton, not two, giving the square root of the normal variation.
- The final option implemented is picking a completely new $b_{\mathbb{P}\mathrm{p}}$, ie. not specifying the size of the $\mathbb{P}$ or its matter profile.

The variation of the size gives different distributions, the latter definition giving much larger multiplicities than the two first (and probably more realistic) cases.

Also the gap survival probability is highly sensitive to the parameters of the MPI model, highlighted in Table 2, where the regulator of the $2 \to 2$ QCD cross section, $p_{\perp 0}^{\mathrm{ref}}$ has been varied with ± 0.5 GeV, resulting in a factor of two in the diffractive fractions. This major $p_{\perp 0}^{\mathrm{ref}}$ dependence holds also for many nondiffractive event properties, however; keeping everything else fixed even a variation of ± 0.1 GeV would be unacceptable. It would be possible to perform a dedicated diffractive tune, but the same parameters would not be able to describe nondiffractive data. Such a retune is considered.

4 Conclusions

We have presented the two frameworks for diffraction present in the general-purpose event generator Pythia 8. The soft diffraction framework allows for QCD processes at various $p_\perp$ scales, although primarily intended for low-$p_\perp$ interactions. It gives a decent description of soft diffraction, although comparisons to data show room for improvement. The new model for hard diffraction allows for non-QCD and very high-$p_\perp$ QCD processes in diffractive systems. The effects of the $\mathbb{P}$ parametrizations as well as the free parameters of the MPI framework have been explored. Future work includes a retuning of the free parameters in both models and developing a new parametrization of the $\mathbb{P}$ flux, in order to obtain a better description of data.

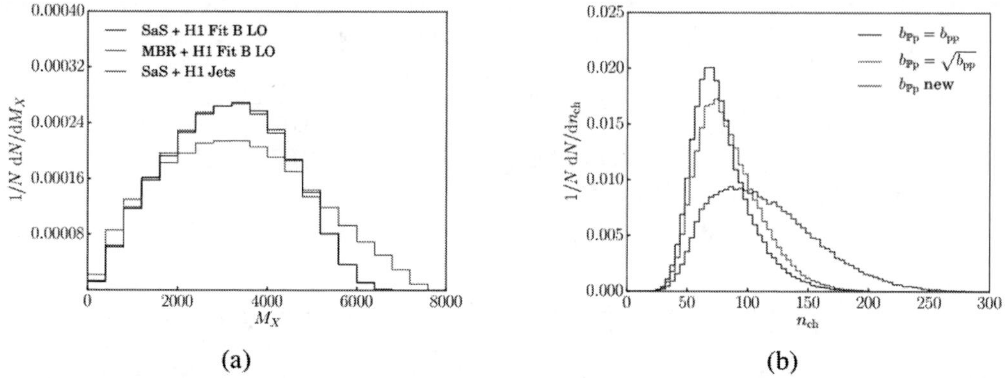

Figure 2. (a) The mass of the diffractive system, M_X, when changing $\mathbb{P}$ parametrization. (b) The number of charged particles in the $\mathbb{P}$p system, when changing the size of the $\mathbb{P}$.

Acknowledgements

Work supported by the MCnetITN FP7 Marie Curie Initial Training Network, contract PITN-GA-2012-315877.

References

[1] T. Sjöstrand *et al.*, Comput. Phys. Commun. **191** (2015) 159 [arXiv:1410.3012 [hep-ph]].
[2] G. Ingelman and P. E. Schlein, Phys. Lett. B **152** (1985) 256.
[3] T. Sjöstrand, S. Mrenna and P. Z. Skands, JHEP **0605** (2006) 026 [hep-ph/0603175].
[4] S. Navin, arXiv:1005.3894 [hep-ph].
[5] A. Donnachie and P. V. Landshoff, Phys. Lett. B **296** (1992) 227 [hep-ph/9209205].
[6] G. A. Schuler and T. Sjöstrand, Phys. Rev. D **49** (1994) 2257.
[7] http://home.thep.lu.se/Pythia/pythia82html/Welcome.html
[8] B. Andersson, G. Gustafson, G. Ingelman and T. Sjöstrand, Phys. Rept. **97** (1983) 31.
[9] C. O. Rasmussen and T. Sjöstrand, in preparation.
[10] R. Ciesielski and K. Goulianos, PoS ICHEP **2012** (2013) 301 [arXiv:1205.1446 [hep-ph]].
[11] A. Aktas *et al.* [H1 Collaboration], Eur. Phys. J. C **48** (2006) 715 [hep-ex/0606004].
[12] A. Aktas *et al.* [H1 Collaboration], JHEP **0710** (2007) 042 [arXiv:0708.3217 [hep-ex]].
[13] A. Aktas *et al.* [H1 Collaboration], Eur. Phys. J. C **48** (2006) 749 [hep-ex/0606003].

EPJ Web of Conferences **120**, 02003 (2016) DOI: 10.1051/epjconf/201612002003
ISMD 2015

Recent forward physics and diffraction results from CMS

Gábor Veres on behalf of the CMS Collaboration[1,a]

[1] *CERN, Geneva, Switzerland*

Abstract. Recent CMS results on forward physics and diffraction are reviewed. The differential diffractive cross section is measured as a function of $\xi = M_X^2/s$ in the region dominated by single dissociation (SD) and double dissociation (DD), where M_X is the mass of one of the two final-state hadronic systems separated by the largest rapidity gap in the event. The total SD and DD cross sections are extracted. The observation of a hard color-singlet exchange process in events with a large rapidity gap between two leading jets (jet-gap-jet) is reported. The fraction of jet-gap-jet to all dijet events is measured as a function of the second leading jet transverse momentum and the size of the pseudorapidity gap. The measured fractions are compared with predictions as well as Tevatron data.

1 Introduction

Diffractive interactions account for about a quarter of the total inelastic proton-proton cross section at high energies. These interactions are characterized by at least one large rapidity gap (LRG) that is not suppressed exponentially. Depending on the topology of the event, one can differentiate between single and double dissociation (SD and DD), and central diffraction (CD). An LRG is presumed to be mediated by a color-singlet exchange (CSE). Measurements of diffractive processes at the LHC provide valuable input to phenomenological models, as these cross sections cannot be calculated within perturbative QCD, and extrapolations in $\sqrt{s}$ from lower energies vary depending on model parameters.

Another interesting and related area of research is the study of events with a LRG between two jets at high transverse momentum p_T. Normally dijet production is the consequence of an exchange of a colored object, and the color field associated with the exchanged parton calls for additional soft parton emissions, populating the η interval between the jets. For dijet events with central LRG, the BFKL approach is expected to describe the data [1]. The LRG is a signature of CSE, this time involving a hard scale, with a much higher momentum transfer compared to soft diffractive events. Thus, studies of CSE events as a function of the rapidity gap width helps disentangle the BFKL dynamics from the DGLAP evolution. The results are also sensitive to rescattering processes, as those can destroy the LRG.

The detailed description of the CMS detector can be found in [2]. For measurements of diffraction, the extensive forward instrumentation is relevant: the forward component of the hadron calorimeter, HF ($2.9 < |\eta| < 5.2$), the CASTOR calorimeter [3] ($-6.6 < \eta < -5.2$), the Zero Degree Calorimeter (ZDC) and the Forward Shower Counters (FSC). The TOTEM experiment, including Roman Pots and

[a] e-mail: gabor.veres@cern.ch

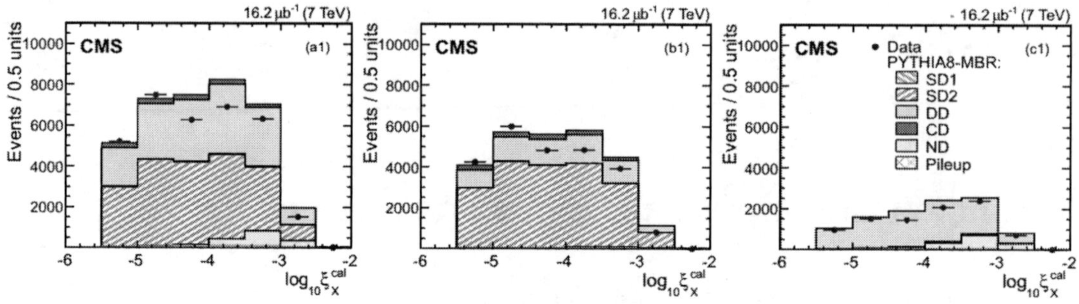

Figure 1. Detector-level distributions of the reconstructed and calibrated ξ_X for (left) the entire FG2 sample, and the FG2 subsamples with (middle) no CASTOR tag, and (right) a CASTOR tag (statistical errors only). The data are compared to the predictions of the PYTHIA 8 MBR simulations, which are normalized to the integrated luminosity of the data sample. The contribution of each of the generated processes is shown separately.

tracker detectors very close to the beamline and sharing collision point with CMS, presents a unique future opportunity to extend the studies discussed here.

2 Diffractive dissociation cross sections

In the experimental analysis presented here, the SD and DD events are separated using the CASTOR calorimeter [3], covering the $-6.6 < \eta < -5.2$ region. A special data set of 16.2 μb^{-1} at $\sqrt{s} = 7$ TeV was used, where the probability of more than one pp collisions occurring coincidentally was only 7%, collected in 2010. Monte Carlo (MC) simulations are used to correct the measured distributions for the geometrical acceptance and efficiency of the detector [4].

The pp collisions are selected with a very inclusive minimum bias trigger. No reconstructed collision vertex or charged particle tracks are required, but at least two reconstructed particle-flow (PF) objects above 4 GeV total energy in the forward region. The acceptance of this selection is 90% for events where any of the diffractive M_X and M_Y masses exceed 12.6 GeV.

Various event topologies are defined: FG1 (FG2) events contain an η-gap on the positive (negative) η-side. In case of FG1 (FG2) topologies, the η-gap extends to $\eta_{\max}$ and $\eta_{\min}$ which are the highest (lowest) η of the PF objects in the central detector. In order to select samples of FG1 and FG2 events with the forward LRG signature in the detector, the requirements $\eta_{max} < 1$ and $\eta_{min} > -1$ are imposed, respectively. The detection of the low-mass dissociated system, Y, and thereby the separation of DD and SD events, is performed by using a CASTOR tag, defined as the presence of a signal above the energy threshold (1.48 GeV) in the CASTOR calorimeter.

The forward rapidity gap cross section is expressed as a function of $\xi_X = M_X^2/s$. Since ξ_X is not directly measurable, it is approximated experimentally by $\xi^{\pm} = \Sigma(E^i \mp p_z^i)/\sqrt{s}$, where E^i and p_z^i refer to PF objects, where the dissociated system is on the positive (negative) side. Good correlation between ξ_X and ξ_X^+ was found using single diffractive events simulated by the PYTHIA 8 Minimum Bias Rockefeller (MBR) tune. Based on this correlation, the measured ξ^+ values can be calibrated (translated into ξ_X). Figure 1 shows the distribution of this calibrated $\log_{10} \xi_X^{cal}$ for the FG2 sample, compared to model predictions, including the SD, DD and ND components separately. The separation of SD and DD processes by the CASTOR tag is clearly seen in these figures.

The PYTHIA 8 MBR tune describes the data well (using a Pomeron trajectory with $\epsilon = 0.08$), and is subsequently used to extract the diffractive cross sections. The differential cross sections are mea-

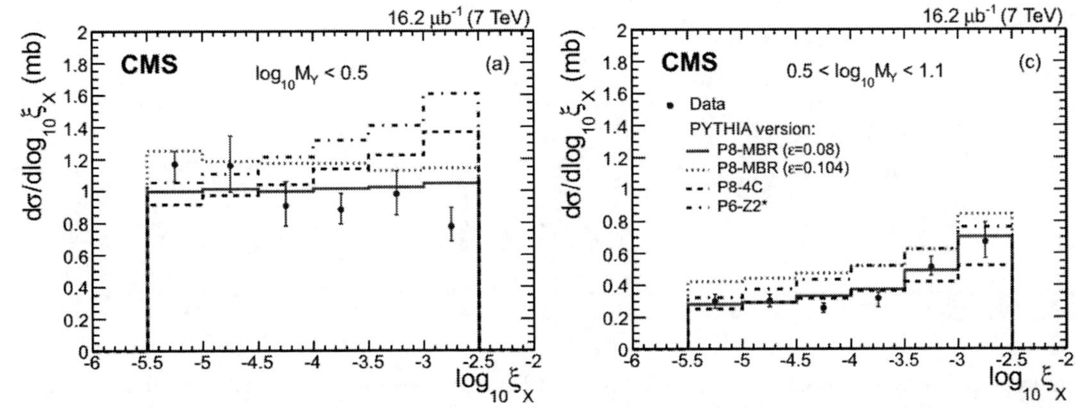

Figure 2. Cross sections $d\sigma/d\log_{10}\xi_X$ for $\log_{10} M_Y < 0.5$ (dominated by single diffraction, left panel) and $0.5 < \log_{10} M_Y < 1.1$ (dominated by double diffraction, right panel) compared to MC predictions: (top) PYTHIA 8 MBR, PYTHIA 8 4C, PYTHIA Z2*, and (bottom) PHOJET, QGSJET-II 03, QGSJET-II 04, EPOS. Error bars are dominated by systematic uncertainties.

sured in ξ_X bins, separately for events with no particles depositing energy in the CASTOR calorimeter (this selection corresponds approximately to $\log_{10} M_Y < 0.5$ and the cross section is corrected to this generator level interval) and for events with CASTOR tag ($0.5 < \log_{10} M_Y < 1.1$). The cross sections are corrected for the small fraction of simultaneous pp collisions in the data, as well as for acceptance and migration effects, evaluated with the iterative Bayesian method, using the response matrix obtained from the PYTHIA 8 MBR model (with $\epsilon = 0.08$).

Figure 2 presents the measured cross sections, compared to the PYTHIA 8 MBR, PYTHIA 8 4C, and PYTHIA 6 Z2* predictions. The error bars are dominated by systematic uncertainties. The predictions of PYTHIA 8 MBR are shown for the values of the parameter ϵ of the Pomeron trajectory $\epsilon = 0.08$ as well as $\epsilon = 0.104$. The data with CASTOR tag, $0.5 < \log_{10} M_Y < 1.1$, favor $\epsilon = 0.08$, especially when both of the M_Y and M_X masses are low. The PYTHIA 6 Z2* tune, using the Schuler-Sjöstrand model, overestimates the data; the predicted rising behavior of the cross section as a function of ξ_X is not confirmed by this measurement.

The cross sections have been integrated over the measured kinematic range $-5.5 < \log_{10}\xi_X < -2.5$ for events with and without a CASTOR tag, corresponding to $0.5 < \log_{10} M_Y < 1.1$ and $\log_{10} M_Y < 0.5$, respectively. The result of the integration is $\sigma_{\text{no-CASTOR}} = 2.99 \pm 0.02(\text{stat})^{+0.32}_{-0.29}(\text{syst})$ mb and $\sigma_{\text{CASTOR}} = 1.18 \pm 0.02(\text{stat}) \pm 0.13(\text{syst})$ mb.

The $\sigma_{\text{no-CASTOR}}$ cross section is dominated by SD events, with background mainly originating from DD processes. The latter is well understood via the CASTOR-tagged events. The PYTHIA 8 MBR simulation was used to correct for the small non-diffractive (ND) fraction and other diffractive processes to obtain the visible SD cross section ($-5.5 < \log_{10}\xi_X < -2.5$), which is found to be $4.06 \pm 0.04(\text{stat})^{+0.69}_{-0.63}(\text{syst})$ mb. The main contributors to σ_{CASTOR} are DD events, with a dominant background from ND events. The corrected visible DD cross section in the CASTOR-tagged sample ($-5.5 < \log_{10}\xi_X < -2.5$ and $0.5 < \log_{10} M_Y < 1.1$) is $1.06 \pm 0.02(\text{stat}) \pm 0.12(\text{syst})$ mb.

In order to compare the SD measurement with results of other experiments and theoretical models, an extrapolation from ($-5.5 < \log_{10}\xi_X < -2.5$) to $\xi < 0.05$ was performed. Similarly, the DD cross section was extrapolated to $\Delta\eta > 3$. This was performed using extrapolation factors determined with PYTHIA 8 MBR, which describes all aspects of our data well. The multiplicative factor of

DOI: 10.1051/epjconf/201612002003

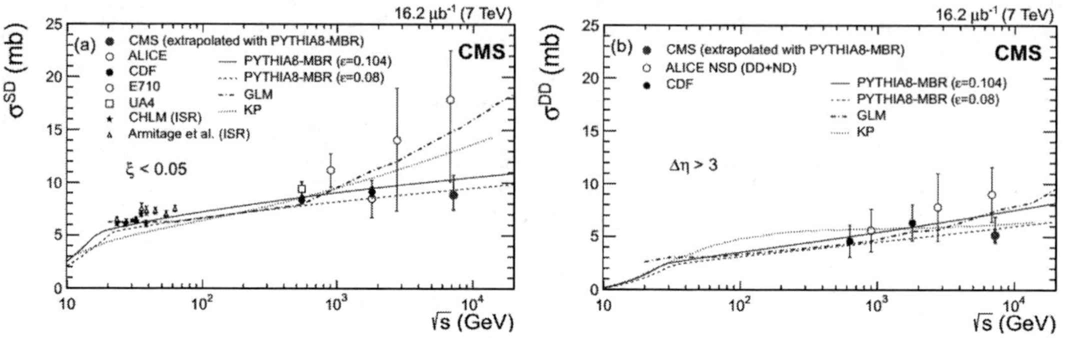

Figure 3. Diffractive cross sections as a function of collision energy measured in pp and $p\overline{p}$ collisions compared to PYTHIA 8 MBR (ϵ = 0.08, 0.104) and other model predictions [5–7]. Left: total single-diffractive cross section for $\xi < 0.05$. Right: total double-diffractive cross section for $\Delta\eta > 3$. The inner (outer) error bars of the CMS data points correspond to the statistical and systematic (and the additional extrapolation) uncertainties added in quadrature.

approximately 2 was obtained for this extrapolation. The final extrapolated SD and DD cross sections were found to be $\sigma^{SD} = 8.84 \pm 0.08(\text{stat})^{+1.49}_{-1.38}(\text{syst})^{+1.17}_{-0.37}(\text{extrap})$ mb ($\xi < 0.05$) and $\sigma^{DD} = 5.17 \pm 0.08(\text{stat})^{+0.55}_{-0.57}(\text{syst})^{+1.62}_{-0.51}(\text{extrap})$ mb ($\Delta\eta > 3$), respectively. Figure 3 presents these extrapolated cross sections compared to the ALICE results and a collection of measurements at lower $\sqrt{s}$, as well as predictions by various models.

3 Dijet events with a rapidity gap

The fraction of jet-gap-jet events in $p\overline{p}$ collisions was measured to be around 1% at the Tevatron. Here we present the first observation of such events at the LHC, along with the CSE fraction at $\sqrt{s}$ = 7 TeV, using two jets with $p_T > 40$ GeV and $1.5 < |\eta| < 4.5$. The CSE signal is extracted from the charged particle multiplicity distribution in the central region $|\eta| < 1$ between the jets, using tracks with $p_T > 0.2$ GeV. For this analysis it is essential that the probability of overlapping pp collisions is low. The amount of data used correspond to $8pb^{-1}$ integrated luminosity collected in 2010 [8].

Jets are clustered with the anti-k_T algorithm, with a radius parameter R=0.5. The jet energy corrections are derived from simulation and confirmed with measurements of dijet and γ+jet energy balance. Jet-gap-jet events are simulated with the HERWIG 6 generator, which takes into account the reactions involving hard CSE via elastic parton-parton scattering according to the Mueller-Tang model, which is based on BFKL evolution [9]. Multi-parton interactions (MPI) in the underlying event is provided by the JIMMY package [10].

The data are divided into three bins of the p_T of the lower-energy jet in the dijet system; p_T^{jet2}: 40–60 GeV, 60–100 GeV and 100–200 GeV. At most one primary collision vertex is allowed in each event, and the jets have to satisfy the conditions $|\eta^{\text{jet1}}| > 1.5$, $|\eta^{\text{jet2}}| > 1.5$ and $\eta^{\text{jet1}} \cdot \eta^{\text{jet2}} < 0$, in order to ensure the necessary separation between the jets and the region (gap) where the charged particle multiplicity is measured.

The left panel of Figure 4 shows the measured track multiplicity distribution for the $60 < p_T^{jet2} <$ 100 GeV bin. The data are well described by the PYTHIA 6 simulation (LO DGLAP), except the lowest multiplicity bins, where a large excess is observed, indicating CSE events. This excess is however well described by the HERWIG 6 model (LL BFKL). It was verified that the kinematical

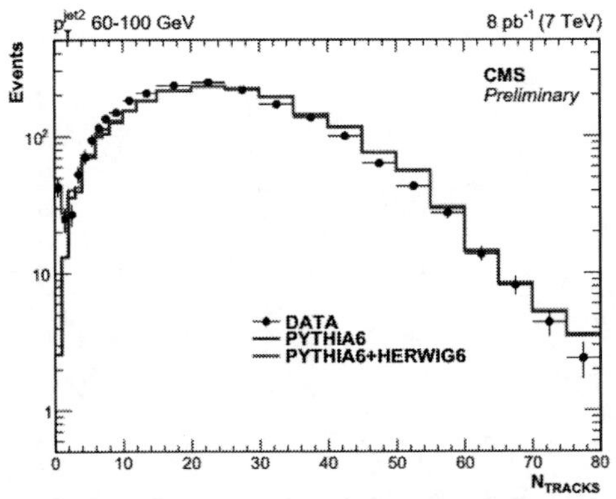

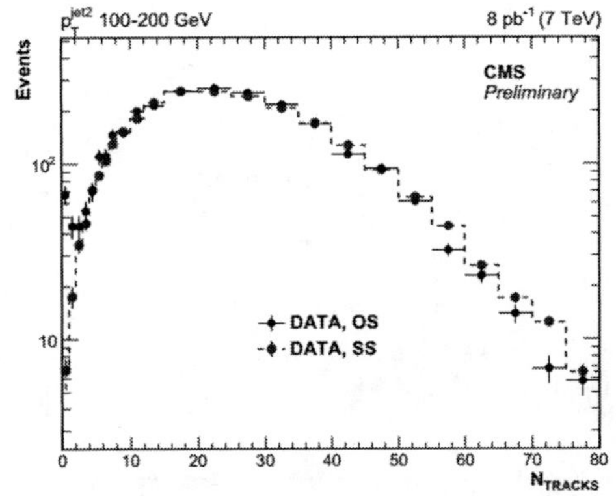

Figure 4. Left: Charged particle multiplicity distributions for p_T^{jet2} between 60 and 100 GeV, compared to predictions of PYTHIA 6 and HERWIG 6. The normalization of the MC distribution is obtained from a fit to the data. Right: Charged particle multiplicity distribution in the highest p_T^{jet2} sample (100–200 GeV), shown together with the charged multiplicity distribution from the same-sign sample.

distributions of jets in the zero-track bin agree well with MC predictions. It was also found that the CSE dijets are more balanced, both in azimuthal angle and transverse momentum, compared to non-CSE dijets.

In order to quantify the contribution of CSE events in inclusive the dijet sample and measure the CSE fraction, f_{CSE}, one has to subtract the background events first, estimated to have a non-CSE origin. The f_{CSE} fraction is not sensitive to trigger inefficiencies and jet reconstruction uncertainties, as they cancel in the ratio.

Two methods are used to estimate the number of background events. First, the shape of the charged particle multiplicity distribution is obtained from a sample with the two leading jets being on the same side of the CMS detector (SS sample, $|\eta^{\rm jet1}| > 1.5$, $|\eta^{\rm jet2}| > 1.5$ and $\eta^{\rm jet1} \cdot \eta^{\rm jet2} > 0$). The region in which the multiplicity was measured was widened to be $|\eta| < 1.2$ (for details, see [8]) and normalized to the nominal opposite-sign (OS) sample. The number of events in the first multiplicity bins is taken as the estimate of the background. An example of the SS and OS multiplicity distributions is shown in the right panel of Fig. 4 for the $100 < p_T^{jet2} < 200$ GeV bin. The second method involves a negative binomial fit to the charged particle multiplicity distribution restricted to the multiplicity region 3–35. The number of background events with these two methods agree within statistical uncertainties. The non-CSE background amounts to about half of the events in the first three bins of the multiplicity distribution.

The left panel of Figure 5 presents the CSE fraction in the three p_T^{jet2} regions, compared to results from the Tevatron ($\sqrt{s}$ = 1.8 TeV). The CSE fraction shows a slight increase with p_T^{jet2}, while a suppression of the gap fraction is observed as $\sqrt{s}$ increases. This trend is in agreement with that observed at different Tevatron energies ($\sqrt{s}$ = 0.63 and 1.8 TeV), and can be explained by a changing rescattering contribution, in which the interactions between spectator partons destroy the rapidity gap. The measurement is also presented on the right panel of Fig. 5, which shows the dependence of f_{CSE} on the η-separation of the two jets, $\Delta\eta_{jj}$, for each p_T^{jet2} bin. The gap fraction increases with $\Delta\eta_{jj}$, not reproduced by the Mueller-Tang model.

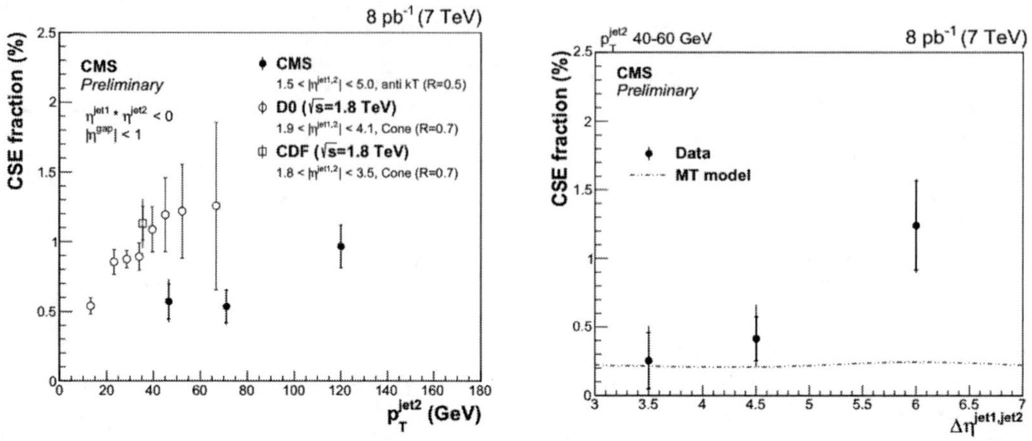

Figure 5. Left: The fraction f_{CSE} as a function of p_T^{jet2} as measured by CMS for $40 < p_T^{jet2} < 60$ GeV, $60 < p_T^{jet2} < 100$ GeV, and $100 < p_T^{jet2} < 200$ GeV at $\sqrt{s} = 7$ TeV, and by D0 [11] and CDF [12] at $\sqrt{s} = 1.8$ TeV. The details of the jet selections are given in the legend. The results are plotted at the mean value of p_T^{jet2} in the bin. Right: The fraction f_{CSE} as a function of $\Delta\eta_{jj}$ measured in the three bins of p_T^{jet2}: 40-60 GeV, 60-100 GeV, and 100-200 GeV. The results are plotted at the center of the $\Delta\eta_{jj}$ bin. Inner and outer error bars correspond to the statistical, and the statistical and systematic uncertainties added in quadrature, respectively.

In Summary, we have presented various measurements concerning color-singlet exchange processes, in particular, inclusive single- and double-diffraction cross sections (those visible by our apparatus as well as extrapolations to a larger phase space), and dijet events with a large rapidity gap between them, that need to invoke CSE to explained their presence in the data. Comparisons to earlier measurements and phenomenological models contribute to the further scrutiny of these important processes in hadron-hadron interactions.

References

[1] J.D. Bjorken, Phys. Rev. **D47**, 101 (1993)
[2] S. Chatrchyan et al. (CMS), JINST **3**, S08004 (2008)
[3] V. Andreev et al., Eur. Phys. J. **C67**, 601 (2010)
[4] V. Khachatryan et al. (CMS), Phys. Rev. **D92**, 012003 (2015), arXiv:1503.08689
[5] E. Gotsman, E. Levin, U. Maor, Phys. Lett. **B716**, 425 (2012), arXiv:1208.0898
[6] A.B. Kaidalov, M.G. Poghosyan, Eur. Phys. J. **C67**, 397 (2010), arXiv:0910.2050
[7] A.B. Kaidalov, M.G. Poghosyan (2011), arXiv:1109.3697
[8] V. Khachatryan et al. (CMS) (2015), CMS-PAS-FSQ-12-001
[9] A.H. Mueller, W.K. Tang, Phys. Lett. **B284**, 123 (1992)
[10] J.M. Butterworth, J.R. Forshaw, M.H. Seymour, Z. Phys. **C72**, 637 (1996), hep-ph/9601371
[11] B. Abbott et al. (D0), Phys. Lett. **B440**, 189 (1998), hep-ex/9809016
[12] F. Abe et al. (CDF), Phys. Rev. Lett. **80**, 1156 (1998)

EPJ Web of Conferences **120**, 02004 (2016) DOI: 10.1051/epjconf/201612002004
ISMD 2015

Evidence for Non-Exponential Differential Cross-Section of pp Elastic Scattering at Low $|t|$ and $\sqrt{s} = 8$ TeV by TOTEM

T. Csörgő[1,2,a] for the TOTEM Collaboration

[1] *Wigner Research Centre for Physics, Budapest, Hungary*
[2] *KRF, Gyöngyös, Hungary*

Abstract.
Recently published and preliminary results of the TOTEM experiment are presented, emphasizing a recent discovery of a non-exponential behaviour of the differential cross-section of elastic proton-proton scattering, that TOTEM measured with an unprecedented precision at the centre-of-mass energy $\sqrt{s} = 8$ TeV based on a high-statistics data sample obtained with the $\beta_* = 90$ m optics of CERN LHC. Both the statistical and systematic uncertainties remained below 1% , except for the t-independent contribution from the overall normalisation. This measurement allowed TOTEM to exclude a purely exponential differential cross-section in the range of four-momentum transfer squared $0.027 < |t| < 0.2$ GeV^2 with a significance greater than 7 σ. In this context we also highlight the innovative TOTEM recalibration of LHC optics, that used elastic scattering data measured by the world's largest and most complex Roman Pot detector system, and discuss recent preliminary TOTEM data on the Coulomb-Nuclear interference region with its physics implications.

1 Introduction

The TOTEM experiment at CERN LHC specializes in forward physics measurements, that deal with colorless exchange, including elastic proton-proton (pp) scattering, single and double diffractive pp scattering [1, 2], and has a program to measure central exclusive production in collaboration with the CMS experiment, the CMS-TOTEM Precision Proton Spectrometer (CT-PPS) project [3].

The main focus of this conference contribution is recent TOTEM the discovery of a non-exponential behaviour of the differential cross-section of elastic pp scattering [4] at low $|t|$ and at $\sqrt{s} = 8$ TeV energy of CERN LHC, which is put into context of other recent TOTEM results, in particular the recalibration of LHC optics from TOTEM data [5] and preliminary results on the observation of Coulomb-nuclear interference at very low four-momentum transfer at LHC [6].

2 Earlier results on the low-$|t|$ behaviour of $d\sigma/dt$ of elastic pp scattering

Let us first summarize what has been known before 2015 about this low-|t| region of elastic scattering in pp and $p\overline{p}$ scattering. Traditionally, the differential cross-section of elastic scattering is characterized in the low-|t| region, before the diffractive minimum, by an approximately exponential behaviour,

[a] e-mail: tcsorgo@cern.ch

$\exp(-B|t|)$, where $B \equiv B(s,t)$ is called the slope parameter, where $s = (p_1+p_2)^2$ and $t = (p_1-p_3)^2$ are Mandelstam variables called squared center of mass energy and the squared four-momentum transfer.

Experimental data reported before 2015 on pp and $p\overline{p}$ elastic scattering were consistent with a simple exponential behaviour, $B(s,t) = B(s)$ independent of t in the low $|t|$ region, before the vicinity of t_{dip}, the region of the diffractive minimum. The slope parameter $B = B(s)$ was measured to increase with increasing $\sqrt{s}$, corresponding to the shrinking of the diffractive cone and the increase of the total elastic cross-section of proton-proton scattering with increasing center of mass energies. Nevertheless, some of the earlier experiments reported hints of deviations from such a simple exponential behaviour. For example, small angle pp elastic scattering at $460 < \sqrt{s} < 2900$ GeV energies indicated a marked change of the slope parameter as a function of t in the $|t| = 0.1$ GeV^2 region as early as in 1972 [7]. A decade later, an indication of a break in the slope parameter B was observed in $p\overline{p}$ reactions, as a function of t, at the ISR energy of 52.8 GeV in ref. [8]. Other measurements at ISR found a simple exponential behaviour, $B \neq B(t)$ at 53 and 62 GeV ISR energies, however, found an indication of a change of the slope parameter at $|t = 0.15$ GeV^2 at the lower $\sqrt{s} = 31$ GeV colliding energies [9], and found that the data are better described by a Gaussian $\exp(-B|t| - Ct^2)$ than a simple exponential $\exp(-B|t|)$ shape. However, these early results were limited by statistics and systematic effects and the significance of these indications of non-exponential behaviour did not reach the discovery, 5σ level. In measurements of elastic $p\overline{p}$ scattering at Tevatron energies, no indications of a possible non-exponential behaviour were found. The first results on the differential cross-section of elastic pp scattering at LHC energies were reported by TOTEM in the dip region [10], but the TOTEM measurement was soon extended to the low-$|t|$ region using a special LHC run [11]. Both measurements were consistent with an exponential behaviour in the diffraction cone, and the measured slope parameter in the low-$|t|$ region was consistent with the B value determined from the data closer to the dip region. However, a high statistics low-$|t|$ data-set from a special LHC run allowed TOTEM to establish the non-exponential behaviour of the differential cross-section of elastic pp scattering with a greater than 7 σ significance at $\sqrt{s} = 8$ TeV LHC colliding energies [4].

The experimental observation of non-exponential low-$|t|$ behavior of the differential cross-section of elastic pp scattering was actually anticipated by a number of theoretical models. To explain the first experimental indications of a non-exponential behaviour at ISR energies, ref. [12] suggested that the corresponding non-linearity of the vacuum or Pomeron Regge trajectory is due to a nearby singularity, the two-pion threshold, restricted by t-channel unitarity requirements. This idea has been applied recently [13] to explain the non-exponential behaviour of preliminary TOTEM data at 8 TeV extrapolating fit parameters from ISR to LHC energies [14]. An independent line of argumentation was based on multiple diffraction theory of elastic proton-proton scattering [15] where the electromagnetic form-factor of the colliding protons and a simple phenomenological parameterizaton of the cluster-averaged parton-parton forward scattering amplitude was shown to lead naturally to a non-exponential behaviour or a t-dependent slope-parameter $B(t)$. This general quantum-optical scattering theory is also able to describe the first TOTEM data if the proton is modelled as a quark-diquark composite object, $p = (q,d)$ even if the internal structure of the quark and the diquark is a featureless Gaussian distribution as far as elastic scattering of protons is concerned. The resulting multiple diffraction theory formulas for elastic pp scattering indicate a non-exponential feature both in the "springy Pomeron" picture of Grichine [16] and the Bialas-Bzdak quark-diquark model of pp elastic scattering, [18], in particular when the non-vanishing nature of the real part of the forward scattering amplitude and the unitarity constraints are both taken into account [17].

3 Non-exponential behaviour of low-$|t|$ elastic pp scattering

In this section, the main results of ref. [4] are summarized and the method of the analysis is outlined.

Detector setup. The TOTEM experiment [1, 2] is located at the LHC interaction point (IP) 5, together with the CMS experiment. The full experimental apparatus of TOTEM consists of two inelastic telescopes, T1 and T2, based on Cathode Strip Chamber and GEM technology, respectively, and a system of Roman Pot detectors. Roman Pots are movable beam-pipe insertions that approach the LHC beam very closely in order to detect particles scattered at very small angles. Currently, 26 Roman Pot detectors belong to TOTEM and the CT-PPS project: this is the largest and most complex Roman Pot detector system that has ever been operated at a collider. In the measurement of the non-exponential shape of the differential cross-section of elastic pp scattering, only the Roman Pot (RP) system, the sub-detector relevant for elastic scattering measurement, was utilized.

The RP-s are organised in two stations placed symmetrically around the IP: one on the left side (in LHC sector 45), one on the right (sector 56). Each station is formed by two units: near (214 m from the IP) and far (220 m). Each unit includes three RPs: one approaching the beam from the top, one from the bottom and one horizontally. Each RP hosts 10 "edgeless" silicon strip sensors with a pitch of 66 μm, each having a strongly reduced insensitive edge of a few tens of micrometers facing the beam. The sensors are equipped with trigger-capable electronics. Since elastic scattering events consist of two anti-parallel protons, the detected events can have two topologies, called diagonals: 45 bottom – 56 top and 45 top – 56 bottom.

Data-set. The measurement presented here is based on data taken in July 2012, during the LHC fill number 2836 at the centre-of-mass energy $\sqrt{s} = 8$ TeV. The vertical RPs were inserted at a distance of 9.5 times the transverse beam size, $\sigma_{\rm beam}$. Initially two, later three colliding bunch-pairs were used, each with a typical population of $8 \cdot 10^{10}$ protons, corresponding to an instantaneous luminosity of about $10^{28}\,\mathrm{cm^{-2}s^{-1}}$ per bunch. The main trigger required a coincidence between the RPs in both arms, combining the near and far units of a station in *OR* to ensure maximal efficiency. During the about 11 h long data-taking, a luminosity of 735 μb^{-1} was accumulated, giving $7.2 \cdot 10^6$ tagged elastic events.

LHC Optics Determination at $\beta_* = 90$ m. For the determination of the total cross-section and the differential cross-section of elastic pp collisions with unprecedented precision by TOTEM, a precise experimental control of the LHC optics was necessary, which was achieved from the recalibration of the LHC optics at IP5 using measurement of elastic pp scattering with RP-s. For the measurement of the non-exponential shape of $d\sigma/dt$, a special LHC optics with $\beta^* = 90$ m was used, with essentially the same characteristics as at $\sqrt{s} = 7$ TeV. In the vertical plane, this optics features parallel-to-point focusing ($v_y \approx 0$) and large effective length L_y. In the horizontal plane, the almost vanishing effective length L_x simplifies the separation of elastic and diffractive events: any sizeable horizontal displacement must be due to a momentum loss ξ. Figures 1 indicate, that TOTEM achieved a precise control of LHC imperfections with perturbed LHC optics and recalibration from data at IP5: Monte-Carlo simulations indicate that the nominal uncertainties of LHC optics have been reduced by factors of 2 - 10, see Refs. [5], [4] for details of this innovative TOTEM method.

Analysis steps. The precise control of LHC optics, together with kinematic cuts that are based on the physics of elastic scattering were needed to measure precisely the differential cross-section. The applied analysis steps included detector alignment, optics recalibration, elastic event selection based on kinematic cuts, resolution unfolding, acceptance correction, background subtraction, detection and efficiency correction, corrections for angular resolution, normalization and binning.

Compatible results have been obtained from the data originating from different bunches, different diagonals and different time periods. In addition, the complete analysis chain has been applied in two independent analysis implementations, yielding compatible results. The details of these steps of the analysis are described in ref. [4].

Results on the non-exponential behaviour. The differential cross-section of elastic pp scattering at 8 TeV is indicated on the top panel of Figure 2, tabulated in Table 3 of ref. [4]. The binning

 DOI: 10.1051/epjconf/201612002004

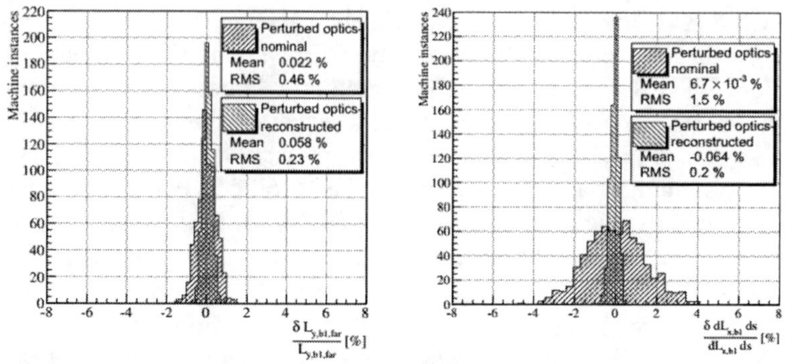

Figure 1. Monte-Carlo error distribution of $\beta_* = 90$ m LHC Optical Functions L_y and dL_x/ds for E = 4 TeV before and after the estimation of the LHC Optics from TOTEM pp elastic scattering data [5].

of this plot is optimized so that the bin size corresponds to 1σ of the resolution in t. In order to visualise small deviations from the leading pure-exponential behaviour, the bottom panel of Figure 2 shows also the relative difference of the cross-section from a reference exponential (pure exponential fit using statistical uncertainties only). This plot indicates the non-exponentiality of the data: pure exponentials would look like (nearly) linear functions in this kind of representation.

To study the detailed behaviour of the differential cross-section, a series of fits has been made using the parametrisation:

$$\frac{d\sigma}{dt}(t) = \frac{d\sigma}{dt}(t=0) \times \exp\left(\sum_{i=1}^{N_b} b_i\, t^i\right), \tag{1}$$

which includes the exponential fit as a special case for ($N_b = 1$) and its straight-forward extensions ($N_b = 2, 3$). The fits have been performed by the standard least-squares method, using the covariance matrix including both the statistical and systematic components. For a detailed description of the determination of the systematic errors, as well as for a discussion on the significance of these observations, see ref. [4]. Two independent lines of analysis indicate that the significance of the non-exponential behaviour in these TOTEM measurements is greater than 7σ. Using the refined, non-exponential parameterizations to extrapolate to the optical point, $t = 0$, yields total cross-section valued compatible with earlier TOTEM determinations, in all cases neglecting the effects due to the Coulomb interactions.

Preliminary TOTEM Results on the Coulomb-Nuclear Interference. In a preliminary analysis, TOTEM investigated if the origin of the non-exponentiality at low-$|t|$, still dominated by the hadronic interactions, can be due to Coulomb interference effects or not. This was achieved with a measurement of the elastic scattering in the Coulomb-Nuclear Interference (CNI) region. The preliminary TOTEM data, shown in Fig. 3, extend the four momentum transfer range of the TOTEM measurements down to $|t| \approx 6x10^{-4}$ GeV^2. These results indicate that the non-exponential behaviour of $d\sigma/dt$ around 0.1 GeV^2 is not due to Coulomb interference effects, but is a property of the hadronic amplitude [6], which allows TOTEM to exclude the simplified West-Yennie interference formula [22]. The constraining power of this TOTEM preliminary analysis on the hadronic phase allows us to study the impact parameter dependence of elastic pp scattering and to measure the parameter ρ. The results of this TOTEM preliminary analysis confirm earlier total pp scattering cross-sections measurements of TOTEM, that neglected the Coulomb-Nuclear Interference effects.

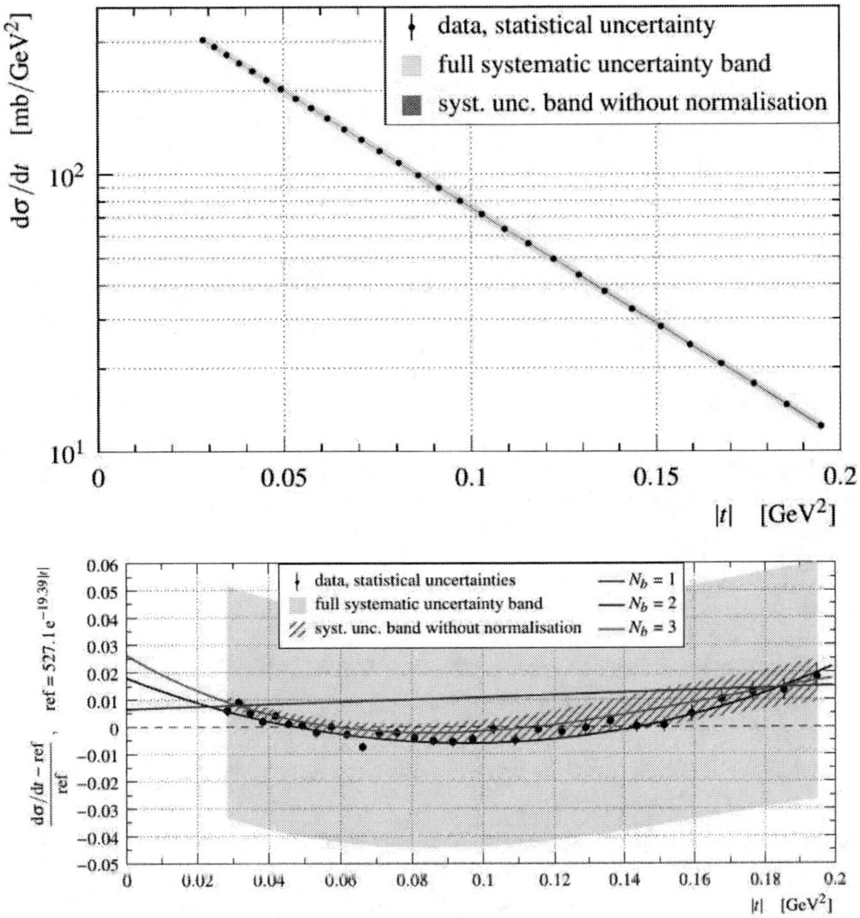

Figure 2. The *top panel* shows the differential cross-section of elastic *pp* scattering at 8 TeV as measured by TOTEM at CERN LHC. The black dots represent TOTEM data with statistical uncertainty bars. The widest error band (yellow) corresponds to full systematic uncertainty, including the normalization error. The narrow band around the data points indicate all systematic errors except the overall normalization uncertainty. Both bands are centered around the fit curve with $N_b = 3$. The *bottom panel* shows the same data points, however, now the relative deviation from an exponential reference distribution is shown (see the vertical axis). From ref. [4].

Acknowledgments: T. Cs. would like to thank the Organizers of this Symposium for their support and for an inspiring and useful meeting. This work was supported by the Magnus Ehrnrooth foundation (Finland), the Waldemar von Frenckell foundation (Finland), the Academy of Finland, the Finnish Academy of Science and Letters (the Vilho, Yrjö and Kalle Väisälä Fund), and the OTKA grant NK 101438 (Hungary).

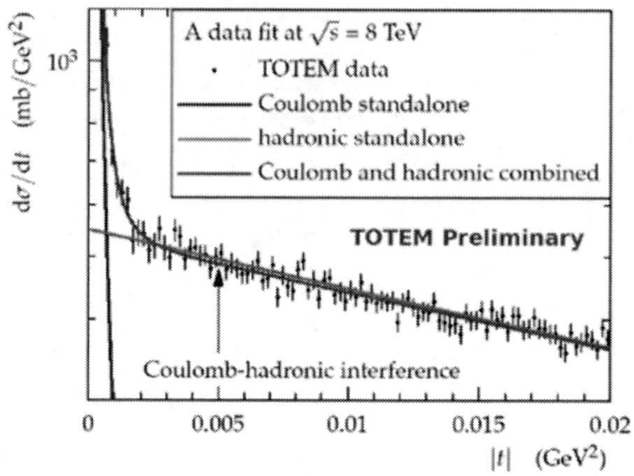

Figure 3. TOTEM preliminary results with a special $\beta_* = 1000$ m run, Roman Pots approaching the beam as close as 3.5 σ probe the Coulomb-Nuclear Interference (CNI) region. From [6].

References

[1] G. Anelli *et al.* [TOTEM Collaboration], JINST **3**, S08007 (2008).
[2] G. Antchev *et al.* [TOTEM Collaboration], Int. J. Mod. Phys. A **28**, 1330046 (2013).
[3] M. Albrow, *et al.* [CMS and TOTEM Collaborations] preprint CERN-LHCC-2014-021, TOTEM-TDR-003, CMS-TDR-13, Sept. 2014, `http://cds.cern.ch/record/1753795`.
[4] G. Antchev *et al.* [TOTEM Collaboration], Nucl. Phys. B **899**, 527 (2015).
[5] G. Antchev *et al.* [TOTEM Collaboration], New J. Phys. **16**, 103041 (2014).
[6] M. Deile for the TOTEM Collaboration and the CT-PPS Project, *Status Report of TOTEM and the CT-PPS Roman Pot Operations* talk at the LHCC Open Session, September 23, 2015.
[7] G. Barbiellini *et al.*, Phys. Lett. B **39**, 663 (1972).
[8] M. Ambrosio *et al.* Phys. Lett. B **115**, 495 (1982).
[9] A. Breakstone *et al.* Nucl. Phys. B **248**, 253 (1984).
[10] G. Antchev *et al.* [TOTEM Collaboration], Europhys. Lett. **95**, 41001 (2011).
[11] G. Antchev *et al.* [TOTEM Collaboration], Europhys. Lett. **101**, 21002 (2013).
[12] G. Cohen-Tannoudji, V. V. Ilyin and L. L. Jenkovszky, Lett. Nuovo Cim. **5S2**, 957 (1972).
[13] L. Jenkovszky and A. Lengyel, Acta Phys. Polon. B **46**, no. 4, 863 (2015).
[14] D. A. Fagundes, L. Jenkovszky, E. Q. Miranda, G. Pancheri and P. V. R. G. Silva, arXiv:1509.02197 [hep-ph].
[15] R. J. Glauber and J. Velasco, Phys. Lett. B **147**, 380 (1984).
[16] V. M. Grichine, Eur. Phys. J. Plus **129**, 112 (2014).
[17] F. Nemes, T. Csörgő and M. Csanád, Int. J. Mod. Phys. A **30**, no. 14, 1550076 (2015).
[18] A. Bialas and A. Bzdak, Acta Phys. Polon. B **38**, 159 (2007).
[19] J. R. Cudell *et al.* [COMPETE Collaboration], Phys. Rev. Lett. **89**, 201801 (2002).
[20] S. Chatrchyan *et al.* [CMS and TOTEM Collaborations], Eur. Phys. J. C **74**, no. 10, 3053 (2014).
[21] G. Antchev *et al.* [TOTEM Collaboration], Eur. Phys. J. C **75**, no. 3, 126 (2015).
[22] G. B. West and D. R. Yennie, Phys. Rev. **172**, 1413 (1968).

EPJ Web of Conferences **120**, 02005 (2016) DOI: 10.1051/epjconf/201612002005
ISMD 2015

Utilizing the CGC/saturation approach for soft scattering at high energies

E. Gotsman[1,a]

[1] *School of Physics and Astronomy, Tel Aviv University, Tel Aviv, 69978, Israel*

Abstract. We propose a consistent approach to describe soft interactions at LHC and higher energies based on the BFKL Pomeron and CGC/saturation. The Green function of the Pomeron is calculated in the framework of CGC/saturation approach. Having eight parameters we obtained a good description of the experimental data at high energies ($W \geq 0.546\,TeV$).

1 Introduction

Strong interactions at high energy are considered as a typical example of processes that occur at long distances, where the unknown confinement of quark and gluons plays a crucial role making all theoretical efforts to treat these processes, fruitless. As a result the description of these soft processes, is the arena of high energy phenomenology based on Pomeron calculus [1].

To overcome this, we suggest an approach which attempts to describe soft interactions by adopting well established aspects of QCD in the saturation region [2], and extending these to the region of soft interactions (low Q^2). Our basic assumption, is that the BFKL Pomeron [3], at high energies, describes both hard and soft processes.

2 Theoretical Input

In this section we briefly review our model which succesfully describes elastic, diffractive [4, 5], inclusive cross sections [6] and long range rapidity correlations [7] The main ingredient of our model is the BFKL [3] Pomeron Green function, that we obtained using a CGC/saturation approach [8]. We determined this function from the solution of the non-linear Balitsky-Kovchegov equation [9], using the MPSI approximation [10] to sum enhanced diagrams, shown in Fig. 1-a. It has the following form:

$$
\begin{aligned}
G^{\text{dressed}}(T) \quad &= \quad a^2(1-\exp(-T)) + 2a(1-a)\frac{T}{1+T} + (1-a)^2 G(T) \\
\text{with} \quad & G(T) = 1 - \frac{1}{T}\exp\left(\frac{1}{T}\right)\Gamma_0\left(\frac{1}{T}\right)
\end{aligned}
\tag{2.1}
$$

[a] e-mail: gotsman@post.tau.ac.il

DOI: 10.1051/epjconf/201612002005

$$T(s,b) \;=\; = \;\phi_0 S(b,m)\, e^{0.63\lambda \ln(s/s_0)} \text{ with } S(b,m) \;=\; \frac{m^2}{2\pi} e^{-mb} \tag{2.2}$$

See [5] for the role of the variables.

We use a two channel model [5], based on the Good-Walker [11] formalism, which allows us to calculate the diffractive production in the region of small masses.

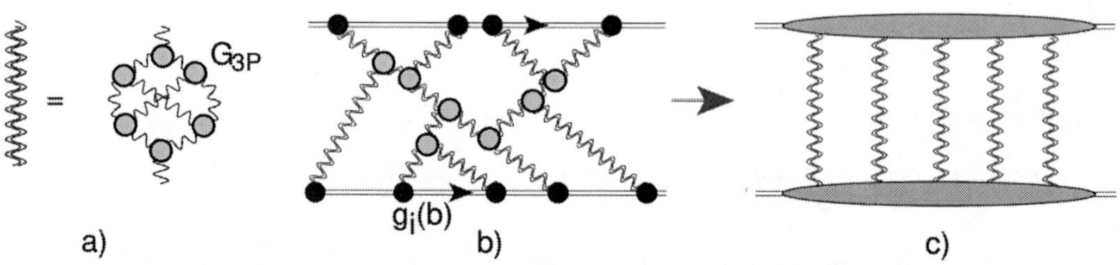

Figure 1. Fig. 1-a shows the set of the diagrams in the BFKL Pomeron calculus that produce the resulting (dressed) Green function of the Pomeron in the framework of high energy QCD. In Fig. 1-b the net diagrams which include the interaction of the BFKL Pomerons with colliding hadrons are shown. The sum of the diagrams reduces to Fig. 1-c after integration over positions of G_{3P} in rapidity.

In the eikonal approach [5] we parametrize the arbitrary functions $\Omega_{ik}(s,b)$ (opacity) in the form

$$\Omega_{ik}(s,b) \;=\; \int d^2b' d^2b'' g_i(m_i,b')\, g_k(m_k b'')\, G^{\rm dressed}\left(T\left(Y,\vec{b}-\vec{b}'-\vec{b}''\right)\right), \tag{2.3}$$

The summation of the 'net' diagram, is then given by the following simplified expression (see Ref. [12] for details)

$$\Omega(s,b)_{i,k} \;=\; \int d^2b' \; \frac{g_i\left(\vec{b}'\right) g_k\left(\vec{b}-\vec{b}'\right) \widetilde{G}^{\rm dressed}(T)}{1 + 1.29\, \widetilde{G}^{\rm dressed}(T)\left[g_i\left(\vec{b}'\right) + g_k\left(\vec{b}-\vec{b}'\right)\right]}, \tag{2.4}$$

where

$$\widetilde{G}^{\rm dressed}(T) \;=\; \int d^2b\, G^{\rm dressed}(T(Y,b)). \tag{2.5}$$

The coefficient 1.29 results from the extraction of the value of G_{3P} in the CGC/saturation approach.

Note that our choice of impact parameter dependence i.e. S(b) $\propto$ $exp(-mb)$ leads to the correct Froissart limit at large b, and at large momentum transfer (Q_T) it decreases as a power of Q_T, as required in perturbative QCD.

3 Diffraction production in the region of large mass

In this section we also include, in the process of diffraction production, the mechanism of production that originates from the dressed Pomeron, and has been discussed in section 2.

For single diffraction the large mass contribution can be written

$$\sigma_{sd}^{\text{large mass}} = 2\int d^2b \tag{3.6}$$

$$\Big\{\alpha^6 A^{sd}_{1;1,1}\, e^{-2\,\Omega^D_{1,1}(Y;b)} + \alpha^2\beta^4 A^{sd}_{1;2,2}\, e^{-2\,\Omega^D_{1,2}(Y;b)} + 2\,\alpha^4\beta^2 A^{sd}_{1;1,2}\, e^{-\left(\Omega^D_{1,1}(Y;b)+\Omega^D_{1,2}(Y;b)\right)}$$

$$+\,\beta^2\alpha^4 A^{sd}_{2;1,1}\, e^{-2\,\Omega^D_{1,2}(Y;b)} + 2\beta^4\alpha^2 A^{sd}_{2;1,2}\, e^{-\left(\Omega^D_{1,2}(Y;b)+\Omega^D_{2,2}(Y;b)\right)} + \beta^6 A^{sd}_{2;2,2}\, e^{-2\,\Omega_{2,2}(Y;b)}\Big\}.$$

where

$$\Omega^D_{i,k}(Y;b) = \int d^2b'\, \frac{g_i\left(\vec{b}'\right) g_k\left(\vec{b}-\vec{b}'\right) \bar{G}^{\text{dressed}}(T)}{\left(1 + 1.29\,\bar{G}^{\text{dressed}}(T)\left[g_i\left(\vec{b}'\right) + g_k\left(\vec{b}-\vec{b}'\right)\right]\right)^2} \tag{3.7}$$

$$A^{sd}_{i;k,l}(Y, Y_{max}, Y_{min}; b) = \tag{3.8}$$

$$\int d^2b'\, \sigma_{diff}(Y, Y_{max}, Y_{min}, 1/m)\, d^2b'\, g_i g_k g_l S_{\mathbf{P}}(b', m_i)\, S_{\mathbf{P}}\left(\vec{b}-\vec{b}', m_k\right) S_{\mathbf{P}}\left(\vec{b}-\vec{b}', m_l\right),$$

where $Y = \ln(s/s_0)$, $Y_{max} = \ln\left(M^2_{max}/s_0\right)$ and $Y_{min} = \ln\left(M^2_{min}/s_0\right)$. M_{max} and M_{min} are the largest and smallest mass produced in the diffractive processes. Eq. (3.6) has simple physical meaning: each term is the product of probability to produce a mass diffractively from the dressed Pomeron (term $\exp(-\Omega_{i,k})$), and the probability of the process of single diffraction, from the dressed Pomeron ($A_{i;k,l}$).

For the double diffraction production at large mass we have

$$\sigma_{dd}^{\text{large mass}} = \int d^2b\, \left\{\alpha^4 A^{dd}_{1,1}\, e^{\,2\Omega^D_{1,1}(Y;b)} + 2\alpha^2\beta^2 A^{dd}_{1,2}\, e^{-2\Omega^D_{1,2}(Y;b)} + \beta^4 A^{dd}_{2,2}\, e^{-2\Omega^D_{2,2}(Y;b)}\right\}. \tag{3.9}$$

$$A^{dd}_{i,k} = \int d^2b\, g_i\, g_k\, S^{i,k}_{DD}(b)\, \sigma_{dd}(Y) \quad \text{where } S^{i,k}_{DD}(b) = \int d^2b'\, S_p(b', m_i)\, S_p\left(\vec{b}-\vec{b}', m_k\right), \tag{3.10}$$

The expressions for $\sigma_{diff}(Y, Y_{max}, Y_{min}, r)$ and σ_{dd} are given in [5]

4 Fitting to Experimental data

There are eight phenomenological parameters, which need to be determined by fitting to the experimental data. ϕ_0, λ and m, g_i and m_i (i = 1,2), as well as β. We determine these parameters by fitting to the experimental data on total, inelastic and elastic cross sections, single and double diffractive production cross sections, and the slope of the forward elastic differential cross section. The value of the minimal energy for data that we use is $W = 0.546$ TeV, as starting from this energy the CGC/saturation approach, is able to describe the data on inclusive production in proton-proton collisions (see Ref. [13]).

The values of the fitted parameters are given in Table I. Our results are shown in Fig. 2, and we obtain very good agreement with the experimental data, except for σ_{dd}, where our results are below the experimental points.

model	λ	ϕ_0	g_1 (GeV^{-1})	g_2 (GeV^{-1})	$m(GeV)$	$m_1(GeV)$	$m_2(GeV)$	β
2 channel	0.38	0.0019	110.2	11.2	5.25	0.92	1.9	0.58

Table 1. Fitted parameters of the model.

5 Conclusions

We have shown that a consistent model, based on the BFKL Pomeron and the CGC/saturation approach, can be built. We have demonstrated that this model successfully describes data for high energy hadron scattering usually classified as "soft" i.e elastic and diffractive cross sections [4, 5], inclusive cross sections [6]and long range correlations [7]. In addition, we hope that this paper provides credence to the arguments, that the matching with long distance physics, (where the confinement of quarks and gluons is essential), can be reached within the CGC/saturation approach, this, without requiring that the soft Pomeron should appear (as a Regge pole).

Acknowledgements

This work was done in collaboration with Evengy Levin and Uri Maor. I would like to thank my colleagues for a fruitful and enjoyable joint effort.

References

[1] P.D.B. Collins, *An introduction to Regge theory and high energy physics* (Cambridge University Press, 1977).

[2] Yuri V Kovchegov and Eugene Levin, *Quantum Choromodynamics at High Energies"*, (Cambridge Monographs on Particle Physics, Nuclear Physics and Cosmology, Cambridge University Press, 2012) .

[3] E. A. Kuraev, L. N. Lipatov, and F. S. Fadin, *Sov. Phys. JETP* **45**, 199 (1977); Ya. Ya. Balitsky and L. N. Lipatov, *Sov. J. Nucl. Phys.* **28**, 22 (1978).

[4] E. Gotsman, E. Levin and U. Maor, Eur. Phys. J. C **75**, 18 (2015).

[5] E. Gotsman, E. Levin and U. Maor, Eur. Phys. J. C **75**, 179 (2015).

[6] E. Gotsman, E. Levin and U. Maor, Phys. Lett. **B741**, 154 (2015).

[7] E. Gotsman, E. Levin and U. Maor, Eur. Phys. J. C **75**, 518 (2015).

[8] E. Levin, JHEP **1311**, 039 (2013).

[9] I. Balitsky, [arXiv:hep-ph/9509348]; *Phys. Rev.* **D60**, 014020 (1999) Y. V. Kovchegov, *Phys. Rev.* **D60**, 034008 (1999), [arXiv:hep-ph/9901281].

[10] A. H. Mueller and B. Patel, Nucl. Phys. **B425**, 471 (1994). A. H. Mueller and G. P. Salam, Nucl. Phys. **B475**, 293 (1996). G. P. Salam, Nucl. Phys. **B461**, 512 (1996); E. Iancu and A. H. Mueller, Nucl. Phys. **A730**, 460 (2004)

[11] M. L. Good and W. D. Walker, Phys. Rev. **120** 1857 (1960).

[12] E. Gotsman, E. Levin and U. Maor, Int.J.Mod.Phys. **A30**, 98 (2015). arXiv:1403.4531 [hep-ph].

[13] E. Levin and A. H. Rezaeian, Phys. Rev. D **83**, 114001 (2011). [arXiv:1102.2385 [hep-ph]], AIP Conf. Proc. **1350** (2011) 243 [arXiv:1011.3591 [hep-ph]], Phys. Rev. D **82**, 054003 (2010).

[14] J. Beringer et al. (Particle Data Group), Phys. Rev. **D86**, 010001 (2012)

[15] F. Ferro [TOTEM Collaboration], AIP Conf. Proc. **1350**, 172 (2011) ; G. Antchev *et al.* [TOTEM Collaboration], Europhys. Lett. **96**, 21002 (2011), Europhys. Lett. **96**, 21002 (2011), Europhys. Lett. **96**, 21002 (2011), **95**, 41001 (2011)

[16] M. G. Poghosyan, J. Phys. G G **38**, 124044 (2011) ALICE Collaboration, J. Phys. G G **38**, 124044 (2011) ALICE Collaboration, arXiv:0911.5430 [hep-ex].

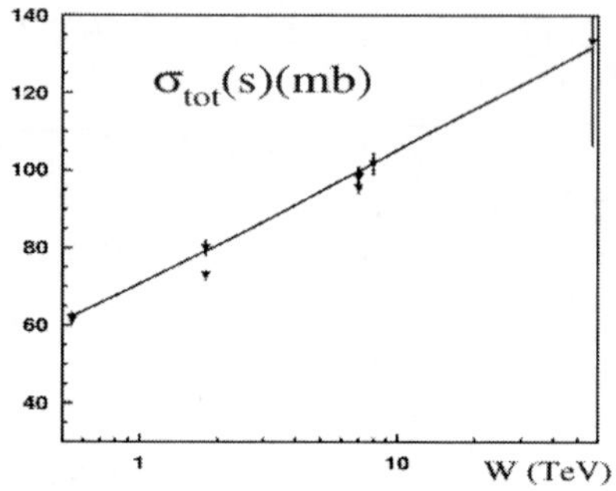

Fig. 2-a

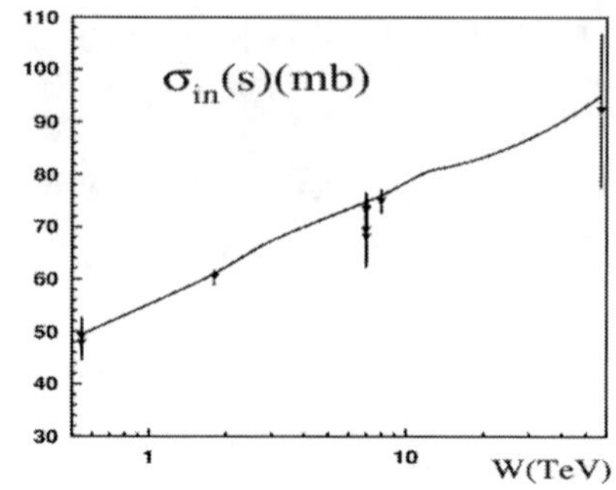

Fig. 2-b

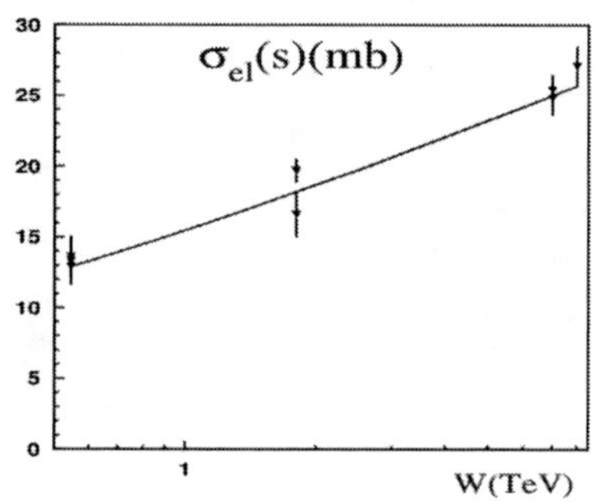

Fig. 2-c

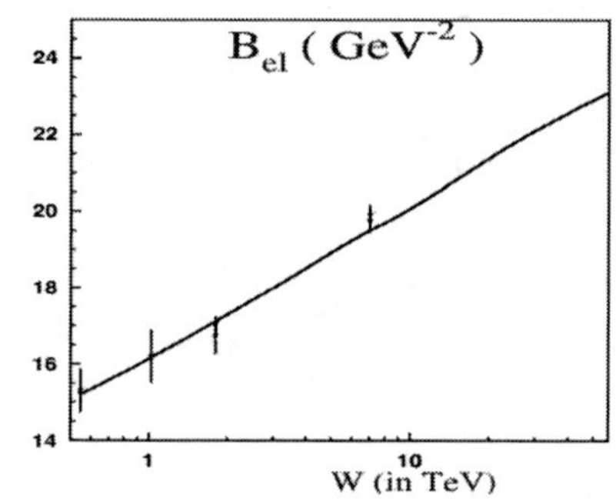

Fig. 2-d

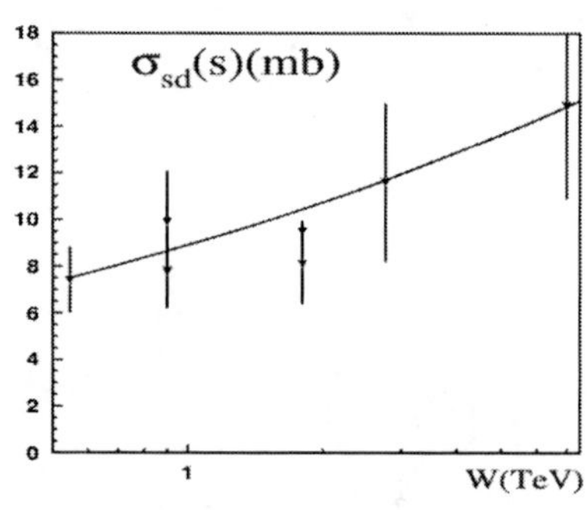

Fig. 2-e

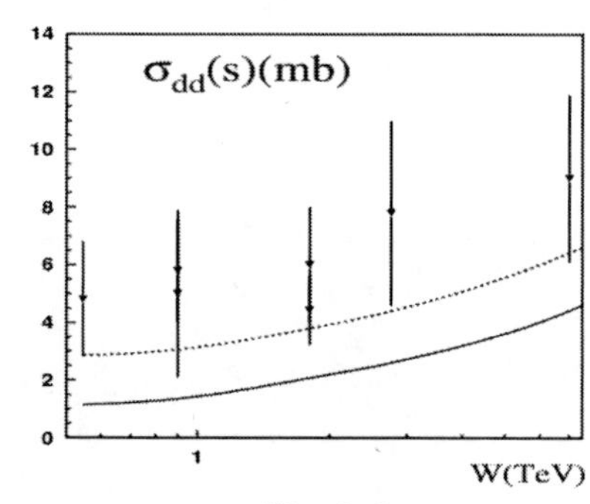

Fig. 2-f

Figure 2. Comparison with the experimental data: the energy behaviour of the total (Fig. 2-a), inelastic (Fig. 2-b), elastic cross sections (Fig. 2-c), as well as the elastic slope (B_{el},Fig. 2-d) and single diffraction (Fig. 2-e) and double diffraction (Fig. 2-f) cross sections. The solid lines show our present fit. The data has been taken from Ref. [14] for energies less than the LHC energy. At the LHC energy for total and elastic cross section we use TOTEM data [15] and for single and double diffraction cross sections are taken from Ref. [16]. The dotted line in (Fig. 2-f) which is obtained assuming factorization, is discussed in [5].

EPJ Web of Conferences **120**, 02006 (2016) DOI: 10.1051/epjconf/201612002006
ISMD 2015

Central diffractive resonance production at the LHC

Roberto Fiore[1,a], Laszlo Jenkovszky[2,b], and Rainer Schicker[3,c]

[1] *INFN and Dipartimento di Physica, Universita della Calabria, I-87036 Arcavacata di Rende, Cosenza, ITALY*
[2] *Bogolyubov ITP, National Academy of Sciences of Ukraine, 03680 Kiev, UKRAINE*
[3] *Physikalisches Institut, Im Neuenheimer Feld 226, Heidelberg University, 69120 Heidelberg, GERMANY*

Abstract. Central production of resonances resulting from the scattering of Pomerons in the central rapidity region of proton-proton scattering is studied. Estimates for relevant cross sections are presented.

1 Introduction

Inelastic diffraction in the triple Regge limit was studied in quite a number of papers, see e.g. [1, 2]. With the advent of the LHC, the interest shifted to the low missing-mass region, below the asymptotic triple Regge region. Progress in theory became possible due to a dual-Regge approach, successfully applied, in the low missing mass region [3] to single- (SD) and double (DD) diffraction dissociation, diagrams 2 and 3 of Fig. 1. By the present contribution we make one more step by applying the above dual-Regge technique to central diffractive production of resonances at the LHC [4].

The total cross section is a sum of the integrated elastic, $\sigma_{el}(s)$ and inelastic cross sections $\sigma_{inel}(s)$, the latter being a sum of single, double and central diffractive cross sections, plus the non-diffractive cross section. The proton-proton total cross section to a good approximation is a sum of a Pomeron and f Reggeon contribution, symbolically, $\sigma_t = P + f$. A number of equally good fits to the total cross section exist, see e.g. [5], most of them being compatible with the negligible contribution of the secondary Reggions (here, that of the f Reggeon) to total cross section at the LHC, $\sqrt{s} \geq 7$ TeV.

We consider:

SD $pp \longrightarrow Xp$ Single Diffraction Dissociation (or Single Diffraction),
or $pp \longrightarrow pY$ Single Diffraction Dissociation (symmetric),
DD $pp \longrightarrow XY$ Double Diffraction Dissociation (Double Diffraction),
CED $pp \longrightarrow pXp$ Central (Exclusive) Diffraction (or DPE) (or Double Pomeron Exchange).

The "symmetric" $pp \longrightarrow Xp$ and $pp \longrightarrow pY$ cross sections, for simplicity, are assumed to be equal.

For large rapidity gaps, $\Delta y \geq 3$, or, equivalently, when Pomeron exchange dominates, in the triple Pomeron limit, the above cross sections can be written as product of the Pomeron flux (first factor in brackets) times the Pomeron-proton total cross section:

---

[a] e-mail: fiore@fis.unical.it
[b] e-mail: jenk@bitp.kiev.ua; L.J. gratefully acknowledges an EMMI visiting Professorship at the University of Heidelberg for completion of this work. He is grateful to the organizers of this meeting for their hospitality and support. His work was supported also by DOMUS, Hungarian Academy of Sciences
[c] e-mail: schicker@physi.uni-heidelberg,de

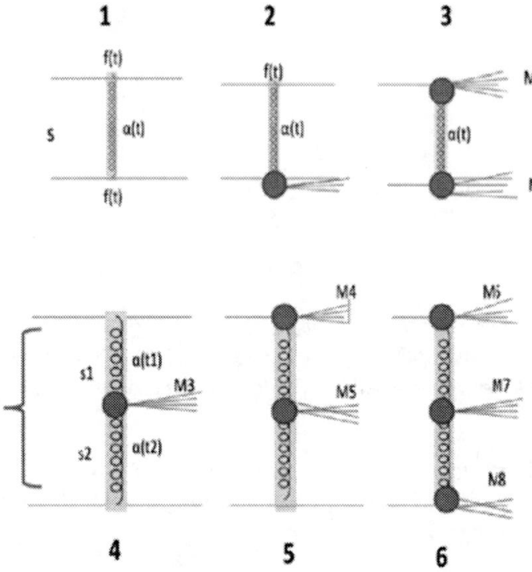

Figure 1. Regge-pole diagrams for elastic, single-, double- and central diffraction.

$$\frac{d\sigma_{SD}}{dt\Delta y} = \Big[\frac{\beta^2(t)}{16\pi}e^{2[\alpha(t)-1]\Delta y}\Big] \times \Big[k\beta^2(0)\Big(\frac{s'}{s_0}\Big)^{\epsilon}\Big], \tag{1}$$

$$\frac{d\sigma_{DD}}{dt\Delta y y_0} = \Big[k\frac{\beta^2(t)}{16\pi}e^{2[\alpha(t)-1]\Delta y}\Big] \times \Big[k\beta^2(0)\Big(\frac{s'}{s_0}\Big)^{\epsilon}\Big], \tag{2}$$

$$\frac{d\sigma_{DPE}}{dt_1 dt_2 \Delta y dy_c} = \Big[\prod_i\Big(\frac{\beta^2(t_i)}{16\pi}e^{2[\alpha(t_i)-1]\Delta y_i}\Big)\Big] \times k\Big[k\beta^2(0)\Big(\frac{s'}{s_0}\Big)^{\epsilon}\Big], \tag{3}$$

where t is the square of the four-momentum transfer at the proton vertex and Δy is the rapidity gap width. The subscript $i = 1, 2$ enumerates Pomerons in DPE; $\Delta y = \Delta y_1 + \Delta y_2$ is the sum of two rapidity-gap width in the same event, and c_i is the center in η of the centrally-produced hadronic system. Since our units below are in GeV, we use 1 GeV^{-2} = 0.389 mb in the final cross sections. We use the relations $\alpha(t) = 1.104 + 0.25t$; $\beta^2(t) \approx 6.566^2(0.9e^{4.6t} + 0.1e^{0.6t})$; $k\beta(0) = \sigma_0 = 7.249GeV^{-2}$; $s_0 = 1GeV^2$; Furthermore, in SD $s' = M^2$, $e^{\Delta y} = \xi^{-1} = s/M^2$ while in DD: $s' = ss_0/(M_1^2 M_2^2)$, $s' = M_1^2 M_2^2/s_0^2$.

The above formalism, based on the triple Regge limit (high missing masses), does not account for the low-mass resonances. In SD and DD they were included [3] by means of a dual-Regge model for the inelastic pPM vertex. We extend the above formalism to low-mass central systems by replacing the asymptotic expression of Pomeron-Pomeron total Pomeron-Pomeron cross section by its low missing mass extension [4] that includes also resonances.

Experimentally, central production in proton-proton collisions has been studied from the low energy range $\sqrt{s}$ = 12.7-63 GeV at the ISR at CERN up to the presently highest energy of $\sqrt{s}$ = 13 TeV available in Run II at the LHC. Ongoing data analysis of central production events include data taken by the COMPASS collaboration at the SPS [6], the CDF collaboration at the TEVATRON [7], the STAR collaboration at RHIC [8], and the ALICE and LHCb collaborations at the LHC [9, 10].

DOI: 10.1051/epjconf/201612002006

2 Low missing masses, nuclear resonances. Basic formulae for single (SD), double (DD) and central (CED) diffraction

- Total cross section: $pp \to X: \quad \sigma^t(s) = g_1(0)g_2(0)(s/s_0)^{\alpha(0)-1}$; here $\beta_1 = \beta_2, \alpha(t)$ is the Pomeron trajectory, whose form will be specified;
- Elastic scattering: $pp \to pp$,
$$\frac{d\sigma}{dt}(s,t) = \frac{F(t)}{16\pi}(s/s_0)^{2[\alpha(t)-1]}, \tag{4}$$
with the form factor $F(t) = g_1(t)g_2(t)$ to be specified;
- single diffraction dissociation, SD:
$$16\pi s\frac{d^2\sigma_{SD}}{dtdM^2}(s,t,M^2) = [g_{Pp}(t)\left(\frac{M^2}{s}\right)^{1-2\alpha(t)}][(g_{PPp}(t)g^2_{Pp}(0)\left(\frac{M^2}{s_0}\right)^{\alpha(0)-1}]. \tag{5}$$

With a simple, smooth Regge behaved total cross section for both Pp and PP scattering one has:
$$\sigma^t(M^2) = (M^2)^{0.1} + R(M^2)^{-0.5}. \tag{6}$$

In what follows we will include also resonances in M^2.

Pomeron-Pomeron (PP) scattering is similar to photon-photon scattering, the only main difference being in C-parities, respectively positive and negative, of the Pomeron and photon. High-virtuality $\gamma^*\gamma^*$ scattering is a favourite process in the framework of perturbative QCD, where the total cross section was calculated in Ref. [11]. In the leading-order BFKL
$$\sigma_{tot}^{\gamma^*\gamma^*} = \frac{\sigma_0}{\sqrt{Q_1^2Q_2^2Y}}(s/s_0)^\lambda, \tag{7}$$
where $Q_i^2 = -q_i^2, \quad i = 1,2$ is the photon virtuality and $Y = \ln(\frac{s}{s_0})$. The quantity σ_0 is a free parameter and the exponent $\lambda \equiv \alpha_P^{BFKL}$ is the familiar BFKL eigenvalue $N_c\alpha_s 4\ln 2/\pi$. We recall that the transition from photon-photon to central Pomeron-Pomeron scattering is accompanied by the change of variables $t \to -Q^2$ and $s \to M^2$.

The one-by-one account of single resonances is possible, but not economic for the calculation of cross section, where a sequence of many resonances contributes at low masses. These resonances overlap and gradually disappear in the continuum at higher masses. An approach to account for many resonances, based on the idea of duality with a limited number of resonances lying on non-linear Regge trajectories, was suggested in Ref. [12]. Later on, this approach was used in Ref. [3] to calculate low mass single- and double-diffractive dissociation at the LHC.

For $s \to \infty$ and fixed t it is Regge-behaved. Contrary to the Veneziano model, DAMA not only allows for, but rather requires the use of non-linear complex trajectories providing the resonance widths via the imaginary part of the trajectory.
$$A(M_X^2,t) = a\sum_{i=f,P}\sum_J \frac{[f_i(t)]^{J+2}}{J-\alpha_i(M_X^2)}. \tag{8}$$

The pole decomposition of the dual amplitude $A(M_X^2,t)$ is that of Eq. (8). The index i sums over the trajectories which contribute to the amplitude. Within each trajectory, the second sum extends over the bound states of spin J.

The pole residue $f(t)$ appearing in the $PP \to PP$ system is fixed by the dual model, in particular by the compatibility of its Regge asymptotics with Bjorken scaling and reads

$$f(t) = (1 - t/t_0)^{-2}, \tag{9}$$

where $t_0 = 0.71$ GeV^2. Note that the residue enters with a power $(J+2)$ in Eq. (8), thereby strongly damping higher spin resonance contributions.

Following Ref. [3], we write the imaginary part of the amplitude $A(M_X^2, t)$, Eq. (8) as

$$\Im m\, A(M_X^2, t) = a \sum_{i=f,P} \sum_J \frac{[f_i(t)]^{J+2} \Im m\, \alpha_i(M_X^2)}{(J - Re\, \alpha_i(M_X^2))^2 + (\Im m\, \alpha_i(M_X^2))^2}, \tag{10}$$

whence the total cross section is

$$\sigma_t^{PP}(M_X^2) = \Im m\, A(M_X^2, t = 0). \tag{11}$$

The Pomeron-Pomeron channel, $PP \to M_X^2$, couples to the Pomeron and f channels. In order to calculate the PP cross section, we therefore take into account the trajectories associated with the $f_0(980)$ and the $f_2(1270)$ resonances, and the Pomeron trajectory.

3 Non-linear, complex Regge trajectories

A non-trivial task for analytic models of Regge trajectories consists in deriving the imaginary part of the trajectory from the seemingly linearly increasing real part [13, 14]. A dispersion relation connects the real and imaginary part of the trajectory.

We follow Ref. [13] (see also [3, 4]) to relate the nearly linear real part of the meson trajectory to its imaginary part

$$\Re e\, \alpha(s) = \alpha(0) + \frac{s}{\pi} PV \int_0^\infty ds' \frac{\Im m\, \alpha(s')}{s'(s' - s)}. \tag{12}$$

The imaginary part is related to the decay width by

$$\Gamma(M_R) = \frac{\Im m\, \alpha(M_R^2)}{\alpha'\, M_R}. \tag{13}$$

The quantity α' in Eq. (13) denotes the derivative of the real part, $\alpha' = \frac{d\Re e\, \alpha(s)}{ds}$. The relation between $\Gamma(M)$ and $\Im m\, \alpha(s)$ requires $\Im m\, \alpha(s) > 0$. In a simple analytical model, the imaginary part is chosen as a sum of single threshold terms [13]

$$\Im m\, \alpha(s) = \sum_n c_n (s - s_n)^{1/2} (\frac{s - s_n}{s})^{|\Re e\, \alpha(s_n)|} \theta(s - s_n). \tag{14}$$

3.1 f mesons and their trajectories

Apart from the Pomeron trajectory, to be discussed below, the direct-channel f trajectories are essential in the PP channel. Guided by conservation of quantum numbers, we include two f trajectories, the first labelled f_1 defined by the resonances $f_0(980)$, $f_1(1420)$, $f_2(1810)$, $f_4(2300)$, and the second labelled f_2 defined by the resonances $f_2(1270)$, $f_4(2050)$ and $f_6(2510)$. The real and imaginary parts of these f trajectories as well as the values of the parameters fitted to the spectroscopic data can be found in Ref. [4]. The $f_0(500)$ resonance is taken as an isolated resonance with a Breit-Wigner parameterisation as described in Ref. [4].

3.2 The Pomeron trajectory

While ordinary meson trajectories can be fitted both in the resonance and scattering region corresponding to positive and negative values of the argument, the parameters of the Pomeron trajectory can only be determined in the scattering region $M^2 < 0$. A fit of the nonlinear Pomeron trajectory to high-energy pp and $p\bar{p}$ data is discussed, e.g. in Ref. [5].

$$\alpha_P(M^2) = 1. + \varepsilon + \alpha' M^2 - c\sqrt{s_0 - M^2}, \tag{15}$$

with $\varepsilon = 0.08$, $\alpha' = 0.25$ GeV^{-2} and s_0 the two pion threshold $s_0 = 4m_\pi^2$. The value of c is set $c = \alpha'/10 = 0.025$.

In order to be consistent with the mesonic trajectories shown above, the linear term in Eq. (15) is replaced by a heavy threshold mimicking linear behaviour in the mass region of interest (M < 5 GeV),

$$\alpha_P(M^2) = \alpha_0 + \alpha_1(2m_\pi - \sqrt{4m_\pi^2 - M^2}) + \alpha_2(\sqrt{M_H^2} - \sqrt{M_H^2 - M^2}), \tag{16}$$

where M_H is an effective heavy threshold set at $M = 3.5$ GeV. The coefficients α_0, α_1 and α_2 are chosen such that the Pomeron trajectory of Eq. (16) has a low energy behaviour as defined by Eq. (15).

The above trajectory is an oversimplified ansatz. An alternative to Eq. (15) is:

$$\alpha_P(M^2) = \frac{1 + \epsilon + \alpha' M^2}{1 - c\sqrt{s_0 - M^2}}. \tag{17}$$

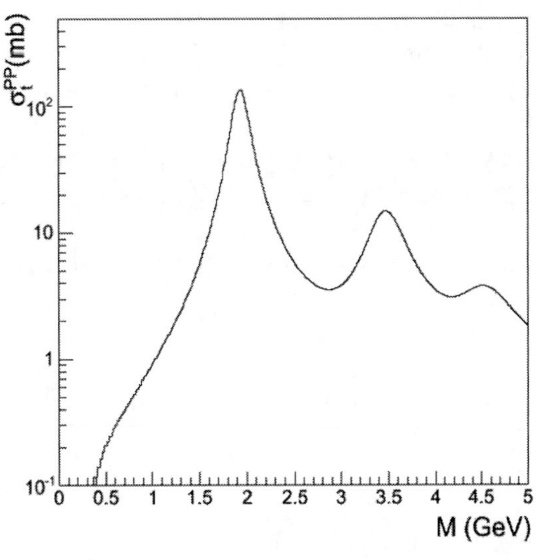

Figure 2. Direct-channel Pomeron contribution to PP total cross section.

Its contribution to the Pomeron-Pomeron total cross section is shown in Fig. 2. Contrary to Eq. (15), used in Ref. [4], it predicts glueballs whose widths increase with their masses, as one expects.

4 Pomeron-Pomeron total cross section

The Pomeron-Pomeron cross section was calculated from the imaginary part of the amplitude by use of the optical theorem

$$\sigma_t^{PP}(M^2) = \Im m\, A(M^2, t=0) = \sum_{i=f,P} \sum_J \frac{[f_i(0)]^{J+2}\, \Im m\, \alpha_i(M^2)}{(J - \Re e\, \alpha_i(M^2))^2 + (\Im m\, \alpha_i(M^2))^2}. \tag{18}$$

The index i in Eq. (18) sums over the trajectories which contribute to the cross section, in our case the f_1, f_2 and the Pomeron trajectory discussed above. Within each trajectory, the summation extends over the resonances of spin J.

To summarize, a formalism to calculate central diffractive production of mesons is presented. Details on SD and DD can be found in Ref. [3], while those on meson production the collisions of Pomerons are in Ref. [4].

References

[1] K.A. Goulianos, Phys. Rep. **101**(1983) 169.

[2] Robert Ciesielski, Konstantin Goulianos, *MBR Monte Carlo Simulation in PYTHIAS*, hep-ph/1205.1446.

[3] L. Jenkovszky, O. Kuprash, J.W. Lamsa, V.K. Magas, and R. Orava, Phys.Rev.D83 (2011) 056014, arXiv:1011.0664.

[4] R. Fiore, L. Jenkovszky, and R. Schicker, *Resonance production in Pomeron-Pomeron collisions at the LHC*, to be published.

[5] L. Jenkovszky, Rivista Nuovo Cim., **10** N12, (1987), p.1.

[6] A. Austregesilo, for the COMPASS Collab., Proceedings 15th Conference on Elastic and Diffractive Scattering, Saariselka, Sept 2013, arXiv:1310.3190.

[7] M. Albrow, for the CDF Collab., Int. J. Mod.Phys. **A29** (2014) 28, 1446009, arXiv:1409.0462.

[8] L. Adamczyk, W. Guryn and J. Turnau, Int. J. Mod.Phys. **A29** (2014) 28, 1446010, arXiv:1410.5752.

[9] R. Schicker, Int. J. Mod. Phys. **A29** (2014) 28, 1446015, arXiv:1411.1283.

[10] R. McNulty, Int. J. Mod.Phys. **A29** (2014) 28, 1446003, arXiv:1409.8113. arXiv:1402.2170.

[11] J. Bartels, A. De Roeck, and H. Lotter, Phys. Lett. **B389** (1996) 742; F. Caporale, D.Y. Ivanov, A. Papa, Eur. Phys. J. C **58** (2008) 1.

[12] R. Fiore, A. Flachi, L.L. Jenkovszky, A. Lengyel and V. Magas, Phys. Rev. D69 (2004), 014004, hep-ph/0308178.

[13] R. Fiore, L.L. Jenkovszky, V. Magas, F. Paccanoni, A. Papa, Eur.Phys.J. A10 (2001) 217-221, hep-ph/0011035.

[14] R.Fiore, L.L. Jenkovszky, F. Paccanoni, A. Prokudin, Phys.Rev.D70 (2004)054003, hep-ph/0404021.

[15] R. Fiore, L.L. Jenkovszky, V. Magas, F. Paccanoni and A. Papa, Phys. Part. Nucl.31 (2000) 46, hep-ph/9911503.

EPJ Web of Conferences **120**, 02007 (2016) DOI: 10.1051/epjconf/201612002007
ISMD 2015

Recent progress in some exclusive and semi-exclusive processes in proton-proton collisions

Antoni Szczurek[1,2,a], Anna Cisek[2,b], Marta Łuszczak[2,c], and Wolfgang Schäfer[1,d]

[1] *Institute of Nuclear Physics, Kraków, ul. Radzikowskiego 152*
[2] *Rzeszòw University, Rzeszów, ul. Rejtana 16*

Abstract. We present the main results of our recent analyses of exclusive production of vector charmonia (J/ψ and ψ') in k_t-factorization approach and for $\gamma\gamma$ production of charged dilepton pairs in exclusive and semiinclusive processes in a new approach, similar in spirit to k_t-factorization.
The results for charmonia are compared with recent results of the LHCb collaboration. We include some helicity flip contributions and quantify the effect of absorption correction. The effect of $c\bar{c}$ wave function is illustrated.
We present uncertainties related to F_2 structure function which are the main ingredient of the approach. Our results are compared with recent CMS data for dilepton production with lepton isolation cuts imposed.

1 Introduction

Here we briefly summarize our recent results for exclusive production of J/ψ mesons obtained in [1, 2] and for photon-photon production of charged dileptons obtained in [5, 6].

2 Exclusive production of vector charmonia

The Born diagrams for the exclusive production of J/ψ mesons are shown in Fig.1. In actual calculations in [1, 2] we include also absorption effects due to soft proton-proton interactions.

All details of the formalism can be found in Refs.[1] and [2]. Here we only sketch the main points.

Imaginary part of the forward $\gamma p \to J/\psi p$ amplitude

$$\Im m\, \mathcal{M}_T(W, \Delta^2 = 0, Q^2 = 0) = W^2 \frac{c_v \sqrt{4\pi\alpha_{em}}}{4\pi^2} 2 \int_0^1 \frac{dz}{z(1-z)} \int_0^\infty \pi dk^2 \psi_V(z, k^2)$$
$$\int_0^\infty \frac{\pi d\kappa^2}{\kappa^4} \alpha_S(q^2) \mathcal{F}(x_{\rm eff}, \kappa^2) \Big(A_0(z, k^2)\, W_0(k^2, \kappa^2) + A_1(z, k^2)\, W_1(k^2, \kappa^2) \Big). \quad (1)$$

[a] e-mail: antoni.szczurek@ifj.edu.pl
[b] e-mail: acisek@univ.rzeszow.pl
[c] e-mail: luszczak@univ.rzeszow.pl
[d] e-mail: Wolfgang.Schafer@ifj.edu.pl

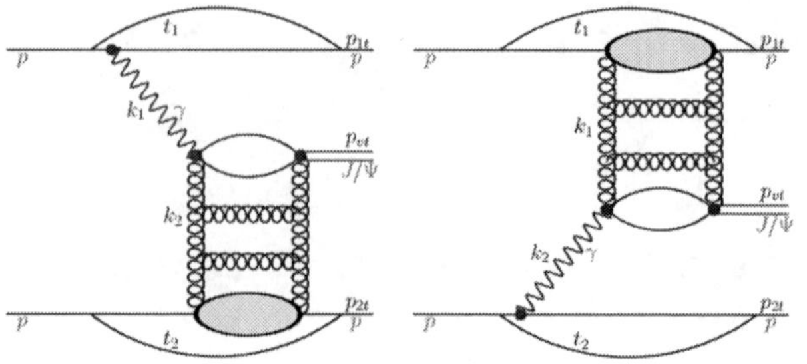

Figure 1. Born diagrams for exclusive production of J/ψ mesons.

The full amplitude, at finite momentum transfer is parametrized as:

$$\mathcal{M}(W, \Delta^2) = (i + \rho)\, \Im m \mathcal{M}(W, \Delta^2 = 0, Q^2 = 0)\, \exp(-B(W)\Delta^2/2)\,, \tag{2}$$

Then the amplitude for the $pp \to ppJ/\psi$ can be written somewhat formally as:

$$\begin{aligned} &\mathcal{M}_{h_1 h_2 \to h_1 h_2 V}^{\lambda_1 \lambda_2 \to \lambda_1' \lambda_2' \lambda_V}(s, s_1, s_2, t_1, t_2) = \mathcal{M}_{\gamma \mathbf{P}} + \mathcal{M}_{\mathbf{P}\gamma} \\ &= \langle p_1', \lambda_1' | J_\mu | p_1, \lambda_1 \rangle \epsilon_\mu^*(q_1, \lambda_V) \frac{\sqrt{4\pi\alpha_{em}}}{t_1} \mathcal{M}_{\gamma^* h_2 \to V h_2}^{\lambda_{\gamma^*} \lambda_2 \to \lambda_V \lambda_2}(s_2, t_2, Q_1^2) \\ &+ \langle p_2', \lambda_2' | J_\mu | p_2, \lambda_2 \rangle \epsilon_\mu^*(q_2, \lambda_V) \frac{\sqrt{4\pi\alpha_{em}}}{t_2} \mathcal{M}_{\gamma^* h_1 \to V h_1}^{\lambda_{\gamma^*} \lambda_1 \to \lambda_V \lambda_1}(s_1, t_1, Q_2^2)\,. \end{aligned} \tag{3}$$

The auxiliary amplitude in Eq. (3) for the emission of a photon of transverse polarization λ_V, and transverse momentum $\boldsymbol{q}_1 = -\boldsymbol{p}_1$ can be written as:

$$\langle p_1', \lambda_1' | J_\mu | p_1, \lambda_1 \rangle \epsilon_\mu^*(q_1, \lambda_V) = \frac{(\boldsymbol{e}^{*(\lambda_V)} \boldsymbol{q}_1)}{\sqrt{1 - z_1}} \frac{2}{z_1} \chi_{\lambda'}^\dagger \Big\{ F_1(Q_1^2) - \frac{i\kappa_p F_2(Q_1^2)}{2m_p} (\sigma_1 \cdot [\boldsymbol{q}_1, \boldsymbol{n}]) \Big\} \chi_\lambda\,. \tag{4}$$

Above F_1 and F_2 are Dirac and Pauli electromagnetic form factors, respectively. In Ref.[2] we have included the Pauli form factors for a first time.

3 Selected results for vector quarkonia

Rapidity distributions of J/ψ mesons are show in Fig.2. In this calculation Gaussian wave functions were used. Only results with UGDFs that include nonlinear effects describe the LHCb experimental distributions [3]. In our opinion it is too preliminary to conclude that we observe nonlinear effects or onset of gluon saturation. In Ref.[2] we obtained similar distributions for ψ' [2].

Very interesting quantity is the ratio of the cross sections for ψ' and J/ψ production. As shown in Fig.3 such a ratio is very sensitive to the functional form of the $c\bar{c}$ light-cone wave function. We conclude that the Gauss $c\bar{c}$ wave function much better describes the LHCb data [3] than Coulomb WF. In some calculation in the literature the wave functions are not included explicitly and only point-like coupling is used [4].

 DOI: 10.1051/epjconf/201612002007

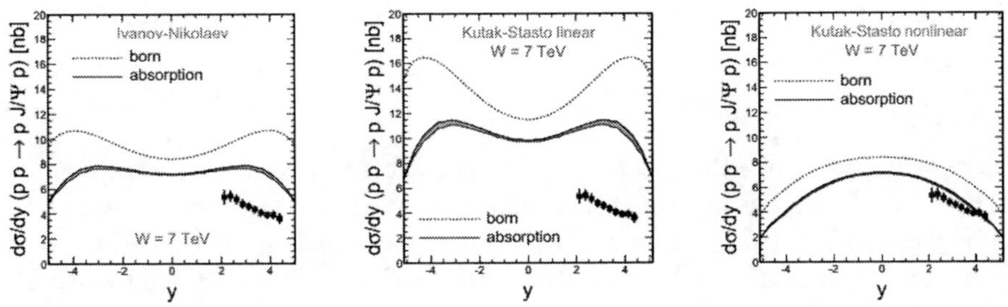

Figure 2. Rapidity distributions for three different unintegrated gluon distributions. The upper curves were obtained in the Born approximation and the lower band includes absorption effects.

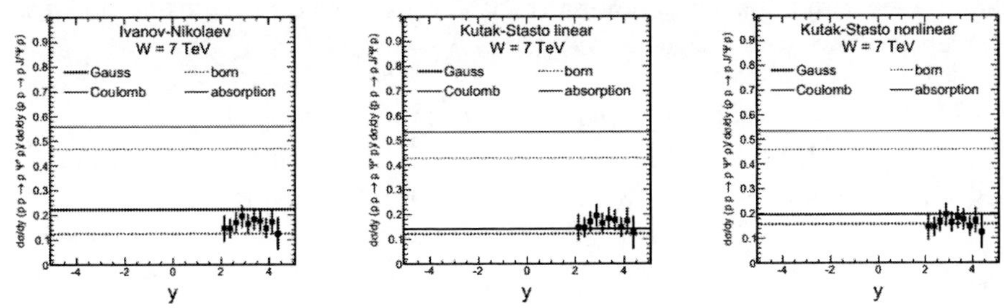

Figure 3. The ratio of the cross section for ψ' and J/ψ.

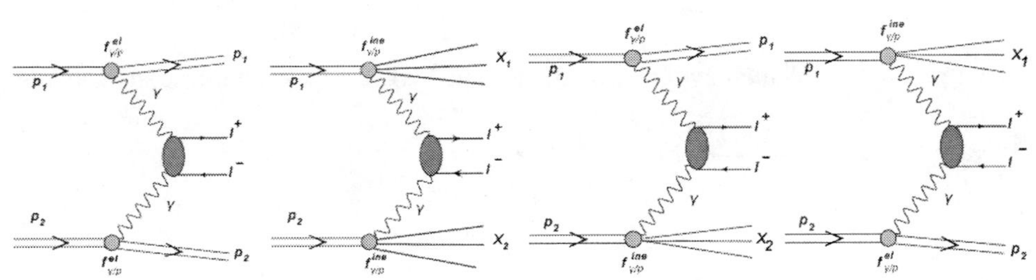

Figure 4. The different categories of $\gamma\gamma$ processes.

4 $\gamma\gamma$ production of dileptons

The $\gamma\gamma$ processes can be categorize according to the final state (see Fig.4). All the processes were discussed in [5, 6].

Two different approaches were discussed in the context of $\gamma\gamma$ production of dileptons.

In the collinear approach the corresponding cross sections are calculated respectively as:

$$\frac{d\sigma^{(k,l)}}{dy_1 dy_2 d^2p_t} = \frac{1}{16\pi^2 \hat{s}^2} x_1\gamma_k(x_1,\mu^2)\, x_2\gamma_l(x_2,\mu^2)\, \overline{|\mathcal{M}_{\gamma\gamma\to l^+l^-}|^2}\,, \tag{5}$$

where $k, l \in \{\text{el}, \text{in}\}$ stand for the processes with intact (el) or dissociated in proton at the photon vertex. The $\gamma_k(x,\mu^2)$ are the corresponding photon distributions in a proton, with or without the condition of breakup. The elastic photon distributions are calculated with the help of the nucleon electromagnetic form factors, while the inelastic distributions can be obtained by using a combined QCD/QED evolution [7].

In the k_t-factorization, proposed recently in [5, 6], the fully unintegrated photon flux can be written as (below incoming particles are denoted as A, B):

$$\frac{d\mathcal{F}_{\gamma^*\leftarrow A}(z,\boldsymbol{q},M_X^2)}{dM_X^2} = \frac{\alpha_{\text{em}}}{\pi}(1-z)\Big(\frac{\boldsymbol{q}^2}{\boldsymbol{q}^2 + z(M_X^2 - m_A^2) + z^2 m_A^2}\Big)^2 \cdot \frac{p_B^\mu p_B^\nu}{s^2} W_{\mu\nu}(M_X^2, Q^2)\,. \tag{6}$$

Information on the virtual photon-proton interaction is contained in the hadronic tensor, which is obtained in terms of the electromagnetic currents as:

$$W_{\mu\nu}(M_X^2,Q^2) = \sum_X (2\pi)^3\, \delta^{(4)}(p_X - p_A - q)\, \langle p|J_\mu|X\rangle\langle X|J_\nu^\dagger|p\rangle\, d\Phi_X\,, \tag{7}$$

The virtual photoabsorption cross sections are related to hadronic tensors as:

$$\sigma_T(\gamma^* p) = \frac{4\pi\alpha_{em}}{4\sqrt{X}}\Big(-\frac{\delta_{\mu\nu}^\perp}{2}\Big) 2\pi W^{\mu\nu}(M_X^2,Q^2)\,,\ \sigma_L(\gamma^* p) = \frac{4\pi\alpha_{em}}{4\sqrt{X}} e_\mu^0 e_\nu^0\, 2\pi W^{\mu\nu}(M_X^2,Q^2)\,. \tag{8}$$

It is customary to introduce dimensionless structure function $F_i(x_{\text{Bj}}, Q^2)$, $i = T, L$ as

$$\sigma_{T,L}(\gamma^* p) = \frac{4\pi^2\alpha_{em}}{Q^2} \frac{1}{\sqrt{1 + \frac{4x_{\text{Bj}}^2 m_A^2}{Q^2}}} F_{T,L}(x_{\text{Bj}}, Q^2)\,, \tag{9}$$

At high energies, in the calculation of photon fluxes, the contribution of the structure function

$$F_2(x_{\text{Bj}}, Q^2) = \frac{F_T(x_{\text{Bj}},Q^2) + F_L(x_{\text{Bj}},Q^2)}{1 + \frac{4x_{\text{Bj}}^2 m_A^2}{Q^2}} \tag{10}$$

dominates.

The unintegrated fluxes enter the cross section for dilepton production as

$$\frac{d\sigma^{(i,j)}}{dy_1 dy_2 d^2\boldsymbol{p}_1 d^2\boldsymbol{p}_2} = \int \frac{d^2\boldsymbol{q}_1}{\pi \boldsymbol{q}_1^2} \frac{d^2\boldsymbol{q}_2}{\pi \boldsymbol{q}_2^2} \mathcal{F}^{(i)}_{\gamma^*/A}(x_1,\boldsymbol{q}_1)\, \mathcal{F}^{(j)}_{\gamma^*/B}(x_2,\boldsymbol{q}_2) \frac{d\sigma^*(p_1,p_2;\boldsymbol{q}_1,\boldsymbol{q}_2)}{dy_1 dy_2 d^2\boldsymbol{p}_1 d^2\boldsymbol{p}_2}\,, \tag{11}$$

where $i, j \in \{\text{el}, \text{in}\}$ again refer to elastic and inelastic processes on the proton sides. For brevity we integrated over invariant masses of the possible dissociated system.

The longitudinal momentum fractions carried by photons can be obtained from rapidities and transverse momenta of the charged leptons as:

$$x_1 = \sqrt{\frac{\boldsymbol{p}_1^2 + m_l^2}{s}} e^{y_1} + \sqrt{\frac{\boldsymbol{p}_2^2 + m_l^2}{s}} e^{y_2}\,,\ x_2 = \sqrt{\frac{\boldsymbol{p}_1^2 + m_l^2}{s}} e^{-y_1} + \sqrt{\frac{\boldsymbol{p}_2^2 + m_l^2}{s}} e^{-y_2}\,. \tag{12}$$

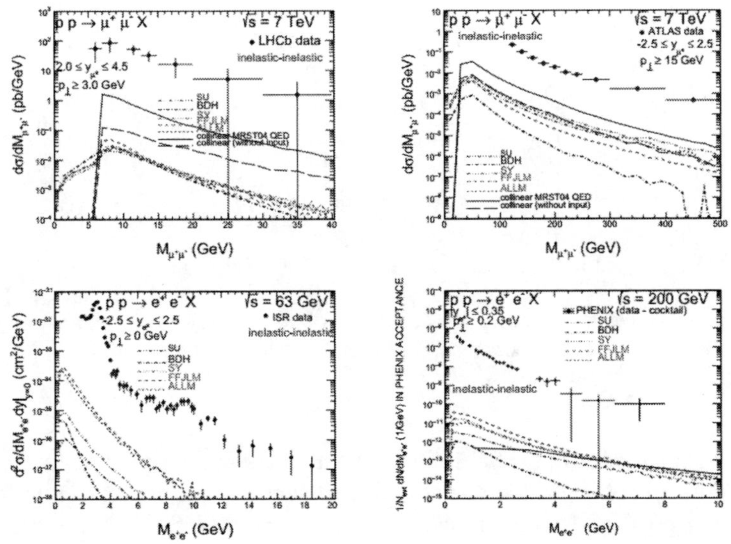

Figure 5. Dilepton invariant mass distributions for different experimental conditions (see the four panels) and for different structure functions (see different lines).

5 Selected results for l^+l^- production

As an example in Fig.5 we show invariant mass distributions for double dissociative processes (both protons undergo electromagnetic dissociation). The calculations have been performed for different experimental conditions specified in the figure legend. We show results for different proton structure functions from the literature. The different structure functions lead to quite different results. The presented results strongly depend on the kinematical regions of longitudinal momentum fraction and photon virtuality (x,Q^2) which were not sufficiently well studied experimentally and theoretically in which an interplay of perturbative and nonperturbative effects takes place. We observe that the $\gamma\gamma$ contribution constitutes only a small fraction of the cross section compared to the experimental data and Drell-Yan contribution (not shown explicitly here). Similar results for elastic-inelastic channel were shown in [5].

Now we wish to show results which can be directly compared to experimental data [8]. The data were obtained by imposing a kinematical constraint on lepton isolation. The results are shown in Fig.6 A relatively good agreement has been achieved. A better quality results are expected from Run II at the LHC.

6 Conclusions

Our results and open problems for charmonia production can be summarized as follows. We have found some model dependent indication of presence of nonlinear effects in the small-x gluon distribution of a proton. It is important to remember, that the present experiments are not fully exclusive and rather semi-inclusive processes have to be studied. Experimentally so far proton dissociation has been "extracted" in a model dependent way assuming some functional form in p_t. From HERA we

DOI: 10.1051/epjconf/201612002007

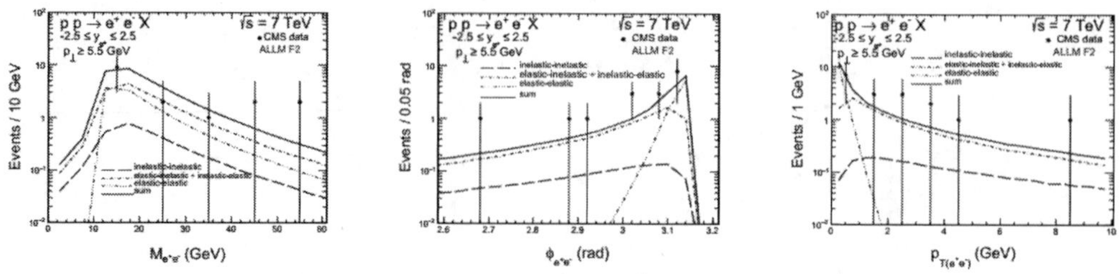

Figure 6. Dilepton invariant mass distributions (left) distribution in relative azimuthal angle between leptons (middle) and distribution in transverse momentum of the pair of leptons (right) for a selected structure function.

have some limited knowledge only about diffractive dissociation. Compared to exclusive production of J/ψ at HERA, in pp collisions there is also electromagnetic dissociation. Interference effects due to the two diagrams in Fig. 1 were predicted. It would be nice to see a modulation in ϕ_{pp} due to interference effects between the two diagrams. In the future, the CMS+TOTEM and ATLAS+ALFA experiments could measure purely exclusive reaction and study dependence on many more variables.

Our results for photon-photon induced processes can be summarized as: We discussed two different approaches for $\gamma\gamma$ processes – collinear and k_t-factorization, they are not completely equivalent. We have found strong dependence on the structure function input to the photon fluxes in the k_t-factorization approach. Semi-exclusive contributions with dissociation are large which is an interesting lesson for $pp \to ppJ/\psi$. The photon-photon contributions are, however, rather small compared to the Drell-Yan contribution. A reasonable description of the CMS data with isolated electrons was obtained (recently also the ATLAS collaboration obtained similar results). So far only the collinear approach has been applied e.g. to the $pp \to (\gamma\gamma) \to W^+W^-XY$ processes. Such a reaction is important in searches for Beyond Standard Model effects. A large cross section was found [9]. A calculation in k_t-factorization approach would be very valuable.

References

[1] W. Schäfer and A. Szczurek, Phys. Rev. **D76** (2007) 094014.

[2] A. Cisek, W. Schäfer and A. Szczurek, JHEP **1504** (2015) 159,

[3] R. Aaij *et al.* [LHCb Collaboration], J. Phys. G **40** (2013) 045001 [arXiv:1301.7084 [hep-ex]]; R. Aaij *et al.* [LHCb Collaboration], J. Phys. G **41** (2014) 055002 [arXiv:1401.3288 [hep-ex]].

[4] S. P. Jones, A. D. Martin, M. G. Ryskin and T. Teubner, JHEP **1311** (2013) 085 [arXiv:1307.7099].

[5] M. Łuszczak, W. Schäfer and A. Szczurek, arXiv:1510.00294,

[6] G. Gil da Silveira, L. Forthomme, K. Piotrzkowski, W. Schäfer, A. Szczurek, JHEP 1502 (2015) 159,

[7] A. D. Martin, R. G. Roberts, W. J. Stirling and R. S. Thorne, Eur. Phys. J. C **39**, 155 (2005) [hep-ph/0411040].

[8] S. Chatrchyan *et al.* [CMS Collaboration], JHEP **1211**, 080 (2012) [arXiv:1209.1666 [hep-ex]].

[9] M. Łuszczak, A. Szczurek and C. Royon, JHEP **1502** (2015) 098 [arXiv:1409.1803 [hep-ph]].

EPJ Web of Conferences **120**, 02008 (2016) DOI: 10.1051/epjconf/201612002008
ISMD 2015

Central Exclusive Production in Proton-Proton Collisions with the STAR Experiment at RHIC

Włodek Guryn[1,a]

[1] *Brookhaven National Laboratory*
Upton, NY 11973 USA

Abstract. We shall describe the physics program with tagged forward protons, focusing on Central Exclusive Production in polarized proton-proton collisions at the Relativistic Heavy Ion Collider (RHIC), with the STAR detector at $\sqrt{s}$ = 200 GeV. Preliminary results in CEP of two oppositely charged pions and kaons produced in the processes $pp \rightarrow pp\pi^+\pi^-$ and $pp \rightarrow ppK^+K^-$ shall be presented. Those Double Pomeron Exchange (DPE) processes, allow the final states to be dominated by gluonic exchanges. Silicon strip detectors placed in Roman Pots were used for measuring forward protons. The preliminary results are based on the measurement of the recoil system of charged particles in the STAR experiment's Time Projection Chamber (TPC). Ionization energy loss, dE/dx, of charged particles was used for particle identification (PID).

1 Introduction

Diffractive processes at high energies occur mostly via the exchange of a color singlet object (the "Pomeron") with internal quantum numbers of the vacuum [1]. Even though properties of diffractive scattering at high energies are described by the phenomenology of Pomeron ($\boldsymbol{P}$) exchange in the context of Regge theory, the exact nature of the Pomeron still remains elusive. The main theoretical difficulties in applying QCD to diffraction are due to the intrinsically non-perturbative nature of the process in the kinematic and energy ranges of the data currently available. In terms of QCD, Pomeron exchange consists of the exchange of a color singlet combination of gluons.

One of the diffractive processes of interest is shown in Fig. 1, a process with tagged forward protons $pp \rightarrow pM_Xp$, in which two protons emerge intact after the scattering and a recoil system M_X is produced mostly around pseudorapidity $\eta \approx 0$ (midrapidity). This process belongs to a class of Double Pomeron Exchange (DPE) processes and is commonly called a Central Production process. The case when all the products of the interaction are measured and the balance of momentum for all the products of the reaction is satisfied in the reaction, including forward protons, is called Central Exclusive Production (CEP) process.

Many other processes are of interest in DPE and CEP: resonance production, jet production and also diffractive Higgs production at the LHC are examples. For a most recent review of CEP see [2] and references therein.

[a] e-mail: guryn@bnl.gov

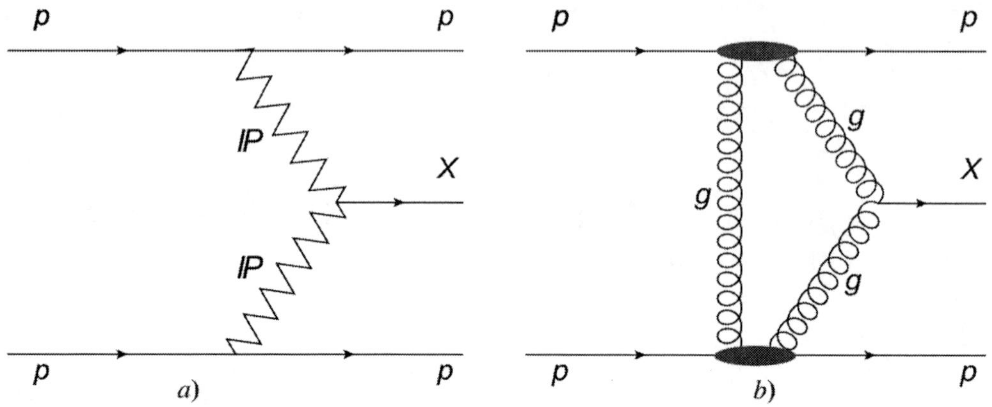

Figure 1. a) Central Production diagram in DPE; and b) pQCDPicture.

For the CEP process at high energy the DPE constraint selects processes mediated by the gluonic matter, see Fig. 1. In the DPE mechanism $pp \to pM_Xp$, as shown in Fig. 1, the two protons stay intact after the interaction, but they lose momentum to the Pomeron and the Pomeron-Pomeron interaction produces a system M_X at mid-rapidity of the colliding protons. Hence, triggering on forward protons at high energies in this central production process allows selection of interactions for which gluonic exchanges are dominant.

One of the important motivations for the inelastic diffraction program at the high energy colliders, to which DPE belongs, is searching for a gluonic bound state (glueball) whose existence is allowed in pure gauge QCD. The idea that glueballs might be preferentially produced in the DPE process due to high gluon density in such process can be traced back to [3]. Two of the gluons in the DPE process could merge into a mesonic bound state without a constituent quark, forming a glueball in the central production process $pp \to pXp$.

QCD predicts the existence of mesons which contain only gluons, the glueballs. These states are a consequence of the non-Abelian nature of the gauge fields which allows that gluons couple to themselves and hence may bind. Despite the theoretical predictions of glueballs, no glueball state has been unambiguously established to date [4–6]. Lattice QCD calculations have predicted the lowest-lying scalar glueball state in the mass range of 1500-1700 MeV/c^2, and tensor and pseudo-scalar glueballs in 2000-2500 MeV/c^2 [7]. Experimentally measured glueball candidates for the scalar glueball states are the $f_0(1500)$ and the $f_0(1710)$ [8] in central production, $pp \to pM_Xp$, as well as other gluon-rich reactions such as $\bar{p}p$ annihilation, and radiative J/ψ decay [5].

Because of the nature of the Pomeron, the central DPE process has been regarded as one of the potential channels of glueball production [7]. Because of the constraints provided by the double Pomeron interaction, the glueballs, and other states coupling preferentially to gluons, are expected to be produced with much reduced backgrounds compared to standard hadronic production processes [7]. It is imperative to cover a wide kinematic range to extract information of the production of glueball candidates at an energy regime where DPE is expected to be a dominant process in Central Production.

However, the energy regime where centrally produced glueball candidates have been identified so far is estimated to be not DPE dominated [6]. The experiments at CERN ISR Collider [9–11] and CERN SPS [12, 13] have provided measurements of many CEP-type processes, however their interpretation in terms of Pomeron-Pomeron interactions is not fully justified [14] at these rather low center-of-mass energies (62 GeV for ISR and 30 GeV for SPS).

2 Experimental Setup

Since the CEP process requires tagging froward protons, those diffraction processes are triggered using Roman Pots as shown in Fig. 2 while the recoil system X is measured in the Central Detector. We shall use an example from the RHIC program to describe how to construct a experiment to search for resonances in the CEP process, including the glueballs. First one needs an accelerator with colliding protons at a high enough energy so that DPE process is dominant. This could be, for example, RHIC where collisions of polarized protons are realized in the $\sqrt{s}$ range up to 510 GeV. We also need a suitable detector, with good charged particle ID to measure the central recoil system, which at RHIC is the STAR detector [15], with its Time Projection Chamber (TPC) which measures charged particle momenta and ionization energy loss dE/dx of particles in azimuth range $0 < \phi < 2\pi$ in pseudorapidity range $-1 < \eta < 1$. In addition the Time-of-Flight (ToF) system extends the momentum range of π/K separation in momentum range up to 1.6 GeV$/c$.

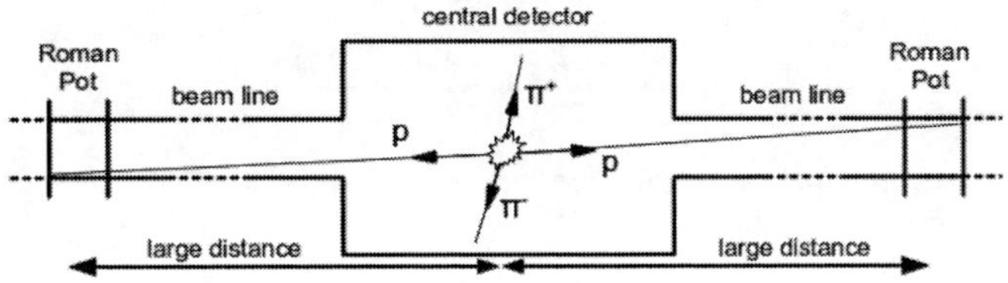

Figure 2. The layout of the general experimental setup. Main detector in the center and forward proton taggers (Roman Pots).

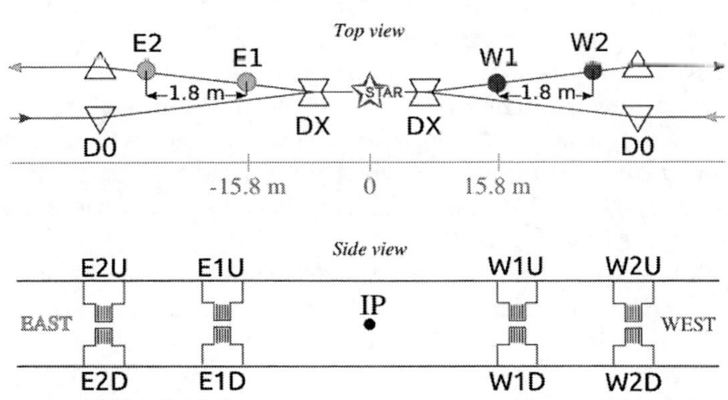

Figure 3. The layout of the RPs with the STAR detector (not to scale). The Roman Pot setup at STAR for measuring forward protons with high-t. Two sets of RPs will be positioned between DX and D0 magnets, at 15.8 m and 17.6 m from the IP. Top and side views are shown.

Finally, to detect forward protons the Roman Pot (RP) system of the pp2pp experiment [16] was installed downstream of the STAR detector at RHIC, see Fig. 3, where the location of the Roman Pots, top view, and schematically Si detectors and scintillation counters in the Roman Pots are shown.

The location is such that no special accelerator conditions, like large β^* are needed to operate Roman Pots together with the rest of the physics program allowing acquiring of large data samples needed for glueball searches.

3 Data Taking and Preliminary Results from Run 15 at RHIC

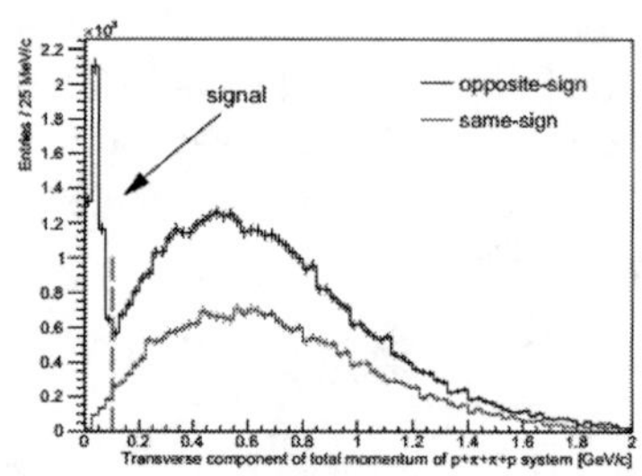

Figure 4. Transverse momentum balance (Δp_T) between centrally produced $\pi\pi$ system and the outgoing protons detected in the Roman Pots

With the setup described in the previous section the Central Production data were collected during Run 15 at RHIC. Roman Pots operated very efficiently through the whole data taking period. The events were required to have two outgoing protons in the RPs, and the inclusive charged tracks in the central region were reconstructed with STAR Time Projection Chamber (TPC). In addition hits in the Time of Flight (ToF) system were required. The pions and kaons were identified by the ionization energy loss dE/dx and the particle momentum measured in the TPC.

Selecting exclusive central reactions requires energy-momentum conservation constraint between the central system and the forward protons. As an example, the balance of the transverse momentum Δp_T between the central system and the forward protons was required, as shown in Fig. 4. The exclusivity cut required $\Delta p_T \leq 0.1$ GeV/c. A small background from lsame-sign pions is shown in red. As one can see the background level is very small, which to large extent is due to the momentum conservation constraint in the CEP process.

Consequently, STAR experiment's preliminary effective mass distributions of two charged pion and kaon states from RHIC-Run 15 at $\sqrt{s} = 200$ GeV with the RP set-up (See Fig. 3) is shown in Fig. 5. Extrapolating from the above preliminary data set we expect about 100k $\pi^+\pi^-$ meson pairs in the mass range above 1 GeV/c^2 and about 10k K^+K^- meson pairs. The features of the $\pi^+\pi^-$ mass distribution are very similar to those obtained by other collider experiments [9, 17]. Namely a sharp drop around 1 GeV/c^2 mass, attributed to the negative interference with f0(980) wave, and a peak structure around 1.270 GeV/c^2.

Acknowledgement. This work was supported in part by the Office of NP within the U.S. DOE Office of Science and by the Polish National Science Centre under contract UMO-2011/01/M/ST2/04126.

References

[1] For a review, see S. Donnachie *et al.*, "Pomeron Physics and QCD", Cambridge University Press.

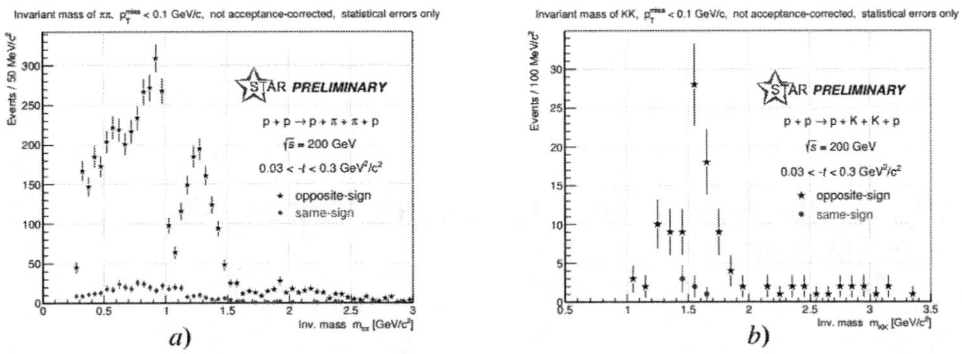

Figure 5. a) Reconstructed, uncorrected mass distributions, dN/dt, for two charged pions in the CEP process at $\sqrt{s}$ = 200 GeV. Asterix points are for neutral states and solid circles represents charged states ; b)Reconstructed, uncorrected mass distributions, dN/dt, for two charged kaons in the CEP process at $\sqrt{s}$ = 200 GeV. Asterix points are for neutral states and solid circles represents charged states. Errors are statistical only.

[2] M. Albrow, V. Khoze and C. Royon, Int J Mod Phys A, **Vol. 29**, No 28 (2014).
[3] D.Robson, Nucl.Phys. **B130**, 328 (1977).
[4] W. Ochs, "The Status of Glueballs", J. Phys. **G 40**, 043001 (2013).
[5] V. Crede and C.A. Meyer, Prog. Part. Nucl. Phys.**63**, 74 (2009).
[6] E. Klempt and A. Zaitsev, Phys. Rept. **454**, 1 (2007).
[7] For a review, see F.E. Close, Rep. Prog. Phys. **51**, 833 (1988), and C. Amsler and N.A. Tornqvist, Phys. Rept.**389**, 61 (2004).
[8] S. Abatziz *et al.*, Phys. Lett. **B324**, 509 (1994).
[9] T.Akesson *e*t al. AFS Collaboration, Nucl.Phys. **B264**, 154 (1986).
[10] A.Breakstone *e*t al. ABCDHW Collaboration, Z.Phys. **C48**, 569 (1990).
[11] A.Breakstone *e*t al. ABCDHW Collaboration, Z.Phys. **C58**, 251 (1993).
[12] F. Antinori *e*t al. WA91 Collaboration, Phys.Lett. **B353**, 589 (1995).
[13] D.Barberis *e*t al. WA102 Collaboration, Phys.Lett. **B474**, 423 (2000).
[14] P.Lebiedowicz, A.Szczurek, Phys.Rev. **D81**, 036003 (2010).
[15] K.H. Ackermann *et al.*, Nucl. Instr. and Meth. **A499**, 624 (2003).
[16] S. Bueltman *et al.*, Nucl. Instrum. and Meth. in Phys. Research **A535**, 415 (2004).
[17] CDF Collaboration: T. Aaltonen *et al.*, Phys.Rev. **D91**, 091101 (2015).

EPJ Web of Conferences **120**, 02009 (2016) DOI: 10.1051/epjconf/201612002009
ISMD 2015

Exclusive Central $\pi^+\pi^-$ Production in Proton Antiproton Collisions at the CDF

Maria Zurek[1,2,a]
on behalf of the CDF Collaboration

[1] *Forschungszentrum Jülich, 52425 Jülich, Germany*
[2] *University of Cologne, 50937 Cologne, Germany*

Abstract. Exclusive $\pi^+\pi^-$ production in proton-antiproton collisions at $\sqrt{s} = 0.9$ and 1.96 TeV in the Collider Detector at Fermilab has been measured. We select events with two particles with opposite charge in pseudorapidity region $-1.3 < \eta < 1.3$ with no other particles detected in $-5.9 < \eta < 5.9$. Particles are assumed to be pions. The $\pi^+\pi^-$system is required to have rapidity $-1.0 < y < 1.0$. The data are expected to be dominated by the double pomeron exchange mechanism. Therefore, the quantum numbers of the central state are constrained. The data extend up to dipion mass $M(\pi^+\pi^-) = 5000$ MeV/c^2. Resonance structures consistent with f_0 and $f_2(1270)$ mesons are visible. The results are valuable for light hadron spectroscopy and for providing information about the nature of the pomeron in a region between non-perturbative and perturbative quantum chromodynamics.

1 Introduction

Diffraction processes with low momentum-transfer squared, i.e., elastic scattering of protons at high energies, cannot be described by perturbative quantum chromodynamics, and the nonperturbative methods have to be involved. Elastic scattering and other diffractive interactions are described in the regime of Regge theory [1, 2] by the exchange of a pomeron, $\mathbb{P}$, which is a strongly interacting color-singlet. At leading order it is a pair of gluons [1]. Such processes are characterized by a region of large rapidity gaps, Δy, with no hadron production. With the requirements of two large rapidity gaps and central hadrons production, it is expected that the observed processes are dominated by double pomeron exchange, D$\mathbb{P}$E [1, 2].

In our work, central exclusive production is defined as a reaction $p\bar{p} \rightarrow p^{(*)}+GAP+X+GAP+\bar{p}^{(*)}$, where X is a specific central state and GAP is a region of rapidity where no particles are detected ($1.3 < |\eta| < 5.9$). We do not detect outgoing (anti)protons, thus events where they dissociate into hadrons, $p \rightarrow p^*$, with $|\eta| > 5.9$ are also included. The highlights of our analysis have been recently published in [3].

The exclusive central hadronic systems resulting primarily from double pomeron exchange have very restrictive quantum numbers: $I^G J^{PC} = 0^+(\text{even})^{++}$. This "quantum number filter" is a powerful tool for meson spectroscopy in the isoscalar sector, especially for glue-rich states. In addition it provides information on the nature of the pomeron.

[a] e-mail: m.zurek@fz-juelich.de

2 Experimental Setup

The CDF II detector setup, located at the Fermi National Accelerator Laboratory, is a general purpose detector designed to study $p\bar{p}$ collisions at the Tevatron ring. The setup is described with details in [4]. Here, a short overview of the detectors used in the analysis is given. The $p\bar{p}$ interaction point is surrounded by silicon microstip detectors and a drift chamber in a 1.4 Tesla magnetic field provided by a solenoid. These detectors compose together a tracking system with almost 100% track reconstruction efficiency for particles with $p_T > 0.4$ GeV/c and $|\eta| < 1.3$. The drift chamber is surrounded by a time-of-flight (TOF) counters barrel, consisting of plastic scintillator bars, read out at both ends by photomultipliers. The TOF detectors cover $|\eta| < 0.9$. Electromagnetic (EM) and hadronic (HAD) calorimeters are installed just after the solenoid. Both the central part of the calorimeters ($|\eta| < 1.32$) and forward plugs ($1.32 < |\eta| < 3.64$) have pointing tower geometry. To monitor the luminosity and veto events with particles in the region $3.7 < |\eta| < 4.7$, gas Cherenkov counters (CLC) were used. Beam shower counters (BSC) covering $5.4 < |\eta| < 5.9$ consist of 1.7 radiation lengths of lead and scintillation counters. The region between $4.7 < |\eta| < 5.4$ stays uninstrumented, giving a contribution to the nonexclusive background.

The data were taken with two different center-of-mass energies, with $\sqrt{s} = 1.96$ TeV and during a special low pile-up run with $\sqrt{s} = 0.9$ TeV. The beam proton rapidities at two center-of-mass energy values are $y_{\text{beam}} = 6.87$ and 7.64 respectively. Since we do not register outgoing (anti)protons, data include diffraction dissociation processes. At $\sqrt{s} = 1.96$ TeV higher dissociation masses are allowed than at 0.9 TeV.

3 Event Selection

The dedicated trigger requires two or more central ($|\eta| < 1.3$) calorimeter towers (EM + HAD) with $E_T > 0.5$ GeV and a veto on BSC, CLC and forward plug calorimeter. The trigger was activated when the mean pile-up was low.

In our analysis we use events with no pile-up and exactly two charged particles in a central region with $|\eta| < 1.3$ and $p_T > 0.4$ GeV/c, with no other particles in the detector, to $|\eta| < 5.9$. We assume that there is no particle in the detector when its readout signal is consistent with noise. To define the noise level unbiased bunch-crossing trigger data (zero-bias) are used. The zero-bias events are divided into two subgroups (a) in which no tracks or CLC hits are detected, (b) all other events, which are dominated by one or more interactions. Comparing the readout for the noise and signal-dominated samples, the noise level is determined for each subdetector. We apply a cut both on the sum of all signals $\sum(EM + HAD)$ and on the highest $E_T(EM)$ tower in each event, for each of the subdetectors. In the central region, tracks are extrapolated to the calorimeters. We require all calorimeter elements to have the pulse height consistent with the noise, apart from the cones $\sqrt{(\Delta\eta)^2 + (\ Delta\phi)^2} < 0.3$ around the track extrapolation point.

An opening-angle cut and a requirement of zero muons eliminate the small background from cosmic ray tracks. Charged-particle track quality cuts are applied. The tracks are required to have at least 25 hits in both stereo and axial layers of the drift chamber, and a fit with χ^2/DoF < 2.5. They both have to pass within 0.5 mm of the beam line in the transverse plane and to be within 1 cm of each other in z at that point. The measured particles are required to have opposite charge. We assume both particle to be pions and we put the requirement of $|y(\pi\pi)| < 1.0$.

4 Effective luminosity

In our analysis, we use only events with no other visible interaction to spoil the exclusivity. To obtain the effective integrated luminosity L_{eff}, the probability of having no pile-up, called the exclusive efficiency ε_{eff}, has to be calculated. Using the zero-bias data, we calculate the probability that the full detector is empty, as a function of the individual bunch integrated luminosity L_{bunch}. The distribution is exponential $\varepsilon_{\text{eff}} = e^{-\sigma_{\text{vis}} L_{\text{bunch}}}$ with the intercept consistent with 1.0. The slope, σ_{vis}, is the total observed ($|\eta| < 5.9$) cross-section. We find $\sigma_{\text{vis}} = 55.9 \pm 0.4$ mb at 1.96 TeV. It is consistent with an expectation from global fits [5] of total inelastic cross-section $\sigma_{\text{inel}} = 61.0 \pm 1.8$ mb, corrected for the estimated part visible in the region $|\eta| < 5.9$, $\sigma_{\text{vis}}/\sigma_{\text{inel}} = 0.85 \pm 0.05$ [6], which gives $\sigma_{\text{vis}}(\text{expected}) = 51.8 \pm 3.4$ mb. We find $L_{\text{eff}} = 1.16$ pb^{-1}, for $\sqrt{s} = 1.96$ TeV, with about 6.7% uncertainty.

For $\sqrt{s} = 0.9$ TeV the CLC counters were not calibrated to the luminosity. Therefore, we use σ_{vis} to calibrate the overall luminosity. The expected observed cross-section $\sigma_{\text{vis}}(\text{expected})$ is obtained from the total inelastic cross-section [5], $\sigma_{\text{inel}} = 52.7 \pm 1.6$ mb, multiplied by an estimate [6] of $\sigma_{\text{vis}}/\sigma_{\text{inel}} = (0.90 \pm 0.05)$. Comparing the value of σ_{vis} from the fit to the zero-bias data to expected one, the correction factor for the $\sqrt{s} = 0.9$ TeV luminosity is determined. The final value of the effective integrated luminosity equals $L_{\text{eff}} = 0.059$ pb^{-1} with about 10% uncertainty.

5 Acceptance calculation

All calculated differential cross-sections are required to be in a certain kinematic region, namely $|\eta(\pi)| < 1.3$, $p_T(\pi) > 0.4$ GeV/c, and $|y(\pi\pi)| < 1.0$. The transverse momentum and pseudorapidity requirements allow to accept only well-reconstructed tracks, and the y cut defines the rapidity gap. The acceptance combined with cut efficiency is a function of single track $p_T(\pi)$, $\eta(\pi)$; the invariant mass of two pions $M(\pi\pi)$; and p_T of the central $\pi\pi$ state.

The trigger efficiency is based on a data-driven procedure, using well-measured isolated tracks from minimum-bias data. Probabilities of triggering 0, 1, 2 or more towers with $\sum E_T \geq 0.5$ GeV in the region around the track extrapolation are calculated. The total trigger efficiency as a function of p_T and η of the track is composed of those three contributions.

The acceptance and cut efficiency as a function of $p_T(\pi)$ and $\eta(\pi)$ is calculated using simulated single pions. After detector simulation using a GEANT-3 Monte Carlo program [7], the track reconstruction probability is determined. As a next step, the simulated track is checked to pass all the quality requirements. The single track acceptance is fitted with a smooth empirical function.

Since the acceptance is dependent not only on single track features, but also on correlations between two tracks, the additional acceptance component as a function of $M(\pi\pi)$ and $p_T(\pi\pi)$ is determined. For that reason, a parent state is generated in mass bins from $2M(\pi)$ to 5000 MeV/c^2 and p_T bins from 0 to 2.5 GeV/c. The central state decays isotropically (S-wave, J=0). A bilinear interpolation to compute the acceptance at every point is used, to avoid fake structures from statistical fluctuations in the Monte Carlo.

6 Background estimation

A background from hadron pairs that are not $\pi^+\pi^-$ is determined using the information from the TOF counters. The time-of-flight information is available when both particles are in a region $|\eta| < 0.9$. Consequently, only 67% of all pairs have both particles identified. For this sample of events, (89±1)% are $\pi^+\pi^-$. When both tracks are in a region $|\eta| < 0.7$, then about 90% of the hadron pairs are identified. For two different regions of η, $|\eta| < 0.7$ and $|\eta| < 0.9$, no significant change in the composition of the identified particle types is observed. All the spectra include non-$\pi^+\pi^-$ background. We assign pion masses for all measured hadron pairs.

The events with two same-charged hadrons are 6.1% and 7.1% at 0.9 TeV and 1.96 TeV respectively, and are rejected. They are an indication of nonexclusive background with at least two untagged charged particles, either below the p_T-threshold, or because of inefficiency with no energy above the noise level in the calorimeters or no reconstructed track. We expect there to be a similar number of same charge events with missed tracks.

7 Cross-sections

The differential cross-section as a function of $M(\pi\pi)$ and $p_T(\pi\pi)$ is presented in Fig. 1. To avoid the region with zero acceptance we present the total cross-section as a function of invariant mass of two pions in the range 1000 MeV/c^2 < $M(\pi\pi)$ < 5000MeV/c^2 integrated over all $p_T(\pi\pi)$ (Figs. 2, 3), and in the range 300 MeV/c^2 < $M(\pi\pi)$ < 5000 GeV/c^2 for $p_T(\pi\pi)$ > 1 GeV/c (Fig. 4).

In Figs. 2, 3 a peak centered at 1270 MeV/c^2 with a full-width at half-maximum ~200 MeV/c^2 is visible. It is consistent with the $f_2(1270)$ state. The shoulder on the high-mass side of the $f_2(1270)$ can be evidence of the $f_0(1370)$ meson. A change of slope in the cross-section distribution at 1500 MeV/c^2 is seen. In this region, at lower $\sqrt{s}$, a dip was observed. It is possibly caused by interference between resonances [10, 11]. At higher masses up to 2400 MeV/c^2, there are structures in the mass distribution, suggesting the production of other resonances. As shown in Fig. 3, the data fall monotonically with $M(\pi\pi)$ from 2000 to 5000 MeV/c^2. The small peak at 3100 MeV/c^2 is consistent with $J/\psi \to e^+e^-$ photoproduction [12]. In Fig. 3 the predictions from the DIME Monte Carlo [13] of the differential cross-section for the mass region $M(\pi\pi)$ > 2600 MeV/c^2 are presented. The large uncertainties of the predictions have their origin in the unknown $\pi\pi\mathbb{P}$ form factor. The MC with an exponential form factor agrees with the data at 3000 MeV/c^2. However, a steeper $M(\pi\pi)$ dependence is predicted.

The ratio of the differential cross-section for two different $\sqrt{s}$, $R(0.9 : 1.96)$, presented in Fig. 2, rises from about 1.2 at 1000 MeV/c^2 to about 2.0 at 4000 MeV/c^2 with no significant structures. The ratio expected by Regge phenomenology when both protons are intact should be equal to approximately 1.3 [2, 8, 9]. For 1000 < $M(\pi\pi)$ < 2000 MeV/c^2 it is $R(0.9 : 1.96) = 1.284 \pm 0.039$. Our data include dissociation, with higher masses allowed at $\sqrt{s}$ = 1.96 TeV. For 2000 < $M(\pi\pi)$ < 3000 MeV $R(0.9 : 1.96) = 1.560 \pm 0.056$.

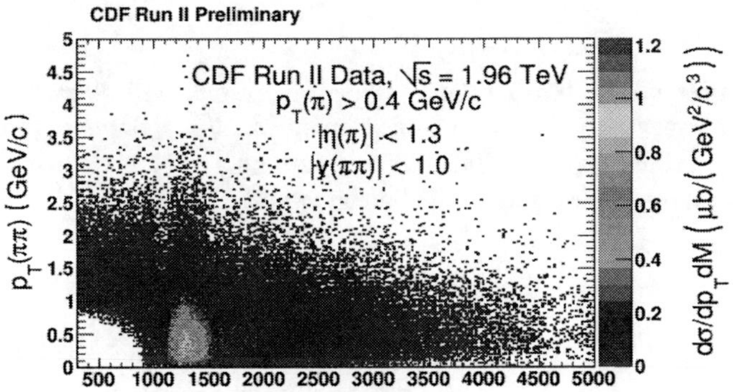

Figure 1. Differential cross-section $d\sigma/dM(\pi\pi)dp_T(\pi\pi)$ for two charged particles, assumed to be $\pi^+\pi^-$, with p_T > 0.4 GeV/c, $|\eta|$ < 1.3 and $|y(\pi\pi)|$ < 1.0 between two rapidity gaps 1.3 < $|\eta|$ < 5.9. (Colour online)

The differential cross-section for $p_T(\pi\pi)$ > 1000 MeV/c^2 is presented in Fig. 4. The distribution is flat up to a sharp drop at $M(\pi\pi)$ = 1000 MeV/c^2, where the $f_0(980)$ and the K^+K^- threshold occur. The drop was seen in previous experiments [2].

The distribution of mean $p_T(\pi\pi)$ as a function of $M(\pi\pi)$, shown in Fig. 5, shows a significant local rise at $M(\pi\pi) \approx$ 1500 MeV/c^2, coinciding with the change in slope of the differential cross-section (Figs. 2 and 3), followed by a trend to larger values at higher mass.

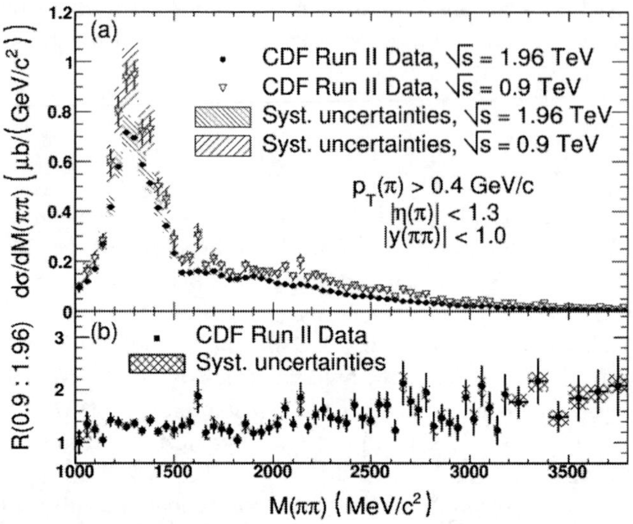

Figure 2. (a) Differential cross-section $d\sigma/dM(\pi\pi)$ for two charged particles, assumed to be $\pi^+\pi^-$, with $p_T > 0.4$ GeV/c, $|\eta| < 1.3$ and $|y(\pi\pi)| < 1.0$ between two rapidity gaps $1.3 < |\eta| < 5.9$. Red open circles for $\sqrt{s} = 0.9$ TeV and black points for $\sqrt{s} = 1.96$ TeV. (b) Ratio of cross-sections at $\sqrt{s} = 0.9$ and 1.96 TeV. Source [3].

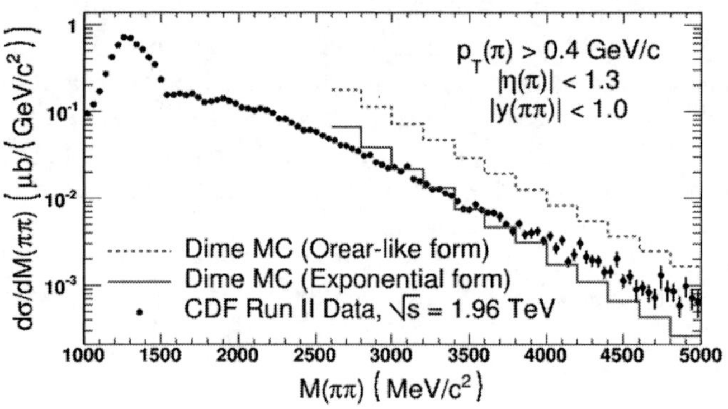

Figure 3. Differential cross-section $d\sigma/dM(\pi\pi)$ at $\sqrt{s} = 1.96$ TeV for two charged particles, assumed to be $\pi^+\pi^-$, with $p_T > 0.4$ GeV/c, $|\eta| < 1.3$ and $|y(\pi\pi)| < 1.0$ between two rapidity gaps $1.3 < |\eta| < 5.9$. Only statistical errors are shown; systematic uncertainties contribute approximately 10% at all masses. The lines show predictions of Ref. [13] with two different pion form factors. Source [3].

Summing up, the exclusive $\pi^+\pi^-$ differential cross-section has been measured, with $|y(\pi\pi)| < 1.0$ and rapidity gaps out to $|\eta(\pi\pi)| < 5.9$, at two $p\bar{p}$ centre of mass energies, $\sqrt{s} = 0.9$ and 1.96 TeV. The differential cross-section shows a sharp drop at 1000 MeV/c^2, a strong $f_2(1270)$ resonance enhancement, and other features at higher mass, the origin of which is uncertain.

Acknowledgments

I would like to thank my CDF colleagues, especially Mike Albrow (Fermilab). I thank the ISMD2015 Organizing Committee and Bonn-Cologne Graduate School for Physics and Astronomy for the student allowance.

References

[1] See, e.g., J.R. Forshaw and D.A. Ross, *Quantum Chromodynamics and the Pomeron*, Cambridge Lecture Notes in Physics (Cambridge University Press, Cambridge, England, 1997); S. Donnachie, G. Dosch, P. Landshoff, and O. Nachtmann, *Pomeron Physics and QCD* (Cambridge University Press, Cambridge, England, 2002).

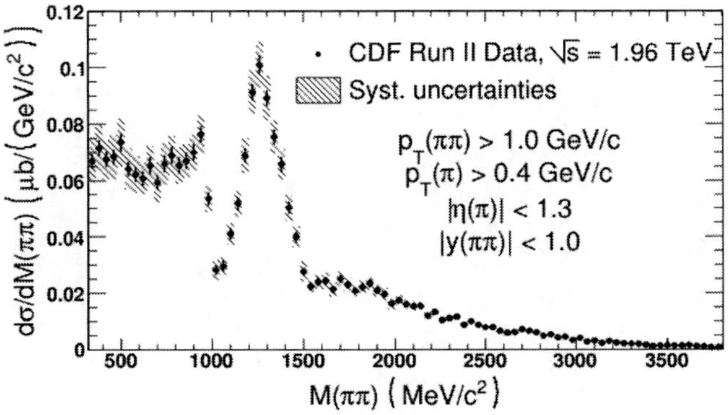

Figure 4. As Fig. 2 at $\sqrt{s} = 1.96$ TeV, but with $p_T(\pi\pi) > 1.0$ GeV/c for which the acceptance extends to low $M(\pi\pi)$. Source [3].

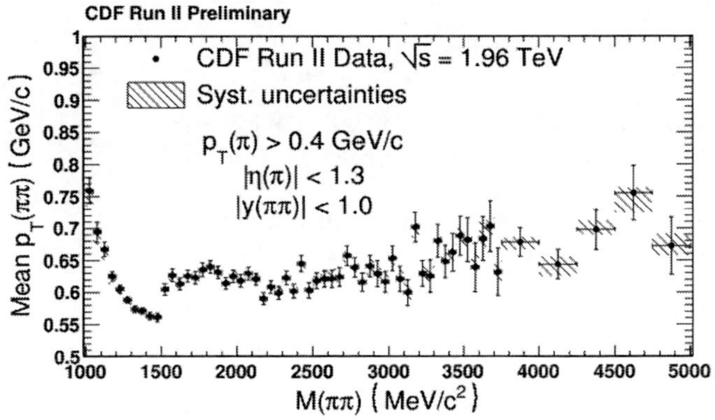

Figure 5. Mean value of the p_T distribution of the central state decaying to two pions, corrected for acceptance, as a function of $M(\pi\pi)$ at $\sqrt{s} = 1.96$ TeV.

[2] See, e.g., M.G. Albrow, T.D. Coughlin, and J.R. Forshaw, Prog. Part. Nucl. Phys. **65**, 149 (2010); M.G.Albrow, Int. J. Mod. Phys. A **29**, 1402006 (2014) and references therein.

[3] T. Aaltonen *et al.* (CDF Collaboration) Phys. Rev. D **91**, 091101 (2015).

[4] D. Acosta *et al.* (CDF Collaboration), Phys. Rev. D **50**, 2966 (1994) and references therein; F. Abe *et al.* (CDF Collaboration), Nucl. Instrum. Methods Phys. Res., Sect. A **271**, 387 (1988).

[5] G. Antchev *et al.* (TOTEM Collaboration), Europhys. Lett. **101**, 21002 and 21004 (2013); Phys. Rev. Lett. **111**, 012001 (2013).

[6] R. Ciesielski and K. Goulianos, Proc. Sci., ICHEP **2012**, 301 (2013).

[7] R. Brun, R. Hagelberg, M. Hansroul, and J.C. Lassalle, GEANT version 3.15, Report No. CERN-DD-78-2-REV.

[8] Y.I. Azimov, V.A. Khoze, E.M. Levin and M.G. Ryskin, Sov. J. Nucl. Phys. **21**, 215 (1975).

[9] B.R. Desai, B.C. Shen, and M. Jacob, Nucl. Phys. **B142**, 258 (1978).

[10] See, e.g., T. Åkesson *et al.* (AFS Collaboration), Nucl. Phys. **B264**, 154 (1986).

[11] A. Breakstone *et al.*, Zeit. Phys. C **31**, 185 (1986).

[12] T. Aaltonen *et al.* (CDF Collaboration) Phys. Rev. Lett. **102**, 242001 (2009).

[13] L.A. Harland-Lang, V.A. Khoze, and M.G. Ryskin (DIME MC), Eur. Phys. J. C **74**, 2848 (2014).

 DOI: 10.1051/epjconf/201612002010

Recent diffractive results from HERA

Alice Valkárová[1]
on behalf of H1 and ZEUS Collaborations·e-mail: alice.valkarova@mff.cuni.cz

[1] *Institute of Particle and Nuclear Physics of the Faculty of Mathematics and Physics of Charles University, V Holešovičkách 2, 180 00 Praha 8, Czech Republic*

Abstract. The diffractive dijet cross sections for photoproduction and deep inelastic scattering were studied and compared with theoretical NLO QCD predictions. The results of exclusive dijet production were compared to predictions from models which are based on different assumptions about the nature of diffractive exchange. Isolated prompt photons in diffractive photoproduction produced inclusively or together with a jet were studied for the first time.

1 Introduction

The first observation of diffractive electron-proton deep-inelastic scattering (DIS) at HERA in 1993 has renewed the interest in understanding such processes. Indeed at HERA a substantial part of the ep cross section (up to 10%) is represented by diffractive scattering processes initiated by a virtual photon. The diffractive deep inelastic scattering process $ep \to eXY$ factorises in QCD and it was proven to hold for inclusive and dijet diffractive processes, assuming high enough photon virtuality such that higher twist effects can be neglected. Diffractive parton distribution functions (DPDFs) have to be determined from QCD fits to the measured inclusive DIS diffractive cross sections. The concept of DPDFs plays an important role in the study of diffractive reactions in DIS and can be an essential input to calculations of hard diffractive processes at the LHC.

Experimentally, diffractive $ep \to eXY$ scattering is characterised by the presence of a leading proton in the final state, retaining most of the initial state proton energy, and by a lack of any other hadronic activity in the outgoing proton direction. System X is cleanly separated from a system Y which is an outgoing intact proton or its low mass excitation. In many analyses diffractive events have been selected on the basis of a large rapidity gap (LRG) between these two systems. The main advantage of this method is its high acceptance. A complementary way to select diffractive events is by direct measurements of the outgoing proton in forward proton spectrometers. This method has a disadvantage of lower acceptance but gives an opportunity to distinguish between the case where the scattered proton remains intact or dissociates into a system of low mass M_Y. These two methods of selection differ partially in the accessible kinematic ranges and substantially in their dominant sources of systematic uncertainties.

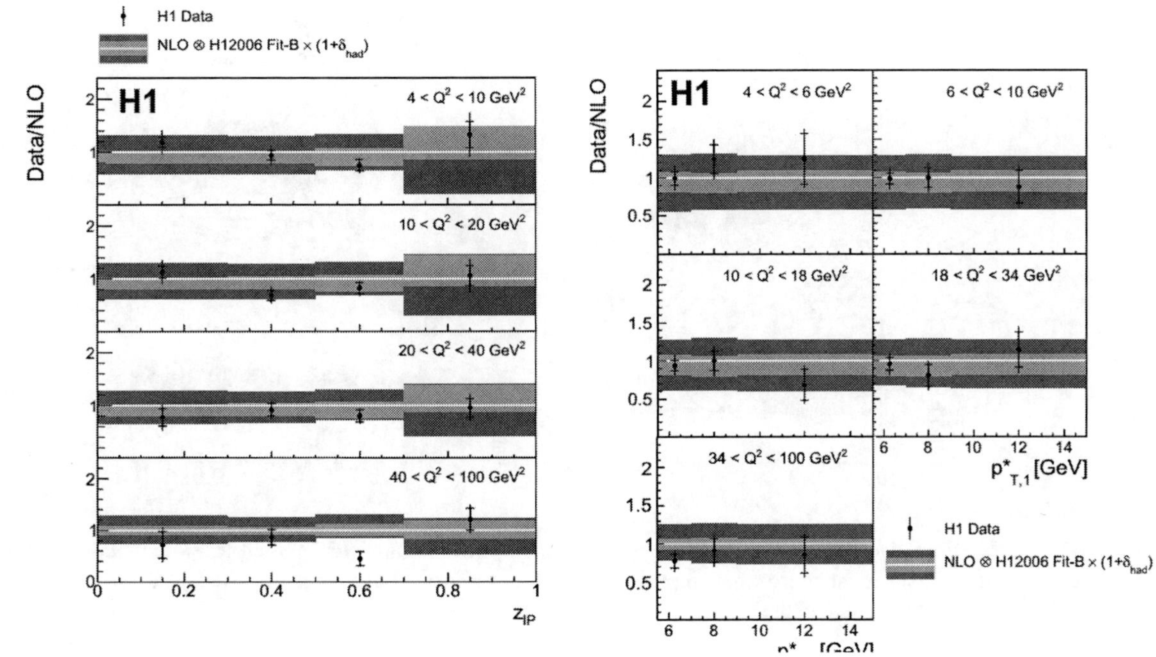

Figure 1. a) Ratio of the double-differential cross section to the NLO QCD predictions as a function of $z_{\mathbb{P}}$ and Q^2 and b) $p^*_{T,1}$ and Q^2. The inner error bars on the data points represent the statistical uncertainties, outer error bars with systematic uncertainties added in quadrature.

2 Diffractive dijets in DIS and photoproduction

A new measurement with about six times higher statistics than previous measurements was provided by H1 using the data of HERA II and LRG method of selection of diffractive events. The measurement was performed for $x_{\mathbb{P}} < 0.03$, $4 < Q^2 < 80$ GeV^2 and events with jets with transverse momenta $p^*_{\mathrm{T},1} > 5.5$ GeV and $p^*_{\mathrm{T},2} > 4$ GeV [1]. The cross sections are determined using a regularised unfolding procedure which fully accounts for efficiencies, migrations and correlations among the measurements. The integrated diffractive dijet cross section is found to be well described by the NLO QCD predictions using the H1 2006 Fit-B DPDF set. Both shapes and normalisation of the single-differential cross sections are reproduced by the theory within the experimental and theory uncertainties, as shown in Fig. 1. The measured dijet cross sections are used to extract the strong coupling constant in diffractive DIS processes for the first time. The result $\alpha_s = 0.119 \pm 0.004(exp) \pm 0.012(DPDF, theo)$ is consistent within the uncertainties with the world average.

Models based on QCD collinear and proton vertex factorisation work well to describe all diffractive processes in DIS, they however fail spectacularly when DPDFs extracted from HERA data are applied to diffractive scattering at the Tevatron [2]. For example, predictions for diffractive dijet production at the Tevatron exceed the data by about one order of magnitude. These results were also recently confirmed at the LHC [3]. This breaking of factorisation was theoretically predicted to be present also in *ep* diffractive dijet photoproduction due to contributions of a resolved component to photoproduction in LO QCD which resemble hadron-hadron collisions [4]. A variable which is commonly used to separate direct and resolved processes in LO is x_γ, measuring the fraction of the photon

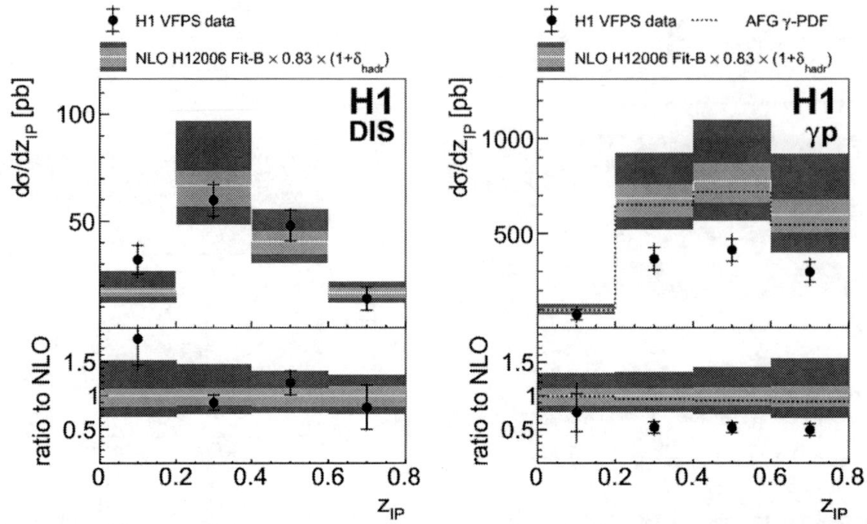

Figure 2. Diffractive dijet a) DIS and b) photoproduction cross sections differential in z_P compared to NLO QCD predictions. The inner error bars represent the statistical errors, outer error bars statistical and systematic uncertainties added in quadrature.

momentum participating in the production of the dijet system. The resolved (direct) processes dominate at low (high) x_γ values, respectively. Due to hadronisation, detector resolution and acceptances the unambiguous separation between direct and resolved components of photoproduction is one of the limiting factors in studies of possible factorisation breaking in photoproduction.

Previous H1 measurements [5, 6] suggested a suppression of the dijet photoproduction cross section by a factor of around 0.5-0.6 relative to NLO QCD predictions. ZEUS results [7] correspond to a smaller suppression and are within uncertainties consistent with no suppression whatsoever. Neither collaboration sees any evidence for the expected x_γ dependence of the survival probability.

To clarify the situation a new measurement is provided by H1 for both DIS diffractive dijets and dijets produced in photoproduction. In contrast to the previous measurements where the LRG method was used, diffractive events were selected using a leading proton spectrometer VFPS [9] placed at a distance of 220 m from the central detector. The VFPS spectrometer covered the kinematic region $0.010 < x_P < 0.024$ and $|t| < 0.6$ GeV^2. The phase space of DIS and photoproduction differs only in momentum transfer Q^2 between the incoming and outcoming electron which is defined as $Q^2 < 2$ GeV^2 for photoproduction and $4 < Q^2 < 80$ GeV^2 for DIS events. Jets were required to have the transverse energy of leading (subleading) jet $E_{T1} > 5.5$ GeV and $E_{T2} > 4$ GeV, respectively. The dijet cross sections were evaluated at the level of stable hadrons using a matrix unfolding method. The measured differential cross sections as a function of z_P are for DIS and photoproduction dijets shown and compared to NLO QCD predictions in Fig. 2. It is seen that theoretical predictions describe properly the DIS data but for photoproduction the measurements are suppressed in comparison with NLO QCD by a factor about 0.5. In Fig. 3a) the differential cross section is shown and compared to NLO QCD as a function of x_γ for dijet photoproduction. Within uncertainties there is clearly no indication of the resolved-part ($x_\gamma < 0.8$) being more suppressed than the direct part ($x_\gamma > 0.8$), which confirms previous measurements at HERA. In a refined method for studying deviations of the NLO QCD predictions from photoproduction data the cross sections measured in this regime are divided

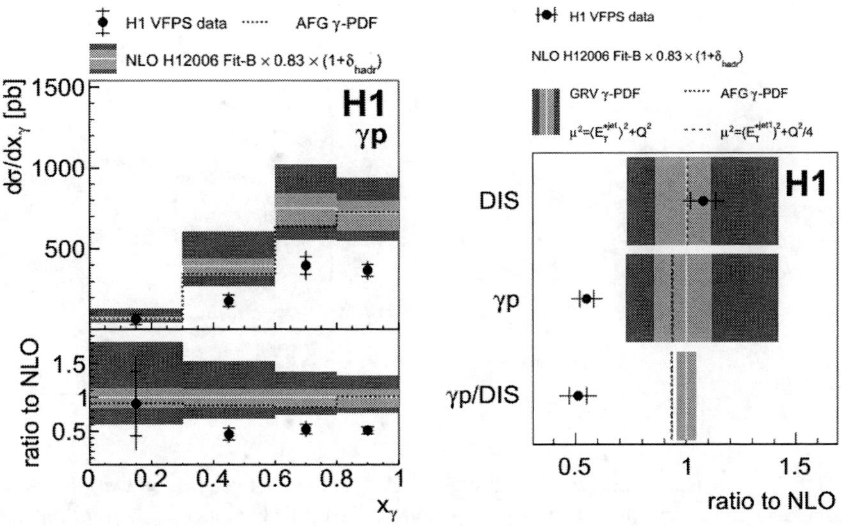

Figure 3. a) Diffractive dijet *ep* cross section in photoproduction differential in x_γ compared to NLO QCD predictions. b) Diffractive dijet DIS and photoproduction cross sections normalised to NLO QCD calculation and double ratio of photoproduction to DIS cross sections, normalised to the corresponding ratio of NLO QCD predictions.

by the corresponding cross sections in DIS. In such ratios most of the data and theoretical systematic uncertainties are reduced. The double-ratio of photoproduction to DIS, data to NLO QCD, is shown in Figure 3 b). The observation of factorisation breaking by factor of about 0.5 is in agreement with previous H1 measurements [5, 6], where complementary experimental methods of diffraction selection were used.

The production of exclusive dijets $e + p \rightarrow e + p + jet_1 + jet_2$ in DIS has been measured with the ZEUS detector [10]. The production of exclusive dijets is sensitive to the nature of the object exchanged between the virtual photon and the proton. The two models - Resolved–Pomeron model [11] and Two–Gluon–exchange model [12] - predict different shapes in the distribution of Φ, the azimuthal angle between lepton and jet planes (definition see Fig. 4 a)). It was pointed out for the first time in [13] that the parameter A in the function $1 + Acos\Phi$ is positive if the quark - antiquark pair is produced via the interaction of a single gluon with the virtual photon and negative if a system of two gluons takes part in the interaction. The measurement was performed in the range of $90 < W < 250$ GeV and for photon virtualities $Q^2 > 25\,\mathrm{GeV}^2$. The shapes of the Φ distributions were parameterised in different intervals of β (momentum fraction of the struck parton with respect to the pomeron) with the function $1 + Acos\Phi$. The dependence of the parameter A in intervals of β is compared with predictions of two versions of MC predictions for the Resolved–Pomeron model and Two–Gluon–Exchange model as implemented in Rapgap [14] in Fig. 4b). The Two-Gluon-Exchange model predicts reasonably well the measured value of A for $\beta > 0.3$, whereas the Resolved–Pomeron model exhibits a very different trend. In terms of absolute normalisation, both models are below the data and thus fail to describe the measurement [10]. It could indicate that the NLO QCD corrections are large or the cross-section enhancement arising from the evolution of the off-diagonal gluon distribution is significant.

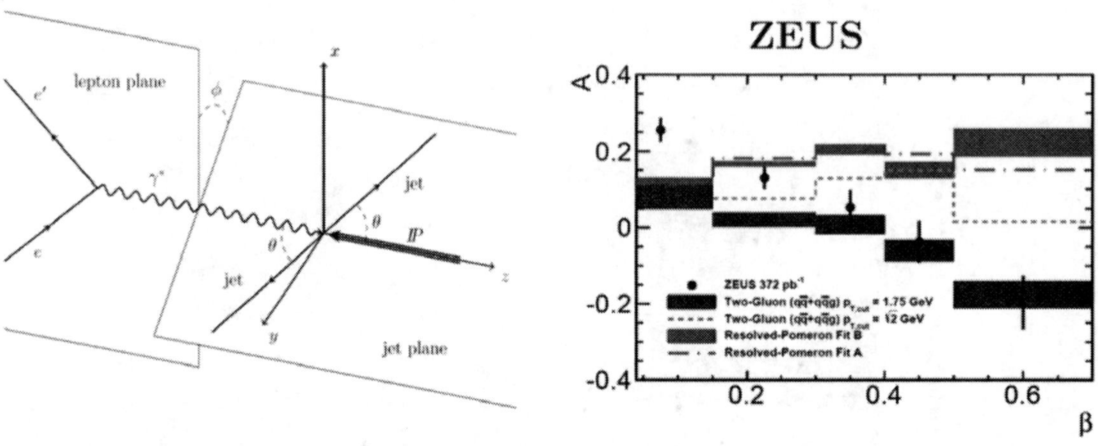

Figure 4. a) Definition of lepton and jet planes, angle Φ is the angle between these two planes. b) The shape parameter A as a function of β in comparison with two models, the bands in model predictions represent statistical uncertainties. The normalisation uncertainty originating from the proton-dissociation background ($\sim 27\%$) is not shown.

3 Hard photons in diffraction

Prompt photon emission in hadronic interactions is a sensitive probe of QCD dynamics and partonic structure, providing additional information to the study of jet production. Although cross sections are small in this case, an isolated photon at large transvers energy can be related directly to the partonic event structure. This is in contrast with jet measurements, where the partonic structure is obscured by hadronization effects.

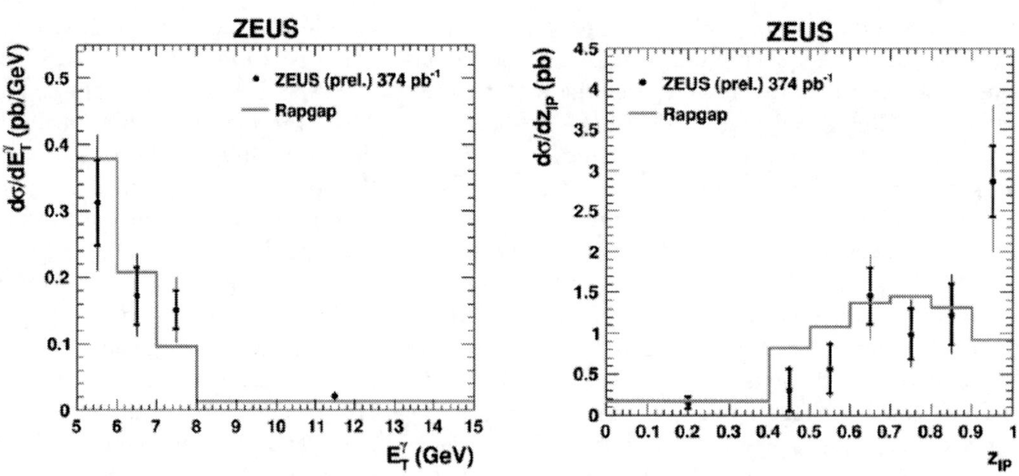

Figure 5. Differential cross sections as functions of a) E_T^{γ} and b) z_P for events with isolated photon accompanied by a jet compared to normalised prediction of RAPGAP.

A new measurement was provided by ZEUS collaboration for diffractive photoproduction *ep* events in which a hard isolated prompt photon is detected in the central region of the ZEUS detector and may be accompanied by a jet [15]. Cross sections are evaluated in the photon transverse energy range $5 < E_T^\gamma < 15$ GeV, inclusively and with a jet with transverse-energy in the range $4 < E_T^{jet} < 35$ GeV. In Fig. 5a) differential cross sections are shown as a function of the photon transverse energy E_T^γ for the photon+jet configuration. The shape of this distribution and also other distributions of inclusive and photon+jet configurations, see [15], are satisfactorily described by the Rapgap Monte Carlo model, normalised to the data. The shape of the z_P distribution, see Fig. 5b), however is not well described. In particular, a prominent peak near $z_P = 1$ is seen and requires further study. In future, comparison with NLO QCD calculations would give another handle on factorisation in *ep* diffractive photoproduction.

References

[1] V. Andreev *et al.*, H1 Collab. *JHEP* **1503**, 092 (2015).
[2] T. Affolder *et al.*, CDF Collab., *Phys. Rev. Lett.* **84**, 5043 (2000).
[3] S. Chatrchyan et al., CMS Collab., *Phys.Rev.* **D87** 012006, (2013).
[4] A. Kaidalov, V. Khoze, A. Martin and M. Ryskin, *Eur. Phys.J.* **C66**, 373 (2010).
[5] A. Aktas *et al.*, the H1 Collab. *Eur. Phys. J.* **C51**, 549 (2007).
[6] F.D. Aaron *et al.*, H1 Collab. *Eur.Phys.J.* **C70**, 15 (2010).
[7] S. Chekanov *et al.*, ZEUS Collab. *Eur.Phys. J.* **C55**, 177 (2008).
[8] V. Andreev *et al.*, H1 Collab., *JHEP* **1505**, 056 (2015).
[9] A. Astvatsatourov, K. Cerny, J. Delvax, L. Favart, T. Hreus, *et al.*, *Nucl.Instrum.Meth.*, **A736**, 46 (2014).
[10] H. Abramowicz *et al.*, ZEUS Collab. DESY-15-070 (May 2015), submitted to EPJ.
[11] G. Ingelman and P.E. Schlein, *Phys. Lett.B* **152**, 256 (1985).
[12] J. Bartels, H. Jung and M. Wuesthoff, *Eur.Phys. J.* **C11**, 111 (1999).
[13] J. Bartels et al., *Phys. Lett.* **B 386**, 389 (1996).
[14] H. Jung, RAPGAP version 3.1, *Comput. Phys. Commun.* **86**, (1995) 147.
[15] P. Bussey for ZEUS Collaboration, Isolated photons in diffraction, PoS (DIS2015) 068.

EPJ Web of Conferences **120**, 02011 (2016) DOI: 10.1051/epjconf/201612002011
ISMD 2015

Breakdown of QCD factorization in hard diffraction

B.Z. Kopeliovich[1,a]

[1] *Departamento de Física, Universidad Técnica Federico Santa María; and Centro Científico-Tecnológico de Valparaíso; Casilla 110-V, Valparaíso, Chile*

Abstract. Factorization of short- and long-distance interactions is severely broken in hard diffractive hadronic collisions. Interaction with the spectator partons leads to an interplay between soft and hard scales, which results in a leading twist behavior of the cross section, on the contrary to the higher twist predicted by factorization. This feature is explicitly demonstrated for diffractive radiation of abelian (Drell-Yan, gauge bosons, Higgs) and non-abelian (heavy flavors) particles.

1 QCD factorization in diffraction

QCD factorization in inclusive processes is nowadays one of the most powerful and frequently used theoretical tools [1]. In spite of lack of understanding of the soft interaction dynamics, the contributions of the soft long-distance and hard short-distance interactions factorise. Making a plausible (not proven) assumption about universality of the former, one can measure it with electro-weak hard probes (DIS, Drell-Yan process) and apply to hard hadronic processes. Although it is tempting to extend this factorization scheme to diffractive, large rapidity gap processes, it turns out to be heavily broken [2, 3], as is demonstrated below.

Ingelman-Schlein picture of diffraction [4].

It looks natural that on analogy of DIS on a hadronic target, DIS on the Pomeron probes its PDF (parton distribution function), like is illustrated in Fig. 1 . Once the parton densities in the Pomeron

Figure 1. DIS on a hadron taget (left) and on the Pomeron, treated as a target (right).

were known, one could predict any hard diffractive hadronic reaction.

The Good-Walker mechanism of diffraction [6–8].

[a] e-mail: boris.kopeliovich@usm.cl

According to this quantum mechanical treatment of diffraction, the diffractive amplitude is given by the difference between the elastic amplitudes of different Fock components in the projectile particle. In the dipole representation hard diffraction of a hadron comes from the difference between elastic amplitudes of hadronic states with and without a hard fluctuation,

$$A^h_{diff} \propto \sigma_{\bar{q}q}(R+r) - \sigma_{\bar{q}q}(R) \propto rR \sim 1/Q, \tag{1}$$

where $\sigma_{\bar{q}q}(R)$ is the total dipole-nucleon cross section [9]; R characterises the hadronic size, while small $r \sim 1/Q \ll R$ is related to the hard process [10, 11]. Apparently such a mild Q-dependence contradicts factorization prediction, based on the DIS relation,

$$A_{diff} \propto \sigma_{\bar{q}q}(r) \propto r^2 \sim 1/Q^2, \tag{2}$$

which is a higher twist effect.

2 Drell-Yan reaction: annihilation or bremsstrahlung?

Parton model is not Lorentz invariant, interpretation of hard reactions varies with reference frame. E.g. DIS is treated as a probe for the proton structure in the Bjorken frame, but looks differently in the target rest frame, as interaction of hadronic components of the photon. Only observables are Lorentz invariant.

The Drell-Yan reaction in the target rest frame looks like radiation of a heavy photon (or Z, W), rather than q-qbar annihilation [12, 13], as is illustrated in Fig. 2

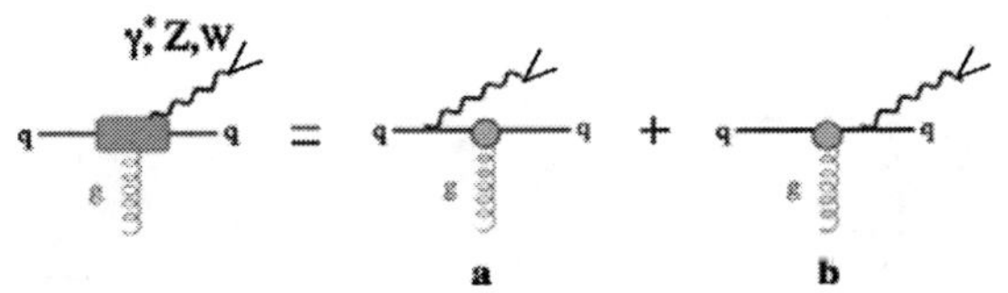

Figure 2. Radiation of a heavy photon, or gauge bosons in the target rest frame, corresponds to $\bar{q}q$ annihilation in the boson rest frame.

The cross section, expressed via the dipoles [12, 13], looks similar to DIS,

$$\frac{d\sigma^{DY}_{inc}(qp \to \gamma^* X)}{d\alpha\, dM^2} = \int d^2r \left|\Psi_{q\gamma^*}(\vec{r},\alpha)\right|^2 \sigma(\alpha r, x_2), \tag{3}$$

where $\Psi_{q\gamma^*}(\vec{r},\alpha)$ is the distribution function for the $|\gamma^* q\rangle$ Fock component of the quark; $\alpha = p^+_{\gamma^*}/p^+_q$ is the fractional ligh-cone momentum of the heavy photon.

In DY diffraction the Ingelman-Schlein factorization is broken. Indeed, diffractive radiation of an abelian particle vanishes in the forward direction [13], due to cancellation of the graphs a, b and c depicted in Fig. 3,

$$\left.\frac{d\sigma^{DY}_{inc}(qp \to \gamma^* qp)}{d\alpha\, dM^2\, d^2p_T}\right|_{p_T=0} = 0. \tag{4}$$

In both Fock components of the quark, $|q\rangle$ and $|q\gamma^*\rangle$ only quark interacts, so they interact equally, and according to the Good-Walker picture cancel in the forward diffractive amplitude. This conclusion holds for any abelian diffractive radiation of γ^*, W, Z bosons, Higgs.

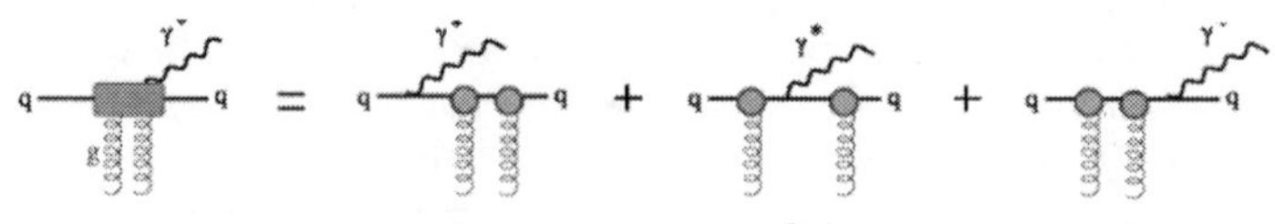

Figure 3. Feynman graphs for diffractive radiation of a heavy photon by a quark.

Diffractive DIS is dominated by soft interactions [2, 14]. On the contrary, diffractive Drell-Yan gets the main contribution from the interplay of soft and hard scales [10, 11] (see Eq. (2)).

The saturated shape of the dipole cross section, $\sigma(R) \propto 1 - \exp(-R^2/R_0^2)$, leads to the unusual features of diffractive Drell-Yan cross section (compare with (2)),

$$\frac{\sigma_{sd}^{DY}}{\sigma_{incl}^{DY}} \propto [\sigma(R+r) - \sigma(R)]^2 \propto \frac{\exp(-2R^2/R_0^2)}{R_0^2} \tag{5}$$

As a result, the fractional diffractive Drell-Yan cross section is steeply falling with energy, but rises with the scale, because of saturation, as is shown in Fig. 4.

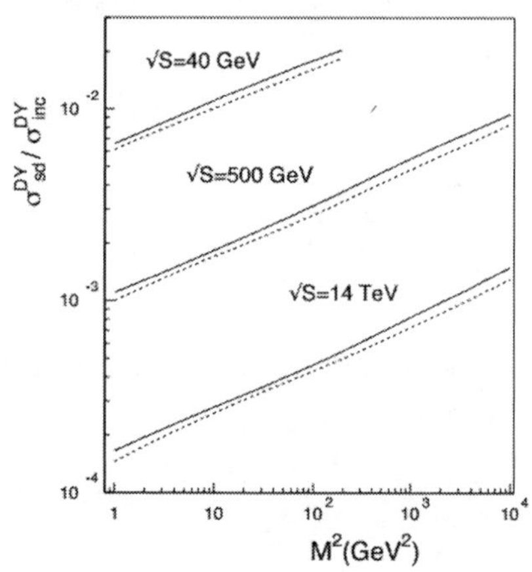

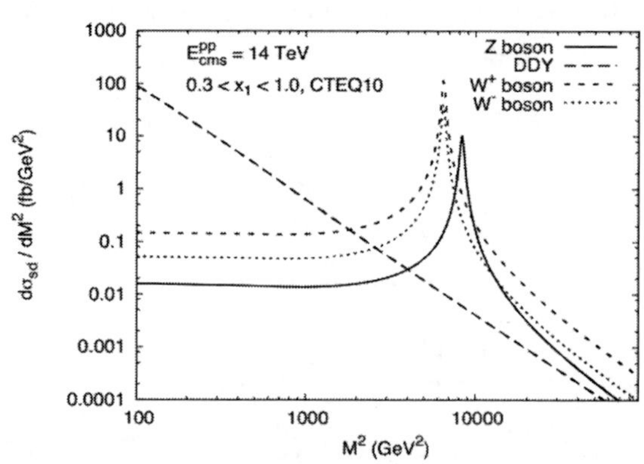

Figure 4. *Left:* Fractional DY cross section vs the dilepton mass squared and at different c.m. energies. *Right:* Diffractive gauge boson production cross section as function of di-lepton invariant mass squared.

3 Diffractive Z and W production

Abelian diffractive radiation of any particle is described by the same Feynman graphs, only couplings and spin structure are different [15]. In Fig. 4 (right panel) we present the results for the single diffractive cross sections for Z^0, γ^* (diffractive DY) and $W^\pm$ bosons production, differential in the di-lepton mass squared, $d\sigma_{sd}/dM^2$.

The single diffractive process $pp \to Xp$ at large Feynman $x_F \to 1$ of the recoil proton is described by the triple Regge graphs, as is illustrated in Fig. 5. The results for the fractional diffractive cross sections of Z and W production is compared with the CDF measurements in Fig. 6 (left panel).

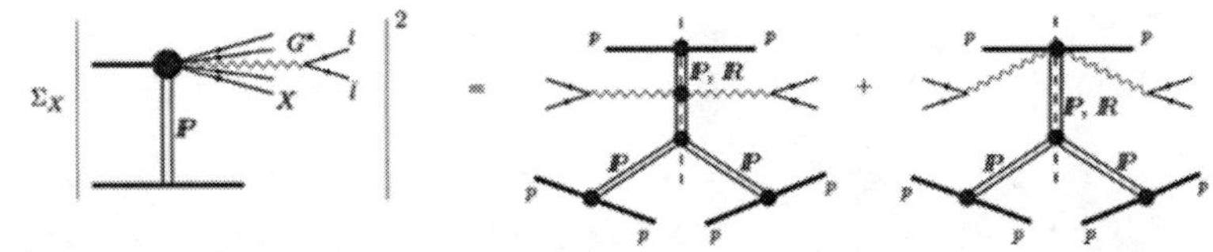

Figure 5. Triple-Regge description of the process $pp \to Xp$, where the diffractively produced state X contains a gauge boson decaying to a lepton pair.

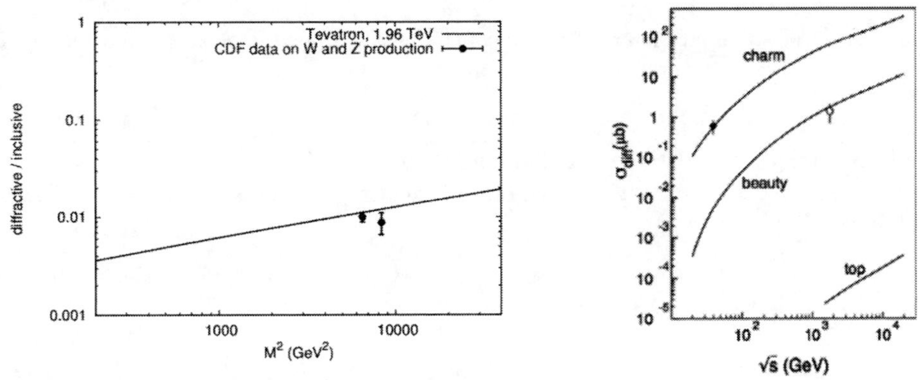

Figure 6. *Left:* The diffractive-to-inclusive ratio vs dilepton invariant mass squared in comparison with the CDF measurements. *Right:* Cross section of diffractive production of heavy flavors in comparison with the CDF data for charm and beauty (see details in [16]).

4 Diffractive heavy flavor production

The dynamics of inclusive heavy flavor production can be classified as : (i) bremsstrahlung (like in DY); and (ii) production mechanisms [16], in accordance with the Feynman graphs presented in Fig. 7. The amplitudes for the two mechanisms are expressed via the amplitudes A_i corresponding to

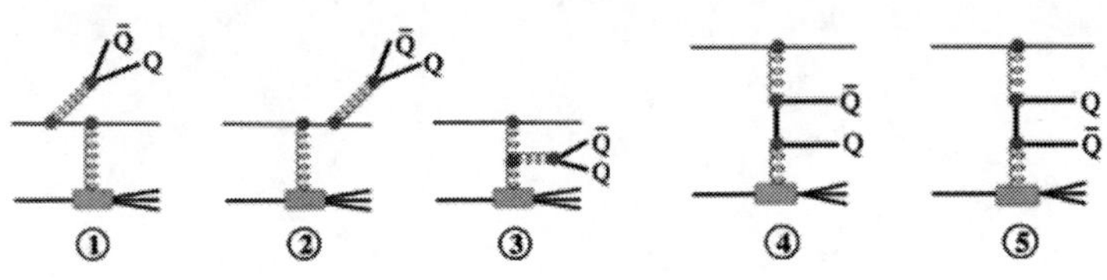

Figure 7. Feynman graphs contributing to inclusive production of a heavy quark pair.

the graph numbering in Fig. 7. The bremsstrahlung and production amplitudes have the form,

$$A_{Br} = A_1 + A_2 + \frac{Q^2}{M^2+Q^2} A_3; \tag{6}$$

$$A_{Pr} = \frac{M^2}{M^2+Q^2} A_3 + A_4 + A_5. \tag{7}$$

For diffractive production one has to provide a colorless two-gluon exchange. Diffractive excitation of a quark turns out to be a higher twist effect, as is depicted in Fig. 8, left, similar to diffractive

Drell-Yan. In this case interaction with spectators again plays crucial role, providing a leading twist contribution, as is shown in Fig. 8, right [16]. Numerically, the leading twist production mechanism

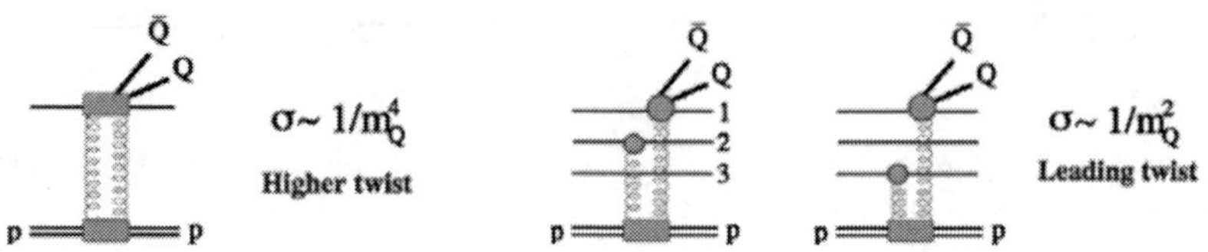

Figure 8. *Left:* Diffractive production of a heavy quark pair in a quark-proton collision. *Right:* Diffractive production of heavy quarks in a proton-proton collision.

is much larger compared with the bremsstrahlung mechanism.

The leading twist behavior $1/m_Q^2$ of the diffractive cross section is confirmed by CDF data, as is demonstrated in Fig. 6, right panel.

5 Diffractive Higgsstrahlung

Diffractive Higgsstrahlung is similar to diffractive DY. Z, W, since in all cases the radiated particle does not participate in the interaction. However, the Higgs decouples from light quarks, due to smallness of the coupling, so the cross section of higgsstrahlung by light hadrons is small. Although light quark do not radiate Higgs directly, they can do it via production of heavy flavors. Therefore the mechanism is closely related to non-abelian diffractive quark production, presented in the previous section. The rapidity dependent cross section of diffractive Higgs production, evaluated in [17], is plotted in Fig. 9, left. The cross section is rather small, below 1fb.

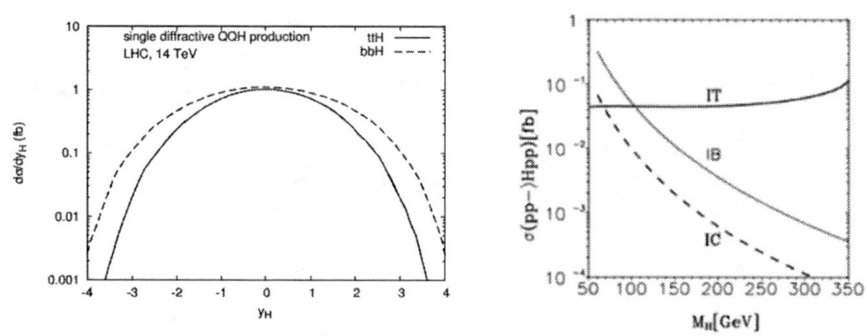

Figure 9. *Left:* The differential cross section of single diffractive Higgs boson production in association with heavy quark pair vs Higgs rapidity (see details in [17]). *Right:* The cross section of diffractive exclusive Higgs production, vs its mass, for different intrinsic heavy flavors [18].

Higgs boson can also be diffractively produced due to intrinsic heavy flavors in light hadrons. Exclusive Higgs production, $pp \to Hpp$, via coalescence of heavy quarks was evaluated in [18, 19].

The cross section of Higgs production was calculated fixing 1% of intrinsic charm in the proton, and assuming that heavier flavors scale as $1/m_Q^2$ [20]. At the Higgs mass 125 GeV intrinsic bottom and top give comparable contributions, as is demonstrated in Fig. 9, right.

6 Summary

Factorization of short and long-distance interactions is heavily broken in hard diffractive hadronic collisions. In particular, forward diffractive radiation of direct photons, Drell-Yan dileptons, and gauge bosons Z, W, by a parton is forbidden. Nevertheless, a hadron can diffractively radiate in the forward direction due to a possibility of soft interaction with the spectators. This property of abelian radiation breaks down diffractive factorization resulting in a leading twist dependence on the boson mass, $1/M^2$.

Non-abelian forward diffractive radiation of heavy flavors is permitted even for an isolated parton. However, interaction with spectators provides the dominant contribution to the cross section. It comes from the interplay between large and small distances. Data well confirm the leading twist behavior.

Diffractive higgsstrahlung is possible due to a double-step process, via heavy quark production. Therefore, the main contribution comes for Higgs production in association with a heavy quark pair. Another important contribution to diffractive Higgs production comes from coalescence of intrinsic heavy quarks in the proton. For $M_H = 125\,\text{GeV}$ dominance of intrinsic bottom and top is expected.

Acknowledgement: This work was supported in part by Fondecyt (Chile) grant 1130543 and Conicyt (Chile) grant Anillo ACT 1406.

References

[1] J. C. Collins, D. E. Soper and G. F. Sterman, Adv. Ser. Direct. High Energy Phys. **5**, 1 (1989).
[2] B. Z. Kopeliovich, I. K. Potashnikova, I. Schmidt, Braz. J. Phys. **37**, 473 (2007).
[3] R. Pasechnik, B. Kopeliovich and I. Potashnikova, Adv. High Energy Phys. **2015**, 701467 (2015).
[4] G. Ingelman, P. E. Schlein, Phys. Lett. **B152** 256 (1985).
[5] UA8 Collaboration, A. Brandt et al., Phys. Lett. **B297**, 417 (1992); **B421**, 395 (1998).
[6] R. J. Glauber, Phys. Rev. 100, 242 (1955).
[7] E. Feinberg and I. Ya. Pomeranchuk, Nuovo. Cimento. Suppl. 3 (1956) 652.
[8] M. L. Good and W. D. Walker, Phys. Rev. 120 (1960) 1857.
[9] B. Z. Kopeliovich, L. I. Lapidus and A. B. Zamolodchikov, JETP Lett. **33**, 595-597 (1981) [Pisma Zh. Eksp. Teor. Fiz. **33**, 612 (1981)].
[10] B. Z. Kopeliovich, I. K. Potashnikova, I. Schmidt, A. V. Tarasov, Phys. Rev. **D74**, 114024 (2006).
[11] R. S. Pasechnik and B. Z. Kopeliovich, Eur. Phys. J. C **71**, 1827 (2011).
[12] B. Z. Kopeliovich, proc. of the workshop Hirschegg 95: Dynamical Properties of Hadrons in Nuclear Matter, Hirschegg January 16-21, 1995, ed. by H. Feldmeyer and W. Nörenberg, Darmstadt, 1995, p. 102 [hep-ph/9609385].
[13] B. Z. Kopeliovich, A. V. Tarasov and A. Schafer, Phys. Rev. C **59**, 1609 (1999).
[14] B. Z. Kopeliovich and B. Povh, Z. Phys. A **356**, 467 (1997).
[15] R. Pasechnik, B. Kopeliovich and I. Potashnikova, Phys. Rev. D **86**, 114039 (2012).
[16] B. Z. Kopeliovich, I. K. Potashnikova, I. Schmidt and A. V. Tarasov, Phys. Rev. D **76**, 034019 (2007).
[17] R. Pasechnik, B. Z. Kopeliovich and I. K. Potashnikova, Phys. Rev. D **92**, no. 9, 094014 (2015).
[18] S. J. Brodsky, B. Kopeliovich, I. Schmidt and J. Soffer, Phys. Rev. D **73**, 113005 (2006).
[19] S. J. Brodsky, A. S. Goldhaber, B. Z. Kopeliovich and I. Schmidt, Nucl. Phys. B **807**, 334 (2009).
[20] M. Franz, M. V. Polyakov and K. Goeke, Phys. Rev. D **62**, 074024 (2000).

EPJ Web of Conferences **120**, 03001 (2016) DOI: 10.1051/epjconf/201612003001
ISMD 2015

Flavor tagging TeV jets for physics beyond the Standard Model

Zack Sullivan[1,a] and Keith Pedersen[1]

[1] *Department of Physics, Illinois Institute of Technology, Chicago, Illinois 60616-3793, USA*

Abstract. We present a new scheme for tagging boosted heavy flavor jets called "μ_x tagging." At the LHC, the primary method to tag b-jets relies on tracking their charged constituents. However, when highly boosted, track-based b-tags lose efficiency, and the probability to mistag light jets rises dramatically. Using muons from B hadron decay and defining a particular combination "x" of angular information and boost estimation, we find fairly flat efficiencies to tag b-jets, c-jets, light-quark jets, and light-heavy jets (containing B hadrons from gluon splitting) of $\epsilon_b = 14\%$, $\epsilon_c = 6.5\%$, $\epsilon_{\text{light-light}} = 0.1\%$, and $\epsilon_{\text{light-heavy}} = 0.5\%$, respectively. We demonstrate the usefulness of this new scheme by showing the reach for discovery of a leptophobic $Z' \to b\bar{b}$ in the dijet channel.

1 Introduction

As searches for W' and Z' bosons at the CERN Large Hadron Collider (LHC) shift to TeV-scale energies, observation of their decay products becomes challenging. Observation of dijet resonances above QCD background is hampered by falling b-tagging efficiencies (28–15% around 1–2 TeV) and large light-jet fake rates of 1–2% [1]. In addition to the low purity ($\epsilon_{\text{fake}}/\epsilon_b \sim 1/10$), large uncertainties in the tagging efficiencies affect the mass limits; e.g., the ATLAS b-tag uncertainty is 35% for $p_T > 500$ GeV [2]. In order to discover multi-TeV physics beyond the Standard Model (BSM), we need a better b tag with good efficiency and purity.

At this conference, we presented a new method for flavor tagging at TeV-scale energies called "μ_x boosted-bottom-jet tagging" [3]. This method is derived from kinematic first principles, and provides both a well-determined 14% efficiency for b-tagging, and a factor of 10 improvement in fake rejection over existing tags ($\epsilon_{\text{fake}}/\epsilon_b \sim 1/100$). In Sec. 2 we summarize the algorithm and cuts for the μ_x tag, show why it works, and plot its transverse momentum p_T- and pseudorapidity η-dependent efficiencies. In Sec. 3 we briefly describe the application of μ_x boosted-b tagging to an analysis for discovery of a leptophobic $Z' \to b\bar{b}$. We summarize our results in Sec. 4.

2 μ_x boosted-b tag

Consider a jet containing a semi-muonic decay of a B hadron. In the center-of-momentum (CM) frame, the muon is emitted with a speed $\beta_{\mu,\text{cm}}$ and at an angle θ_{cm} with respect to the beam axis (see

[a] e-mail: Zack.Sullivan@IIT.edu

Fig. 1). In the lab frame, the boost γ_B of the B hadron compresses its decay products into a narrow subjet at high energy. We define a lab frame observable

$$x \equiv \gamma_B \tan\theta_{\text{lab}} = \frac{\sin\theta_{\text{cm}}}{\kappa + \cos\theta_{\text{cm}}}, \tag{1}$$

where $\kappa \equiv \beta_B/\beta_{\mu,\text{cm}}$.

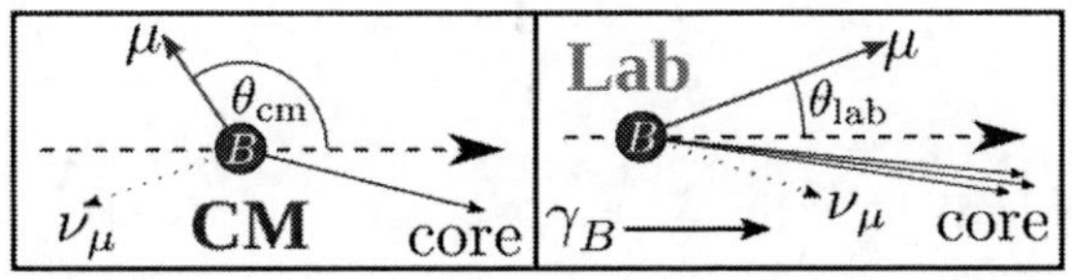

Figure 1. Nomenclature for the center-of-momentum frame and boosted lab frame.

While κ is unobservable, for sufficiently boosted B hadrons ($\gamma_B \gg \gamma_{\mu,\text{cm}} \geq 3$) the lab frame distribution of the muon count N vs. x is effectively independent of κ,

$$\frac{dN}{dx} \approx \frac{2x}{(x^2+1)^2}. \tag{2}$$

This leads to a universal shape in x for highly boosted jets containing B hadrons. Using this shape we define the μ_x boosted-b tag as a cut on two variables: We capture 90% of muons from B decay by demanding $x < 3$. To further isolate b decays, we note the hard fragmentation function for b quarks leads to the B hadron subjet carrying a large fraction f_{subjet} of the total jet momentum. Hence, we demand

$$f_{\text{subjet}} \equiv \frac{p_{T\text{subjet}}}{p_{T\text{jet}}} \geq 0.5. \tag{3}$$

There are two challenges in applying the μ_x tag to real events: we must identify the correct decay remnant of the B hadron to reconstruct its four-vector p_{subjet}, and we must deal with the missing muon neutrino. Most of the neutrino energy in the lab frame comes from the boost, so we use the measured four-vector of the muon as a proxy $p_\nu = p_\mu > 10$ GeV. In order to find the non-leptonic remnant "core" of the subjet, we need a more sophisticated algorithm.

In order to reconstruct the boosted subjet we first cluster the jet using the anti-k_T algorithm with a $R = 0.4$. We then search for the core (generally the charm hadron remnant) by reclustering the muon and calorimeter towers with total jet energy fraction $f_{\text{tower}}^{\text{min}} > 0.05$ using a smaller $R_{\text{core}} = 0.04$. We assume $m_{\text{core}} = 2$ GeV (a typical charm hadron mass), and identify the "correct" core as the one which comes closest to $\sqrt{p^2_{\text{subjet}}} = 5.3$ GeV. Since mismeasurements smear out the reconstructed energy of the subjet, if $m_{\text{subjet}} > 12$ GeV we constrain the subjet mass to be 12 GeV. The parameters of the μ_x tag are summarized in Table 1.

Table 1. A summary of parameters chosen for μ_x boosted bottom jet tagging.

R	0.4	m_{core}	2 GeV	$p_{T\mu}^{\text{min}}$	10 GeV
R_{core}	0.04	m_B	5.3 GeV	x_{max}	3 ($x_{90\%}$)
$f_{\text{tower}}^{\text{min}}$	0.05	$m_{\text{subjet}}^{\text{max}}$	12 GeV	$f_{\text{subjet}}^{\text{min}}$	0.5

In spite of its non-trivial reconstruction, x is effectively a dynamic angular cut on the muon. Defining ξ, the lab frame angle between the muon and the core, it is possible to calculate ξ_{max}, the maximum μ-to-core angle which produces $x \leq 3$. For "soft" muons ($E_\mu \ll E_{\text{core}}/18$), this angular cut is relatively tight

$$\xi_{\max}^{\text{soft}} \approx 3\frac{m_{\text{core}}}{E_{\text{core}}}. \tag{4}$$

Once the muons become "hard" ($E_\mu \geq E_{\text{core}}/18$), the cut loosens significantly

$$\xi_{\max}^{\text{hard}} \approx 3\frac{m_{\text{subjet}}^{\max}}{E_{\text{core}}}. \tag{5}$$

While the transition between these limits depends explicitly on the muon's p_T, this dependence is small until just below the hard threshold. Thus, not only is x a *smart* angular cut — scaling with the energy of the core — it is a *dual* angular cut; tight for soft muons, looser for hard muons, and sensitive to the p_T resolution of the muon system only within the narrow transition region.

The separation of reconstructed b jets from light-quark-initiated jets can be seen in Fig. 2. Bottom jets (b-quarks hadronized as B hadrons) above 500 GeV produce large f_{subjet} and $x \sim 0.8$. Light jets (mostly π and K) produce either incompatible values of $x > 3$, or random subjet recombinations that lead to small f_{subjet}. A small fraction of b jets is not well-reconstructed (represented by the low-f_{subjet} tail), but it has little effect on the total efficiency.

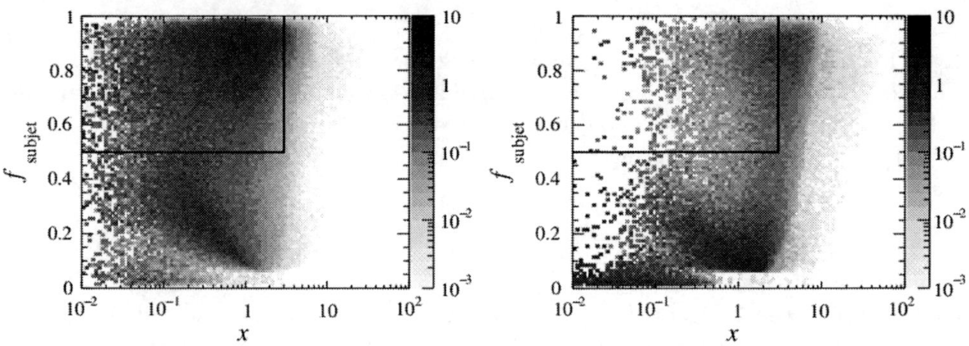

Figure 2. Density of reconstructed candidate tags with $\mu = 40$ pileup events as a function of f_{subjet} vs. x for (left) bottom and (right) light-quark-initiated jets.

We extract the standalone μ_x tagging efficiencies using PYTHIA 8.210 [4, 5] fed into an ATLAS-like version of DELPHES 3.2 [1], and a custom μ_x tagging module `MuXboostedBTagging` (available on GitHub [6]). In Fig. 3 we show separate efficiencies as a function of p_T and η for bottom jets, charm jets, light-light jets (where the muon came from a light-flavor hadron), and light-heavy jets (where a gluon split to $b\bar{b}/c\bar{c}$ — producing heavy-flavor hadrons in the final state). The kinematic nature of the tagging variables leads to fairly flat efficiencies in pseudorapidity, and when $p_T > 500$ GeV. The exception is the η distribution for B hadrons from gluon splitting. This leads to the intriguing possibility that the $g \to b\bar{b}$ contribution to jets in the Monte Carlo could be calibrated using the rapidity dependence of these highly-boosted jets.

3 A search for leptophobic $Z' \to b\bar{b}$

Very massive Z' bosons are expected to exist in many BSM models. We test the μ_x boosted-bottom tag by examining the reach at a 13 TeV LHC for a leptophobic Z' decaying to $b\bar{b}$ or $c\bar{c}$. For this study

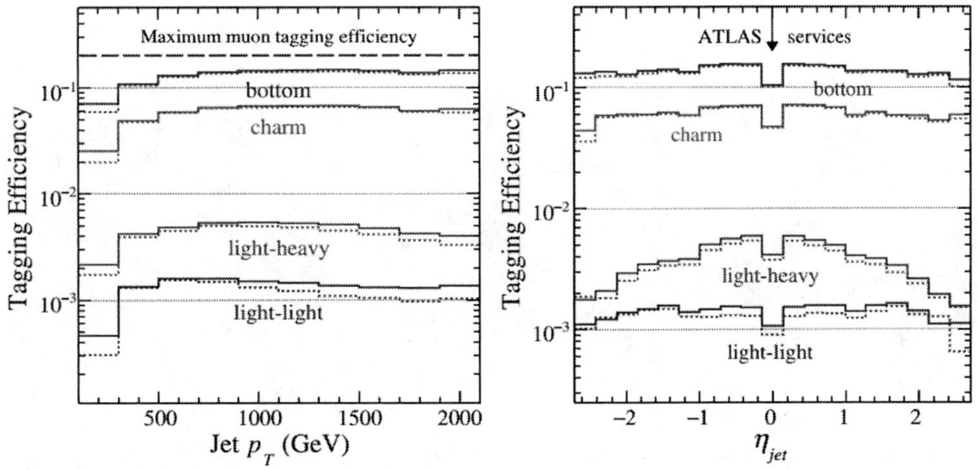

Figure 3. μ_x tagging efficiency vs. (left) jet p_T and (right) $\eta_{\rm jet}$. Solid (dashed) lines include $\mu = 0$ (40) pileup events.

we choose a $U(1)'_B$ Lagrange density

$$\mathscr{L} = \frac{g_B}{6} Z'_{B\mu} \bar{q}\gamma^\mu q, \tag{6}$$

with a flavor-independent coupling to quarks [7, 8].

We simulate the signal and backgrounds using a MLM-matched MadEvent sample [9] and CT14llo PDFs [10] fed through PYTHIA into DELPHES. In addition to demanding one or two μ_x tags (as defined in Sec. 2), we require $|\eta_j| < 2.7$, and $\Delta\eta_{jj} < 1.5$. We reconstruct a dijet mass out of the two leading-p_T jets, and look for a resonance in the mass window $[0.85, 1.25] \times M_{Z'_B}$.

The results for 5σ discovery of this leptophobic Z' are shown in Fig. 4 for a two-tag, and one-tag inclusive sample, compared to current exclusion limits from Ref. [7]. In 100 fb^{-1} of integrated luminosity at 13 TeV, a two b-tag analysis could discover a Z' of 3 TeV if the universal coupling $g_B \sim 2.5$. For this particular model, the single-tag inclusive search would be more effective — allowing for discovery up to nearly 1 TeV above current mass limits. Should a discovery not be made, the two-tag search (not shown) would set a 95% C.L. exclusion comparable to the one-tag discovery reach; while the one-tag search would set a 95% C.L. exclusion that can access g_B couplings a factor of 2 smaller than current limits, and masses up to 2 TeV higher.

4 Conclusions

In this paper we discuss the new μ_x boosted-bottom-jet tag. Combining angular information x from B hadron decay with jet substructure $f_{\rm subjet}$ in TeV-scale jets allows for tagging efficiencies of $\epsilon_b = 14\%$, $\epsilon_c = 6.5\%$, $\epsilon_{\rm light-light} = 0.1\%$, and $\epsilon_{\rm light-heavy} = 0.5\%$, respectively. The results here focused on ATLAS because their standalone non-isolated muon tagging efficiency is publicly available. We expect that if CMS has similar non-isolated muon tagging capability this tag will be just as effective, since it is kinematically driven and not sensitive to fine details of the detector.

When applying the μ_x tag to a search for leptophobic Z' bosons, we find that the reach for discovery at a 13 TeV LHC is about 1 TeV higher than current limits. If a Z' is not found, 95% C.L.

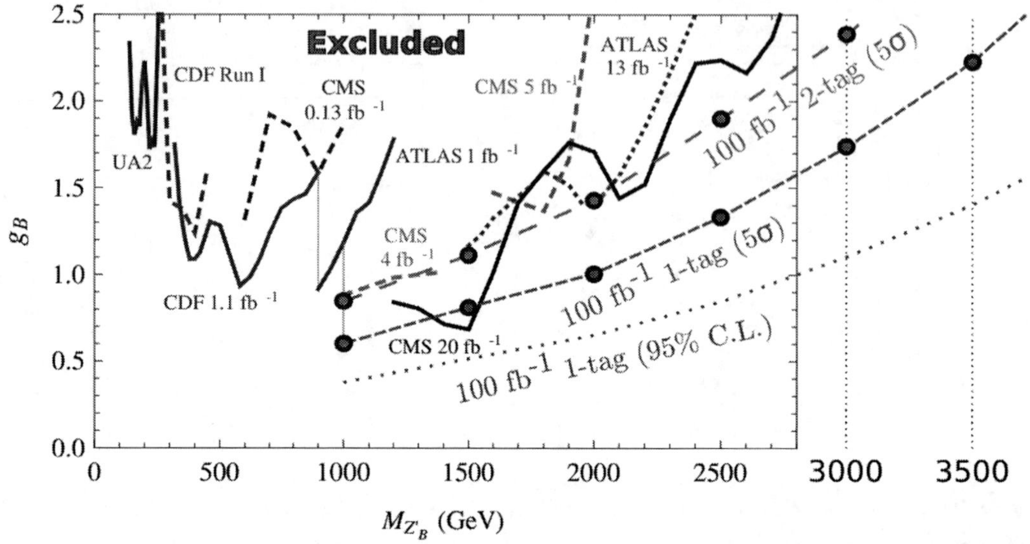

Figure 4. 5σ discovery reach for a leptophobic Z' with universal coupling in the with one or two boosted-b tags at a 13 TeV LHC compared to exclusion limits from Ref. [7]. Also shown is the 95% C.L. exclusion reach of the one-tag analysis.

exclusion limits can be set up to 2 TeV higher, or for g_B couplings a factor of 2 smaller, than the current limits. In addition to $Z' \to b\bar{b}$, the μ_x tag should be of immediate use in the search for $W' \to t\bar{b}$ in the boosted-top and boosted-bottom channel [11] conducted by the ATLAS Collaboration [2].

Acknowledgments

Z.S. would like to thank the ISMD organizers for an exceptional symposium and environment for lively discussions. This work was supported by the U.S. Department of Energy under award No. DE-SC0008347.

References

[1] J. de Favereau *et al.* (DELPHES 3 Collaboration), J. High Energy Phys. **02**, 057 (2014) [arXiv:1307.6346 [hep-ex]].

[2] G. Aad *et al.* (ATLAS Collaboration), Phys. Lett. B **743**, 235 (2015) [arXiv:1410.4103 [hep-ex]].

[3] K. Pedersen and Z. Sullivan, arXiv:1511.05990 [hep-ph].

[4] T. Sjostrand, S. Mrenna, and P.Z. Skands, J. High Energy Phys. **05**, 026 (2006) [hep-ph/0603175].

[5] T. Sjostrand, S. Mrenna, and P.Z. Skands, Comput. Phys. Commun. **178**, 852 (2008) [arXiv:0710.3820 [hep-ph]].

[6] K. Pedersen, [https://github.com/keith-pedersen/delphes/tree/MuXboostedBTagging].

[7] B.A. Dobrescu and F. Yu, Phys. Rev. D **88**, 035021 (2013); Erratum: Phys. Rev. D **90**, 079901 (2014) [arXiv:1306.2629 [hep-ph]].

[8] B.A. Dobrescu, arXiv:1506.04435 [hep-ph].
[9] J. Alwall *et al.*, J. High Energy Phys. **07**, 079 (2014) [arXiv:1405.0301 [hep-ph]].
[10] S. Dulat *et al.*, arXiv:1506.07443 [hep-ph].
[11] D. Duffty and Z. Sullivan, Phys. Rev. D **90**, 015031 (2014) [arXiv:1307.1820 [hep-ph]].

EPJ Web of Conferences **120**, 03002 (2016) DOI: 10.1051/epjconf/201612003002
ISMD 2015

Combined analysis of charm-quark fragmentation-fraction measurements

Mykhailo Lisovyi[1,3,a], Andrii Verbytskyi[2,b], and Oleksandr Zenaiev[3,c]

[1] *Physikalisches Institut der Universität Heidelberg*
[2] *Max-Planck-Institut für Physik*
[3] *on leave from DESY*

Abstract. A summary of measurements of the fragmentation of charm quarks into a specific hadron is given. Measurements performed in photoproduction and deep inelastic scattering in $e^{\pm}p$, pp and $e^{+}e^{-}$ collisions are compared, using up-to-date branching ratios. Within uncertainties, all measurements agree, supporting the hypothesis that fragmentation is independent of the specific production process. Averages of the fragmentation fractions over all measurements are presented. The average has significantly reduced uncertainties compared to individual measurements.

1 Introduction

The fragmentation process is one of the most important phenomena in the high-energy physics. Every new-born quark participates in this process producing more quarks and consequently forms a hadron.

The fragmentations is difficult for modelling, so it is important to improve the knowledge about the process and study the most basic properties of the fragmentation. One of these is the probability of the fragmented quark q to create a given hadron H_q. This is called the fragmentation fraction and is denoted below as $f(q \to H_q)$.

An important question is whether the fragmentation fractions are universal, i.e. independent of the hard production mechanism. If true, once precisely measured in one environment, these can be applied in any other. This is especially important for the studies of heavier c and b quarks, as the precise knowledge of $f(c \to H_c)$ and $f(b \to H_b)$ can help for many measurement in the modern high-energy physics experiments.

The goal of this analysis [1] is to provide the most precise and model-independent values for fragmentation fractions of the charm quark. In this context it is also important to check that the sum of fragmentation fractions to all known ground states of charm hadrons is equal to unity, thus, all these states are known.

To achieve these goals, we extract the fragmentation fractions from all measurements simultaneously and separately for different environments and compare the results. To perform the desired checks we utilise the Standard Model predictions.

[a] e-mail: mikhaylo.lisovyi@desy.de
[b] e-mail: andrii.verbytskyi@mpp.mpg.de
[c] e-mail: oleksandr.zenaiev@desy.de

2 Selection of measurements for the extraction of fragmentation fractions

The selection of the measurements for the extraction of fragmentation fractions was done according to a set of criteria explained below. First, the selection is limited to the measurements obtained in the collisions of particle beams as these environments have well know production mechanisms of charm quarks and assure an absence of possible matter effects. Thus, only the measurements from high-energy ep, pp and e^+e^- collisions satisfy the requirements. The second criterion of the selection is the precision of the measured quantities. Basically, only the experiments on LEP, TEVATRON, HERA, LHC and B-factories. The third criterion of the selection is the sufficient number of measurements in the given physical environment. Several analyses of charm production (e.g. Ref. [2]) do not contain enough simultaneous measurements of hadron production and, therefore, cannot be treated independently and/or constrain the fragmentation fractions. The fourth criterion of the selection is a minimal model dependence of the results. Finally, only results that were published, or, at least submitted to peer-reviewed journals were considered.

The measurements selected according to the criteria described above form five groups. The first group is provided by B-factories. The results of the CLEO [3, 4] and ARGUS [5–7] experiments are represented as a product of the charm-hadron cross-sections times decay branching ratios, $\sigma(e^+e^- \to H_c) \cdot \mathcal{B}(H_c \to X)$. The BELLE experiment [8] provided measurements of the charm-hadrons cross-sections $\sigma(e^+e^- \to H_c)$. The BABAR experiment [9] provided a measurement of an average number of $\Lambda_c^+ \to pK\pi$ decays per hadronic event. The full list of the measured quantities used in the analysis is given in Ref. [1]. The second group of measurements is provided by the OPAL [10, 11], ALEPH [12] and DELPHI [13] experiments. Of those, the most valuable for the studies of fragmentation are results obtained from the studies of hadronic Z decays. These are represented in the form of fraction of charm events multiplied by branching ratios $\frac{\Gamma(Z\to c\bar{c})}{\Gamma(Z\to \text{hadrons})} \cdot f(c \to H_c) \cdot \mathcal{B}(H_c \to X)$ [10–13], or $\frac{\Gamma(Z\to c\bar{c})}{\Gamma(Z\to \text{hadrons})} \cdot f(c \to H_c)$ [12, 14]. In addition, ALEPH [12], DELPHI [14] and OPAL [11] provided measurements of $f(c \to H_c)$ and $f(c \to H_c) \cdot \mathcal{B}(H_c \to X)$ from the fits of fragmentation functions. The full list of the measured quantities used in the analysis is given in Ref. [1]. The third group of measurements are the charm-hadron production measurements in DIS in $e^{\pm}p$ collisions done by ZEUS [15, 16] and H1 [17] collaboration. Those experiments provide the cross-sections of the charm-hadron production in the restricted kinematic space. To extract the fragmenation fractions a procedure described in Ref. [1] is used. The fourth group of measurements are the charm-quark production measurements in PHP in $e^{\pm}p$ collisions done by ZEUS [18, 19]. The experiment provides the cross-sections of the charm hadron production in the restricted kinematic space and the fragmentation fractions. To extract the fragmenation fractions a procedure described in Ref. [1] is used. The last group of measurements comes from the studies of pp collisions at LHCb. The experiment provided measurements of charm-hadron cross-sections at $\sqrt{s}$ = 7TeV [20] and at $\sqrt{s}$ = 13TeV [21][1].

To make separate inputs consistent, the original measurements are corrected to the same up-to-date world averages of branching ratios of the charm-hadron decays. Most of the values were taken from Ref. [22]. Exceptions are made for the branching ratios of Λ_c^+ and D^{*0} decays, calculated from Refs. [23–25] as described in Ref. [1].

2.1 Calculation of the fragmentation fractions

In this analysis the charm-quark fragmentation fraction to a specific hadron assumed to be equal to the ratio of the production cross-section of the hadron via charm quark over the production cross-section of the charm quark

$$f(c \to H_c) = \sigma(H_c)/\sigma(c), \tag{1}$$

[1]The latter was presented on this conference.

i.e. the charm hadrons produced from beauty decays are excluded. With the available precise predictions for the total charm cross-section in e^+e^- collisions it is possible to calculate $f(c \to H_c)$ according to Eq. (1). However, sufficiently precise predictions for the charm-quark production in pp and $e^\pm p$ collisions are not available and other approaches are needed. Hereby,we make an assumption that the sum of charm-quark fragmentation fractions to all known ground states of charm hadrons is unity, the charm-quark fragmentation fraction to a specific hadron can be calculated as the ratio of the hadron-production cross-section over the sum of cross-sections of all known known ground states of charm hadrons

$$f(c \to H_c) = \sigma(H_c)/\Sigma_i \sigma(H_{c,i}). \quad (2)$$

The fragmentation fractions calculated according to Eq. (1) for the e^+e^- collisions and Z decays allow an independent check that

$$S = f(c \to D^0) + f(c \to D^+) + f(c \to D_s^+) + f(c \to \Lambda_c^+) + \sum f(c \to \Xi_c^{+,0}, \Omega_c^0) \quad (3)$$

is close to unity with sufficient accuracy. To perform the check we combine the available measurements of D^0, D^+, D_s^+, Λ_c^+ and assume $\sum f(c \to \Xi_c^{+,0}, \Omega_c^0) = \lambda f(c \to \Lambda_c^+)$, where $\lambda \ll 1$ is estimated from the production rates of strange hadrons.

2.2 Extraction procedure

The extraction of $f(c \to H_c)$ from the measurements in the present analysis is done separately for each group of measurements and then from all of them together regardless of the affiliation.

The procedure is based on numerical fit with respect to observables of interest. The details on the implementation are in Ref. [1] and only a brief description is given below. Basically, for the set of measurements the corresponding expectation values calculated from fit parameters together with the residuals R, the differences between the measurements and the corresponding expectation. The covariance matrix, V is calculated from the uncertainties of the residuals and takes into account all of the known correlations, including these for branching ratios, theoretical predictions etc. The full description of the correlations taken into account is given in Ref. [1]. The fit is performed with a minimisation of the $\chi^2 = R^T V^{-1} R$ quantity. The uncertainties on the fit parameters (e.g. fragmentation fractions) are determined using the Hessian method.

The obtained fragmentation fractions and the corresponding correlation matrix are used to calculate, the quantities commonly used as Monte Carlo generator parameters:

$$R_{u/d} = \frac{f(c \to D^0) - f(c \to D^{*+})\mathcal{B}_{D^{*+}\to D^0}}{f(c \to D^+) + f(c \to D^{*+})\mathcal{B}_{D^{*+}\to D^0}},$$

$$\gamma_{s(1)}^{(*)} = \frac{2f(c \to D_{s(1)}^{(*)+})}{f(c \to D^{(*)+}) + f(c \to D^{(*)0})} \text{ and } P_V^d = \frac{f(c \to D^{*+}) + f(c \to D^{*0})}{f(c \to D^+) + f(c \to D^0)}.$$

3 Results

The results of the fragmentation fraction extraction procedure for each of the groups introduced above is shown in Fig. 1. The numerical values are given in Ref. [1].

To check the consistency of the data from different production environments and also to extract the charm-quark fragmentation fractions with high precision, all input measurements are used together to produce a global combination. The constraint on the sum of the cross-sections of the weakly decaying

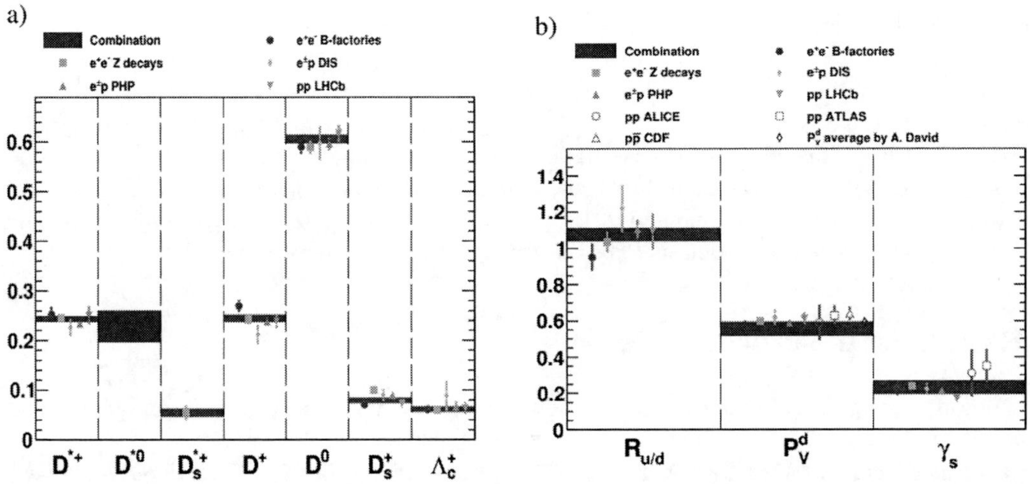

Figure 1. The values of a)charm-quark fragmentation fractions, $f(c \to H_c)$, and b) $R_{u/d}$, P^d_V, γ_s in different experiments with the S constraint. The global combination with the S constraint is shown with the shaded band. Averages of included data in different production environments are shown with various full symbols. Data that were not included in the combination [26–31] are shown with open symbols. Note, that the latter are quoted from the original papers, i.e. without correction to the up-to-date branching ratios and with no branching ratio uncertainty, if not given in the source.

Table 1. Results of the global combination.

	Constrained S	Constrained S, fix $\sigma(e^+e^- \to c\bar{c})$, $\frac{\Gamma_{c\bar{c}}}{\Gamma_{\mathrm{hadrons}}}$.
$f(c \to D^{*+})$	0.2436 ± 0.0050	0.2411 ± 0.0048
$f(c \to D^{*0})$	0.2286 ± 0.0313	0.2270 ± 0.0304
$f(c \to D_s^{*+})$	0.0548 ± 0.0076	0.0549 ± 0.0076
$f(c \to D^{+})$	0.2449 ± 0.0065	0.2451 ± 0.0064
$f(c \to D^{0})$	0.6058 ± 0.0079	0.6130 ± 0.0075
$f(c \to D_s^{+})$	0.0794 ± 0.0047	0.0803 ± 0.0048
$f(c \to \Lambda_c^{+})$	0.0615 ± 0.0046	0.0542 ± 0.0030
χ^2	60.0	74.6
n_{dof}	57	60
S	1.0000 ± 0.0005	1.0000 ± 0.0004
$R_{u/d}$	1.0757 ± 0.0341	1.1017 ± 0.0335
P^d_V	0.5551 ± 0.0372	0.5455 ± 0.0357
γ_s	0.1866 ± 0.0120	0.1872 ± 0.0123
γ_s^*	0.2321 ± 0.0356	0.2344 ± 0.0361

charm states, S, is imposed in the combination, i.e. the prediction for the total charm cross-sections in e^+e^- collisions is not used, in order to minimise model dependence of the averaging procedure. The result of averaging e^+e^-, $e^\pm p$ and pp data, with the constraint $S = 1$ is presented in the middle column of Tab. 1 and is shown in Fig. 1. The input data are in very good agreement with $\chi^2/n_{\rm dof} = 64/61$. The result of the combination has significantly reduced uncertainties compared to individual measurements. The combined results are also compared to recent measurements [26–31] (see Fig. 1) that were not included in the combination. In particular, $R_{u/d} = 1.078 \pm 0.033$ is in fair agreement with the isospin invariance hypothesis $R_{u/d} = 1$ within 2.4 standard deviations.

4 Results for the excited states

In addition to the average fragmentation fractions for the ground, $L = 0$, states, some fragmentation fractions for the excited, $L = 1$ charm hadrons are calculated.

The measurements used for the averaging are obtained by ZEUS [32, 33], OPAL [34] and ALEPH [35] experiments. For the detailed list of those see Ref. [1]. The unpublished measurement of $f(c \to D_{s1}^+)$ from Ref. [36] was rejected. As the measurements have limited precision, the correction to the most recent branching ratios cannot improve the results and was not done. The averages, as well as the γ_{s1}, are calculated with an assumption of fully uncorrelated statistical and systematical uncertainties. The results are given in Tab. 2.

Table 2. Results of the global combination for $L = 1$ states.

	Average
$f(c \to D_1^+)$	$0.0460^{+0.0269}_{-0.0182}$
$f(c \to D_2^{*+})$	$0.0320^{+0.0094}_{-0.0082}$
$f(c \to D_1^0)$	0.0297 ± 0.0038
$f(c \to D_2^{*0})$	0.0394 ± 0.0068
$f(c \to D_{s1}^+)$	0.0109 ± 0.0014
γ_{s1}	$0.287^{+0.079}_{-0.109}$

5 Summary

A summary of measurements of the fragmentation of charm quarks into a specific charm hadron is given. The analysis includes data collected in photoproduction and deep inelastic scattering in $e^\pm p$ collisions and well as e^+e^- and pp data. Measurements in different production regimes agree within uncertainties, supporting the hypothesis that fragmentation proceeds independent of the specific production process. Averages of the fragmentation fractions are presented. The global average has significantly reduced uncertainties compared to individual measurements. In addition, the hypothesis that the sum of fragmentation fractions of all known weakly decaying charm hadrons is equal to unity is checked to hold within 3 standard deviations using the e^+e^- data. The relation $\gamma_{s1} > \gamma_s^* > \gamma_s$ indicates smaller charm-strange hadron production suppression for the heavier charm states in comparison to the ground ones.

The obtained results should be applied for the best precision of relevant measurements and theoretical predictions.

Acknowledgements

We thank Erich Lohrmann for his major contribution to the development of this analysis. We thank Uri Karshon and Stefan Kluth for useful discussions and help in the work with the bibliography. We also thank Alexander Glazov and Ian Brock for the critical reading of the analysis manuscript [1] and useful suggestions on text improvement.

References

[1] M. Lisovyi, A. Verbytskyi and O. Zenaiev (2015), 1509.01061v1
[2] S. Chekanov et al. (ZEUS Collaboration), Eur.Phys.J. **C63**, 171 (2009), 0812.3775
[3] D. Bortoletto et al. (CLEO Collaboration), Phys.Rev. **D37**, 1719 (1988)
[4] P. Avery et al. (CLEO Collaboration), Phys.Rev. **D43**, 3599 (1991)
[5] H. Albrecht et al. (ARGUS Collaboration), Z.Phys. **C52**, 353 (1991)
[6] H. Albrecht et al. (ARGUS Collaboration), Z.Phys. **C54**, 1 (1992)
[7] H. Albrecht et al. (ARGUS Collaboration), Phys.Lett. **B207**, 109 (1988)
[8] R. Seuster et al. (Belle Collaboration), Phys.Rev. **D73**, 032002 (2006), hep-ex/0506068
[9] B. Aubert et al. (BaBar Collaboration), Phys.Rev. **D75**, 012003 (2007), hep-ex/0609004
[10] G. Alexander et al. (OPAL Collaboration), Z.Phys. **C72**, 1 (1996)
[11] K. Ackerstaff et al. (OPAL Collaboration), Eur.Phys.J. **C1**, 439 (1998), hep-ex/9708021
[12] R. Barate et al. (ALEPH Collaboration), Eur.Phys.J. **C16**, 597 (2000), hep-ex/9909032
[13] P. Abreu et al. (DELPHI Collaboration), Eur.Phys.J. **C12**, 225 (2000)
[14] P. Abreu et al. (DELPHI Collaboration), Eur.Phys.J. **C12**, 209 (2000)
[15] S. Chekanov et al. (ZEUS Collaboration), JHEP **0707**, 074 (2007), 0704.3562
[16] H. Abramowicz et al. (ZEUS Collaboration), JHEP **1011**, 009 (2010), 1007.1945
[17] A. Aktas et al. (H1 Collaboration), Eur.Phys.J. **C38**, 447 (2005), hep-ex/0408149
[18] S. Chekanov et al. (ZEUS Collaboration), Eur.Phys.J. **C44**, 351 (2005), hep-ex/0508019
[19] H. Abramowicz et al. (ZEUS Collaboration), JHEP **1309**, 058 (2013), 1306.4862
[20] R. Aaij et al. (LHCb Collaboration), Nucl.Phys. **B871**, 1 (2013), 1302.2864
[21] R. Aaij et al. (LHCb Collaboration), Submitted to JHEP (2015), 1510.01707
[22] K.A. Olive et al. (Particle Data Group), Chin.Phys. **C38**, 090001 (2014)
[23] A. Zupanc et al. (Belle Collaboration), Phys.Rev.Lett. **113**, 042002 (2014), 1312.7826
[24] M. Ablikim et al. (BESIII Collaboration), Phys.Rev. **D91**, 031101 (2015), 1412.4566
[25] B. Aubert et al. (BaBar Collaboration), Phys.Rev. **D72**, 091101 (2005), hep-ex/0508039
[26] G. Aad et al. (ATLAS Collaboration), ATLAS preliminary ATLAS-CONF-2011-017 (2011)
[27] B. Abelev et al. (ALICE Collaboration), JHEP **1201**, 128 (2012), 1111.1553
[28] B. Abelev et al. (ALICE Collaboration), JHEP **1207**, 191 (2012), 1205.4007
[29] B. Abelev et al. (ALICE Collaboration), Phys.Lett. **B718**, 279 (2012), 1208.1948
[30] D. Acosta et al. (CDF Collaboration), Phys.Rev.Lett. **91**, 241804 (2003), hep-ex/0307080
[31] A. David, Phys.Lett. **B644**, 224 (2007)
[32] H. Abramowicz et al. (ZEUS Collaboration), Nucl.Phys. **B866**, 229 (2013), 1208.4468
[33] S. Chekanov et al. (ZEUS Collaboration), Eur.Phys.J. **C60**, 25 (2009), 0807.1290
[34] K. Ackerstaff et al. (OPAL Collaboration), Z.Phys. **C76**, 425 (1997)
[35] A. Heister et al. (ALEPH Collaboration), Phys.Lett. **B526**, 34 (2002), hep-ex/0112010
[36] A. Verbytskyi, Ph.D. thesis, Universität Hamburg (2013)

EPJ Web of Conferences **120**, 03003 (2016) DOI: 10.1051/epjconf/201612003003
ISMD 2015

Perspective study of exotics and flavour baryons in antiproton-proton annihilation and proton-proton collisions

Mikhail Barabanov, Alexander Vodopyanov

Joint Institute for Nuclear Research, 141980 Joliot-Curie 6 Dubna, Russia

Abstract. The spectroscopy of exotic states with hidden charm is discussed. Together with charmonium, these provide a good tool for testing theories of the strong interactions including both perturbative and non-perturbative QCD, lattice QCD, potential and other phenomenological models. An elaborated analysis of exotics spectrum is given, and attempts to interpret recent experimentally observed states with masses above the $D\bar{D}$ threshold region are considered. Experimental results from different collaborations (BES, BaBar, Belle, LHCb) are analyzed with special attention given to recently discovered hidden charm states. Some of these states can be interpreted as higher-lying charmonium states and others as tetraquarks with hidden charm. It has been shown that charged/neutral tetraquarks must have their neutral/charge partners with mass values differ by at most a few MeV/c^2, hypotheses that tend to coincide with those proposed by Maiani and Polosa. However, measurements of different decay modes are needed before firm conclusions can be made. These data can be derived directly from the experiments using ahigh quality antiproton beam with momentum up to 15 GeV/c and proton-proton collisions with momentum up to 26 GeV/c.

1 Introduction

The study of strong interactions and hadron matter in the process of antiproton-proton annihilation and proton-proton collisions seems to be a challenge nowadays. One of the main goals of contemporary physics is to search for new exotic forms of matter, some of which will be manifest in the form of charmed hybrids and multiquark states, such as meson molecules and tetraquarks [1, 2]. Studies of the spectra of charmed hybrids $c\bar{c}g$ and tetraquarks containing hidden charm ($cq\bar{c}\bar{q}'$, q and $q'= u, d, s$),together with completion of the charmonium spectrum are promising approaches for the development of an understanding of the dynamics of quark interactions at small distances. They provide a good testing tool that challenges theories of strong interactions: QCD in both perturbative and non-perturbative regimes, QCD inspired potential models, phenomenological models, and non-relativistic QCD and LQCD.

Charmed hybrids $c\bar{c}g$ are hypothesized $c\bar{c}$ states with an excited gluonic degree of freedom. These states are described by different models and calculational schemes (LQCD, bag model, flux tube model) [1, 2]. All model predictions and calculations tend to agree that the masses of the lowest-lying charmonium hybrids are between 3.9 and 4.6 GeV/c^2 and that the state with $J^{PC} = 1^{-+}$ has the lowest mass. Until now, discussions have been focused only around the lowest-lying charmonium hybrids.

Four of these states $J^{PC} = 2^{-+}, 1^{-+}, 1^{--}, 0^{-+}$ correspond to a $c\bar{c}$ pair with $J^{PC} = 0^{-+}$ or 1^{--}, coupled to a gluon in the lightest mode with $J^{PC} = 1^{+-}$. Four other states with a $c\bar{c}$ pair with $J^{PC} = 2^{+-}, 1^{+-}, 1^{++}, 0^{+-}$ coupled with a $J^{PC} = 1^{-+}$ mode are, probably, a bit heavier. The expected mass splitting between the 1^{-+} and 0^{+-}states is about 150 - 250 MeV/c^2. Three of these eight low-lying charmonium hybrids have spin-exotic quantum numbers $1^{-+}, 0^{+-}, 2^{+-}$, and for these mixing effects with nearby $c\bar{c}$ states are excluded, making their experimental identification especially unambiguous. The next possible hybrid states with quantum numbers $2^{++}, 2^{+-}, 1^{++}, 1^{+-}, 0^{+-}, 0^{++}$ correspond to $c\bar{c}$ pair with quantum numbers $J^{PC} = 1^{+-}$ or $J^{PC} = (0,1,2)^{++}$ coupled to a gluon that, in the lightest mode, has $J^{PC} = 1^{+-}$. The states with quantum numbers $2^{--}, 2^{-+}, 1^{--}, 1^{-+}, 0^{-+}, 0^{--}$ correspond to $c\bar{c}$ pair with quantum numbers $J^{PC} = 1^{+-}$ or $J^{PC} = (0,1,2)^{++}$ that are coupled to a gluon mode with $J^{PC} = 1^{-+}$. One can find a possibility of the existence of hybrid state with the exotic quantum numbers $J^{PC} = 0^{--}$.The most interesting and promising decay channels for charmed hybrids are as follows: $\bar{p}p \rightarrow \tilde{\eta}_{c0,1,2}(0^{-+}, 1^{-+}, 2^{-+})\, \eta \rightarrow \chi_{c0,1,2}(\eta, \pi\pi;\ldots)$; $\bar{p}p \rightarrow \tilde{h}_{c0,1,2}(0^{+-}, 1^{+-}, 2^{+-})\, \eta \rightarrow \chi_{c0,1,2}(\eta, \pi\pi;\ldots)$; $\bar{p}p \rightarrow \tilde{\Psi}(1^{--}, 2^{--}) \rightarrow J/\Psi(\eta, \omega, \pi\pi, \ldots)$; $\bar{p}p \rightarrow \tilde{\eta}_{c0,1,2}, \tilde{h}_{c0,1,2}, \tilde{\chi}_{c1}(0^{-+}, 1^{-+}, 2^{-+}, 0^{+-}, 1^{+-}, 2^{+-}, 1^{++}, 2^{++})\eta \rightarrow D\bar{D}_J^{*}\eta$.

In the last few years we have witnessed the discovery of a number of narrow hadron resonances with hidden charm that do not match the standard quark-antiquark interpretation, and have, thereby, been named exotic hadrons [2 – 5]. This has called for alternative interpretations of their inner structure. One of the possible explanations is that these particles are loosely bound molecules of open charm mesons. Another possibility is that new aggregation patterns of quarks in matter are possible. We follow the suggestion of using diquarks as building blocks. Light diquarks have been an object of several lattice studies. The idea that the coloured diquark can be handled as a constituent building block is at the core of this approach. Two generic types of multiquark states have been described in the literature [3 – 5]. The first one, the molecular state, is comprised of two charmed mesons bound together to form a molecule. These states are, by nature, loosely bound. Molecular states bound through two mechanisms: quark/colour exchange interactions at short distances and pion exchange at a large distance. Since the mesons inside the molecule are weakly bound, they tend to decay as if they are free. The second type is a tightly bound four-quark state, a so-called tetraquark that is predicted to have properties that are distinct from those of a molecular state. In the model of Maiani [6], for example, the tetraquark is described as a diquark-diantiquark structure in which the quarks group into the colour-triplet scalar and vector clusters with interactions that are dominated by a simple spin-spin interaction. Here, strong decays are expected to proceed via rearrangement processes followed by dissociation that gives rise, for example, to processes such as: $\bar{p}p \rightarrow X \rightarrow J/\Psi\rho \rightarrow J/\Psi\pi\pi$; $\bar{p}p \rightarrow X \rightarrow J/\Psi\omega \rightarrow J/\Psi\pi\pi\pi$, $\bar{p}p \rightarrow X \rightarrow \chi_{cJ}\pi$; $\bar{p}p \rightarrow X \rightarrow D\bar{D}^{*} \rightarrow D\bar{D}\gamma$; $\bar{p}p \rightarrow X \rightarrow D\bar{D}^{*} \rightarrow D\bar{D}\eta$. A prediction that distinguishes tetraquarks containing a $c\bar{c}$ pair from conventional charmonia is possible existence of multiplets that include members with non-zero charge $cu\bar{c}\bar{d}$, strangeness $cd\bar{c}\bar{s}$, or both $cu\bar{c}\bar{s}$.

2 The model of confinement for resonance decay products

The actual calculation of the dynamics of a resonance is an impracticable task, but we considered the particular moment when two decay products are out of range for a mutual interaction and no potential energy is involved in calculations. This happens at a distance R for a given group of resonances, the distance at which the resonance ceases to exist and two or more decay products are only the objects to consider. At that moment there is no translational energy of the decay products and only the orbital kinetic energy is relevant for each of them. The spin rotational energy is described with the relativistic spherical symmetric top model with quantized kinetic energy. So the resonance decay products are described as a collection of two tops in the same rotational state. Therefore the decay channels may be joint to the groups with the properties of quantum objects. Inside each group the decay products momenta are quantized relatively characteristic length R. Thus we can consider the resonances as the

states of relativistic quantum system. It was shown that the dynamic of the system is derived from SU(2)-group generators [7, 8]. Hence SU(2)-group plays the fundamental role in quantization of energy-momentum.
Let us define the set of generators of *SO*(4) group

$$\vec{M} = [\vec{r} \times \vec{p}], \vec{N} = R\vec{p} + \vec{r}(\vec{r}\,\vec{p})/R, \tag{1}$$

The linear combinations of these orthonormal operators

$$\mu_{\pm} = (\vec{M} \pm \vec{N}), \tag{2}$$

contribute two sets of generators of the SU(2) group. Thus the *SU*(2) group generates the action on a three-dimensional sphere S^3.
This action consists of the translation with whirling around the direction of translation.
We get:

$$H = \frac{1}{2mR^2}\{2\hbar + (\vec{\mu}_{\pm}, \vec{\sigma})\}\{2\hbar + (\vec{\mu}_{\pm}, \vec{\sigma})\}, \tag{3}$$

where $\vec{\sigma}$ is the spin operator, $\hbar$ is the Plank constant, m is the top mass.

The spectrum is:

$$H\Psi_n = \frac{\hbar^2}{2mR^2}(n+1)^2\Psi_n, n = 0, 1, 2..., \tag{4}$$

The wave function $\Psi_n = |LSJM_J\rangle$ was taken as eigenfunction of whole momentum $J^2 = (\mu^{\pm} + s^2)$ of the top.
Let us generalize this concept to the relativistic case:

$$M(R) = \sqrt{m_a^2 + \frac{1}{R^2}((\vec{\sigma}\vec{\mu}_{\pm}) + 2\hbar)^2} + \sqrt{m_b^2 + \frac{1}{R^2}((\vec{\sigma}\vec{\mu}_{\pm}) + 2\hbar)^2}, \tag{5}$$

where m_a and m_b are the masses of resonance decay products.
The spectrum is:

$$E = \sqrt{m_a^2 + \frac{\hbar^2(n+1)^2}{R^2}} + \sqrt{m_b^2 + \frac{\hbar^2(n+1)^2}{R^2}}, n = 0, 1, 2..., \tag{6}$$

Finally, the formula for resonance mass spectrum has the form:

$$E = \sqrt{m_a^2 + P_n^2} + \sqrt{m_b^2 + P_n^2} = \sqrt{m_a^2 + [n * P_0]^2} + \sqrt{m_b^2 + [n * P_0]^2} \tag{7}$$

where $R \to a + b$ - a binary decay channel (we use the system where $\hbar = c = 1$), m_a, m_b – the masses of resonance binary decay products, P_0 – basic momentum, P_n – asymptotic momentum of their relative motion.

3 Calculation the exotics spectrum

For these purposes we have performed an elaborated analysis of the spectrum of charmed hybrids and tetraquarks with the hidden charm with masses above the $D\bar{D}$ threshold region. An analysis of spectrum of charmonium states was carried out earlier [9, 10]. In particular, different decay modes of charmed hybrids and tetraquarks, such as decays into charmonium and light mesons and decays into $D\bar{D}_J^*$ and $D\bar{D}^*$ pairs, were analyzed. Special attention was given to recently discovered states with hidden charm [2 – 5]. Experimental results from different collaborations like Belle, BaBar, LHCb, BES were used. It was demonstrated [11, 12] that the dynamics of the decaying resonant system (charmonium, charmed hybrid and tetraquark) is determined by a set of *SU(2)* group generators. Mathematical aspects of spin particle motion on the spherical surface S^3 defined in four-dimensional Euclidean space were considered [7, 8]. The radius of the sphere is a constant. Conformity with the three-dimensional actual space is achieved by transition to the spherical surface S^3. The authors shown [11, 12] that two manifolds of *SU(2)* group generators specify the equation of motion of particles with spin ½ on the tree-dimensional sphere S^3 in four-dimensional Euclidean space. Subsequently, they generalized the aforesaid to particles with spin equals 1. In this confinement model, a geometrical point, lying on the sphere, is associated with relativistic spherically symmetric top. The totality of resonant states with identical quantum numbers is interpreted as radially excited states of the same quantum system (in our case charmonium, charmed hybrids or tetraquark). The proposed model was developed in this paper. According to a combined approach based on the quarkonium potential model and a confinement model [11, 12], more than twenty charmed hybrids are expected to exist in the mass region under consideration (see Figs. 1,2). Charmed hybrids with exotic quantum numbers are marked with dark shading and charmed hybrids with nonexotic quantum numbers are indicated with light shading. The results of calculations for the hybrids are in good agreement with the well accepted picture that the quartet $1^{--},(0,1,2)^{-+}$ is lower in mass than $1^{++},(0,1,2)^{+-}$.The mass splitting between the states 1^{-+} and 0^{+-} is about 200 MeV/c^2. More than twenty tetraquarks with hidden charm (see Fig. 3) are expected to exist in the mass region above the $D\overline{D}$ threshold. The black-white boxes correspond to the recently revealed *XYZ* states with the hidden charm that may be interpreted as tetraquarks. White boxes correspond to the tetraquark states that have yet to be found. The possible existence of these states is predicted in the framework of the combined approach. It has been shown that charge/neutral tetraquarks with hidden charm must have neutral/charged partners with mass values that differ at most by a few MeV/c^2. This assumption coincides with that proposed earlier by Maiani and Polosa [13] and can shed light on the nature of neutral $X(3872)$, X(4350) and charged $Z_c(3885)^{\pm}, Z_c(3900)^{\pm}, Z_c(4020)^{\pm}, Z_c(4025)^{\pm}, Z_c(4200)^{\pm}, Z_c(4050)^{\pm}, Z_c(4250)^{\pm}$ $Z_c(4430)^{\pm}$ states. The quantum numbers J^{PC} of the $X(3872)$ meson have been recently determined by LHCb [14]. We find that the $X(3872)$ can be interpreted as tetraquark state with J^{PC} = 1^{++}, and the $X(4350)$ as the tetraquark state with $J^{PC} = 2^{++}$. The charged $Z_c(3900)^{\pm}$ state that was recently observed by BES [15, 16] and Belle [17], the $Z_c(4020)^{\pm}$ seen by BES [18, 19], the $Z_c(4050)^{\pm}$ and $Z_c(4250)^{\pm}$ seen by Belle decaying to $\chi_{c1}\pi^{\pm}$ [20] and the $Z_c(4430)^{\pm}$ charged state seen by Belle decaying to $\Psi'\pi^{\pm}$ [21] can be identified as charge tetraquarks with $J^{PC} = 1^{+}$. (The proposed approach doesn't distinguish the states $Z_c(3900)^{\pm}$ [15, 17] and $Z_c(3885)^{\pm}$ [16] as well the $Z_c(4020)^{\pm}$ [18] and $Z_c(4025)^{\pm}$ [19] states.) The values of their masses and widths coincide in the framework of the combined approach. Two states (one charged and one neutral) with $J^{PC} = 1^{++}$ are expected to exist in the mass range of 4200 – 4300 MeV/c^2. The $Z_c(4200)^{\pm}$ charged state recently reported by Belle [22] may be a good candidate for the one of them.

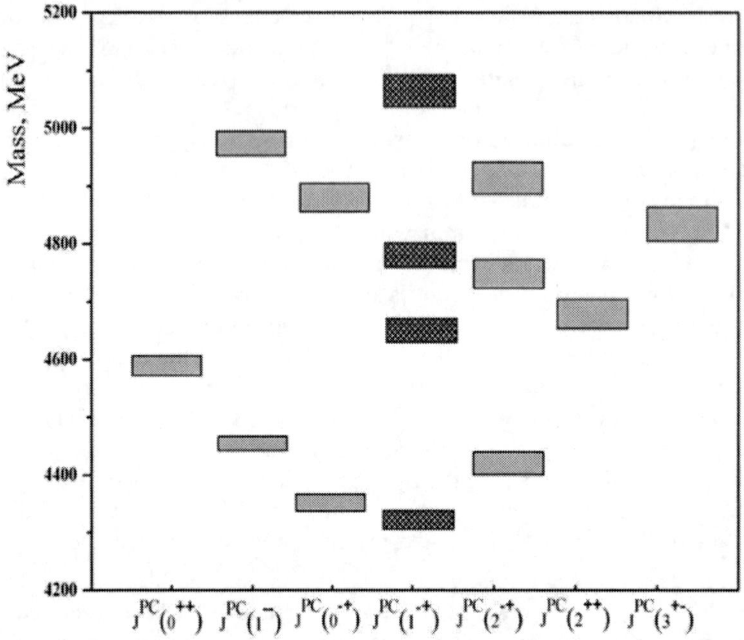

Figure 1. The spectrum of charmed hybrids with quantum numbers $J^{PC} = 3^{+-}, 2^{++}, 2^{-+}, 1^{-+}, 1^{--}, 0^{-+}, 0^{++}$. Charmed hybrids with exotic quantum numbers are marked with dark shading and charmed hybrids with nonexotic quantum numbers are indicated with light shading.
All these states are predictions in the combined approach.

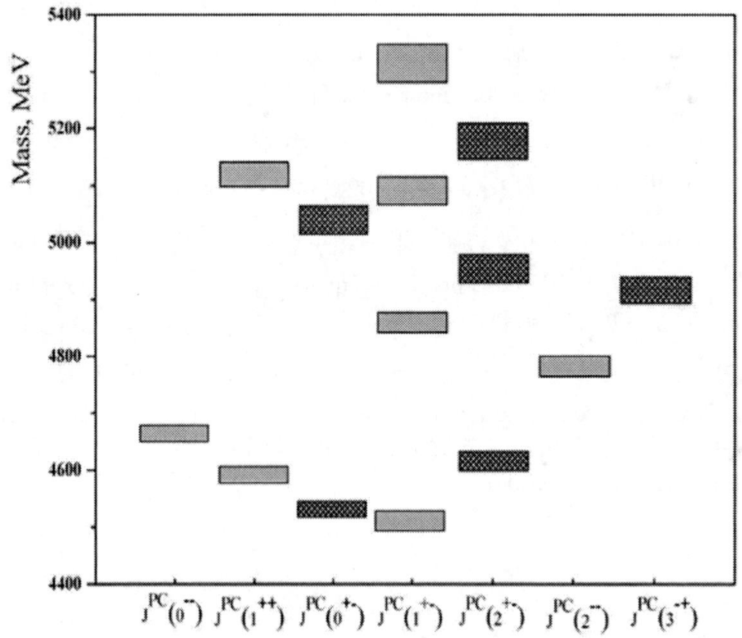

Figure 2. The spectrum of charmed hybrids with quantum numbers $J^{PC} = 3^{-+}, 2^{--}, 2^{+-}, 1^{+-}, 1^{++}, 0^{+-}, 0^{--}$. Charmed hybrids with exotic quantum numbers are marked with dark shading and charmed hybrids with nonexotic quantum numbers are indicated with light shading.
All these states are predictions in the combined approach.

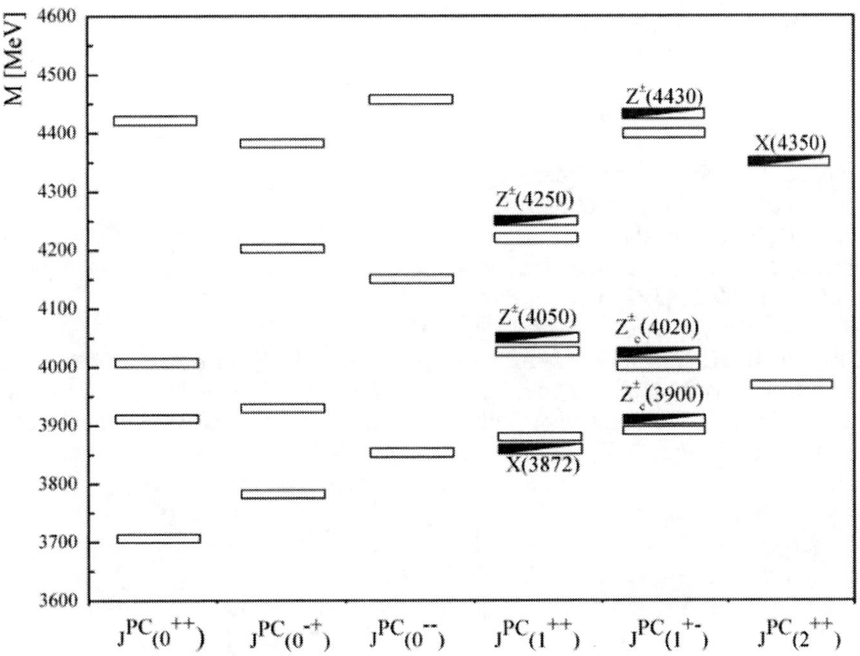

Figure 3. The spectrum of tetraquarks with hidden charm.The black-white boxes correspond to the recently revealed *XYZ* states with the hidden charm that may be interpreted as tetraquarks. White boxes correspond to the tetraquark states that have yetto be found. The existence of these states is predicted in the framework of the combined approach.

To confirm that the predicted states actually exist and can be found experimentally, their widths and branching ratios were calculated [10, 12]. The feature of the considered states is their narrowness compared with light unflavored mesons, baryons and hybrids. The states we find in this model have small widths; their values are of the order of several tens of MeV. This fact facilitates experimental searches. The values of the calculated widths coincide (within the experimental error) with the experimentally determined values for the *XYZ* particles; the correspondence of the mass values has been discussed above. This fact strongly suggests that some of the *XYZ* particles may be interpreted as higher-lying charmonium states [9, 10] and tetraquarks as it can be verified by the experiments with antiproton beams with momentum up to 15 GeV/c and proton-proton collisions with momentum up to 26 GeV/c. The values of branching ratios in the considered decay channels of charmonium and exotics are of the order of $10^{-1} - 10^{-2}$ dependent of their decay mode. From this one can conclude that the branching ratios are significant and searches for charmonium and exotics, and studies of the main characteristics of their spectrum seem to be promising.

4 Conclusion

The prospects for future exotics research are related with the results obtained below:
A combined approach has been employed to study charmonium and exotics on the basis of the quarkonium potential model and a confinement model that uses a three-dimensional sphere embedded into the four-dimensional Euclidian space of the decay products.
The most interesting and promising decay channels of charmed hybrids and tetraqurks with the hidden charm have been analyzed. Many new states above $D\overline{D}$ threshold are expected to exist in the framework of this model.

The recently discovered states with the hidden charm above the $D\overline{D}$ threshold (*XYZ* particles) have been analyzed. Ten of these states can be interpreted as higher-lying tetraquark states with hidden charm. The necessity of further studies of the *XYZ* particles and improved measurements of their main characteristics has been demonstrated.

References

1. W. Erni *et al.*, arXiv:0903.3905v1 [hep-ex] (2009)
2. N. Brambilla *et al.*, Eur. Phys. J. **C 71**: 1534, 1 (2011)
3. J. Beringer *et al.*, Reviewof Particle Physics, Phys. Rev. **86** (2012)
4. S. Godfrey and S. Olsen, Ann. Rev. Nucl. Part. Sci. **58**, 51 (2008)
5. S. Olsen, arXiv : 0909.2713v1 [hep-ex] (2009)
6. L. Maiani, F. Piccinini, A.D. Polosa, V. Riquer, Phys. Rev. **D 71**, 014028 (2005)
7. J. Keller, R. Yamaleev, A. Rodrigues, Adv. in Appl. Clifford Algebras **8**, N.2, 235 (1998)
8. F. Gareev, J. Keller, R. Yamaleev, Adv. in Appl. Clifford Algebras **8**, N.2, 255 (1998)
9. M.Yu. Barabanov, A.S. Vodopyanov, Fizika Elementarnyh Chastits Atomnogo Yadra Pisma, **8**, N.10, 63 (2011) / Phys. Part. Nucl. Lett. **8**, N.10, 1069 (2011).
10. M.Yu. Barabanov, A.S. Vodopyanov, S. L. Olsen, YadernayaFizica, **77**, N.1, 1 (2014) / Phys. At. Nucl., **77**, N.1,1 26 (2014).
11. M.Yu. Barabanov *et al.*, Russ. Phys. J., **50**, 1243 (2007)
12. M.Yu. Barabanov *et al.*, Hadronic J., **32**, N.2, 159 (2009)
13. L. Maiani, F. Piccinini, A.D. Polosa, V. Riquer, Phys. Rev. Lett. **99**, 182003 (2007)
14. R. Aaij *et al.* (LHCb Collaboration), Phys. Rev. Lett. **110**, 222001 (2013)
15. M. Ablikim *et al.* (BESIII Collaboration), Phys. Rev. Lett. **110**, 252001 (2013)
16. M. Ablikim *et al.* (BESIII Collaboration), Phys. Rev. Lett. **112**, 022001 (2014)
17. Z.Q. Liu*et al.* (Belle Collaboration), Phys. Rev. Lett. **110**, 252002 (2013)
18. M. Ablikim *et al.* (BESIII Collaboration), Phys. Rev. Lett. **111**, 242001 (2013)
19. M. Ablikim *et al.* (BESIII Collaboration), Phys. Rev. Lett. **112**, 132001 (2014)
20. R. Mizuk *et al.* (Belle Collaboration), Phys. Rev. **D 78**, 072004 (2008)
21. S.-K. Choi*et al.* (Belle Collaboration), Phys. Rev. Lett. **100**, 142001 (2008)
22. K. Chilikin*et al.* (Belle Collaboration), arXiv: 1408.6457v3 [hep-ex] (2014)

EPJ Web of Conferences **120**, 03004 (2016) DOI: 10.1051/epjconf/201612003004
ISMD 2015

The Weighted GMD Model for multiplicity distributions

Probing gluon production at LHC energies

W.Y.Wang[1,a], S.Seah[1], J.Setianegara[1], A.H.Chan[1], and C.H.Oh[1]

[1] *Physics Department, National University of Singapore, Singapore*

Abstract. A new distribution, the Weighted GMD (WGMD) is obtained from the Generalised Multiplicity Distribution (GMD), describing charged-particle multiplicity distributions as the hadronisation products of quark and gluon branching with fluctuations in the initial gluon numbers produced from the collision. The WGMD is shown to describe charged-particle multiplicity distributions in *pp* collisions at the Large Hadron Collider (LHC), and the average initial gluon number is obtained for Poisson distributed gluon multiplicities.

1 Introduction

Multiplicity distributions reveal useful information on the characteristics of particle production processes in high energy collisions. The shape of multiplicity distributions reveals whether the production of the particles within the emission region are correlated. A Poisson distributed (PD) multiplicity for example, indicates the independent emission of single particles. Cascades of particles originating from an initial number of ancestor particles produced after the collision, lead to a broader distribution like the Negative Binomial Distribution (NBD) [1].

Like the NBD, the Generalised Multiplicity Distribution (GMD) [2–4] can be derived from a very similar cascade process. The GMD is a solution of a stochastic branching equation [5] describing the evolution of quarks and gluons as Markov branching processes. This intuitive picture has been shown to be equivalent to an algorithm calculating the multi-parton distributions within a QCD jet in a leading logarithmic approximation [6], and yields the GMD and NBD as solutions.

Given m and n number of quarks (q) and gluons (g) respectively, the dominant processes of quark bremsstrahlung ($q \to q{+}g$) and gluon fission ($g \to g{+}g$) result in a shower of quarks and gluons which eventually hadronise into the observed particles. In this picture, the stochastic branching equation [5]

$$\frac{\partial P_{m,n}(t)}{\partial t} = -\tilde{A}mP_{m,n}(t) + \tilde{A}mP_{m,n-1}(t) - AnP_{m,n}(t) + A(n-1)P_{m,n-1}(t), \tag{1}$$

relates the probability of obtaining m quarks and n gluons ($P_{m,n}$) from m quarks and n or $n-1$ gluons at each branching step. $A\Delta t$, and $\tilde{A}\Delta t$ are the probabilities of quark bremsstrahlung and gluon fission

[a] e-mail: w.y.wang@u.nus.edu

within the infinitesimal interval Δt. The QCD evolution parameter t is given by

$$t = \frac{6}{11N_c - 2N_f} \ln \left[\frac{\ln\left(Q^2/\mu^2\right)}{\ln\left(Q_0^2/\mu^2\right)} \right], \tag{2}$$

where Q is the initial parton (q/g) invariant mass, Q_0 is the hadronisation mass, μ is a QCD mass scale of a few GeV, N_c is the number of colours, and N_f is the number of flavours.

For an initial m and k' number of quarks and gluons, the solution of equation 1 yields the final multiplicity of n partons which hadronise to form n particles distributed according to the GMD:

$$P_{GMD}(n; p, k, k') = \frac{\Gamma(n+k)}{\Gamma(n-k'+1)\Gamma(k'+k)} (1-p)^{n-k'} (p)^{k'+k}, \tag{3}$$

where $k = m\tilde{A}/A$ and $p = \exp(-At) = (k' + k) / (\overline{n} + k)$. The average multiplicity of final state hadrons $\overline{n}$ is controlled by the initial quark and gluon numbers as well as the branching probability through

$$\overline{n} = \frac{k' + k}{p} - k. \tag{4}$$

The GMD is then a more general solution as the name suggests, in the sense that it encompasses to the Fury-Yule Distribution [7] for which $k = 0$, and the NBD where $k' = 0$. Furthermore, the NBD converges to the PD for $k \to \infty$.

The GMD has been successfully applied to describe charged-particle multiplicity distributions from e^+e^-, pp, and $p\overline{p}$ collisions [2–4]. However it increasingly experiences a difficulty due to the emergence of a kink or "shoulder" like structure in charged-particle multiplicity data above $\sqrt{s} = 900$ GeV [8, 9] in both pp and $p\overline{p}$ collisions, as well as a spike in the $n = 0$ probability. Theoretically, the hadron production mechanism and hadronisation in the GMD only allows for $n \geq k'$, and poses a difficulty in the low multiplicity and high energy regime.

In brief, this paper proposes a modified GMD with a more realistic description of the initial state of the particle emission source. In addition, this proposed modified GMD achieves a better fit of data from the CMS collaboration at 0.9, 2.36 and 7 TeV [9] with the same number of parameters, and provides a physical interpretation of the extracted model parameters.

2 The weighted GMD model

The weighted GMD (WGMD) model is obtained from the GMD by introducing event-by-event fluctuations in the production of initial gluons. If each independent collision event emits k' number of gluons distributed according to a distribution $P(k'; x_1, ..., x_r)$ with r parameters, then the WGMD is given by

$$P_{WGMD}(n; x_1, ..., x_r, p, k) = \sum_{k'=0}^{n} P(k'; x_1, ..., x_r) \times P_{GMD}(n; p, k, k'). \tag{5}$$

The observed final state multiplicity distribution is thus a weighted sum of the GMD over the probabilities of the initial states. The summation is done for $k' = 0$ to n as those are the only physical possibilities since the number of gluons cannot exceed the number of final state hadrons. The gluons are assumed to go through a cascade process and then eventually split into quarks and gluons that hadronise (i.e. a 1–1 correspondence between the gluon number right before hadronisation, and the hadron number right after).

DOI: 10.1051/epjconf/201612003004

Without specifying the form of the weight $P(k'; x_1, ..., x_r)$, the WGMD can be shown to be normalised and the mean, $\overline{n} = \sum_{n=0}^{\infty} nP_{WGMD}(n; x_1, ..., x_r, p, k)$, given by

$$\overline{n} = \frac{\langle k' \rangle + k}{p} - k, \tag{6}$$

where $\langle k' \rangle = \sum_{k'=0}^{\infty} k' P(k'; x_1, ..., x_r)$. Equation 6 holds the same form as the mean of GMD expressed by equation 4, except with k' replaced by $\langle k' \rangle$. Detailed calculations are provided in the appendix.

Perhaps the simplest manifestation of the WGMD is where the weight is Poisson distributed, i.e. $P(k'; \overline{k'}) = \overline{k'}^{k'} \exp\left(-\overline{k'}\right)/k'!$, where $\overline{k'}$ is the average number of gluons. From equation 6, the mean of n with the Poisson weighted GMD (PGMD) is simply given by $\overline{n} = \frac{\overline{k'}+k}{p} - k$, where $\langle k' \rangle$ is replaced by the Poisson average $\overline{k'}$. The Poisson weight in the WGMD represents the independent production of single gluons in the initial state of QCD evolution after collision and the parameter $\overline{k'}$ represents the average gluon number over all collision events.

The number of initial gluons k' is often said to be able to take non-integral values if interpreted in an average sense [2–4]. The WGMD describes an ensemble of events with varying gluon numbers described by the weight factor. This model opens up possibilities to probe the average gluon number as a function of the collision centre-of-mass energy and pseudorapidity acceptance. Given a different weight distribution (e.g. NBD), correlations in gluon production can also be studied and compared to the Poisson case.

3 Results

Figure 1 exhibits a comparison of 3 different PGMD with parameters $k = 0.1$, 1, and 5. The other parameters are fixed at $p = 0.1$ and $\overline{k'} = 2$. For a smaller k, the distribution also peaks at smaller n. At $k = 0.01$, the distribution has a spike at $n = 0$. For a relatively larger k, varying the parameters p and $\overline{k'}$ does not reproduce the spike while for a small k, the variation of p and $\overline{k'}$ only modulates the height of the spike.

Figures 2–4 show the multiplicity distributions measured by the CMS collaboration [9] at centre-of-mass energies of 0.9, 2.36, and 7 TeV. Comparison is made with the best fit PGMD and GMD. The interior point algorithm [10, 11] is used to find the parameters that give the minimum χ^2 fit to the data, excluding point $n = 0$. The parameters that give the best fit distributions are shown in table 1.

The χ^2/dof show that the PGMD describes the multiplicity distributions better than the GMD. The PGMD describes the tail ends of the distribution better than the GMD, but not as well at low multiplicities.

From table 1, it is evident that the GMD that best describes the data reduces to the NBD ($k' = 0$). Given the computation of χ^2 which considers the points $n \geq 1$, we have the constrain of $k' \leq 1$. The characteristic feature of the GMD having an initial condition of k' number of gluons that branch and eventually hadronise, ensure that there are at least the same number of n hadrons as the initial number of gluons. This constraint limits the applicability of the GMD in finding an optimised solution.

The PGMD does not have this constraint since it considers a spread of initial conditions that describe Poisson processes which produce the gluons. At 0.9, 2.36, and 7 TeV, the model gives 1.07, 0.741, and 0.865 gluons respectively, on average. For increasing centre-of-mass energies, the parameter p decreases monotonically from 0.135 to 0.0614. From equation 2 and the definition of p, the model describes an increasing initial parton invariant mass Q as a function of the centre-of-mass energy, given a fixed branching probability A. The corresponding mean hadron numbers 18.1, 23.3, and 30.4 calculated using equation 6 is compatible with the experimental values measured by the CMS collaboration [9].

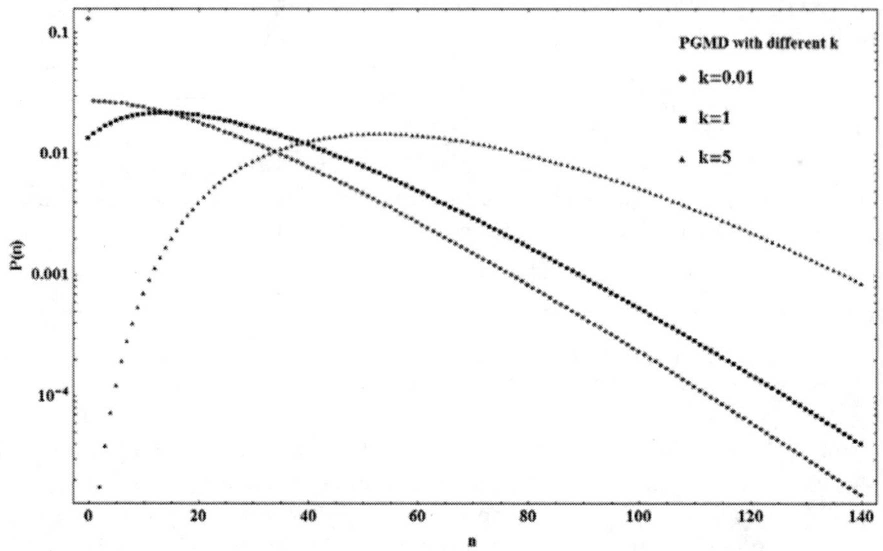

Figure 1. Comparison of PGMD with different *k* values.

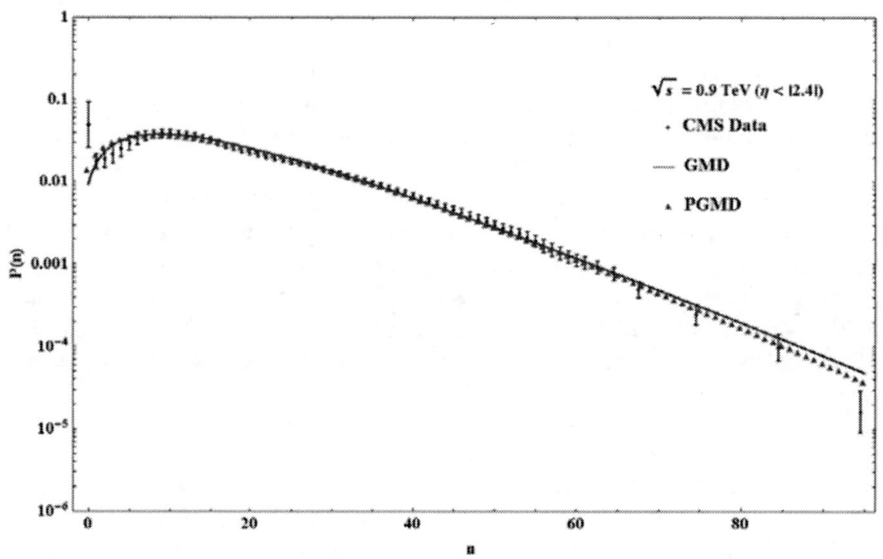

Figure 2. Charged-particle multiplicity distribution at $\sqrt{s} = 0.9$ TeV and $|\eta| < 2.4$ compared to the best fit PGMD (triangle) and GMD (line). The vertical lines in the data points represent the statistical errors and systematic uncertainties added in quadrature.

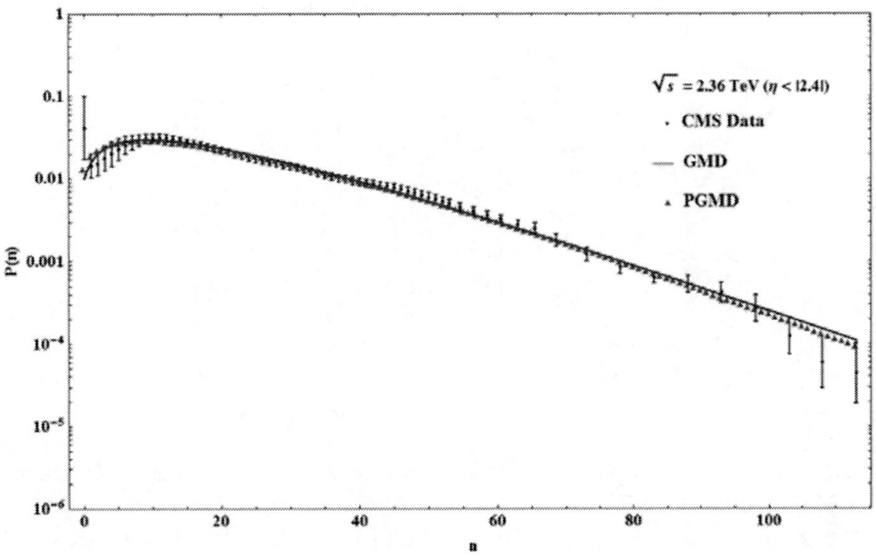

Figure 3. Charged-particle multiplicity distribution at $\sqrt{s}$ = 2.36 TeV and $|\eta| < 2.4$ compared to the best fit PGMD and GMD. The markers used and description of error bars are the same as in Fig. 2.

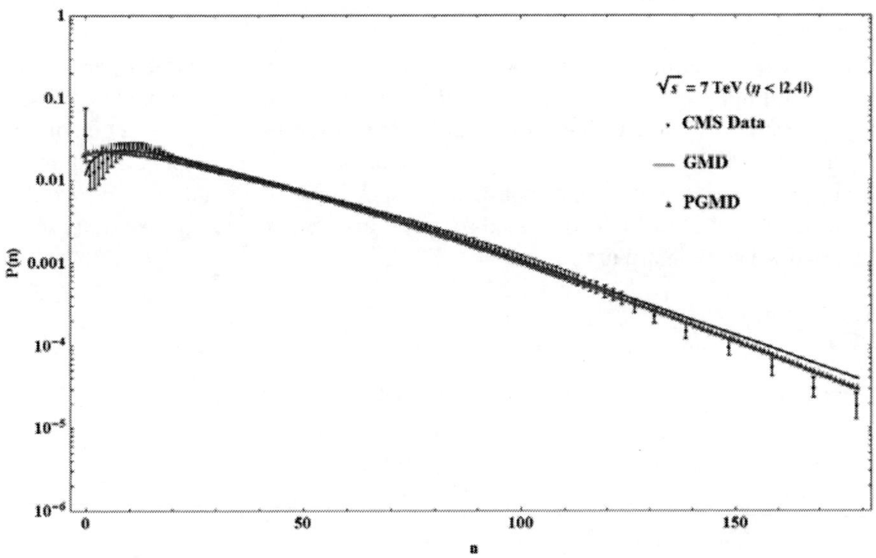

Figure 4. Charged-particle multiplicity distribution at $\sqrt{s}$ = 7 TeV and $|\eta| < 2.4$ compared to the best fit PGMD and GMD. The markers used and description of error bars are the same as in Fig. 2.

Table 1. Summary of the best fit parameter values.

	$\sqrt{s}$ (TeV)	0.9	2.36	7
PGMD	p	0.135	0.0904	0.0614
	k	1.59	1.50	1.07
	k'	1.07	0.741	0.865
	χ^2/dof	36.7/65	36.9/67	93.4/124
GMD	$\bar{n}$	18.4	23.4	31.0
	k	2.02	1.71	1.41
	k'	0.00	0.00	0.00
	χ^2/dof	47.1/65	41.3/67	136/124

4 Conclusion

The weighted Generalised Multiplicities Distribution model is obtained from a weighted sum of the Generalised Multiplicities Distribution. The model describes a mixed distribution with weights corresponding to the likelihood of the initial production of a certain gluon multiplicity. This model provides a more realistic description of collision and production conditions where the initial gluon number fluctuates on an event by event basis. The normalisation of the distribution is shown and the mean is derived.

A specific realisation of the model has weights described by a Poisson distribution of the gluon multiplicity. This Poisson weighted Generalised Multiplicities Distribution is applied to charged-particle multiplicity distributions measured by the CMS collaboration. Although the "shoulder" in the observed multiplicity distribution is still not reflected by the model, it describes the data better than the pure Generalised Multiplicities Distribution. The parameters that give the best fit are obtained, and the mean multiplicity given by the model is compatible with the experimentally observed values.

The application of Poisson weights to describe the gluon number production implies that the sources produce independent and uncorrelated single gluons. The use of other distributions as weights, for example the Negative Binomial Distribution, may be feasible and would suggest a different picture where the gluons are produced in a correlated manner. Insight on gluon production may be obtained by the comparison of such distributions.

References

[1] A. Giovannini, L. Van Hove, A. Phys. Pol. B **19**, 495 (1988)
[2] C.K. Chew, D. Kiang, H. Zhou, Phys. Lett. B **186**, 411 (1987)
[3] A.H. Chan, C.K. Chew, Phys. Rev. D **41**, 851 (1990)
[4] A.H. Chan, C.K. Chew, Z. Phys. C **55**, 503 (1992)
[5] A. Giovannini, Nucl. Phys. B **161**, 429 (1979)
[6] K. Konishi, A. Ukawa, G. Veneziano, Nucl. Phys. B **157**, 45 (1979)
[7] R.C. Hwa, C.S. Lam, Phys. Lett. B **173**, 346 (1986)
[8] UA5 Collaboration, Z. Phys. C **43**, 357 (1989)
[9] CMS Collaboration, JHEP **01**, 79 (2011)
[10] Wolfram Research, Inc., *Mathematica*, version 9.0 edn. (Wolfram Research, Inc., Champaign, Illinois, 2012)
[11] S. Mehrotra, SIAM J. Optim. **2**, 575 (1992)

EPJ Web of Conferences **120**, 03005 (2016) DOI: 10.1051/epjconf/201612003005
ISMD 2015

Particle production beyond the thermal model

Georg Wolschin[1,a]

[1] *Institut für Theoretische Physik der Universität Heidelberg, Philosophenweg 16, D-69120 Heidelberg, Germany, EU*

Abstract. The sources of particle production in relativistic heavy-ion collisions are investigated from RHIC to LHC energies. Whereas charged-hadron production in the fragmentation sources follows a $\ln(s_{NN}/s_0)$ law, particle production in the mid-rapidity low-x gluon-gluon source exhibits a much stronger dependence $\propto \ln^3(s_{NN}/s_0)$, and becomes dominant between RHIC and LHC energies. The equilibration of the three sources is investigated in a relativistic diffusion model (RDM). It agrees with the thermal model only for $t \to \infty$.

1 Introduction

The statistical hadronization or thermal model [1] has consistently provided good descriptions of relative or absolute particle production yields in e^+e^-, pp and relativistic heavy-ion collisions, e.g. [2, 3]. However, a necessary and sufficient condition for statistical equilibrium in the systems under investigation is provided by the distribution functions of the relevant observables rather than the particle yields. An example for a purely thermal distribution may be found in the cosmic microwave background radiation. It has a blackbody spectrum with a temperature of 2.735 K at redshift zero [4], although there are temperature fluctuations on the level of less than 1 part in 10^5 which give rise to structure formation, and have meanwhile been measured with excellent accuracy by the WMAP [5] and Planck [6] collaborations.

In relativistic heavy-ion collisions, the distributions of both transverse momentum p_T (Fig. 1) as well as rapidity y (or pseudorapidity η) of produced charged hadrons (Figs. 2, 4) clearly deviate from thermal distributions. At RHIC and LHC energies, the deviations in a p_T-region of $0.5 < p_T < 8$ GeV/c and the ensuing transition from exponential to power-law p_T-distributions are usually attributed to collective expansion. This transition is conveniently parametrized by the distribution functions shown in Fig 1. Above 8 GeV/c, however, hard processes become visible which require a pQCD treatment. When integrated over p_T to obtain particle yields, their contribution is almost negligible, but decisive as an indicator for nonequilibrium processes.

Indications for nonequilibrium behaviour can also be found in (pseudo)rapidity distributions of produced charged hadrons as measured by ALICE in PbPb [7], and both ATLAS and ALICE in pPb collisions [8, 9]. Here the distribution functions have important contributions from the fragmentation regions that are clearly visible in net-proton rapidity distributions at SPS and RHIC energies [10],

[a] e-mail: g.wolschin@thphys.uni-heidelberg.de

increase in particle content with $\log(s_{NN})$ for produced particles (Fig. 3), and are not in equilibrium with particles produced in the midrapidity source.

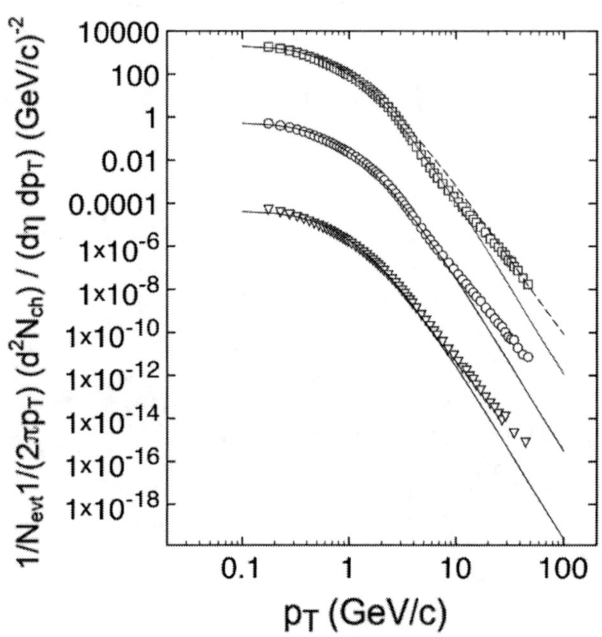

Figure 1. Transverse momentum distributions of produced charged hadrons in 2.76 TeV PbPb collisions as measured by ALICE [3] for 0-5%, 30-40% and 70-80% centralities compared with distribution functions $\propto [1 + (q-1)m_T/T]^{-q/(q-1)}$ for $q = 1.12$ that implicitly account for thermal emission and collective expansion, but not for hard (pQCD) processes at $p_T \geq 8$ GeV/c. The dashed curve is for $q = 1.14$. From GW, unpublished.

2 Relativistic diffusion model

In the Relativistic Diffusion Model (RDM) [11, 12], the (pseudo-)rapidity distribution of produced particles emerges from an incoherent superposition of the beam-like fragmentation components at larger rapidities y arising from valence quark-gluon interactions, and a component centered at (or near) midrapidity due to gluon-gluon collisions. All three time-dependent distribution functions (sources) $R_{1,2,gg}(y,t)$ are broadened in rapidity space as a consequence of diffusion-like processes governed by a Fokker-Planck equation (FPE). The fragmentation sources tend to shift towards midrapidity due to the drift term. Whereas this drift is sizeable at lower (AGS, SPS) energies, it is not pronounced at LHC energies where the interaction times are very short.

Since the FP-equation is linear, a superposition of the distribution functions using the initial conditions $R_{1,2}(y,t=0) = \delta(y \mp y_{max})$ with the absolute value of the beam rapidities $y_{max} = \ln(\sqrt{s_{NN}}/m_p)$ and $R_{gg}(y,t=0) = \delta(y - y_{eq})$ yields the solution. The charged-particle content of the respective sources is $N_{ch}^{qg,1}$ (projectile-like), $N_{ch}^{qg,2}$ (target-like) and N_{ch}^{gg} for the midrapidity low-x gluon-gluon source. In symmetric collisions the equilibrium value is $y_{eq} = 0$, whereas it is calculated from energy and momentum conservation at each centrality for asymmetric systems like pPb. The total rapidity distribution of produced charged particles becomes

$$\frac{dN_{ch}^{tot}(y, t = \tau_{int})}{dy} = N_{ch}^{qg,1} R_1(y, \tau_{int}) + N_{ch}^{qg,2} R_2(y, \tau_{int}) + N_{ch}^{gg} R_{gg}(y, \tau_{int})$$

with the rapidity $y = 0.5 \cdot \ln((E + p)/(E - p))$, and the interaction time τ_{int}. The latter corresponds to the total integration time of the underlying partial differential equation, which is a linear partial differential equation of the Fokker-Planck type, as described in [16].

In the solution, the mean values and variances are obtained analytically from the moments equations. Relaxation time and diffusion coefficient are related through a dissipation-fluctuation theorem (Einstein relation). Due to the collective expansion of the system, however, the effective diffusion coefficient is substantially larger. Hence the partial widths Γ_k (FWHM) are treated here as independent variables, which are related to the standard deviations through $\Gamma_k = \sqrt{8 \ln 2} \cdot \sigma_k$. For symmetric systems, the RDM then has four parameters (five including the total number of produced charged particles). This nonequilibrium-statistical model goes beyond the thermal model: Only for $t \to \infty$ a thermal distribution would be attained. At RHIC and LHC energies, comparison of the model results with data clearly shows that equilibrium among the three sources is not reached.

This finding is indirectly related to the current intense theoretical investigations of local equilibration within the gluonic source, cf. [13, 14] and references therein. These works concern the microscopic equilibration mechanisms and eventually aim at at fully QCD-based nonperturbative description. A direct connection to the current macroscopic investigation of equilibration among the three sources is difficult to perform conceptually and mathematically.

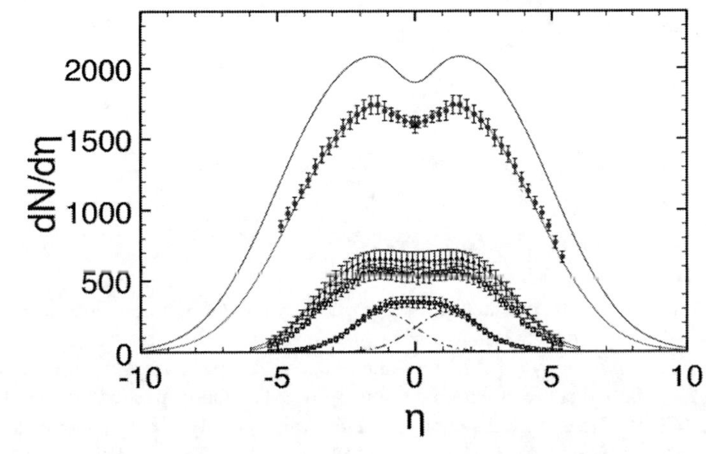

Figure 2. The RDM pseudorapidity distribution functions for charged hadrons in central AuAu (RHIC) and PbPb (LHC) collisions at c.m. energies of 19.6 GeV, 130 GeV, 200 GeV and 2.76 TeV shown here are optimized in χ^2–fits with respect to the PHOBOS [15] (bottom) and ALICE [7] (top) data, with parameters from [16]. The upper distribution function is an extrapolation to the LHC design energy of 5.52 TeV. At the lowest energy, only the fragmentation sources contribute (dash-dotted curves). From [16].

3 Pseudorapidity distributions

If particle identification is not available, one has to convert the results to pseudorapidity, $\eta = -\ln[\tan(\theta/2)]$ with the scattering angle θ. The Jacobian used for the conversion is

$$\frac{dN}{d\eta} = \frac{dN}{dy}\frac{dy}{d\eta} = J(\eta, m/p_T)\frac{dN}{dy},$$

$$J(\eta, m/p_T) = \cosh(\eta)\cdot[1 + (m/p_T)^2 + \sinh^2(\eta)]^{-1/2}$$

with the mass m fixed at the pion mass m_π, and a corresponding effective mean transverse momentum [17] $\langle p_{\mathrm{T,eff}}\rangle = m_\pi J_{y=0}/(1 - J^2_{y=0})^{1/2}$. Here the Jacobian $J_{y=0}$ at midrapidity is taken from experiment (for pions, kaons and protons) as discussed in [17]. The effective transverse momenta are smaller than the mean transverse momenta determined from the p_T-distributions, and the corresponding effect of the Jacobian is therefore larger than that estimated with $\langle p_T\rangle$ taken from the transverse momentum distributions for each particle species. At high RHIC and LHC energies the effect of the Jacobian transformation remains, however, essentially confined to the midrapidity source.

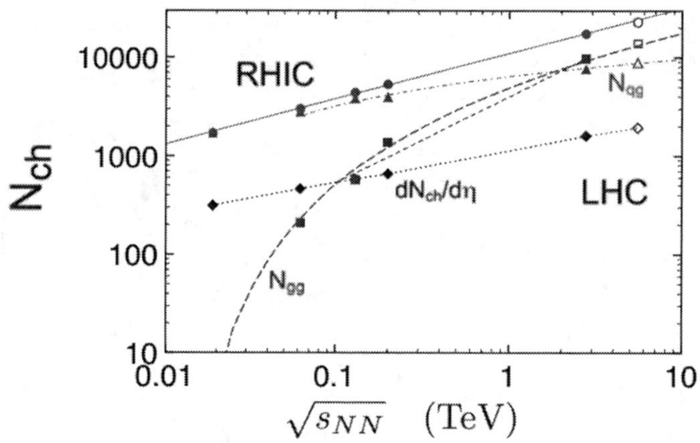

Figure 3. The total charged-hadron production in central AuAu and PbPb collision in the energy region 19.6 GeV to 5.52 TeV is following a power law $N_{tot} \propto (s_{NN}/s_0)^{0.23}$ (solid line), whereas the particle content in the fragmentation sources is $Nqg \propto \ln(s_{NN}/s_0)$, dash-dotted curve. The particle content in the mid-rapidity source obeys $N_{gg} \propto \ln^3(s_{NN}/s_0)$, dashed curve, not too far from a power law (short-dashed line) only in the intermediate energy range 0.1–2.76 TeV. The energy dependence of the mid rapidity yield is shown as a dotted line, with PHOBOS data [15] at RHIC energies, and ALICE data [18] at 2.76 TeV. From [19].

With parameter values for central collisions given in [16], the 3-sources RDM results for PbPb at 2.76 TeV energy are shown in Fig. 2 [19]. Corresponding centrality-dependent results for pPb collisions at 5.02 TeV are shown in Fig. 4 [20]. Here the asymmetric shapes of the distributions are much more sensitive to the details of the model, enhancing the credibility of the nonequilibrium 3-sources approach. The model had previously also been compared with dAu data at RHIC energies in [12].

The total hadron-production yield integrated over η follows $\ln(s_{NN}/s_0)$ for the fragmentation sources in both pPb and PbPb collisions at LHC energies, whereas the midrapidy gluonic source has a cubic log dependence $\propto \ln^3(s_{NN}/s_0)$, see Fig. 3 and the discussion in [19] with theoretical arguments for such an energy dependence. For experimental confirmations at RHIC energies based on STAR results cf. [21] and references therein.

DOI: 10.1051/epjconf/201612003005

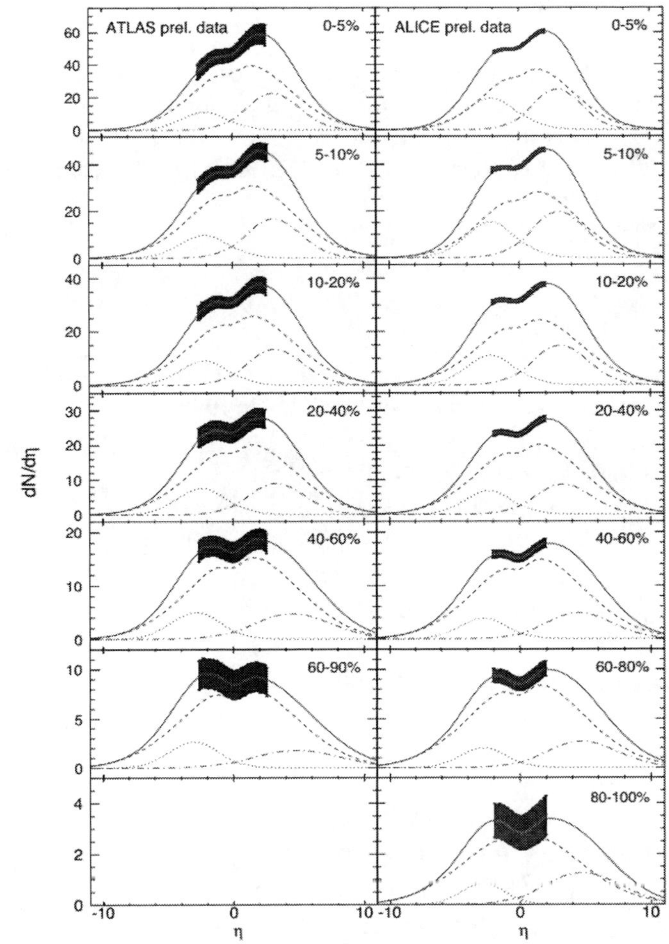

Figure 4. The centrality-dependent RDM pseudorapidity distribution functions for charged hadrons in *p*Pb collisions at LHC c.m. energy of 5.02 TeV are adjusted in the mid-rapidity region through χ^2−minimizations to the preliminary ATLAS (left) [9] and ALICE (right) [8] data (systematic error bars only for ALICE). The underlying distributions in the three–sources RDM are also shown, with the dashed curves arising from gluon-gluon collisions, the dash-dotted curves from valence quark-gluon events in the Pb-like region, and the dotted curves in the proton-like region (fragmentation sources). From [20].

4 Conclusion

Deviations from thermal distribution functions for produced particles in relativistic heavy-ion collisions are sensitive indicators for nonequilibrium processes. These are discussed for both transverse momentum and pseudorapidity distributions of produced charged hadrons.

A three-sources nonequilibrium-statistical relativistic diffusion model (RDM) is used for the description and prediction of pseudorapidity distributions of produced charged hadrons. Two fragmentation sources and a midrapidity gluonic source evolve as functions of time according to a Fokker-Planck equation, but they remain far from the thermal equilibrium distribution due to the short interaction times of about 10^{-23} s at RHIC and LHC energies.

The particle content integrated over pseudorapidity is proportional to $\ln(s_{NN}/s_0)$ only in the fragmentation sources. In the midrapidity source that arises essentially from gluon-gluon collisions, a dependence $\propto \ln^3(s_{NN}/s_0)$ is found and discussed in [19].

Acknowledgments

I am grateful to Jean-Paul Blaizot, Larry McLerran and Raju Venugopalan for conversations about local equilibration during their stays at the Heidelberg Institute for Theoretical Physics, and to Tom Trainor for discussions and suggestions regarding the $\sqrt{s_{NN}}$-dependence of the RDM-sources. This work is supported by DFG through the Transregional Research Center TRR33 at the Universities of Bonn, Heidelberg and LMU Munich.

References

[1] R. Hagedorn, Nuovo Cim. Suppl. **3**, 147 (1965)
[2] P. Braun-Munzinger, J. Stachel, J. Wessels, N. Xu, Phys. Lett. B **344**, 43 (1995)
[3] J. Manninen, F. Becattini, Phys. Rev. C **78**, 054901 (2008)
[4] J.C. Mather et al. (COBE Collaboration), Astrophys. J. **354**, L37 (1990)
[5] G. Hinshaw et al. (WMAP Collaboration), Astrophys. J. Suppl. **180**, 225 (2009)
[6] R. Adam et al. (Planck Collaboration), Astron. Astrophys. (to be published), 1502.01582v2
[7] E. Abbas et al. (ALICE Collaboration), Phys. Lett. B **726**, 610 (2013)
[8] J. Adam et al. (ALICE Collaboration), Phys. Rev. C **91**, 064905 (2015)
[9] B. Cole et al. (ATLAS Collaboration), ATLAS-CONF-2013-096 (2013)
[10] Y. Mehtar-Tani, G. Wolschin, Phys. Rev. Lett. **102**, 182301 (2009)
[11] G. Wolschin, Eur. Phys. J. A **5**, 85 (1999)
[12] G. Wolschin, M. Biyajima, T. Mizoguchi, N. Suzuki, Phys. Lett. B **633**, 38 (2006)
[13] L. McLerran, Nucl. Phys. A **926**, 3 (2014)
[14] R. Venugopalan, Nucl. Phys. A **928**, 209 (2014)
[15] B. Alver et al. (PHOBOS Collaboration), Phys. Rev. C **83**, 024913 (2011)
[16] G. Wolschin, J. Phys. G: Nuclear and Particle Physics **40**, 45104 (2013), 1301.1868
[17] D. Röhrscheid, G. Wolschin, Phys. Rev. C **86**, 024902 (2012)
[18] K. Aamodt et al. (ALICE Collaboration), Phys. Rev. Lett. **106**, 032301 (2011)
[19] G. Wolschin, Phys. Rev. C **91**, 014905 (2015), 1501.03026
[20] P. Schulz, G. Wolschin, Eur. Phys. J. A **51**, 18 (2015)
[21] T. Trainor, D.J. Prindle (2015), 1512.01599

EPJ Web of Conferences **120**, 03006 (2016) DOI: 10.1051/epjconf/201612003006
ISMD 2015

Systematic study of real photon and Drell-Yan pair production in p+A (d+A) interactions

M. Krelina[1,a], E. Basso[2,3], V. P. Goncalves[3,4], J. Nemchik[1,5], and R. Pasechnik[3]

[1] *FNSPE, Czech Technical University in Prague, Brehova 7, 115 19 Prague, Czech Republic*
[2] *Instituto de Fısica, Universidade Federal do Rio de Janeiro, Caixa Postal 68528, Rio de Janeiro, RJ 21941-972, Brazil*
[3] *Department of Astronomy and Theoretical Physics, Lund University, SE-223 62 Lund, Sweden*
[4] *High and Medium Energy Group, Instituto de Fisica e Matematica, Universidade Federal de Pelotas, Pelotas, RS, 96010-900, Brazil*
[5] *Institute of Experimental Physics SAS, Watsonova 47, 04001 Kosice, Slovakia*

Abstract. We study nuclear effects in production of Drell-Yan pairs and direct photons in proton-nucleus collisions. For the first time, these effects are studied within the color dipole approach using the Green function formalism which naturally incorporates the color transparency and quantum coherence effects. The corresponding numerical results for the nuclear modification factor are compared with available data. Besides, we present a variety of predictions for the nuclear suppression as a function of transverse momentum p_T, Feynman variable x_F and invariant mass M of the lepton pair which can be verified by experiments at RHIC and LHC. We found that the nuclear suppression is caused predominantly by effects of quantum coherence (shadowing corrections) and by the effective energy loss induced by multiple initial state interactions. Whereas the former dominate at small Bjorken x_2 in the target, the latter turns out to be significant at large x_1 in the projectile beam and is universal at different energies and transverse momenta.

1 Introduction

The color dipole approach [1] represents a phenomenological framework that effectively takes into account the higher-order and nonlinear QCD effects. There are many studies in the literature demonstrating a reliable agreement of predictions with experimental data, especially at high energies and/or small Bjorken variable x_2 in proton-proton (pp) collisions and DIS (see e.g. Refs. [2–4] and references therein).

The color dipole approach which is formulated in the target rest frame provides a consistent way of studying the nuclear effects, especially the nuclear shadowing, in both proton-nucleus (pA) and nucleus-nucleus (AA) collisions. The dynamics of pA or AA collisions is controlled by the coherence length l_c. When the coherence length is sufficiently large or small, one talks about the long coherence length (LCL) or short coherence length (SCL) approximations, respectively. In the intermediate kinematics when both approximations fail, one should employ the Green function technique which

[a] e-mail: michal.krelina@fjfi.cvut.cz

 DOI: 10.1051/epjconf/201612003006

accounts for the exact coherence length l_c and naturally incorporates the color transparency and quantum coherence effects. Such a kinematic region corresponds e.g. to kinematics at RHIC fixed target experiments or planned experiments such as AFTER@LHC.

In this paper, we present numerical results on the quark-nucleus cross section within the Green function formalism for the Drell-Yan (DY) lepton pair production and production of direct photons. Besides, we include also the gluon shadowing (GS) that dominates at small Bjorken x_2 and the effective energy loss induced by multiple initial state interactions.

2 Coherence length

The rest frame of the nucleus is very convenient for study of coherence effects. The dynamics of Drell-Yan (DY) process is regulated by the coherence length l_c, which controls the interference between amplitudes of the hard reaction occurring on different nucleons and is given by

$$l_c = \frac{1}{x_2 m_N} \frac{(M^2 + p_T^2)(1-\alpha)}{\alpha(1-\alpha)M^2 + \alpha^2 m_f^2 + p_T^2}\,, \tag{1}$$

where α is the fraction of the light-cone momentum of the projectile quark carried out by the photon, and $m_q = 0.2$ GeV is an effective quark mass. Figs. 1 and 2 show the energy dependence of the mean coherence length for $x_F = 0$ and $x_F = 0.6$ corresponding to small x_2 fractions, explicitly separating the regimes with the long coherence length (LCL), $l_c > R_A$, and short coherence length (SCL), $l_c \lesssim 1 \div 2$ fm. For the transition region between both limits we used the Green function formalism as the general case with no restrictions on l_c.

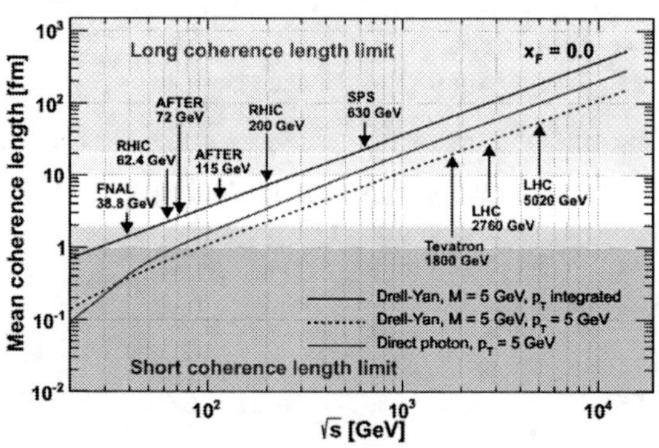

Figure 1. The mean coherence length for Drell-Yan and direct photons production for $x_F = 0$.

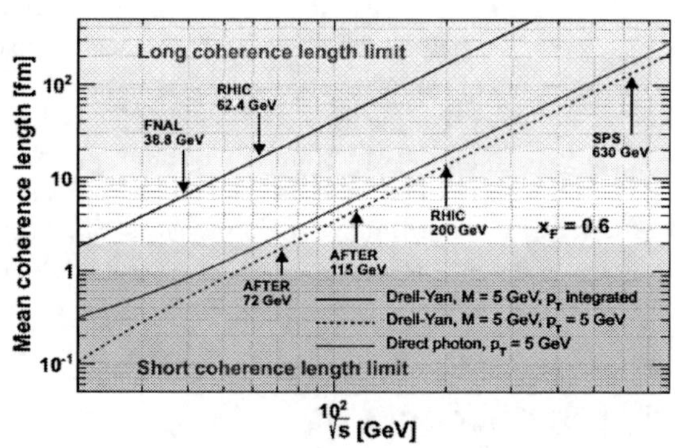

Figure 2. The mean coherence length for Drell-Yan and direct photons production for $x_F = 0.6$.

3 Color dipole approach

The DY process in the target rest frame can be treated as a radiation of a heavy photon or Z^0 boson by a projectile quark. The transverse momentum p_T distribution of photon bremsstrahlung in quark-nucleon interactions reads [5]

$$\frac{d^3\sigma^{(qN\to\gamma^* X)}}{d\ln\alpha d^2 p_T} = \frac{1}{(2\pi)^2}\int d^2\rho_1 d^2\rho_2 e^{i\vec{p}_T\cdot(\vec{\rho}_1-\vec{\rho}_2)}\Psi^*_{\gamma^* q}(\alpha,\vec{\rho}_2)\Psi_{\gamma^* q}(\alpha,\vec{\rho}_1)\Sigma(\alpha,\vec{\rho}_1,\vec{\rho}_2)\,, \tag{2}$$

where

$$\Sigma(\alpha,\vec{\rho}_1,\vec{\rho}_2) = \frac{1}{2}\left(\sigma^N_{q\bar{q}}(\alpha\vec{\rho}_1) + \sigma^N_{q\bar{q}}(\alpha\vec{\rho}_2) - \sigma^N_{q\bar{q}}(\alpha(\vec{\rho}_1 - \vec{\rho}_2))\right) \tag{3}$$

and the light-cone (LC) wave functions of the projectile $q \to q + \gamma$ fluctuation $\Psi_{T,L}(\alpha,\vec{\rho})$ can be found in Ref. [5]. For the dipole cross section $\sigma^N_{q\bar{q}}(\alpha\vec{\rho})$ we used GBW [6], KST [7] and GBWnew [8] parameterisations. The hadron cross section is given by a convolution of the qN cross section with the corresponding parton distribution functions (PDFs) f_q and $\bar{f}_q$ as follows

$$\frac{d^4\sigma^{(pp\to l^+l^-X)}}{d^2p_T dx_F dM^2} = \frac{\alpha_{EM}}{3\pi^2}\frac{x_1}{x_1+x_2}\int_{x_1}^{1}\frac{d\alpha}{\alpha^2}\sum_q Z_q\left(q_f(x_1/\alpha,Q^2) + \bar{q}_f(x_1/\alpha,Q^2)\right)\frac{d^3\sigma^{(qN\to\gamma^* X)}}{d\ln\alpha\, d^2p_T} \tag{4}$$

where Z_q is the fractional quark charge, the (anti)quark PDFs f_q and $\bar{f}_q$ are used with the leading order (LO) parameterisation from Ref. [9] at the scale $Q^2 = p_T^2 + (1 - x_1)M^2$. After integration over the transverse momentum $\vec{p}_T$ we get for hadronic cross section

$$\frac{d^2\sigma^{(pp\to l^+l^-X)}}{dx_F dM^2} = \frac{\alpha_{EM}}{3\pi^2}\frac{x_1}{x_1+x_2}\int_{x_1}^{1}\frac{d\alpha}{\alpha^2}\sum_q Z_q\left(q_f(x_1/\alpha,Q^2) + \bar{q}_f(x_1/\alpha,Q^2)\right)\frac{d\sigma^{(qN\to\gamma^* X)}}{d\ln\alpha} \tag{5}$$

and for quark-nucleon cross section

$$\frac{d\sigma^{(qN\to\gamma^* X)}}{d\ln\alpha} = \int d^2\rho\, |\Psi_{\gamma^* q}(\alpha,\vec{\rho})|^2 \sigma^N_{q\bar{q}}(\alpha\vec{\rho})\,. \tag{6}$$

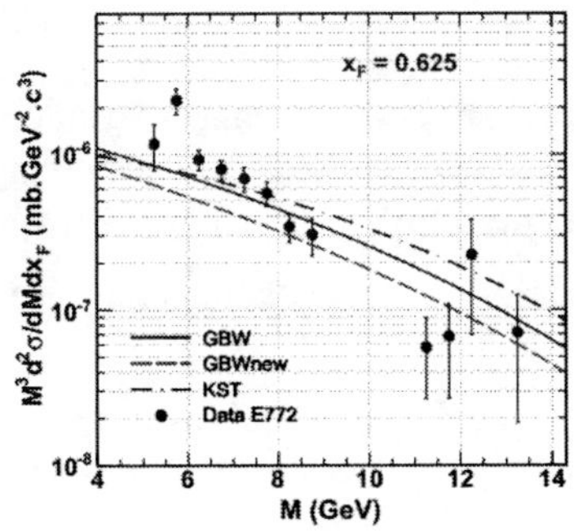

Figure 3. Differential dilepton cross sections in pp collisions vs E772 [13].

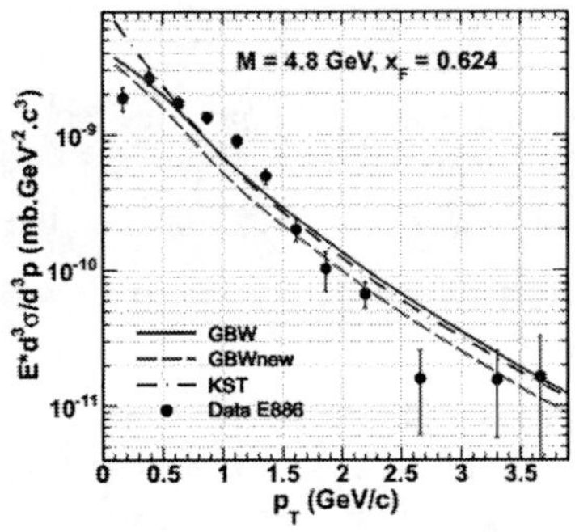

Figure 4. Differential dilepton cross sections in pp collisions vs E886 [14].

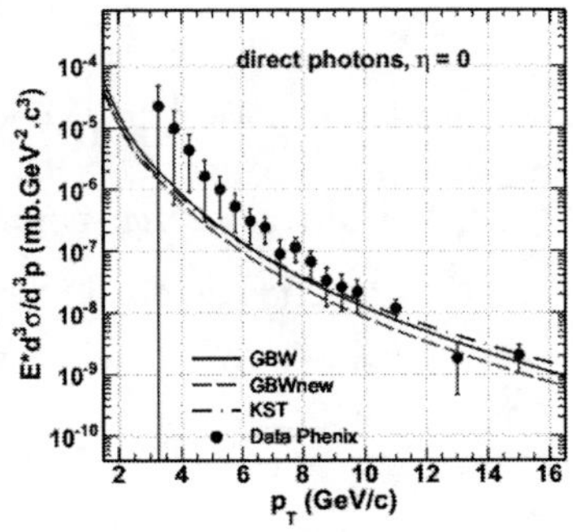

Figure 5. Differential direct photon cross sections in pp collisions vs PHENIX [15].

In Figs. 3, 4 and 5 we compare our predictions for various dipole cross section parameterisations with available DY data from E772 and E886 experiments and for direct photons from the PHENIX experiment where the dipole model predictions agree with the data reasonably well.

4 Transition to nuclear target

Within the Green function formalism the quark-nucleus cross section for DY pair production on nuclear targets reads [5]

$$\frac{d\sigma^{(qA\to\gamma^* X)}}{d\ln\alpha} = A\frac{d\sigma^{(qN\to\gamma^* X)}}{d\ln\alpha} - \frac{1}{2}Re\int_{-\infty}^{\infty}dz_1\int_{-\infty}^{z_1}dz_2\int d^2b d^2\rho_1 d^2\rho_2$$
$$\times\ \Psi^*_{\gamma^* q}(\alpha,\vec{\rho}_2)\rho_A(b,z_2)\sigma^N_{q\bar{q}}(\alpha\vec{\rho}_2)G(\vec{\rho}_2,z_2|\vec{\rho}_1,z_1)\rho_A(b,z_1)\sigma^N_{q\bar{q}}(\alpha\vec{\rho}_1)\Psi_{\gamma^* q}(\alpha,\vec{\rho}_1)\,, \quad (7)$$

where the Green function $G(\vec{\rho}_2, z_2|\vec{\rho}_1, z_1)$ describes the propagation of $|\gamma^* q\rangle$ Fock state between longitudinal positions z_1 and z_2 through the nucleus with initial and final separations $\vec{\rho}_1$ and $\vec{\rho}_2$, respectively. The Green function above satisfies the two-dimensional time-dependent Schroedinger equation (z_2 plays the role of time)

$$\left[i\frac{\partial}{\partial z_2} + \frac{\Delta_T(\vec{\rho}_2) - \eta^2}{2E_q\alpha(1-\alpha)} - V(z_2,\vec{\rho}_2,\alpha)\right]G(\vec{\rho}_2,z_2|\vec{\rho}_1,z_1) = 0 \quad (8)$$

with the boundary condition $G(\vec{\rho}_2, z_2|\vec{\rho}_1, z_1)|_{z_1=z_2} = \delta^2(\vec{\rho}_2 - \vec{\rho}_1)$. The imaginary part of the potential $V(z_2, \vec{\rho}_2, \alpha)$ describes an absorption of the dipole in a nuclear medium and reads

$$V(z_2,\vec{\rho},\alpha) = -\frac{i}{2}\rho_A(b,z_2)\sigma^N_{q\bar{q}}(\alpha\vec{\rho})\,. \quad (9)$$

For the p_T-dependent DY production cross section we solved the Schroedinger equation analytically which is possible for quadratic $\sigma^N_{q\bar{q}}(\rho) = C\rho^2$ and the uniform nuclear density. For p_T-integrated DY production cross section we solved the Schroedinger equation numerically using an algorithm proposed in Ref. [10].

In the LCL limit the Green function formalism naturally leads to a simple modification of the dipole cross section:

$$\sigma^N_{q\bar{q}}(\vec{\rho},x) \to \sigma^A_{q\bar{q}}(\vec{\rho},x) = 2\int d^2b\left(1 - e^{-\frac{1}{2}\sigma^N_{q\bar{q}}(\vec{\rho},x)T_A(b)}\right)\,. \quad (10)$$

Besides the lowest $|qG^*\rangle$ Fock state one should include also the higher Fock components containing gluons $|\gamma^* qG\rangle$, $|\gamma^* q2G\rangle$ etc. They cause an additional suppression known as the gluon shadowing (GS). The corresponding suppression factor R_G [11] calculated as a correction to the total $\gamma^* A$ cross section for the longitudinal photon, $R_G(x, Q^2, b) \approx 1 - \frac{\Delta\sigma_L^{(\gamma^* A)}}{\sigma_{tot}^{(\gamma^* A)}}$, was included in calculations replacing $\sigma^N_{q\bar{q}}(\vec{\rho}, x) \to \sigma^N_{q\bar{q}}(\vec{\rho}, x)R_G(x, Q^2, b)$.

The initial state energy loss (due to ISI effects) is expected to suppress the nuclear cross section significantly towards the kinematical limits, $x_L = \frac{2p_L}{\sqrt{s}} \to 1$ and $x_T = \frac{2p_T}{\sqrt{s}} \to 1$. Correspondingly, the proper variable which controls this effect is $\xi = \sqrt{x_L^2 + x_T^2}$. The magnitude of suppression was evaluated in Ref. [12]. It was found within the Glauber approximation that each interaction in the nucleus leads to a suppression factor $S(\xi) \approx 1 - \xi$. Summing up over the multiple initial state interactions in a pA collision at impact parameter b, one arrives at the nuclear ISI-modified quark PDF

$$q_f(x,Q^2) \Rightarrow q_f^A(x,Q^2,b) = C_v q_f(x,Q^2)\frac{e^{-\xi\sigma_{eff}T_A(b)} - e^{-\sigma_{eff}T_A(b)}}{(1-\xi)(1-e^{-\sigma_{eff}T_A(b)})}\,. \quad (11)$$

Here, $\sigma_{eff} = 20$ mb is the effective hadronic cross section controlling the multiple interactions. The normalisation factor C_v is fixed by the Gottfried sum rule (for more details, see Ref. [12]), $T_A(b)$

is the nuclear thickness function at given impact parameter b normalized to the mass number A. It was found that such an additional nuclear suppression due to the ISI effects represents an energy independent feature common for all known reactions, experimentally studied so far, with any leading particle (hadrons, Drell-Yan dileptons, charmonium etc).

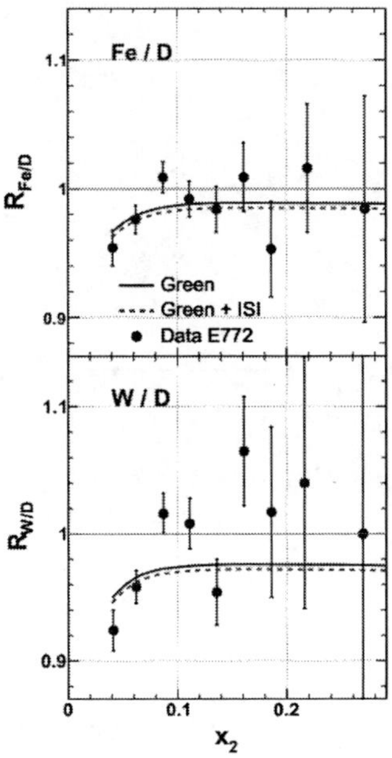

Figure 6. Comparison of the dipole model predictions for $R_{A/B}(x_2)$ with the E772 data at $\sqrt{s}$ = 38.8 GeV [13].

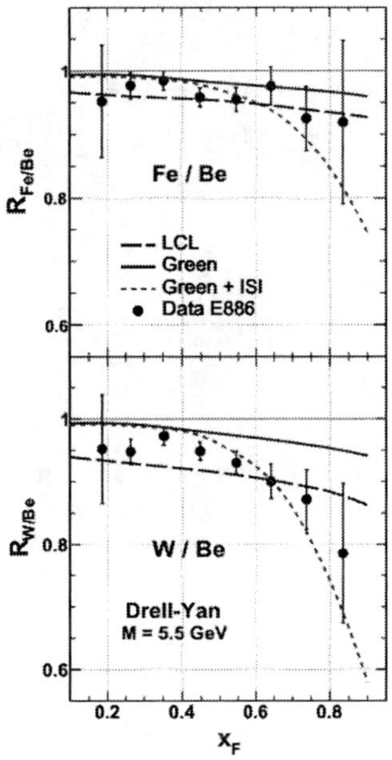

Figure 7. Comparison of the dipole model predictions for $R_{A/B}(x_F)$ with the E886 data at $\sqrt{s}$ = 38.8 GeV [14].

In Figs. 6 and 7 we compare our predictions for ratios $R_{A/B}(x_2)$ and $R_{A/B}(x_F)$ with the E772 and E886 data where the GS is not expected. We obtain a reasonable agreement with the E886 data including the ISI effects. In Fig. 8 we present our predictions for the nuclear suppression of DY pairs production at the future AFTER@LHC experiment demonstrating separate contributions from the GS and ISI effects. Fig. 9 shows the difference between calculations using the Green function formalism and the LCL limit in the RHIC kinematics region for production of direct photons and DY pairs at midrapidity. The RHIC data [15] indicate a strong large-p_T suppression that can be explained only by the ISI effects.

5 Conclusions

For the first time, we use the Green function formalism based on the color dipole approach for description of DY pair and direct photon production on nuclear targets in the kinematic regions where the SCL and LCL limits should not be used. We demonstrate that the GS and ISI energy loss causes a significant nuclear suppression. While the GS dominates at large energies and p_T, the ISI effects are important at large p_T and/or x_F. Our predictions are in a good agreement with FNAL E772 and E886 data as well as with the data from the PHENIX Collaboration. Finally, we predict a strong suppression due to the ISI effects that can be verified by the AFTER@LHC experiment in the future.

DOI: 10.1051/epjconf/201612003006

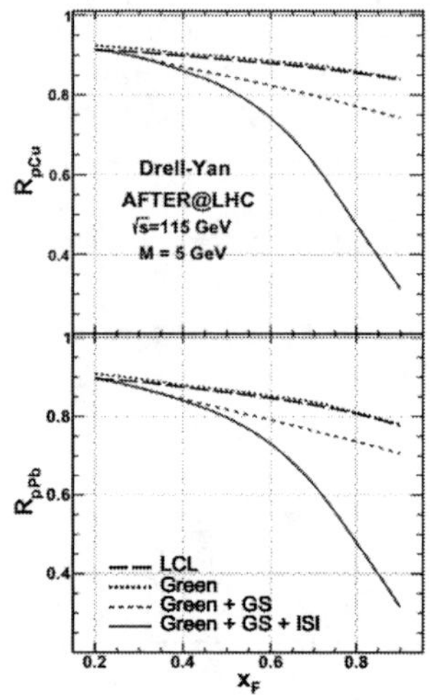

Figure 8. Predictions for the nuclear suppression R_{pA} in the DY process for the AFTER@LHC experiment.

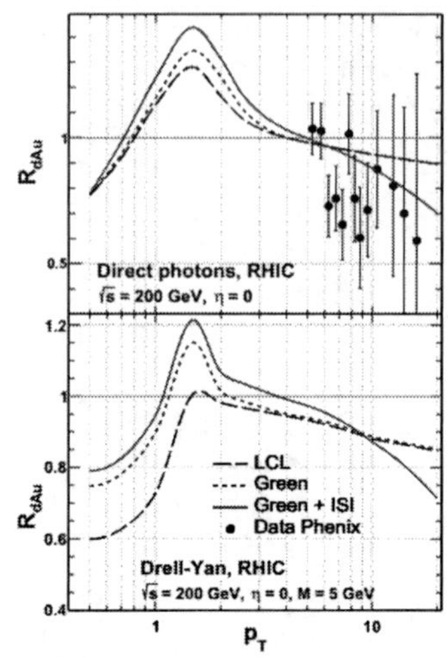

Figure 9. Comparison of R_{pA} with the data from RHIC [15] for direct photons and for the DY process with $M = 5$ GeV.

Acknowledgements

E. B. is supported by CAPES and CNPq (Brazil), contract numbers 2362/13-9 and 150674/2015-5. V. P. G. has been supported by CNPq, CAPES and FAPERGS, Brazil. R. P. is supported by the Swedish Research Council, contract number 621-2013-428. J. N. is partially supported by the grant 13-20841S of the Czech Science Foundation (GAČR), by the Grant MŠMT LG13031, by the Grant MŠMT LG15001, by the Slovak Research and Development Agency APVV-0050-11 and by the Slovak Funding Agency, Grant 2/0020/14.

References

[1] N. N. Nikolaev and B. G. Zakharov, Z. Phys. C**64**, 631 (1994).
[2] E. Basso, V. P. Goncalves, J. Nemchik, R. Pasechnik, M. Sumbera, arXiv:1510.00650 [hep-ph].
[3] B. Z. Kopeliovich, E. Levin, A. H. Rezaeian and I. Schmidt, Phys. Lett. B**675**, 190-195 (2009).
[4] B. Z. Kopeliovich, A. H. Rezaeian, H. J. Pirner and I. Schmidt, Phys. Lett. B**653**, 210-215 (2007).
[5] B. Z. Kopeliovich, A. V. Tarasov and A. Schafer, Phys. Rev. C**59**, 1609-1619 (1999).
[6] K. Golec-Biernat and M. Wüsthoff, Phys. Rev. D**59**, 014017 (1998).
[7] B. Z. Kopeliovich, A. Schafer and A. V.Tarasov, Phys. Rev. D**62**, 054022 (2000).
[8] H. Kowalski, L. Motyka and G. Watt, Phys. Rev. D**74**, 074016 (2006).
[9] A. D. Martin, W. J. Stirling, R. S. Thorne, G. Watt, Eur. Phys. J. C**63**, 189 (2009).
[10] J. Nemchik, Phys. Rev. C**68**, 035206 (2003).
[11] B. Z. Kopeliovich, J. Nemchik and A. Schafer, Phys. Rev. C**65**, 035201 (2002).
[12] B. Z. Kopeliovich *et al.*, Int. J. Mod. Phys. E**23**, 1430006 (2014).
[13] D. M. Alde *et al.* [E772 Collaboration], Phys. Rev. Lett. **64**, 2479 (1990).
[14] M. A. Vasiliev *et al.* [E886 Collaboration], Phys. Rev. Lett. **83**, 2304 (1999).
[15] S. S. Adler *et al.* [PHENIX Collaboration], Phys. Rev. Lett. **98**, 012002 (2007).

Nuclear effects in Drell-Yan production at the LHC

M. Krelina[1,a], E. Basso[2,3], V. P. Goncalves[3,4], J. Nemchik[1,5], and R. Pasechnik[3]

[1] *FNSPE, Czech Technical University in Prague, Brehova 7, 115 19 Prague , Czech Republic*
[2] *Instituto de Fısica, Universidade Federal do Rio de Janeiro, Caixa Postal 68528, Rio de Janeiro, RJ 21941-972, Brazil*
[3] *Department of Astronomy and Theoretical Physics, Lund University, SE-223 62 Lund, Sweden*
[4] *High and Medium Energy Group, Instituto de Fisica e Matematica, Universidade Federal de Pelotas, Pelotas, RS, 96010-900, Brazil*
[5] *Institute of Experimental Physics SAS, Watsonova 47, 04001 Kosice, Slovakia*

Abstract. Using the color dipole formalism we study production of Drell-Yan (DY) pairs in proton-nucleus interactions in the kinematic region corresponding to LHC experiments. Lepton pairs produced in a hard scattering are not accompanied with any final state interactions leading to either energy loss or absorption. Consequently, dileptons may serve as more efficient and cleaner probes for the onset of nuclear effects than inclusive hadron production. We perform a systematic analysis of these effects in production of Drell-Yan pairs in *pPb* interaction at the LHC. We present predictions for the nuclear suppression as a function of the dilepton transverse momentum, rapidity and invariant mass which can be verified by the LHC measurements. We found that a strong nuclear suppression can be interpreted as an effective energy loss proportional to the initial energy universally induced by multiple initial state interactions. In addition, we study a contribution of coherent effects associated with the gluon shadowing affecting the observables predominantly at small and medium-high transverse momenta.

1 Introduction

The Drell-Yan (DY) process provides an important test of the Standard Model as well as a comprehensive tool for studies of strong interaction dynamics in an extended kinematical range of energies and rapidities. In this paper, we focus on dilepton pairs coming from decay of virtual γ^*/Z^0 as a probe of nuclear effects in proton-lead (*pPb*) collisions at LHC. In this case, the DY process represents a cleaner probe than typical hadron production since the dilepton pairs have no final state interactions leading to either energy loss or absorption in the hot medium. For the same reason, no convolution with the jet fragmentation function is required and no nuclear effects are expected besides the saturation effects.

First, we give a short introduction into the color dipole picture as a framework in which the DY looks like a radiation of γ^*/Z^0 boson by a quark. We also compare calculations with the existing DY data in proton-proton (*pp*) collisions at LHC. A more detailed recent study of DY observables for *pp* collisions by some of the authors can be found in Ref. [1]. Then, we demonstrate that the coherence

[a] e-mail: michal.krelina@fjfi.cvut.cz

length l_c is large enough, and the long coherence length (LCL) limit can be safely used in calculations of the DY cross section in pPb collisions.

Besides the quark shadowing which is naturally incorporated in the LCL formula, we take into account for following two important effects. The first one is the gluon shadowing which plays a greater role at LHC energies at very small fractions x and can be estimated as a correction associated with the higher Fock states $|\gamma^* qG\rangle$, $|\gamma^* q2G\rangle$, etc. The second effect is the so-called effective energy loss due to the initial state interactions (ISI). The latter describes a suppression of the cross section at large dilepton p_T which was indicated at midrapidity, $y = 0$, by the PHENIX data [2] on π^0 production in central dAu collisions and on direct photon production in central $AuAu$ collisions [3]. The same mechanism of nuclear attenuation is important, especially at forward rapidities where we expect a much stronger onset of nuclear suppression as was demonstrated by the BRAHMS and STAR data [4].

Finally, we present new results on dilepton-pion azimuthal correlation in proton-lead collisions, where the characteristic double peak structure particularly sensitive to the saturation scale is predicted.

2 Color dipole picture

The color dipole formalism is treated in the target rest frame where the process of DY pair production can be viewed as a radiation of gauge bosons $G^* = \gamma^*, Z^0$ by a projectile quark [1, 5]. Assuming only the lowest $|qG^*\rangle$ Fock component the quark–nucleon differential cross section reads [1, 5, 6]

$$\frac{d\sigma^f_{T,L}(qN \to qG^*X)}{d\ln\alpha d^2p_T} = \frac{1}{(2\pi)^2}\int d^2\rho_1 d^2\rho_2 e^{i\vec{p}_T\cdot(\vec{\rho}_1-\vec{\rho}_2)}\Psi^{V-A}_{T,L}(\alpha,\vec{\rho}_1)\Psi^{V-A,*}_{T,L}(\alpha,\vec{\rho}_2)\Sigma(\alpha,\vec{\rho}_1,\vec{\rho}_2), \quad (1)$$

where

$$\Sigma(\alpha,\vec{\rho}_1,\vec{\rho}_2) = \frac{1}{2}\left(\sigma^N_{q\bar{q}}(\alpha\vec{\rho}_1) + \sigma^N_{q\bar{q}}(\alpha\vec{\rho}_2) - \sigma^N_{q\bar{q}}(\alpha(\vec{\rho}_1 - \vec{\rho}_2))\right) \quad (2)$$

and $\vec{p}_T$ is the transverse momentum of the outgoing gauge boson, and α is a fraction of the quark LC momentum taken by the gauge boson G^*. The vector and axial-vector wave functions are decorrelated in the simplest case of an unpolarized incoming quark [5]. In this work, we take into account the presence of both interfering $G^* = \gamma^*$ and Z^0 contributions. The corresponding wave functions $\Psi^{V-A}_{T,L}(\alpha,\vec{\rho})$ can be found in Ref. [5]. The cross section for inclusive production of a virtual gauge boson in pp collisions is found as follows [1]

$$\frac{d\sigma_{L,T}(pp \to G^*X)}{d^2p_T d\eta dM^2} = J(\eta,p_T)\frac{x_1}{x_1+x_2}\sum_f\int_{x_1}^1\frac{d\alpha}{\alpha^2}\left(q_f(x_q,\mu^2) + \bar{q}_f(x_q,\mu^2)\right)\frac{d\sigma^f_{T,L}(qN \to qG^*X)}{d\ln\alpha d^2p_T} \quad (3)$$

where $J(\eta, p_T) = \frac{2}{\sqrt{s}}\sqrt{M^2 + p_T^2}\cosh\eta$ is the Jacobian of the transformation between the Feynman $x_F = x_1 - x_2$ and the pseudorapidity η variables, with Bjorken fractions x_1 and x_2. Then, $q_f, \bar{q}_f$ denote quark and antiquark PDFs, respectively, for which the CTEQ parameterisations [7] will be used, with the hard scale $\mu^2 = p_T^2 + (1 - x_1)M^2$, where M is a dilepton mass. In practical calculations we use several parametrisations for dipole cross sections such as GBW [8], BGBK [9] and IP-Sat [10] models.

In Figs. 1 and 2 we compare our predictions for the dilepton invariant mass distributions with recent ATLAS data in the high invariant mass range and with recent CMS data covering the Z^0 boson resonance region taking into account its interference with the γ^* contribution. We can conclude that the parametrisation of the dipole cross section including the DGLAP evolution via the gluon PDF (BGBK and IP-Sat) describe the DY data better than naive GBW model as expected.

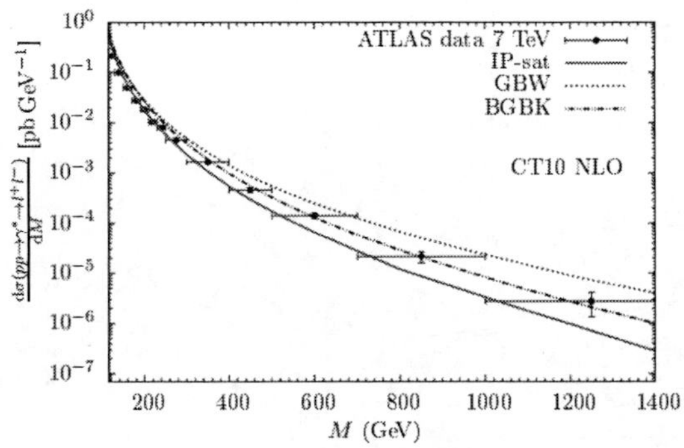

Figure 1. Predictions for the DY dilepton invariant mass distributions in pp vs. data from ATLAS [11].

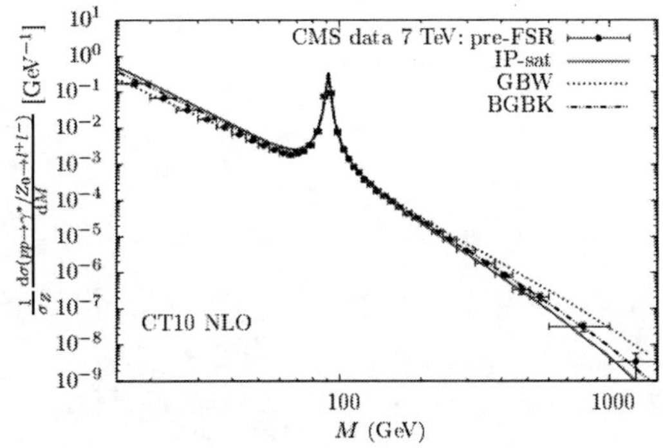

Figure 2. Predictions for DY dilepton mass distributions in pp vs. data from CMS [12].

3 Proton-nucleus interactions

The dynamics of DY production on nuclear targets is controlled by the mean coherence length

$$l_c = \frac{1}{x_2 m_N} \frac{(M^2 + p_T^2)(1-\alpha)}{\alpha(1-\alpha)M^2 + \alpha^2 m_f^2 + p_T^2}, \tag{4}$$

where M is the dilepton invariant mass, m_f is the mass of a projectile quark (we take same values as in Ref. [1]) and α is the fraction of the light-cone momentum of the projectile quark carried out by the gauge boson. The condition for the onset of shadowing is that the coherence length exceeds the nuclear radius R_A, $l_c \gtrsim R_A$. In the LHC kinematic region the long coherence length (LCL) limit can be safely used in practical calculations as is demonstrated in Fig. 3. In particular, this enables us to incorporate the shadowing effects via eikonalization of $\sigma^N_{q\bar{q}}(\vec{\rho}, \alpha)$ [13]

$$\sigma^N_{q\bar{q}}(\vec{\rho}, x) \to \sigma^A_{q\bar{q}}(\vec{\rho}, x) = 2 \int d^2b \left(1 - e^{-\frac{1}{2}\sigma^N_{q\bar{q}}(\vec{\rho},x)T_A(b)}\right) \tag{5}$$

where $T_A(b)$ is the nuclear thickness function at given impact parameter b normalized to the mass number A.

In the LCL limit, besides the lowest $|qG^*\rangle$ Fock state one should include also the higher Fock components containing gluons, e.g. $|\gamma^* qG\rangle$, $|\gamma^* q2G\rangle$, etc. They cause an additional suppression known as the gluon shadowing (GS). The corresponding suppression factor R_G [15] computed as a correction to the total $\gamma^* A$ cross section for the longitudinal photon reads $R_G(x, Q^2, b) \approx 1 - \frac{\Delta\sigma_L^{(\gamma^* A)}}{\sigma_{tot}^{(\gamma^* A)}}$, was included in calculations replacing $\sigma^N_{q\bar{q}}(\vec{\rho}, x) \to \sigma^N_{q\bar{q}}(\vec{\rho}, x) R_G(x, Q^2, b)$.

3.1 Effective energy loss

The initial-state energy loss (due to ISI effects) is expected to suppress noticeably the nuclear cross section when reaching the kinematical limits, $x_L = \frac{2p_L}{\sqrt{s}} \to 1$ and $x_T = \frac{2p_T}{\sqrt{s}} \to 1$. Correspondingly, a proper variable which controls this effect is $\xi = \sqrt{x_L^2 + x_T^2}$. The magnitude of suppression was evaluated in Ref. [16]. It was found within the Glauber approximation that each interaction in the nucleus

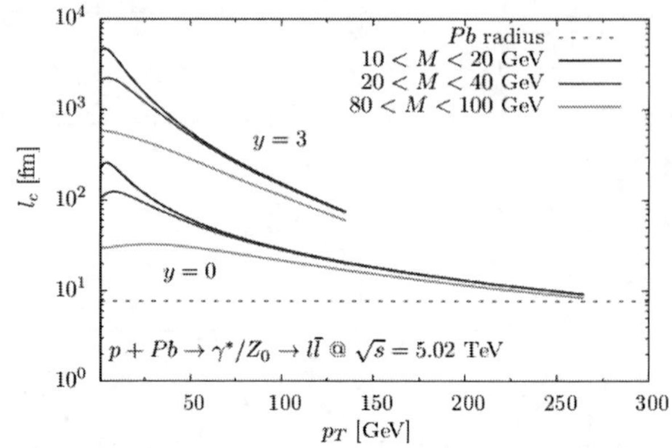

Figure 3. The mean coherence length for dilepton rapidities $y = 0$, 3 for different dilepton invariant mass intervals.

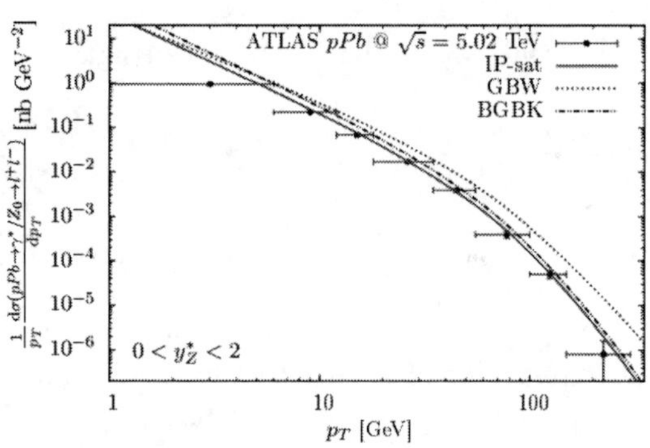

Figure 4. The dilepton p_T distribution in *pPb* collisions vs. ATLAS data [14].

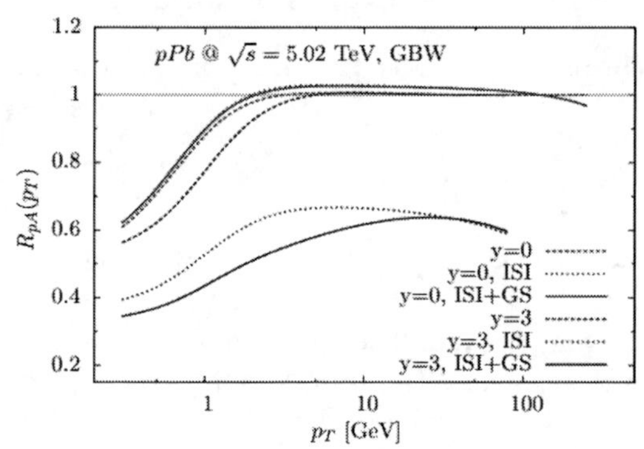

Figure 5. Predictions for the nuclear modification factor R_{pPb}.

leads to a suppression factor $S(\xi) \approx 1 - \xi$. Summing up over the multiple initial state interactions in a *pA* collision at impact parameter *b*, one arrives at a nuclear ISI-modified PDF

$$q_f(x, Q^2) \Rightarrow q_f^A(x, Q^2, b) = C_v q_f(x, Q^2) \frac{e^{-\xi\sigma_{eff}T_A(b)} - e^{-\sigma_{eff}T_A(b)}}{(1-\xi)(1 - e^{-\sigma_{eff}T_A(b)})}. \quad (6)$$

Here, $\sigma_{eff} = 20$ mb is the effective hadronic cross section controlling the multiple interactions. The normalisation factor C_v is fixed by the Gottfried sum rule (for more details, see Ref. [16]). It was found that such an additional nuclear suppression emerging due to the ISI effects represents an energy independent feature common for all known reactions experimentally studied so far, with any leading particle (hadrons, Drell-Yan dileptons, charmonium, etc).

Fig. 4 demonstrates a good agreement of our calculations for DY production in *pPb* collisions with the data from ATLAS experiment [14]. Fig. 5 shows predictions for the nuclear modification factor R_{pPb} as a function of p_T for production of DY pairs at distinct rapidities $y = 0$, 3 and dilepton invariant masses in the interval $66 < M < 122$ GeV typical for the ATLAS measurements. The GS and ISI effects are irrelevant at $y = 0$. However, we expect a strong nuclear suppression in the forward region $y = 3$ and large p_T's caused by the ISI effects. Here, the GS effects give a small contribution to the nuclear suppression at $p_T < 15 \div 20$ GeV.

4 Drell-Yan–Hadron correlations

The correlation function $C(\Delta\phi)$ depends on the azimuthal angle difference $\Delta\phi$ between the trigger and associate particles. The azimuthal correlations are investigated through a coincidence probability defined in terms of a trigger particle which could be either the gauge boson (dilepton) or the hadron. If we assume the former as a trigger particle then the correlation function is written as [1]

$$C(\Delta\phi) = \frac{2\pi \int_{p_T^{G^*}, p_T^h > p_T^{cut}} dp_T^{G^*} p_T^{G^*} dp_T^h p_T^h \frac{d\sigma(pp\to hG^*X)}{d^2p_T^h d\eta^h d^2p_T^{G^*} d\eta^{G^*} d^2b}}{\int_{p_T^{G^*} > p_T^{cut}} dp_T^{G^*} p_T^{G^*} \frac{d\sigma(pp\to G^*X)}{d^2p_T^{G^*} d\eta^{G^*} d^2b}} \tag{7}$$

where $\Delta\phi$ is the angle between the gauge boson and the hadron. The differential cross sections for G^* and G^*h production in momentum representation can be found in Ref. [1]. Fig. 6 demonstrates that a double peak structure emerges around $\Delta\phi = \pi$ in pp collisions considering that the photon and the pion are produced at forward rapidities, close to the limit of the phase space. Taking into account the nuclear dependence of the saturation scale in the GBW model in the LCL limit we calculated the correlation functions for proton-lead collisions as is shown in Fig. 7, where we expect again the characteristic double peak structure at forward rapidity $y = 4$. These results are in agreement with Ref. [17].

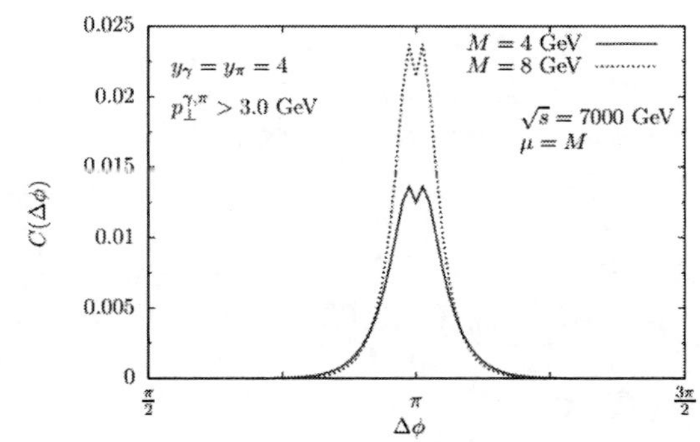

Figure 6. The correlation function for the DY–pion production in pp collisions.

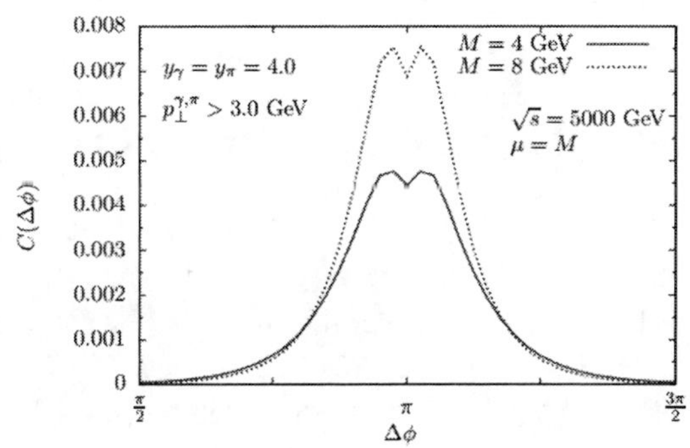

Figure 7. The correlation function for the DY–pion production in pPb collisions.

5 Conclusions

Within the color dipole picture we analyzed the DY pair production process accouning for virtual γ^* and Z^0 contributions. In the case of pp collisions, we found a good agreement of the differential cross section as a function of the dilepton invariant mass M with the ATLAS and CMS data. We demonstrate that for production of DY pairs on nuclear targets, the LCL limit can be safely employed in the LHC kinematic region. Our calculations of differential cross section for DY production in pPb collisions were compared to the ATLAS data and a good agreement has been found. We showed that the main source of a strong nuclear suppression, especially at forward rapidities expected at the LHC, comes mainly from the ISI corrections. A small onset of the GS is visible only at large $y = 3$ for $p_T < 15 \div 20$ GeV. Investigating the anglular correlation function corresponding to associated dilepton and pion production we found a characteristic double peak structure around $\Delta\phi = \pi$ not only for pp but also for pPb collisions.

Acknowledgements

E. B. is supported by CAPES and CNPq (Brazil), contract numbers 2362/13-9 and 150674/2015-5. V. P. G. has been supported by CNPq, CAPES and FAPERGS, Brazil. R. P. is supported by the Swedish Research Council, contract number 621-2013-428. J. N. is partially supported by the grant 13-20841S of the Czech Science Foundation (GAČR), by the Grant MŠMT LG13031, by the Grant MŠMT LG15001, by the Slovak Research and Development Agency APVV-0050-11 and by the Slovak Funding Agency, Grant 2/0020/14.

References

[1] E. Basso, V. P. Goncalves, J. Nemchik, R. Pasechnik, M. Sumbera, arXiv:1510.00650 [hep-ph].
[2] S. S. Adler *et al.* [PHENIX Collab.], Phys. Rev. Lett. **98**, 172302 (2007); Phys. Rev. Lett. **109**, 152302 (2012).
[3] S. Afanasiev *et al.* [PHENIX Collab.], Phys. Rev. Lett. **109**, 152302 (2012); T. Sakaguchi, Nucl. Phys. A**805**, 355 (2008).
[4] I. Arsene *et al.* [BRAHMS Collab.], Phys. Rev. Lett. **93**, 242303 (2004); J. Adams *et al.* [STAR Collab.], Phys. Rev. Lett. **97**, 152302 (2006).
[5] R. Pasechnik, B. Z. Kopeliovich, I. Potashnikova, Phys. Rev. D**86**, 114039 (2012).
[6] B. Z. Kopeliovich, A. V. Tarasov, A. Schafer, Phys. Rev. C**59**, 1609-1619 (1999).
[7] H. Lai, M. Guzzi, J. Huston, Z. Li, P. M. Nadolsky, J. Pumplin, C. P. Yuan, Phys. Rev. D**82**, 074024 (2010).
[8] K. Golec-Biernat, M. Wüsthoff, Phys. Rev. D**59**, 014017 (1998).
[9] J. Bartels, K. Golec-Biernat, H. Kowalski, Phys. Rev. D**66**, 014001 (2002).
[10] H. Kowalski, L. Motyka, G. Watt, Phys. Rev. D**74**, 074016 (2006).
[11] G. Aad *et al.* [ATLAS Collab.], Phys. Lett. B**725**, 223 (2013).
[12] V. Khachatryan *et al.* [CMS Collab.], Eur. Phys. J. **75**, 147 (2015).
[13] B. Z. Kopeliovich, L. I. Lapidus, A. B. Zamolodchikov, JETP Lett. **33**, 595-597 (1981).
[14] G. Aad *et al.* [ATLAS Collab.], Phys. Rev. C**92**, 044915 (2015).
[15] B. Z. Kopeliovich, J. Nemchik, A. Schafer, Phys. Rev. C**65**, 035201 (2002).
[16] B. Z. Kopeliovich, J. Nemchik, I. K. Potashnikova, I. Schmidt, Int. J. Mod. Phys. E**23**, 1430006 (2014).
[17] A. Stasto, B. W. Xiao and F. Yuan, Phys. Lett. B**716**, 430 (2012).

EPJ Web of Conferences **120**, 03008 (2016) DOI: 10.1051/epjconf/201612003008
ISMD 2015

Relativistic second-order dissipative fluid dynamics at finite chemical potential

Amaresh Jaiswal[1,a], Bengt Friman[1], and Krzysztof Redlich[2,3,4]

[1] *GSI, Helmholtzzentrum für Schwerionenforschung, Planckstrasse 1, D-64291 Darmstadt, Germany*
[2] *Institute of Theoretical Physics, University of Wroclaw, PL-50204 Wroclaw, Poland*
[3] *Extreme Matter Institute EMMI, GSI, Planckstrasse 1, D-64291 Darmstadt, Germany*
[4] *Department of Physics, Duke University, Durham, North Carolina 27708, USA*

Abstract. We employ a Chapman-Enskog like expansion for the distribution function close to equilibrium to solve the Boltzmann equation in the relaxation time approximation and subsequently derive second-order evolution equations for dissipative charge current and shear stress tensor for a system of massless quarks and gluons. We use quantum statistics for the phase space distribution functions to calculate the transport coefficients. We show that, the second-order evolution equations for the dissipative charge current and the shear stress tensor can be decoupled. We find that, for large chemical potential, the charge conductivity is small compared to the shear viscosity. Moreover, we demonstrate that the limiting behaviour of the ratio of heat conductivity to shear viscosity is identical to that obtained for a strongly coupled conformal plasma.

1 Introduction

At extreme conditions of temperature and density, quarks and gluons are deconfined to form quark-gluon plasma (QGP). The QGP behaves as a strongly coupled plasma having a very small shear viscosity-to-entropy density ratio, η/s [1–7]. Relativistic hydrodynamics has been quite successful in describing the space-time evolution of the QGP formed in relativistic heavy ion collisions and estimate its transport coefficients [8]. While most of the studies have focused on exploring the effects of shear and bulk viscosity on the QGP evolution, the dissipative charge current has been largely ignored. This may be due to the fact that at very high energies, baryon chemical potential are negligible. On the other hand, at lower energies such as those probed in the RHIC low-energy scan and at the upcoming Facility for Antiproton and Ion Research (FAIR), baryon chemical potential can no longer be ignored and therefore dissipative charge diffusion may play an important role.

Formulation of a causal theory of relativistic dissipative hydrodynamics are due to Grad [9], Müller [10] and Israel and Stewart (IS) [11]. However, the IS formulation of relativistic hydrodynamics from kinetic theory has several approximations and inconsistencies, the resolution of which is an active research area [12–25]. Here we present the derivation of causal (second-order) equations for dissipative charge current and shear stress tensor for a system of massless quarks and gluons. We employ an iterative Chapman-Enskog like expansion [26] to solve the Boltzmann equation in the

[a]e-mail: jaiswal.amaresh@gmail.com

relaxation time approximation [18]. Subsequently, we derive evolution equations for the dissipative quantities and exact expressions for the associated transport coefficients. We employ quantum statistics for the quark and gluon phase-space distribution functions with a non-vanishing quark chemical potential. We show that the evolution equations for the dissipative charge current and the shear stress tensor can be decoupled. We find that, for large chemical potential, the charge conductivity is small compared to the shear viscosity. Finally we demonstrate that in the limiting cases, the ratio of heat conductivity to shear viscosity, obtained here in the relaxation-time approximation, is identical to that of a conformal fluid in the strong coupling limit.

2 Relativistic hydrodynamics

The conserved energy-momentum tensor and the net-quark current for a system of massless quarks and gluons can be expressed in terms of the phase-space distribution function as [27]

$$T^{\mu\nu} = \int dp\, p^\mu p^\nu \left[g_q(f_q + f_{\bar q}) + g_g f_g\right] = \epsilon u^\mu u^\nu - P\Delta^{\mu\nu} + \pi^{\mu\nu}, \tag{1}$$

$$N^\mu = \int dp\, p^\mu \left[g_q(f_q - f_{\bar q})\right] = n u^\mu + n^\mu, \tag{2}$$

where $dp = d\mathbf{p}/[(2\pi)^3|\mathbf{p}|]$, p^μ is the particle four momenta, and g_q and g_g are the quark and gluon degeneracy factor, respectively. Here f_q, $f_{\bar q}$, and f_g are the phase-space distribution functions for quarks, anti-quarks, and gluons. In the tensor decompositions, ϵ, P, n, $\pi^{\mu\nu}$ and n^μ are the energy density, pressure, net quark number density, shear stress tensor and particle diffusion current, respectively. The bulk viscous pressure vanishes in the massless case. The projection operator $\Delta^{\mu\nu} = g^{\mu\nu} - u^\mu u^\nu$ is orthogonal to the hydrodynamic four-velocity u^μ defined in the Landau frame: $T^{\mu\nu}u_\nu = \epsilon u^\mu$. The conservation of energy-momentum, $\partial_\mu T^{\mu\nu} = 0$, and particle four-current, $\partial_\mu N^\mu = 0$, gives

$$\dot\epsilon + (\epsilon + P)\theta - \pi^{\mu\nu}\sigma_{\mu\nu} = 0, \tag{3}$$

$$(\epsilon + P)\dot u^\alpha - \nabla^\alpha P + \Delta^\alpha_\nu \partial_\mu \pi^{\mu\nu} = 0, \tag{4}$$

$$\dot n + n\theta + \partial_\mu n^\mu = 0. \tag{5}$$

Here we use the notation $\dot A = u^\mu\partial_\mu A$ for co-moving derivatives, $\theta \equiv \partial_\mu u^\mu$ for the expansion scalar, $\sigma^{\mu\nu} \equiv \frac{1}{2}(\nabla^\mu u^\nu + \nabla^\nu u^\mu) - \frac{1}{3}\theta\Delta^{\mu\nu}$ for the velocity stress tensor, and $\nabla^\alpha = \Delta^{\mu\alpha}\partial_\mu$ for space-like derivatives.

The equilibrium phase-space distribution functions for quarks, anti-quarks and gluons are given by

$$f_q^{(0)} = \frac{1}{\exp(\beta\, u\cdot p - \alpha) + 1}, \quad f_{\bar q}^{(0)} = \frac{1}{\exp(\beta\, u\cdot p + \alpha) + 1}, \quad f_g^{(0)} = \frac{1}{\exp(\beta\, u\cdot p) - 1}, \tag{6}$$

respectively, where $\beta = 1/T$ is the inverse temperature, $u \cdot p \equiv u_\mu p^\mu$ and $\alpha = \mu/T$ is the ratio of the quark chemical potential to temperature (we assume vanishing chemical potential for gluons). For a dissipative system, the temperature T and chemical potential μ is determined by the matching condition $\epsilon = \epsilon_0$ and $n = n_0$. In equilibrium, the energy density, pressure and net quark number density for a system of massless quarks and gluons is given by

$$\epsilon_0 \equiv u_\mu u_\nu \int dp\, p^\mu p^\nu \left[g_q\left(f_q^{(0)} + f_{\bar q}^{(0)}\right) + g_g f_g^{(0)}\right] = \frac{(4g_g + 7g_q)\pi^2}{120}T^4 + \frac{g_q}{4}T^2\mu^2 + \frac{g_q}{8\pi^2}\mu^4, \tag{7}$$

$$P_0 \equiv -\frac{1}{3}\Delta_{\mu\nu} \int dp\, p^\mu p^\nu \left[g_q\left(f_q^{(0)} + f_{\bar q}^{(0)}\right) + g_g f_g^{(0)}\right] = \frac{(4g_g + 7g_q)\pi^2}{360}T^4 + \frac{g_q}{12}T^2\mu^2 + \frac{g_q}{24\pi^2}\mu^4, \tag{8}$$

$$n_0 \equiv u_\mu \int dp\, p^\mu \left[g_q\left(f_q^{(0)} - f_{\bar q}^{(0)}\right)\right] = \frac{g_q}{6}T^2\mu + \frac{g_q}{6\pi^2}\mu^3, \tag{9}$$

Using the above equations and Eqs. (3)-(5), we obtain expressions for the derivatives of β and α as

$$\dot{\beta} = \frac{\beta}{3}\theta + O(\delta^2), \quad \dot{\alpha} = O(\delta^2), \tag{10}$$

$$\nabla^\mu \beta = -\beta \dot{u}^\mu + \frac{n}{\epsilon + P}\nabla^\mu \alpha - \frac{\beta}{\epsilon + P}\Delta^\mu_\rho \partial_\gamma \pi^{\rho\gamma}, \tag{11}$$

where $O(\delta^2)$ represents terms which are of second order in derivatives. Note that while Eq. (11) is exact, Eq. (10) is terminated at first-order which is sufficient for the present work. For a system close to local thermodynamic equilibrium, $f = f^{(0)} + \delta f$, where $|\delta f|/f^{(0)} \ll 1$. Therefore, using Eqs. (1) and (2), the shear stress tensor $\pi^{\mu\nu}$ and the particle diffusion current n^μ can be written as

$$\pi^{\mu\nu} = \Delta^{\mu\nu}_{\alpha\beta} \int dp\, p^\alpha p^\beta \left[g_q(\delta f_q + \delta f_{\bar{q}}) + g_g \delta f_g\right], \tag{12}$$

$$n^\mu = \Delta^\mu_\alpha \int dp\, p^\alpha \left[g_q(\delta f_q - \delta f_{\bar{q}})\right], \tag{13}$$

where $\Delta^{\mu\nu}_{\alpha\beta} \equiv \frac{1}{2}(\Delta^\mu_\alpha \Delta^\nu_\beta + \Delta^\mu_\beta \Delta^\nu_\alpha) - \frac{1}{3}\Delta^{\mu\nu}\Delta_{\alpha\beta}$ is a traceless symmetric projection operator orthogonal to u_μ and $\Delta_{\mu\nu}$.

3 Dissipative evolution equations

The relativistic Boltzmann equation in the relaxation-time approximation is given by [28],

$$p^\mu \partial_\mu f = -\frac{u \cdot p}{\tau_R}\left(f - f^{(0)}\right), \tag{14}$$

where τ_R is the relaxation time. We employ iterative solution of the above equation to derive expressions for the dissipative quantities [17–19]. We obtain the first-order expressions for shear stress tensor and dissipative charge current as [18],

$$\pi^{\mu\nu} = 2\beta_\pi \tau_R \sigma^{\mu\nu}, \quad n^\mu = \beta_n \tau_R \nabla^\mu \alpha. \tag{15}$$

Here β_π and β_n are the first-order transport coefficients obtained as

$$\beta_\pi = \frac{\epsilon + P}{5}, \quad \beta_n = \frac{J^+_{10}}{3} - \frac{n^2 T}{\epsilon + P}, \tag{16}$$

where,

$$J^+_{10} \equiv g_q \int dp (u \cdot p)\left(f^{(0)}_q \tilde{f}^{(0)}_q + f^{(0)}_{\bar{q}} \tilde{f}^{(0)}_{\bar{q}}\right) = \frac{g_q}{6}T^3 + \frac{g_q}{2\pi^2}T\mu^2 = \frac{\pi^2 + 3\alpha^2}{\alpha(\pi^2 + \alpha^2)}n. \tag{17}$$

Here $\tilde{f}^{(0)} \equiv 1 - rf^{(0)}$, where $r = 1$ for Fermions (quarks and anti-quarks) and $r = -1$ for Bosons (gluons).

To obtain the second-order dissipative equations, we consider the comoving derivative of Eqs. (12) and (13), and rewrite Eq. (14) in favour of $\delta\dot{f}$. Using first-order expressions for δf obtained in Ref. [18] and performing the momentum integrations, we obtain the second-order evolution equations for $\pi^{\mu\nu}$ and n^μ [25],

$$\dot{\pi}^{\langle\mu\nu\rangle} + \frac{\pi^{\mu\nu}}{\tau_\pi} = 2\beta_\pi \sigma^{\mu\nu} + 2\pi_\gamma^{\langle\mu}\omega^{\nu\rangle\gamma} - \frac{4}{3}\pi^{\mu\nu}\theta - \frac{10}{7}\pi_\gamma^{\langle\mu}\sigma^{\nu\rangle\gamma}, \tag{18}$$

$$\dot{n}^{\langle\mu\rangle} + \frac{n^\mu}{\tau_n} = \beta_n \nabla^\mu \alpha - n_\nu \omega^{\nu\mu} - n^\mu \theta - \frac{3}{5}n_\nu \sigma^{\nu\mu} - \frac{3\beta_n}{\epsilon + P}\pi^{\mu\nu}\nabla_\nu \alpha. \tag{19}$$

Here $\omega^{\mu\nu} \equiv (\nabla^\mu u^\nu - \nabla^\nu u^\mu)/2$ is the anti-symmetric vorticity tensor. It is important to note that in the relaxation-time approximation, τ_R is the common time scale for evolution of both $\pi^{\mu\nu}$ and n^μ, i.e., $\tau_\pi = \tau_n = \tau_R$. Comparing the first-order equations, Eq. (15), with the relativistic Navier-Stokes equations, $\pi^{\mu\nu} = 2\eta\sigma^{\mu\nu}$ and $n^\mu = \kappa_n \nabla^\mu \alpha$, the dissipative relaxation times are given by $\tau_\pi = \eta/\beta_\pi$ and $\tau_n = \kappa_n/\beta_n$. It is interesting to note that for dissipative charge current evolution, the coefficients of $\Delta^\mu_\nu \partial_\gamma \pi^{\nu\gamma}$ and $\pi^{\mu\nu}\dot{u}_\nu$ vanish and hence they do not appear in Eq. (19).

The last term in Eq. (19) couples the evolution of dissipative charge current with that of the shear stress tensor. As shown in Ref. [29], this type of coupling leads to disagreement with transport results. However, we observe that using Eq. (15), the last term in Eq. (19) is equivalent to $-(6/5)n_\nu\sigma^{\nu\mu}$ up to second-order in the gradient expansion. Therefore the evolution equation, Eq. (19), for the dissipative charge current becomes [25]

$$\dot{n}^{\langle\mu\rangle} + \frac{n^\mu}{\tau_n} = \beta_n \nabla^\mu \alpha - n_\nu \omega^{\nu\mu} - n^\mu \theta - \frac{9}{5} n_\nu \sigma^{\nu\mu}. \tag{20}$$

This is one of the main results of the present calculation. The compact form of the above equation makes it straightforward to directly implement it in a dissipative hydrodynamic code.

4 Transport coefficients

A measure of the relative importance of charge conductivity and shear viscosity is the dimensionless ratio $\kappa_n T/\eta$. Here, this quantity is given by $\kappa_n T/\eta = \beta_n T/\beta_\pi$, which is, in general, a function of μ/T. In order to quantify this ratio, one still need to specify the appropriate degeneracy factors for quarks (g_q) and gluons (g_g), as

$$\begin{aligned} g_q &= N_s \times N_c \times N_f = 6 N_f, \\ g_g &= N_s \times (N_c^2 - 1) = 16, \end{aligned} \tag{21}$$

where $N_c = 3$ is the number of colours, $N_s = 2$ is the number of spin degrees of freedom, and N_f is the number of flavours. In Fig. 1 (a), we show the ratio $\kappa_n T/\eta$ as a function of μ/T for $N_f = 2$ and $N_f = 3$. We see that, while this ratio is almost constant for small μ/T, it drops rapidly for larger μ/T, indicating that at low temperature and high density the conductivity of the QGP is small relative to the shear viscosity. Moreover, we observe that although the qualitative behaviour is the same for $N_f = 2$ and $N_f = 3$, the drop in $\kappa_n T/\eta$ is more pronounced for $N_f = 3$. Furthermore, we note that for $\mu/T > 1$ the ratio $\kappa_n T/\eta$ is almost independent of N_f.

Another interesting quantity is the ratio of heat conductivity to shear viscosity. Note that the heat flow is related to the dissipative charge current via the relation $q^\mu = -(\epsilon + P)n^\mu/n$, and hence we get [30]

$$q^\mu = -\kappa_q \frac{n T^2}{\epsilon + P} \nabla^\mu \alpha \quad \Rightarrow \quad \kappa_q = \kappa_n \left(\frac{\epsilon + P}{n T}\right)^2 \quad \Rightarrow \quad \frac{\kappa_q}{\eta} = \frac{\beta_n}{\beta_\pi} \left(\frac{\epsilon + P}{n T}\right)^2, \tag{22}$$

where κ_q is the coefficient of heat conductivity. In Fig. 1 (b), we plot κ_q/η scaled by the factor $\mu^2/\pi^2 T$ against μ/T, for a two and three flavor QGP. We see a constant behaviour in the limit of small as well as large μ/T. Moreover, we see that $(\kappa_q/\eta)\mu^2/\pi^2 T$ is independent of number of flavors for large μ/T. These limiting behaviours have interesting consequences.

In the limit of both small and large μ/T, the last equality in Eq. (22) reduces to [25]

$$\frac{\kappa_q}{\eta} = C \frac{\pi^2 T}{\mu^2}, \tag{23}$$

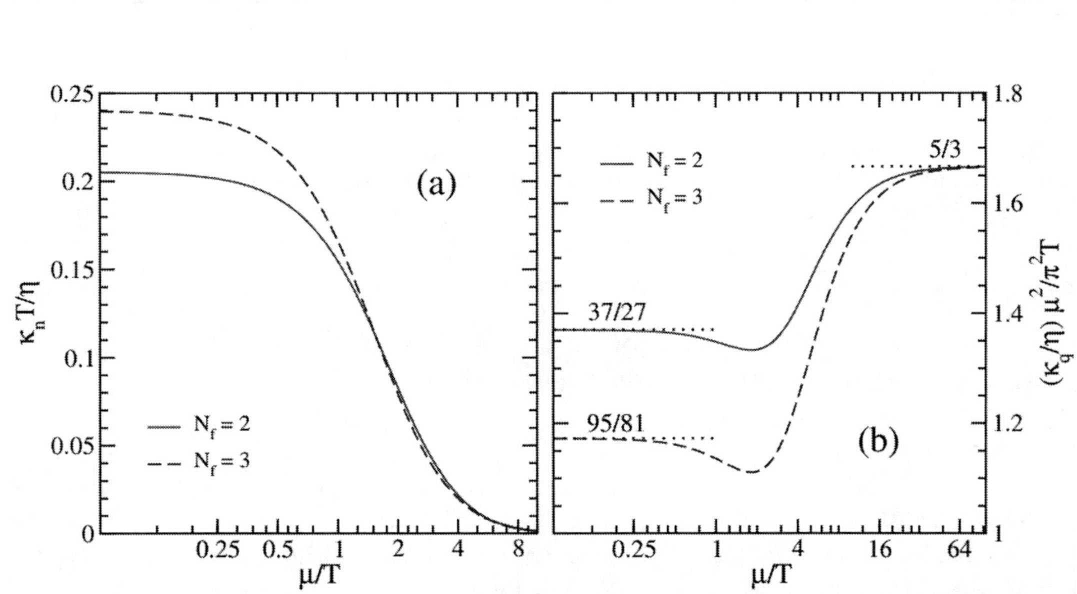

Figure 1. (a): The ratio of charge conductivity to shear viscosity scaled by the temperature, $\kappa_n T/\eta$, and (b): The ratio of thermal conductivity to shear viscosity, κ_q/η, scaled by the factor $\mu^2/\pi^2 T$, for two flavor (solid line) and three flavor (dashed line) massless quarks, plotted against μ/T.

which is analogous to the Wiedemann-Franz law [30]. The factor π^2 in the above equation is entirely due to quantum statistics and does not appear for a classical Boltzmann gas. In the small μ/T limit, the constant C in Eq. (23) becomes $C = (4g_g + 7g_q)/9g_q$. Hence, $C = 37/27$ and $95/81$ for two and three flavor QGP, respectively; see Fig. 1 (b). On the other hand, in the large μ/T limit, we find $C = 5/3$, independent of the number of flavors, as shown in Fig. 1 (b). It is interesting to note that these values of C are comparable to $C = 8/9$ (here a factor of $1/9$ indicates that the quark chemical potential is one-third of the baryon chemical potential) obtained in the calculations for strongly coupled conformal fluid with finite chemical potential [30]. Indeed, it is intriguing that up to a constant of proportionality, the limiting behaviour of the ratio κ_q/η obtained here is identical to that derived in the case of a strongly coupled conformal plasma.

5 Conclusions

In this work we employed the iterative Chapman-Enskog method to solve the Boltzmann equation and then derived the second-order dissipative hydrodynamical equations for a system of massless quarks and gluons. For such a system the bulk viscous pressure vanishes and therefore the dissipation is solely due to shear stress tensor and dissipative charge current. We considered quantum statistics with non vanishing quark chemical potential for the equilibrium distribution function. We obtained novel, exact expressions for the second-order transport coefficients corresponding to the dissipative charge current evolution. We also demonstrated that the evolution equations for dissipative charge current and shear stress tensor can be decoupled. We found that, when the ratio of chemical potential to temperature is large, the charge conductivity is small compared to the shear viscosity. Finally, we showed that the limiting behaviour of the ratio of heat conductivity to shear viscosity is qualitatively identical to that of a conformal fluid in the strong coupling regime.

Acknowledgements:- A.J. thanks Gabriel Denicol for useful discussions. A.J. was supported by the Frankfurt Institute for Advanced Studies (FIAS). The work of B.F. was supported in part by the Extreme Matter Institute EMMI. K.R. acknowledges support by the Polish Science Foundation (NCN), under Maestro grant DEC-2013/10/A/ST2/00106, and of the U.S. Department of Energy under Grant No. DE-FG02- 05ER41367.

References

[1] P. Romatschke and U. Romatschke, Phys. Rev. Lett. **99**, 172301 (2007).
[2] H. Song and U. W. Heinz, Phys. Rev. C **77**, 064901 (2008).
[3] M. Luzum and P. Romatschke, Phys. Rev. Lett. **103**, 262302 (2009).
[4] B. Schenke, S. Jeon and C. Gale, Phys. Rev. Lett. **106**, 042301 (2011).
[5] H. Song, S. A. Bass, U. Heinz, T. Hirano and C. Shen, Phys. Rev. Lett. **106**, 192301 (2011); **109**, 139904(E) (2012).
[6] R. S. Bhalerao, A. Jaiswal and S. Pal, Phys. Rev. C **92**, 014903 (2015).
[7] A. Jaiswal and V. Koch, arXiv:1508.05878 [nucl-th].
[8] U. Heinz and R. Snellings, Ann. Rev. Nucl. Part. Sci. **63**, 123 (2013).
[9] H. Grad, Comm. Pure Appl. Math. **2**, 331 (1949).
[10] I. Muller, Z. Phys. **198**, 329 (1967).
[11] W. Israel and J. M. Stewart, Annals Phys. **118**, 341 (1979).
[12] G. S. Denicol, T. Koide and D. H. Rischke, Phys. Rev. Lett. **105**, 162501 (2010).
[13] A. Jaiswal, R. S. Bhalerao and S. Pal, Phys. Rev. C **87**, 021901(R) (2013).
[14] A. Jaiswal, R. S. Bhalerao and S. Pal, Phys. Lett. B **720**, 347 (2013); J. Phys. Conf. Ser. **422**, 012003 (2013); arXiv:1303.1892 [nucl-th].
[15] A. El, Z. Xu and C. Greiner, Phys. Rev. C **81**, 041901 (2010).
[16] G. S. Denicol, H. Niemi, E. Molnar and D. H. Rischke, Phys. Rev. D **85**, 114047 (2012).
[17] P. Romatschke, Phys. Rev. D **85**, 065012 (2012).
[18] A. Jaiswal, Phys. Rev. C **87**, 051901(R) (2013); arXiv:1408.0867 [nucl-th].
[19] A. Jaiswal, Phys. Rev. C **88**, 021903(R) (2013); Nucl. Phys. A **931**, 1205 (2014); Proc. Indian Natl. Sci. Acad. **81**, no. 1, 62 (2015).
[20] C. Chattopadhyay, A. Jaiswal, S. Pal and R. Ryblewski, Phys. Rev. C **91**, 024917 (2015).
[21] A. Jaiswal, R. Ryblewski and M. Strickland, Phys. Rev. C **90**, 044908 (2014).
[22] W. Florkowski, A. Jaiswal, E. Maksymiuk, R. Ryblewski and M. Strickland, Phys. Rev. C **91**, 054907 (2015).
[23] R. S. Bhalerao, A. Jaiswal, S. Pal and V. Sreekanth, Phys. Rev. C **88**, 044911 (2013).
[24] R. S. Bhalerao, A. Jaiswal, S. Pal and V. Sreekanth, Phys. Rev. C **89**, 054903 (2014).
[25] A. Jaiswal, B. Friman and K. Redlich, Phys. Lett. B **751**, 548 (2015).
[26] S. Chapman and T. G. Cowling, *The Mathematical Theory of Non-uniform Gases* (Cambridge University Press, Cambridge, 1970), 3rd ed.
[27] R. Vogt, *Ultrarelativistic Heavy-Ion Collisions*, (North-Holland, Amsterdam, 2007).
[28] J. L. Anderson and H. R. Witting Physica **74**, 466 (1974).
[29] I. Bouras, E. Molnar, H. Niemi, Z. Xu, A. El, O. Fochler, C. Greiner and D. H. Rischke, Phys. Rev. C **82**, 024910 (2010).
[30] D. T. Son and A. O. Starinets, JHEP **0603**, 052 (2006).

EPJ Web of Conferences **120**, 04001 (2016) DOI: 10.1051/epjconf/201612004001
ISMD 2015

Searches for Dark Matter at the ATLAS experiment

Henso Abreu[1] on behalf of the ATLAS Collaboration[a]

[1] *Technion – Israel Institute of Technology. Haifa, 3200003, Israel*

Abstract. Searches for strongly produced dark matters in events with jets, photons, heavy-flavor quarks or massive gauge bosons recoiling against large missing transverse momentum in ATLAS are presented. These "$E_{\mathrm{T}}^{\mathrm{miss}}$+X" signatures provide powerful probes to dark matter production at the LHC, allowing us to interpret results in terms of effective field theory and/or simplified models with pair production of Weakly Interacting Massive Particles. Recent ATLAS results on dark matter searches at LHC Run 1 and the connection to astroparticle physics are discussed.

1 Introduction

Dark Matter existence in the Universe is well established through numerous astrophysical and cosmological observations, as documented in Refs. [1–4], however little is known of its particle nature or its non-gravitational interactions. At the Large Hadron Colliders, one can search for a weakly interacting massive particle (WIMP), denoted by χ, and for interactions between χ and Standard Model particles, see Ref. [5]. Searches conducted at the Large Hadron Collider are especially sensitive at low Dark Matter masses ($m_\chi \leq 10$ GeV), and therefore provide results complementary to direct Dark Matter searches Refs. [6–9] .

Interaction of particles mediating between Dark Matter and Standard Model particles can be described by contact operators in the framework of an effective field theory (EFT) Refs. [10–12] in cases where they are too heavy to be produced directly in the experiment. In the absence of signal, limits can be placed in terms of the effective mass scale of the interaction, M_* and of the χ–nucleon cross-section, $\sigma_{\chi-N}$, as a function of m_χ. In addition to the investigation with the EFT operators, pair production of WIMPs is also investigated within so-called simplified models, where a pair of WIMPs couples to a pair of Standard Models particles explicitly via a new mediator particle, e.g a new vector boson Z'. In this case, limits on M_* and/or m_χ are placed as a function of the mediator mass M_{med}.

A number of dedicated approaches to finding evidence for Dark Matter have been carried out with the ATLAS detector experiment Ref. [13] during Run 1, using typically 20.3 fb^{-1} of data collected at a centre-of-mass energy of $\sqrt{s} = 8$ TeV. So-called mono-X searches take advantage of a variety of different *tag* objects, X, together with large absolute values of missing transverse momentum, $E_{\mathrm{T}}^{\mathrm{miss}}$, in the final state to constitute a clean and distinctive signature. Heavy-quark searches use events with large $E_{\mathrm{T}}^{\mathrm{miss}}$ in association with high-momentum jets of which one or more are identified as jets containing b or top–quarks.

[a] e-mail: habreu@cern.ch

2 Mono-X Seaches

Tagging events using a variety of recoil objects, mostly stemming from initial state radiation (ISR), gives access to a broad range of EFT operators, as the respective sensitivity depends on the *tag* object in question. All cases require large amounts of $E_{\mathrm{T}}^{\mathrm{miss}}$, coming from the Dark Matter particle, and *tag* objects include single jets or photons as well as electroweak bosons Z and W .

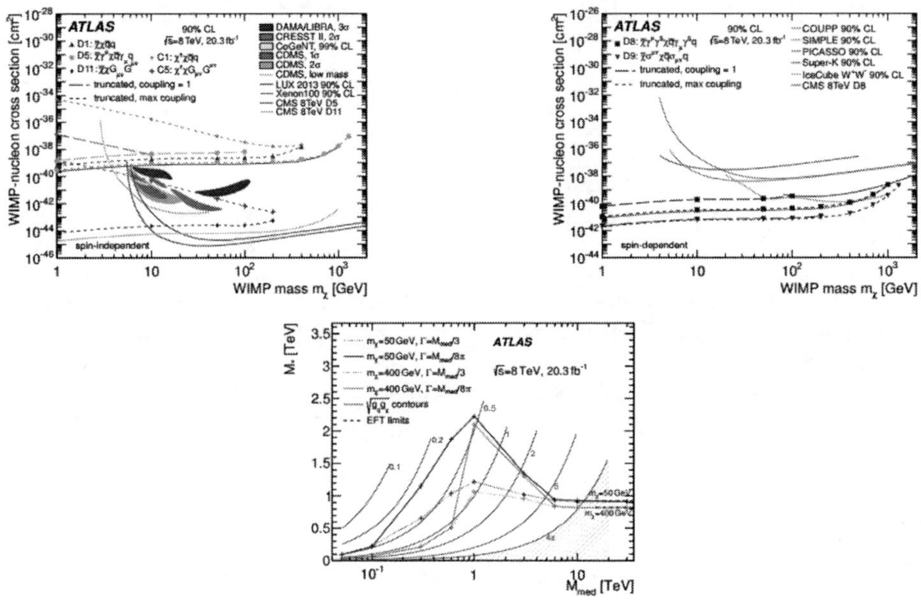

Figure 1. 90% CL limits on spin-independent (top-left) and spin-dependent (top-right) $\sigma_{\chi-N}$ as a function of m_χ for different EFT operators. Observed 95% CL limits on M_* as a function of M_{med} (bottom), assuming a Z'-like bosons in a simplified model and a Dark Matter mass of 50 GeV ad 400 GeV. The width of the mediator is varied between $M_{med}/3$ and $M_{med}/8\pi$. The corresponding limits from EFT models are shown as dashed lines; contour lines indicating a range of values of the product of the coupling constants ($\sqrt{g_q g_\chi}$) are also shown. For details see Refs. [6, 9, 14–30]. Plots taken from Ref. [14]

2.1 Mono-jet+ $E_{\mathrm{T}}^{\mathrm{miss}}$

As introduced in Ref. [14], using a single jet as recoil object gives sensitivity to six EFT operators (D1, D5, D8, D9, D11 and D5). To enhance the expected signal, events are required to contain at least one central jet with transverse momentum p_{T}, larger than both 120 GeV and half of $E_{\mathrm{T}}^{\mathrm{miss}}$. To ensure that the jet is in fact recoiling against the Dark Matter particles, the angle between the jet and the missing transverse momentum in the events is required to be above one. To further suppress background, which is mainly comprised of $Z(\nu\nu)$+jets and $W(\ell\nu)$+jets events, events containing leptons or high-p_{T} isolated tracks are vetoed. As all measurements are consistent with Standard Model expectations, the most sensitive out of nine signal regions, defined by requirements on $E_{\mathrm{T}}^{\mathrm{miss}}$ ranging from 150 GeV to 700 GeV, is used to place limits on $\sigma_{\chi-N}$ for each of the operators under investigation. As an example result of this analysis, inferred 90% confidence level (CL) limits on $\sigma_{\chi-N}$ as a function of m_χ for the spin-independent as well as spin-dependent case for different operators are shown in Fig. 1 (top-left and top-right). To ensure the validity of the EFT approach, the results are also shown after applying a *truncation* procedure as described in Ref. [14]. Limits on M_* as a function of M_{med} in the context of a simplified model are shown in Fig. 1 (bottom).

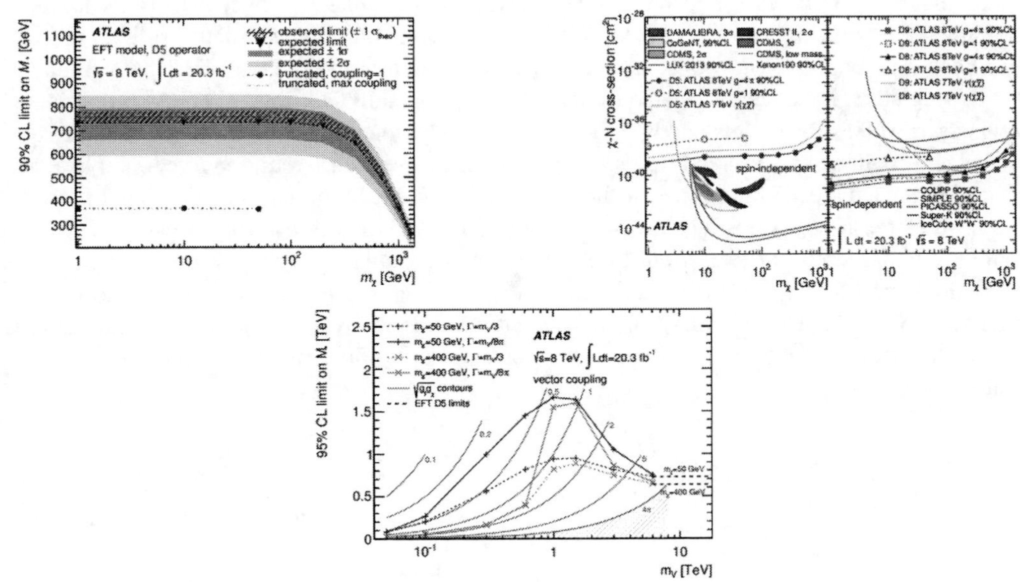

Figure 2. Limits at 90% CL on M_* as a function of m_χ (top-left), for the vector operator D5 . Results where EFT truncation is applied are also shown, assuming coupling values $\sqrt{g_f g_\chi} = 1, 4\pi$. Upper limits at 90% CL on $\sigma_{\chi-N}$ as a function of m_χ spin-independent (top-right-left) and spin-dependent (top-right-right) interactions, for a coupling strength $\sqrt{g_f g_\chi}$ of unity or the maximum value (4π) that keeps the model within its perturbative regime. The truncation procedure is applied for both cases. Observed lower limits at 95% CL on M_* as a function of M_{med} (bottom), for a Z'-like mediator with vector interactions. For an m_χ of 50 GeV or 400 GeV, results are shown for different values of the mediator total decay width Σ and compared to the EFT observed limit results for a D5 (vector) interaction. M_* vs m_V contours for an overall coupling $\sqrt{g_f g_\chi} = 0.1, 0.2, 0.5, 1, 2, 5, 4\pi$ are also shown. The corresponding limits from the D5 operator are shown as a dashed line. For details see Refs. [6, 8, 9, 20, 21, 24, 26, 29, 31–36]. Plots taken from Ref. [31]

2.2 Mono-photon+ E_T^{miss}

A search using a photon as *tag* object, allow to access three EFT operators (D5, D8 and D9), has been shown in Ref. [31]. Using events with a single highly energetic photon, large E_T^{miss}, no leptons and at most one jet; the main backgrounds in this analysis remain $Z(\nu\nu)+\gamma$, diboson, $W\gamma$ and $Z\gamma$ with lost leptons as well as W and Z production with leptons misidentified as photons. As for the mono-jet search, all measurements are consistent with Standard Model expectations and lower (upper) limits on M_* ($\sigma_{\chi-N}$) are presented both with and without applying the EFT truncation procedure mentioned above, as shown in examples in Fig. 2 (top-left and top-right). In addition limits on M_* as a function of M_{med}, as shown in Fig. 2 (bottom), are derived in the context of a simplified model.

2.3 Mono-W/Z+ $E_{\mathrm{T}}^{\mathrm{miss}}$

An analysis using W and Z bosons, respectively their decay products, as *tag* objects gives sensitivity to four EFT operators (C1, D1, D5 and D9) and has been presented in Refs. [37–39], for both hadronic and leptonic decay of the vector bosons. In the analysis aiming at hadronic decays, events are required to contain at least one high-p_{T} *large-radius jet* with reasonably balanced sub-jets, originating from the vector boson; at most one additional *regular* jet; and no leptons and photons. The background yield in the two signal regions, defined by requirements on $E_{\mathrm{T}}^{\mathrm{miss}}$ of 350 GeV and 500 GeV, is dominated by $Z(\nu\nu)$+jets as well as $Z(\ell\ell)$+jets and $W(\ell\nu)$+jets with *lost* leptons. Looking at the leptonic decays, the event selection differs for W or Z bosons. In the first case, events are required to contain exactly one high-p_{T} lepton and a transverse mass of the W boson candidate incompatible with direct production; while in the latter case, events have to contain two leptons giving an invariant mass close to Z peak and no additional lepton or jets. Both cases require large values of $E_{\mathrm{T}}^{\mathrm{miss}}$ and the main background contributions are coming from diboson events in the Z case and in addition $W(\ell\nu)$ and $Z(\ell\ell)$ with *lost* leptons in the W case. All yields are consistent with Standard Model expectations and limits on M_* as a function of m_χ and $\sigma_{\chi-N}$, are presented both for spin-independent and spin-dependent EFT operators, as shown in Fig. 3.

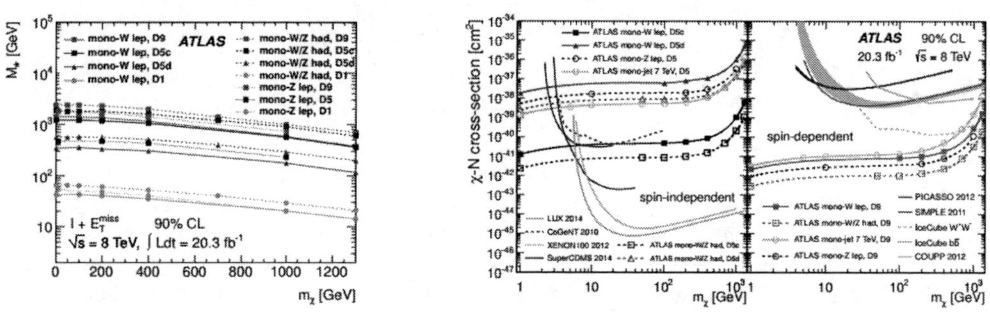

Figure 3. Limits at 90% CL on M_* as a function of m_χ (left), for EFT operators D9, D5 and D1. Observed limits on $\sigma_{\chi-N}$ as a function of m_χ at 90% CL for spin-independent (right-left) and spin-dependent (right-right) operators in the EFT. For details see Refs. [6–9, 29, 30, 32, 34, 35, 37, 38]. Plots taken from Refs. [37–39]

2.4 Heavy-quarks Searches

A search for Dark Matter pair production in association with bottom or top quarks has been presented in Ref. [40]. Aside from being sensitive to three EFT operators (C1, D1 and D9), this analysis also places constraints on the mass of a coloured mediator suitable to explain a possible signal of annihilating Dark Matter, using a simplified model model approach. Several signal regions are defined requiring combinations of increasing jet and b-jet multiplicity, 0 or 1 lepton and values of $E_{\mathrm{T}}^{\mathrm{miss}}$ above 200 GeV to 300 GeV in the events. Applying additional kinematic cuts, the main backgrounds is still coming form $t\bar{t}$-events as well as single-top and W/Z+jets events. All measurements are consistent with Standard Model expectations and lower (upper) limits on $M_*(\sigma_{\chi-N})$ are presented for three EFT operators (C1, D1 and D9), as shown in Fig. 4 for D1operator (top-left and top-right). Due to the proportionality of the scalar operator to the quark mass, limits for D1 are in fact better than those obtained by the above mentioned mono-jet analysis. Constraints on b-flavoured Dark Matter models, using a simplified model, are also presented and shown in Fig. 4 (bottom). For a Dark Matter particle of about 35 GeV, as suggested by an interpretation in Ref. [41] of data recorded by the Fermi–LAT collaboration see Ref. [42], mediator masses between roughly 300 GeV and 500 GeV are excluded at 95% CL.

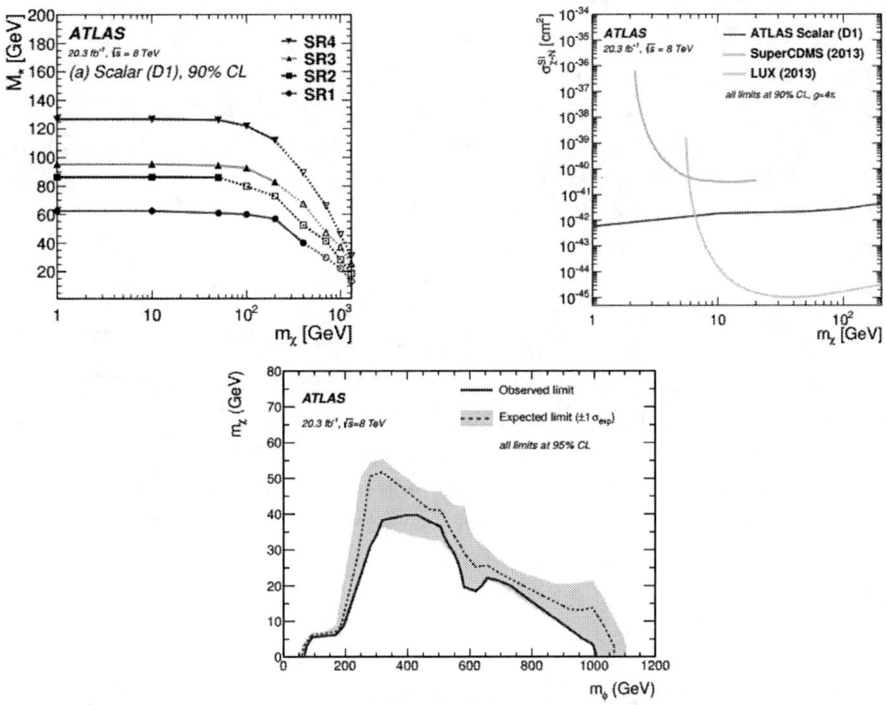

Figure 4. Lower limits on M_* at 90% CL for different signal regions as a function of m_χ (top-left) for the operator D1. Solid lines and markers indicate the validity range of the effective field theory assuming couplings $g_q g_\chi < 4\pi$, the dashed lines and hollow makers represent the full collider constraints. Upper limits at 90% CL on $\sigma_{\chi-N}$ for the scalar operator D1 as a function of m_χ (top-right) compared to other results. The coupling is assumed to be $g_q g_\chi = 4\pi$. Exclusion contour at 95% CL for the b-flavoured Dark Matter model (bottom) from combined results of two signal regions. The expected limit is given by the solid red line. The region beneath the curve indicating the observed limit is excluded. For details see Refs. [6, 7, 40, 43]. Plots taken from Ref. [40]

2.5 Conclusions

The ATLAS Collaboration has performed a broad variety of searches for Dark Matter signatures, using $\sqrt{s} = 8$ TeV Run 1 data and with *tag* objects ranging from single jets, photons, to W/Z bosons and as well as heavy quarks . No signs of Dark Matter have been observed, and stringent limits have set on the different benchmark models, emphasising the complementary nature of collider searches to direct and indirect detection experiments, especially at low Dark Matter masses and for spin-dependent EFT operators.

References

[1] F. Zwicky, Helv. Phys. Acta **6**, 110–127 (1933).

[2] G. Bertone et al. , Phys. Rept. **405**, 279–390 (2005).
[3] G. Jungman et al. , Phys. Rept. **267** , 195–373 (1996).
[4] J. Binney and S. Tremaine, (1993), arXiv:astro-ph/9304010 [astro-ph] .
[5] J. Beringer et al. (Particle Data Group), Phys. Rev. **D86**, p. 010001Jul (2012).
[6] D. S. Akerib et al. (LUX Collaboration), Phys. Rev. Lett. **112** , p. 091303 (2014).
[7] R. Agnese et al. (SuperCDMS Collaboration), Phys. Rev. Lett. **112** , p. 041302 (2014).
[8] E. Behnke et al. (COUPP Collaboration), Phys. Rev. D 86, p. 052001Sep (2012).
[9] S. Archambault et al. (PICASSO Collaboration), Phys. Lett. **B711**, 153?161 (2012).
[10] M. Beltran et al. , JHEP **09** , p. 037 (2010).
[11] P. J. Fox et al. , Phys. Rev. **D85**, p. 056011 (2012).
[12] J. Goodman et al. , Phys. Rev. **D82**, p. 116010 (2010).
[13] ATLAS Collaboration, JINST **3**, p. S08003 (2008).
[14] ATLAS Collaboration, Eur. Phys. J. **C75** , p. 299 (2015).
[15] CMS Collaboration, Eur. Phys. J. **C75**, p. 235 (2015).
[16] P. A. R. Ade et al. (Planck Collaboration), Astron. Astrophys. **571**, p. A16 (2014),
[17] G. Hinshaw et al. , ApJS **208**, p. 19 (2013)
[18] M. Ackermann et al. (Fermi-LAT Collaboration), Phys. Rev. **D89**, p. 042001 (2014)
[19] A. Abramowski et al. (H.E.S.S. Collaboration), Phys. Rev. Lett. **106**, p. 161301 Apr(2011).
[20] G. Angloher et al. , Eur. Phys. J. **C72**, p. 1971 (2012)
[21] R. Agnese et al. (CDMS Collaboration), Phys. Rev. Lett.**111**, p. 251301 (2013).
[22] R. Agnese et al. (SuperCDMS Collaboration), Phys. Rev. Lett. **112**, p. 241302 (2014).
[23] C. E. Aalseth et al., (2014), arXiv:1401.6234 [astro-ph.CO]
[24] R. Bernabei et al. (DAMA Collaboration), Eur. Phys. J. **C56**, 333–355 (2008).
[25] E. Aprile et al. (XENON100 Collaboration), Phys. Rev. Lett.**111** , p. 021301 (2013).
[26] S. Desai et al. (Super-Kamiokande Collaboration), Phys. Rev. **D70**, p. 083523 (2004).
[27] R. Abbasi et al. (IceCube Collaboration), Phys. Rev. Lett. **102** , p. 201302May (2009).
[28] E. Behnke et al. (COUPP Collaboration), Phys. Rev. Lett. **106**, p. 021303Jan (2011).
[29] M. Felizardo et al. (The SIMPLE Collaboration), Phys. Rev. Lett. **108**, p. 201302May (2012) .
[30] ATLAS Collaboration, ATL-PHYS-PUB-2014-007 (2014). https://cds.cern.ch/record/1708859 .
[31] ATLAS Collaboration, Phys. Rev. **D91**, p. 012008 (2015).
[32] E. Aprile et al. (XENON100 Collaboration), Phys. Rev. Lett. **109**, p. 181301Nov (2012).
[33] Z. Ahmed et al. (CDMS Collaboration), Phys. Rev. Lett. **106**, p. 131302Mar (2011).
[34] C. E. Aalseth et al. (CoGeNT Collaboration), Phys. Rev. Lett. **106**, p. 131301Mar (2011).
[35] M. G. Aartsen et al. (IceCube Collaboration), Phys. Rev. Lett. **110**, p. 131302Mar (2013).
[36] A. Abramowski et al. (HESS Collaboration), Phys. Rev. Lett. **110**, p. 041301 (2013).
[37] ATLAS Collaboration, Phys. Rev. Lett. **112**, p. 041802 (2014).
[38] ATLAS Collaboration, JHEP **09**, p. 037 (2014).
[39] ATLAS Collaboration, Phys. Rev. **D90**, p. 012004 (2014).
[40] ATLAS Collaboration, Eur. Phys. J. **C75** , p. 92 (2015).
[41] T. Daylan et al. , (2014), arXiv:1402.6703 [astro-ph.HE] .
[42] W. B. Atwood et al. , The Astrophysical Journal **697** , p. 1071 (2009).
[43] R. Gaitskell et al. , http://dmtools.brown.edu/.

EPJ Web of Conferences **120**, 04002 (2016) DOI: 10.1051/epjconf/201612004002
ISMD 2015

Extensive Air Showers and Cosmic Ray Physics above 10^{17} eV

Mario Bertaina[1,2,a]

[1] *Dipartimento di Fisica, Università degli Studi di Torino - Italy*
[2] *Istituto Nazionale di Fisica Nucleare - Sezione di Torino - Italy*

Abstract. Cosmic Rays above 10^{17} eV allow studying hadronic interactions at energies that can not be attained at accelerators yet. At the same time hadronic interaction models have to be applied to the cosmic-ray induced air-shower cascades in atmosphere to infer the nature of cosmic rays. The reliability of air-shower simulations has become the source of one of the largest systematic uncertainty in the interpretation of cosmic-ray data due to the uncertainties in modeling the hadronic interaction driving the air-shower development. This paper summarises in the first part the recent results on the cosmic ray energy spectrum, composition and anisotropy from the knee region to the GZK cutoff [1, 2] of the spectrum by means of ground-based experiments. Most of the information reported in this contribution is taken from [3–5]. Aspects interconnecting cosmic ray and particle physics are reviewed in the second part of the paper.

1 Introduction

The cosmic ray energy spectrum above 10^{14} eV has a power-law like behaviour ($\propto E^{-\gamma}$, with $\gamma \sim 2.7$) with features which are known as the 'knee' at $3-4\times10^{15}$ eV, where the spectrum steepens to $\gamma \sim 3.0$, the 'ankle' at $2-8\times10^{18}$ eV, which is characterised by a flattening of the spectrum by roughly the same change of the spectral index, i.e. back to $\gamma \sim 2.7$, and the GZK cut-off around 5×10^{19} eV.

The shape and composition of the primary spectrum as well as the large-scale anisotropy in the arrival direction of cosmic rays are key elements to understand the origin, acceleration and propagation of the Galactic radiation. The paradigm of the origin of Galactic cosmic rays (CR) are supernovae, as their shock waves can provide the required power to explain the intensity of the CR radiation at least up to 10^{15} eV [6]. This paradigm has been recently confirmed well below the knee by the observations of AGILE [7] and FERMI satellites [8]. However, the possibility for supernovae to accelerate CRs at energies above 10^{15} eV is quite challenging, therefore, different populations of sources have been envisaged as responsible for the radiation in galactic and extragalactic energy ranges [9, 10]. Those sources would be subject to a rigidity cutoff in the maximum energy at which the various elements are accelerated, as proposed originally by Peters [11]. In this approach, the knee at $\sim 4\times10^{15}$ eV would represent the end of the spectrum of CRs accelerated by supernova remnants in the Milky Way and the ankle at $\sim 4\times10^{18}$ eV the transition to particles from extragalactic sources. However, the ankle structure could be explained also in a completely different way, such as a consequence of the physical process of pair production by protons during propagation through the cosmic microwave

[a] e-mail: bertaina@to.infn.it

background radiation [12]. In this case, the Galactic-extragalactic transition occurs below 10^{18} eV. Around $\sim 5 \times 10^{19}$ eV another break is observed. Its position is compatible with the GZK cutoff, however, it could be explained also as the limit in the UHECRs acceleration by the sources.

A refined study of the CR primary spectrum and composition is, therefore, extremely important to address the above questions. As acceleration and propagation mechanisms in magnetic fields would lead to the same rigidity dependence, the study of large scale anisotropies in the arrival direction could provide relevant information to distinguish source and propagation effects.

The direct study of CRs by means of satellites or balloon-borne detectors is performed only at energies below 10^{15} eV. Close to the knee, the flux becomes of the order of 1 particle m^{-2} sr^{-1} yr^{-1}. This fact prevents the possibility of a direct observation of its structure by currently planned satellites or balloon experiments. Indeed, at least hundred of events above the knee are necessary to determine its existence with enough significance. Around the knee and at higher energies, CRs are studied by means of large arrays located at ground that measure the secondary particles produced by the primary CR cascading in the atmosphere, the so-called Extensive Air Showers (EAS). Typically, the energy is proportional to the total number of secondaries sampled at ground, while the composition is inferred either through a multi-component measurement, such as the electromagnetic and muonic components, or through the measurement of the emitted light (Cherenkov or fluorescence lights) along the longitudinal development of the shower. Despite the fact that shower arrays allow one to collect high statistics, the interpretation of the results is based on the comparison with expectations from simulation describing the EAS development in atmosphere, which are at some level inaccurate. This introduces a systematic uncertainty on the results, especially on the mass composition.

The TeV region allows some partial overlap between direct and indirect measurements. Several techniques have been employed recently on ground detectors that are sensitive to specific components of the CR radiation to overcome those uncertainties [13]. Among them, it is worth mentioning the measurement of the light component (p alone, or p+He) using hadron calorimeters [14], or Cherenkov light measurements in coincidence with TeV muons [15], and RPC counters at high altitude [16]. Those results are in quite good agreement with measurements by CREAM [17] balloon. In particular, the ARGO results allow one to cross-check the fluxes on an extended energy range (5-250 TeV). These results show that, when indirect measurements have the opportunity of selecting almost pure beams, their findings are in reasonable agreement with direct ones and confirm a fair representation of the EAS development in the atmosphere by simulation codes such as CORSIKA [18].

2 The knee region up to the ankle

The 'knee' is a distinct feature of the all-particle CR energy spectrum at $\sim 4\times10^{15}$ eV, where the power index suddenly changes from $\gamma \sim -2.7$ to $\gamma \sim -3.1$. Since its discovery the origin of this feature is still under debate. From the experimental point of view, measurements indicate that such a break is observed in the hadronic, muonic, and electromagnetic components [19–22], as well as in Cherenkov light [23]. These results give a clear indication that the knee is a peculiarity of the primary spectrum, disfavouring a hypothesis based on changes of the interaction characteristics of the primaries with air nuclei. This conclusion has been reinforced by the first comparisons of the predictions from hadronic models and LHC data [24].

Several experimental results associate the knee with the bending of the light component, and are compatible with a rigidity dependent cut-off [20, 21]. Unfortunately, the flux of the different components vary significantly depending on the interaction model used to interpret the data [21]. The recent results from the ARGO experiment, however, are in tension with this interpretation as they indicate that the proton component bends already around 1 PeV [25]. Nevertheless, if the interpretation that

DOI: 10.1051/epjconf/201612004002

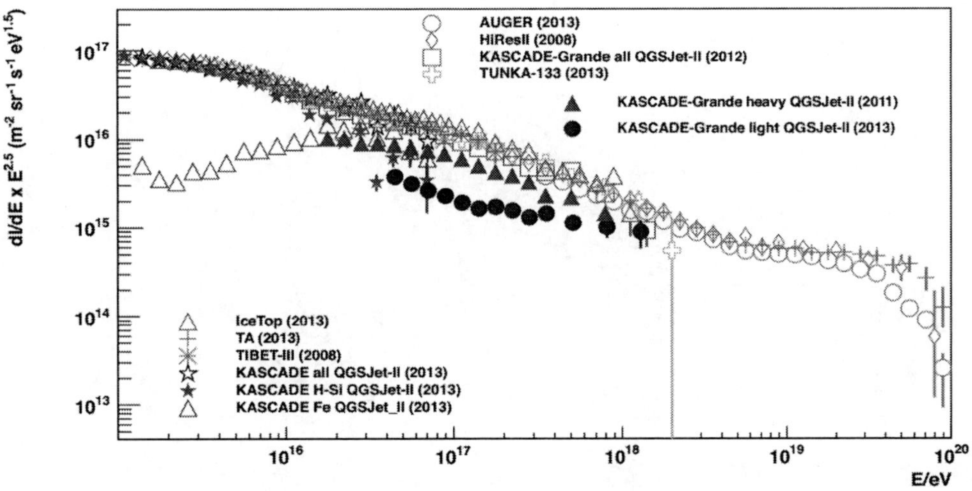

Figure 1. The measurement of the CR spectrum by EAS experiments from the knee till the end of the spectrum. The plot is taken from [3].

the main knee is due to the bending of the light component is correct, the heavy one should show a similar bending in the energy range $5 \times 10^{16} - 10^{17}$ eV. This is indeed the experimental finding of KASCADE-Grande [26] confirmed by TUNKA-133 [27] and IceTop [28] experiments.

In detail, the measured all-particle energy spectrum by KASCADE-Grande exhibits a less pronounced but still clear deviation from a single power law between the knee and the ankle, with a spectral hardening at $\sim 2\times10^{16}$ eV and a steepening at $\sim 10^{17}$ eV [29]. The average mass composition gets heavier after the knee till $\sim 10^{17}$ eV, where a bending of the heavy component is observed [30] . An indication of a hardening of the light component just above 10^{17} eV has been measured as well [31]. The flux of heavy and light components depends on the hadronic interaction model used to interpret the experimental data [32] as previously mentioned. The above results are summarised on Fig. 1.

Fig. 2 compares the results of several experiments in terms of $< lnA >$, as it is often reported in literature to describe the evolution of the composition as a function of energy. Only QGSJet-II model is considered. A detailed description including the role of interaction models in composition is reported in [33]. Use of different models introduces a shift in the average mass comparable to the dispersion of the data in Fig. 2. Despite the large uncertainty in the absolute composition, a common general trend is visible. Composition gets heavier through the knee region and becomes lighter approaching the ankle. The solid line in the plot is used as a guidance line to show how the average $< lnA >$ of data in the plot evolves with energy. This result is compatible with the concept that the galactic component of the cosmic ray radiation reaches an end following a rigidity cut-off at $\sim 10^{17}$ eV and that the ankle indicates a transition to a population dominated by extra-galactic sources.

The search for anisotropies in the arrival direction of CR can provide relevant information to distinguish source and propagation effects. The anisotropy varies with energy but the topological structure remains the same till $\sim 10^{14}$ eV where it has an abrupt change as pointed out by IceCube [34]. Such a change is confirmed till knee energies. This result seems to be inconsistent with the amplitude and phase expected according to the Compton-Getting prediction due to the relative motion of the Earth in the Galaxy. KASCADE-Grande published recently [35] an update on the anisotropy study based on

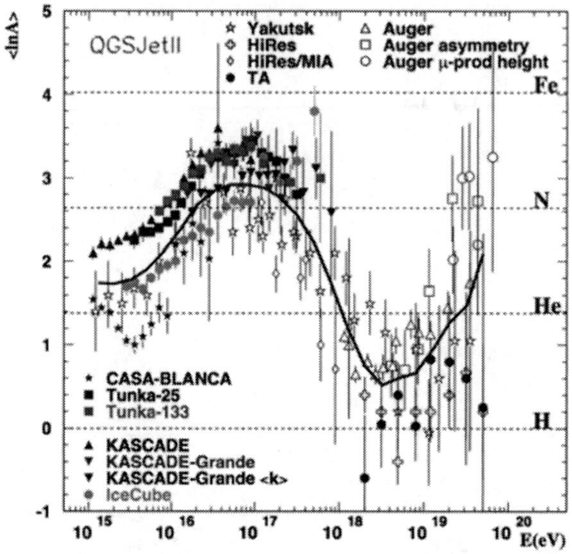

Figure 2. Average logarithmic mass of CRs as a function of energy derived from X_{max} and particle detector measurements using QGSJet-II interaction model. Most of the data are taken from [33] and references therein. The solid line is drawn as a guidance and is obtained by averaging, in each energy bin, the values of $< lnA >$ reported in the figure. The plot is taken from [3].

the East-West method [36]. By investigating the variation of the amplitude as a function of energy it was found that the amplitudes were not significant, however, the phases were in almost all energy bins centred around 250 ± 25 degrees. This is interesting in itself because it points towards the Galactic Center. Moreover, it agrees inside the statistical uncertainties with the results of the anisotropy studies of the Pierre Auger Observatory (PAO) [37] in the range $3\times10^{17} - 10^{18}$ eV.

3 From the ankle to the end of the spectrum

Fig. 1 shows that the features at the Ultra High Energies (UHE), the ankle and the cutoff, have been established beyond doubt. The spectral slopes before and after the ankle have been measured and agree between PAO [38] and Telescope Array (TA) [39]. The positions of the ankle also agree within the quoted errors, and are compatible with the existing model(s). The parameters of the break at the highest energies are known less accurately. There seems to be some discrepancy concerning the shape of the spectrum around the break; however more statistics is needed for a firm conclusion. The position of the break is compatible with the GZK cutoff for protons in TA, but the PAO spectrum fits better the case of a limit in the UHECRs acceleration by the sources.

As far as the mass composition is concerned, the situation is less definite [33]. While PAO sees a change in the composition towards a heavier one at the highest energies, the TA observes no such a trend and is compatible with a pure proton composition. This difference in the data has profound consequences: the Auger data suggest that we see the maximum energy of sources, similarly to what is observed at the knee in the cosmic ray spectrum, while the TA data suggest we observe the GZK effect. Moreover, the different compositions in the GZK- and maximum-energy scenario will affect the level of anisotropies expected to be seen in the data. It should be noticed, however, that a recent analysis of the joint working group on composition of the PAO and TA collaborations, seem to reconcile the two results obtained independently by PAO and TA experiment [40]. In this analysis an adequate comparison was achieved by taking into account that the $< X_{max} >$ published by PAO are corrected for detector effects, whereas those published by TA includes detector effects. From the preliminary comparison the authors conclude that the data of the two observatories are in good agreement. Therefore, the apparent differences seem to be related to the different hadronic model used to compare the data and to the role of detector effects which doesn't allow a direct comparison between the two results.

Despite the major advances in the understanding of UHECRs nature, the current experiments face the limit in the statistics they can accumulate due to the extremely low flux of 1 particle km^{-2} year^{-1} above 5×10^{19} eV. For this reason future space-based observatories, such as JEM-EUSO, that reach much higher and uniform exposures [41], could help in sheding light on the mystery of the most energetic radiation of the Universe.

4 The interconnection between cosmic ray and particle physics

A fundamental ingredient of the CORSIKA simulation is the hadronic interaction model which generates the hadronic cascade at the origin of the electromagnetic one. Since 2009, the Large Hadron Collider (LHC) provides a lot of very precise data which have been used to improve the hadronic models used for air shower simulations, such as EPOS [42] and QGSJetII [43], giving birth to the most updated versions: EPOS-LHC and QGSJetII-04. References [44] and [45] provide, respectively, the most updated review on the measurements at LHC and their relevance for UHECRs and a review on the cosmic ray interaction models.

Remarkably, interaction models employed in air shower simulations provide a reasonable prediction of global observables (multiplicities, p_T-distributions, forward and transverse energy flow, etc.), in some cases even better than typical tunes of HEP models, such as PYTHIA or PHOJET [46]. The cross-section is particularly important for the EAS development and the depth of shower maximum. The proton-proton scattering total cross-section measured by TOTEM [47] at $\sqrt{s} = 7$ TeV allowed to reduce to 20 gr/cm^2 the difference in X_{max} position of the two models, which is comparable to the systematic uncertainties in the measurement by experiments. Detecting UHECRs enables to measure the proton-air inelastic cross section $\sigma^{\mathrm{inel}}_{\mathrm{p-air}}$ at energies that can not be reached with particle acceleerators. The proton-proton cross section $\sigma_{\mathrm{p-p}}$ is subsequently inferred from the proton-air cross section at these energies. UHECRs experiments have been reported results on the proton-air inelastic cross section till $\sqrt{s} \sim 100$ TeV. A recent review on this topic is reported in [48]. The muon number depends on the ratio between particles producing hadronic sub-showers and the total number of particles. LHC data allow constraining the (anti)baryon and strangeness production at mid rapidity as well as the forward production of π^0 in fixed target experiments [49]. Taking into account these data, EPOS-LHC and QGSJetII-04 provide now better results [50] compared to the previous versions (see Fig. 3). Despite improvements obtained using the updated version of the interaction models, shortcomings still exist. As an example PAO estimate the muon number at ground and the muon production depth using measurements of the longitudinal, lateral and temporal disstribution of particles in the EAS. The results, obtained at about 140 TeV center-of-mass energy for proton primaries, are compared to the predictions of LHC-tuned hadronic interaction models for different primary masses. The models exhibit a deficit in the muon content [52].

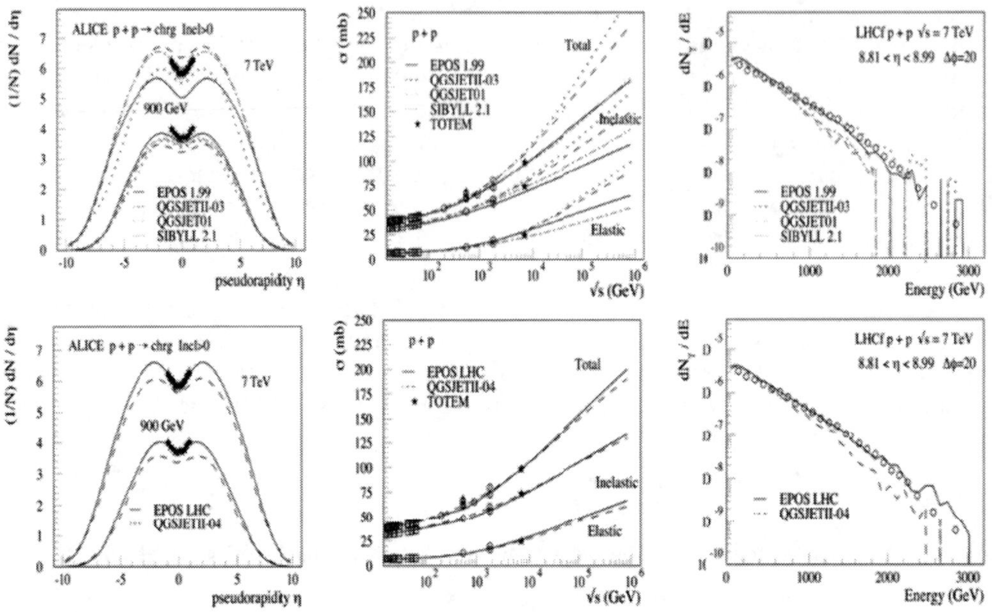

Figure 3. Examples of tuning interaction models to LHC data. Top (bottom) panel shows a comparison between old (new) versions of hadronic interaction models used in CR science and LHC data. Left column refers to pseudorapidities, central one to cross-sections and right one to distributions of particle energy (plot from [51]).

5 Conclusions

A review of the current understanding of cosmic ray data at different energies has been presented. The interpretation of CR data requires the knowledge of the physics of hadronic interactions in atmosphere, but at the same time provides a means to cross-check the validity of the physics principles embedded in the models. Hadronic interaction models do a fairly well job not only in the interpretation of EAS cascades in atmosphere but also of LHC data. However, shortcomings exist. LHC data are extremely helpful in fine tuning the models and give solid bases for the extrapolation at high energies. CR remain the sole mean to test hadronic interactions at energies well beyond those reachable with colliders. CRs and accelerator data provide an excellent mix of information to understand the physics of interactions.

References

[1] Greisen K., Phys. Lett. **16** 148 (1966).
[2] Zatsepin G.T. and Kuz'min V.A., JETP **4** 78 (1966).
[3] Bertaina M., Comptes Rendus Phys. **15/4** 300 (2014).
[4] Kampert K.-H. and Tinyakov P., Comptes Rendus Phys. **15/4** 318 (2014).
[5] Bertaina M., Il Nuovo Cimento **38 C** 29 (2015).
[6] Blasi P., Astron. Astrophys. Rev. **21** 1174 (2013).
[7] Giuliani A. et al. (AGILE Coll.), Astrophys. J. **742** L30 (2011).
[8] Ackermann M. et al. (FERMI-LAT Coll.), Science **339** 807 (2013).

[9] Hillas M., J. Phys. G, Nucl. Part. Phys. **31** R95 (2005).
[10] Gaisser T, Stanev T. and Tilav S., arXiv: 1303.3365v1 (2013).
[11] Peters B., Il Nuovo Cimento **22** 800 (1961).
[12] Berezinsky V., Gazizov A. and Grigorieva S., Phys. Rev. D **74** 043005 (2006).
[13] Bertaina M. et al., J. Phys. Soc. Jpn. Suppl. A **78** 2010 (2009).
[14] Aglietta M. et al. (EAS-TOP Coll.), Astrop. Phys. **19** 329 (2003).
[15] Aglietta M. et al. (EAS-TOP and MACRO Coll.), Astrop. Phys. **21** 223 (2004).
[16] Bartoli B. et al. (ARGO-YBJ Coll.), Phys. Rev. D **85** 092005 (2012).
[17] Yoon Y.S. et al. (CREAM Coll.), Astrophys. J. **728** 122 (2011).
[18] Heck D., Report FZKA 6019 (1998).
[19] Aglietta M. et al. (EAS-TOP Coll.), Astrop. Phys. **10** 1 (1999).
[20] Aglietta M. et al. (EAS-TOP Coll.), Astrop. Phys. **21** 583 (2004).
[21] Antoni T. et al. (KASCADE Coll.), Astrop. Phys. **24** 1 (2005).
[22] Aglietta M. et al. (EAS-TOP and MACRO Coll.), Astrop. Phys. **20** 641 (2004).
[23] Budnev N. et al. (TUNKA Coll.), Astrop. Phys. **50-52** 18 (2013).
[24] Ostapchenko S., Prog. Theor. Phys. Suppl. **193** 204 (2012).
[25] Bartoli B. et al. (ARGO-YBJ Coll.), Phys. Rev. D **92** 092005 (2015).
[26] Apel W.D. et al. (KASCADE-Grande Coll.), Nucl. Instr. Meth. A **620** 202 (2010).
[27] Prosin V.V. et al. (TUNKA-133 Coll.), Nucl. Instr. Meth. A **756** 94 (2014).
[28] Aartsen M.G. et al. (IceTop Coll.), Phys. Rev. D **88** 042004 (2013).
[29] Apel W.D. et al. (KASCADE-Grande Coll.), Astrop. Phys. **36** 183 (2012).
[30] Apel W.D. et al. (KASCADE-Grande Coll.), Phys. Rev. Lett. **107** 171104 (2011).
[31] Apel W.D. et al. (KASCADE-Grande Coll.), Phys. Rev. D **87** 081101(R) (2013).
[32] Apel W.D. et al. (KASCADE-Grande Coll.), Adv. Space Res. **53** 1456 (2014).
[33] Kampert K.-H. and Unger M., Astrop. Phys. **35** 660 (2012).
[34] Abbasi R. et al. (ICECUBE Coll.), Astrophys. J. Lett. **718** L194 (2010).
[35] Chiavassa. et al. (KASCADE-Grande Coll.), Proc. 33rd ICRC, #0093 (2013).
[36] Bonino R. et al., Astrophys. J. **738** 67 (2011).
[37] Abreu P. et al. (Pierre Auger Coll.), Astrop. Phys. **34** 627 (2011).
[38] Abraham J. et al. (Pierre Auger Coll.), Phys. Rev. Lett. **101** 061101 (2008).
[39] Abu-Zayyad T. et al. (Telescope Array Coll.), Astrophys. J. **768** L1 (2013).
[40] Unger M. et al. (PAO and TA Colls.), Proc. 34th ICRC, PoS(ICRC2015) #307 (2015).
[41] Adams Jr. J.H. et al. (JEM-EUSO Coll.), Astrop. Phys. **44** 76 (2013).
[42] Werner K., Liu F.-M. and Pierog T., Phys. Rev. C **74** 044902 (2006).
[43] Ostapchenko S., Phys. Rev. D **74** 014026 (2006).
[44] Ulrich R., These Proceedings (2015).
[45] Ostapchenko S., These Proceedings (2015).
[46] d'Denterria D. et al., Astrop. Phys. **35** 98 (2011).
[47] Antchev G. et al. (TOTEM Coll.), Europhys. Lett. **96** 21002 (2011).
[48] Abbasi R., These Proceedings (2015).
[49] Adriani O. et al. (LHCf Coll.), Phys. Rev. C **89** 065209 (2014).
[50] Pierog T. and Heck D., Proc. 33rd ICRC, #0163 (2013).
[51] Engel R., Proc. 34th ICRC, Invited talk (2015).
[52] Veberic D. (Pierre Auger Coll.), These Proceedings (2015).

Cosmic Ray Interaction Models: an Overview

Sergey Ostapchenko[1,2,a]

[1] *Frankfurt Institute for Advanced Studies (FIAS), 60438 Frankfurt am Main, Germany*
[2] *D.V. Skobeltsyn Institute of Nuclear Physics, Moscow State University, 119992 Moscow, Russia*

Abstract. I review the state-of-the-art concerning the treatment of high energy cosmic ray interactions in the atmosphere, discussing in some detail the underlying physical concepts and the possibilities to constrain the latter by current and future measurements at the Large Hadron Collider. The relation of basic characteristics of hadronic interactions to the properties of nuclear-electromagnetic cascades induced by primary cosmic rays in the atmosphere is addressed.

1 Introduction

Experimental studies of high energy cosmic rays (CR) are traditionally performed using the so-called extensive air shower (EAS) techniques: the properties of the primary CR particles are reconstructed from measured characteristics of nuclear-electro-magnetic cascades induced by their interactions in the atmosphere. This naturally implies the importance of detailed Monte Carlo simulations of EAS development, particularly, of its backbone - the cascade of nuclear interactions of both the primary particles and of the secondary hadrons produced. Thus, the very success of experimental studies depends crucially on the validity of hadronic interaction models used in the analysis.

Typically, one chooses between two main experimental procedures [1]. In the first case, one deals with the information obtained with scintillation detectors positioned at ground. The energy of the primary particle is reconstructed from the measured lateral density of charged particles (mostly, electrons and positrons) while the particle type is inferred from the relative fraction of muons, compared to all charged particles at ground. Alternatively, one may study the longitudinal EAS development by measuring fluorescence light produced by excited air molecules at different heights in the atmosphere, for which purpose dedicated fluorescence telescopes are employed. In the latter case, the primary energy is related to the total amount of fluorescence light emitted. In turn, the particle type may be determined from the measured position of the so-called shower maximum $X_{\rm max}$ - the depth in the atmosphere (in g/cm^2) where the number of ionizing particles reaches its maximal value (which is thus the brightest spot on the shower image).

Not surprisingly, the observables used to determine the primary particle type - the lateral muon density ρ_μ and the EAS maximum position $X_{\rm max}$ - appear to be very sensitive to details of high energy hadronic interactions. More precisely, $X_{\rm max}$ depends strongly on the properties of the primary particle interaction with air nuclei: the inelastic cross section and the forward spectra of secondary hadrons produced. In turn, the EAS muon content is formed in a multi-step cascade process, driven mostly

[a] e-mail: sergei@tf.phys.ntnu.no

by interactions of secondary pions and kaons with air. Hence, ρ_μ depends strongly on the properties of pion-air collisions, most importantly, on the multiplicity and spectral shape for charged hadrons produced.

The discussion above helps one to understand the basic requirements to cosmic ray interaction models: a treatment of various hadron-nucleus and nucleus-nucleus interactions over a wide energy range, including the respective inelastic cross sections and particle production, with a special importance of forward particle spectra. By far, the most crucial requirement is the one for a significant predictive power, which stems from the fact that there is a little possibility for model tuning based on air shower data. Rather, cosmic ray data allow one to check the overall consistency of a model approach in question.

Among the models most frequently used in cosmic ray studies have been QGSJET [2] and SIBYLL [3], as well as more recent ones: QGSJET-II [4, 5] and EPOS [6, 7]. Overall, CR interaction models behaved rather well when confronted with the first set of data from Run 1 of LHC [8]. QGSJET-II and EPOS had been further retuned with LHC data, and an update of the SIBYLL model is in progress [9].

2 Underlying physical approaches

By construction, CR interaction models are designed to cope with both nonperturbative "soft" and hard processes. QGSJET(-II) and EPOS employ the so-called "semihard Pomeron" approach [10, 11], treating both contributions within the Reggeon Field Theory framework [12]. Basically, one introduces a "soft-hard" separation scale Q_0^2, describing parton evolution in the perturbative range of high virtuality $|q^2| > Q_0^2$ by means of DGLAP equations. In turn, for "soft" ($|q^2| < Q_0^2$) parton cascades (and for a "soft pre-evolution" for hard cascades) a phenomenological soft Pomeron amplitude is employed. The total amplitude of a "general Pomeron" which contains both kinds of processes is used as a basic building block to develop a Pomeron calculus and to derive all partial cross sections and probabilities for particular final states [4, 6, 11].

On the other hand, SIBYLL is similar to most of the Monte Carlo generators used in the collider physics, being based on the "minijet" approach [13]. The latter employs an eikonalization of the inclusive cross section $\sigma_{pp}^{\rm jet}(s, p_{\rm t}^{\rm cut})$ for the production of parton jets of transverse momenta $p_{\rm t} > p_{\rm t}^{\rm cut}$. Partial probabilities for having a given number of jet pairs for some impact parameter b are expressed via the eikonal function defined by a product of $\sigma_{pp}^{\rm jet}(s, p_{\rm t}^{\rm cut})$ and a parton overlap function $A(b)$. This is further supplemented by a simplified treatment of soft processes.

There are substantial similarities between the results of the two approaches concerning the description of inclusive hadron spectra in the central rapidity region. This is, first of all, due to a calibration of the models to similar sets of accelerator data. Moreover, in the very high energy limit, the bulk of central particle production comes from a hadronization of (mini-)jets, thus being driven by the input parton momentum distribution functions and DGLAP evolution.

However, the predictions of the two approaches differ significantly in the fragmentation region. In the "semihard Pomeron" scheme, multiple scattering affects both central and forward production of hadrons, creating thus long-range rapidity correlations. This gives rise to a noticeable violation of Feynman scaling in the fragmentation region and of the limiting fragmentation. Also the plateau-like behavior of the pseudorapidity η spectra of secondary hadrons extends to large values of η. In contrast, in minijet-like models, multiple scattering affects mostly central hadron production, being almost decoupled from the fragmentation region. As a consequence, the above-discussed effects are practically absent in that case.

Actually, the minijet-like scheme is somewhat disfavored by recent combined measurements of η-density of charged hadrons $dn/d\eta$ by the CMS and TOTEM experiments [14]. Indeed, the ob-

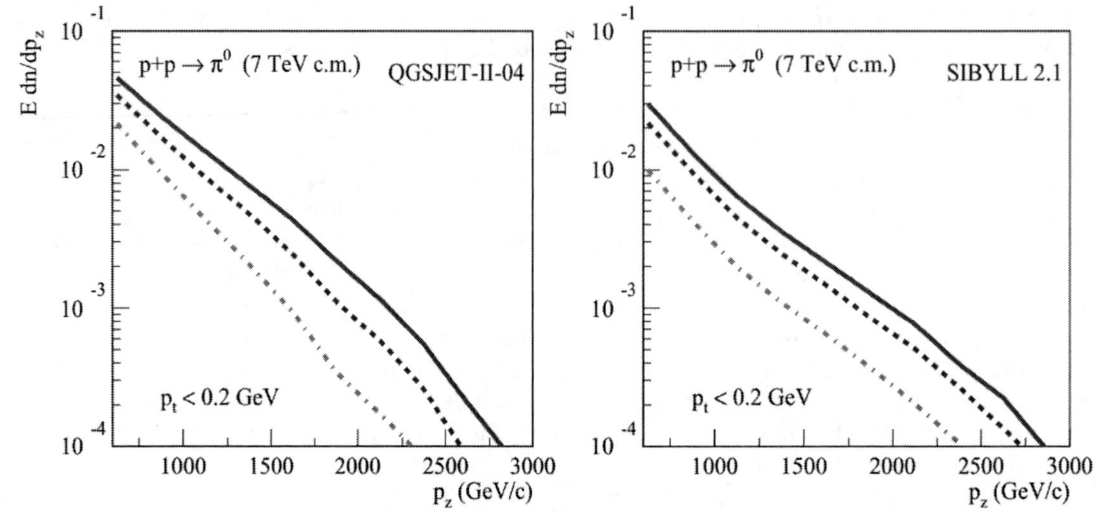

Figure 1. Longitudinal momentum spectra of neutral pions ($p_t < 0.2$ GeV) in proton-proton collisions at $\sqrt{s} = 7$ TeV, calculated using QGSJET-II-04 (left) or SIBYLL 2.1 (right) for different "centrality" triggers: ≥ 1, ≥ 5, and ≥ 10 charged hadrons with $p_t > 0.5$ GeV at $|\eta| < 2.5$ - solid, dashed, and dot-dashed lines respectively.

served η-dependence of $dn/d\eta$ in inelastic and nondiffractive event samples has been correctly reproduced by EPOS and QGSJET-II while other Monte Carlo generators in the study, e.g. various tunes of PYTHIA[15], fell short of describing the high yield of secondary hadrons in the TOTEM detector. Another important piece of evidence comes from studies of forward production of neutral pions by the LHCf experiment [16]: due to their characteristic scaling-like behavior, π^0 spectra of the SIBYLL and PYTHIA models appear to be too hard compared to the experimental observations.

However, a more reliable discrimination between the two basic approaches requires to study correlations between central and forward particle production, which may be realized in combined measurements by e.g. ATLAS and LHCf detectors. Indeed, triggering different levels of activity in ATLAS, for example, measuring at least 1, 5, or 10 charged hadrons of $p_t > 0.5$ GeV at $|\eta| < 2.5$, one could be able to discriminate between predictions of models of the two discussed types, as illustrated in Fig. 1 for a particular case of QGSJET-II-04 [5] and SIBYLL 2.1 [3] models. Similar studies may be performed with the CMS and TOTEM detectors.

3 Uncertainties of model predictions for $X_{\max}$

By far, the most suitable EAS parameter for studying primary CR composition is the shower maximum depth $X_{\max}$. Apart from the possibility to measure it reliably by modern air fluorescence detectors, the uncertainties of the respective model predictions have been greatly reduced with the start of the Large Hadron Collider - primarily, thanks to the precise measurements of the total and elastic proton-proton cross sections by the TOTEM and ATLAS experiments [17, 18].

Yet another potential source of uncertainty for $X_{\max}$ is related to its sensitivity to the rate of inelastic diffraction in proton-proton and proton-nucleus collisions. Meanwhile, there exists presently a serious tension between the results of diffraction measurements at LHC by the TOTEM and CMS experiments, as discussed in [19, 20]. The potential impact of the latter on the calculated shower

 DOI: 10.1051/epjconf/201612004003

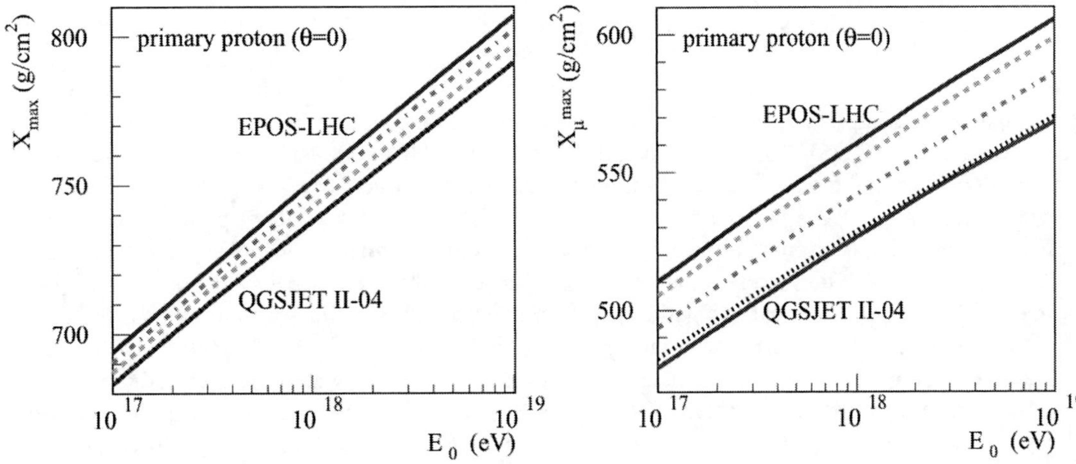

Figure 2. Primary energy dependence of the average shower maximum depth (left) and of the maximum of the muon production profile (right) for proton-initiated EAS, calculated using the EPOS-LHC or QGSJET-II-04 models (respectively blue and red solid lines), or applying mixed model descriptions, as explained in the text (dashed, dot-dashed, and dotted lines).

maximum position has been studied in [19] in the framework of the QGSJET-II-04 model. In more detail, one considered two alternative model parameter tunes which allowed one to reproduce either the TOTEM or CMS results on single and double diffraction cross sections, while leaving essentially unchanged the model results for $\sigma_{pp}^{\mathrm{tot/el}}$ and for central particle production. Applying these alternative model tunes to EAS simulations, one observed that the difference between the respective $X_{\max}$ predictions does not exceed 10 g/cm^2, being thus comparable to the characteristic uncertainty of shower maximum measurements by modern EAS experiments.

However, present differences between various calculations of $X_{\max}$ are substantially bigger, e.g. reaching 20 g/cm^2 in the ultra high energy limit in the particular case of the QGSJET-II-04 [5] and EPOS-LHC [7] models, the corresponding $X_{\max}$ predictions being shown by red and blue solid lines respectively in Fig. 2 (left). This is rather surprising, taking the fact that both models have been recently updated using LHC data. Thus, the question arises if the analysis of Ref. [19] was not general enough or the position of the shower maximum depends on some other characteristics of hadronic interactions, not well constrained by the present LHC data.

To reveal the interaction features which are responsible for the above-discussed differences in $X_{\max}$ predictions, we are going to use a "cocktail" model approach, using QGSJET-II-04 results for some selected interactions of hadrons in the atmospheric cascades or for some particular features of the primary interaction, while treating the rest with the EPOS-LHC model. As the first step, we apply QGSJET-II-04 to determine the position of the primary particle interaction in the atmosphere and to describe the production of secondary nucleons in this interaction; all other characteristics of the first proton-air collision and all the subsequent interactions of secondary hadrons in the cascade are treated using EPOS-LHC. This way we check the sensitivity of the calculated shower maximum depth to the model differences concerning the proton-air cross section and the predicted spectra of leading nucleons (the latter thus comprise the effects of the inelastic diffraction). The obtained $X_{\max}$ shown by the green dot-dashed line in Fig. 2 (left) differs from the original EPOS-LHC results by not more than 5 g/cm^2, which is well within the uncertainty range obtained in [19].

Further, we apply QGSJET-II-04 to describe all the characteristics of the primary interaction while treating the rest of the hadron cascade using EPOS-LHC. The obtained $X_{\max}$ shown by the yellow dashed line in Fig. 2 (left) is shifted further towards the QGSJET-II-04 results by up to 5 g/cm^2. This additional shift is explained by somewhat softer spectra of secondary mesons, most importantly, of secondary pions in QGSJET-II-04, compared to the EPOS-LHC model. Here we actually observe an important change in the physics of the hadronic cascade in the atmosphere. At lower energies, there is a very pronounced "leading nucleon" effect, i.e. the most energetic secondary particles in proton-air collisions are typically protons and neutrons (produced either directly or via decays of hyperons and resonances). On the other hand, in the very high energy limit the energy loss of leading nucleons is noticeably higher and the most energetic secondary hadron may well be a pion or a kaon.

What is the reason for the remaining difference with the QGSJET-II-04 results? One potential explanation could have been a difference between the two models concerning the predicted pion-air cross section or/and concerning the multiplicity and spectral shape for secondary mesons. However, the real reason is a different and a more interesting one. If we apply QGSJET-II-04 to describe both the primary interaction and the production of nucleons and antinucleons in all the secondary hadron-air collisions, while treating the rest with EPOS-LHC, the obtained $X_{\max}$ shown by the black dotted line coincides with the QGSJET-II-04 results. Thus, it is the more copious production of baryon-antibaryon pairs and harder (anti-)baryon spectra in EPOS-LHC[1] [21] that gives rise to the largest differences in $X_{\max}$ predictions of the two models in the very high energy limit.

4 Constraining model predictions for $X_{\max}$ by cosmic ray data

As demonstrated in Section 3, a large part of the model uncertainty for $X_{\max}$ is related to the treatment of proton-air interactions at very high energies, in particular, to the predicted diffraction rate and the forward spectra of secondary mesons. There are serious hopes that this will be strongly constrained by further studies of proton-proton and proton-nucleus collisions at LHC. Of particular importance are the measurements of very forward production of neutral pions by the LHCf experiment [16], which already demonstrated that the spectra predicted by EPOS-LHC are somewhat harder than the observed ones. However, another serious uncertainty is related to the production of baryon-antibaryon pairs in pion-air interactions. While more precise measurements of (anti-)baryon spectra at fixed target energies may be performed by the NA61 experiment [23], the prospects for extending such studies to much higher energies are rather dim.

Nevertheless, the discussed mechanism can be constrained indirectly by characteristics of EAS induced by very high energy cosmic rays. Recently, Pierre Auger collaboration reported measurements of the so-called maximum muon production depth in air showers, $X^{\mu}_{\max}$ - the depth in the atmosphere (in g/cm^2) where the rate of muon production via decays of pions and kaons reaches its maximal value [24]. One observed a strong contradiction between the results of EAS simulations with EPOS-LHC and the experimental data: the measured maximum of the muon production profile appeared to be reached much higher in the atmosphere than predicted by the model, even when the heaviest possible primary cosmic ray nuclei were considered.

There are both similarities and differences concerning the relation of $X_{\max}$ and $X^{\mu}_{\max}$ to the properties of hadron-air interactions. Obviously, both characteristics are sensitive to the position X_0 of the primary particle interaction in the atmosphere, which depends on the respective inelastic cross section: fluctuations of X_0 shift the whole cascade upwards and downwards in the atmosphere and thus do so

[1]Production of baryon-antibaryon pairs in EPOS-LHC has been tuned to reproduce fixed-target data on pion-proton collisions from [22]. There are, however, indications on serious particle misidentification effects in that experiment: e.g., large differences between the reported forward spectra of secondary protons and antiprotons in $\pi^+ + p \to p$ and $\pi^- + p \to \bar{p}$ reactions.

for X_{max} and X^{μ}_{max} for a particular shower. However, in contrast to X_{max}, maximum muon production depth is much less sensitive to the production of secondaries in the primary interaction. EAS muon content rather depends on the multi-step hadronic cascade in which the number of pions and kaons increases in an avalanche way until the probability of their decays becomes comparable to the one for interactions. This happens when energies for most of the hadrons approach the so-called critical energy: around 100 GeV for charged pions and an order of magnitude higher for charged kaons [25]. The position of the maximum of the muon production profile is actually close to this turning point.

From the above-discussion, it is clear that X^{μ}_{max} is very sensitive to the forward spectral shape of secondary mesons in pion-air collisions: producing in each cascade step a meson of a slightly higher energy would mean that a larger number of cascade branchings is required for reaching the critical energy, with the result that the maximum of the muon production profile will be observed deeper in the atmosphere. However, there exists another potential mechanism which may influence model predictions for X^{μ}_{max}. If the production of baryon-antibaryon pairs in pion-air interactions is copious enough, it will also influence the obtained X^{μ}_{max}. Indeed, (anti-)nucleons do not decay, hence, they continue to interact even when their energies fall below 100 GeV, producing additional generations of secondary hadrons in the cascade. Muons emerging from decays of secondary pions and kaons created in interactions of such low energy (anti-)nucleons contribute to the elongation of the muon production profile and give rise to larger values of X^{μ}_{max}. It is noteworthy that the respective effect is noticeable if (and only if) the yield of baryon-antibaryon pairs in pion-air interactions is comparable to the one of secondary pions.

To reveal the reasons for the large difference of the values of X^{μ}_{max}, predicted by QGSJET-II-04 and EPOS-LHC, shown respectively by red and blue solid lines in Fig. 2 (right), we are going to use the same "cocktail" model approach as in the previous Section. First, we apply QGSJET-II-04 to describe all the characteristics of the primary interaction while treating the rest of the hadron cascade using EPOS-LHC [yellow dashed line in Fig. 2 (right)]. As expected, the obtained X^{μ}_{max} deviates only slightly from the pure EPOS-LHC calculation (by less than 7 g/cm^2): the bulk of the difference between QGSJET-II-04 and EPOS-LHC predictions for the maximum of the muon production profile is due to secondary (mainly pion-air) interactions in the cascade. Further, we apply QGSJET-II-04 to describe both the primary interaction and the production of nucleons and antinucleons in all the secondary hadron-air collisions, while treating the rest with EPOS-LHC. The obtained X^{μ}_{max} shown in Fig. 2 (right) by the green dot-dashed line appears to be right in the middle between the QGSJET-II-04 and EPOS-LHC results. Finally, we have to check if the remaining half of the difference between the two models is indeed due to harder spectra of secondary mesons in EPOS-LHC for pion-air and kaon-air collisions. To this end, we use QGSJET-II-04 to describe the production of mesons for $x_F > 0.1$ in all the secondary hadron-air interactions, apart from treating the primary interaction and the production of nucleons and antinucleons in secondary collisions. It is easy to see that the result shown by the black dotted line is indeed very close to the pure QGSJET-II-04 calculation.

5 Conclusions

We have demonstrated that both the maximum of the muon production profile and the average shower maximum depth are very sensitive to properties of very high energy pion-air and kaon-air interactions. In particular, a large part of the difference between the predictions of the QGSJET-II-04 and EPOS-LHC models for the two EAS characteristics is due to the copious production of baryon-antibaryon pairs in pion-air collisions in the latter model. As a matter of the fact, the mechanism is thus disfavored by the measurements of X^{μ}_{max} by the Pierre Auger experiment, which puts serious constraints on potential variations of model predictions for the shower maximum depth.

Yet a large part of the model dependence for calculated X_{max} is related to the properties of primary proton-air interactions. These will be constrained by further studies of proton-proton and proton-nucleus collisions at the Large Hadron Collider, in particular, by measurements of forward particle production. Especially promising for testing the underlying physical mechanisms of the present interaction models seem to be combined studies by central and forward-looking particle detectors.

Acknowledgments

The author acknowledges the support by Deutsche Forschungsgemeinschaft (project OS 481/1).

References

[1] M. Nagano and A. A. Watson, Rev. Mod. Phys. **72**, 689 (2000).
[2] N. N. Kalmykov and S. S. Ostapchenko, Phys. Atom. Nucl. **56**, 346 (1993); N. N. Kalmykov, S. S. Ostapchenko and A. I. Pavlov, Nucl. Phys. Proc. Suppl. **52B**, 17 (1997).
[3] R.S. Fletcher, T. K. Gaisser, P. Lipari, T. Stanev, Phys. Rev. D **50**, 5710 (1994); E.-J. Ahn, R. Engel, T. K. Gaisser, P. Lipari, T. Stanev, Phys. Rev. D **80**, 094003 (2009).
[4] S. Ostapchenko, Nucl. Phys. Proc. Suppl. **151**, 143 (2006); Phys. Rev. D **74**, 014026 (2006).
[5] S. Ostapchenko, Phys. Rev. D **83**, 014018 (2011); EPJ Web Conf. **52**, 02001 (2013).
[6] K. Werner, F.-M. Liu and T. Pierog, Phys. Rev. C **74**, 044902 (2006).
[7] T. Pierog, Iu. Karpenko , J. M. Katzy, E. Yatsenko, K. Werner, Phys. Rev. C **92**, 034906 (2015).
[8] D. d'Enterria, R. Engel, T. Pierog, S. Ostapchenko, K. Werner, Astropart. Phys. **35**, 98 (2011).
[9] F. Riehn, R. Engel, A. Fedynitch, T. K. Gaisser, T. Stanev, arXiv:1510.00568 [hep-ph].
[10] H. J. Drescher, M. Hladik, S. Ostapchenko, K. Werner, J. Phys. G **25**, L91 (1999); S. Ostapchenko, H. J. Drescher, F. M. Liu, T. Pierog, K. Werner, *ibid.*, **28**, 2597 (2002).
[11] H. J. Drescher, M. Hladik, S. Ostapchenko, T. Pierog, K. Werner, Phys. Rep. **333**, 329 (2001).
[12] V. N. Gribov, Sov. Phys. JETP **26**, 414 (1968).
[13] L. Durand and P. Hong, Phys. Rev. Lett. **58**, 303 (1987); H. N. Wang, Phys. Rep. **280**, 287 (1997).
[14] S. Chatrchyan *et al.* (CMS and TOTEM Collaborations), Eur. Phys. J. C **74**, 3053 (2014); G. Antchev *et al.* (TOTEM Collaboration), *ibid.* **75**, 126 (2015).
[15] T. Sjostrand *et al.*, JHEP **0605**, 026 (2006); Comput. Phys. Commun. **178**, 852 (2008).
[16] O. Adriani *et al.* (LHCf Collaboration), Phys. Rev. D **86**, 092001 (2012); arXiv:1507.08764 [hep-ex].
[17] G. Antchev *et al.* (TOTEM Collaboration), Europhys. Lett. **101**, 21002 (2013); *ibid.*, 21003 (2013); *ibid.*, 21004 (2013); Phys. Rev. Lett. **111**, 012001 (2013).
[18] G. Aad *et al.* (ATLAS Collaboration), Nucl. Phys. B **889**, 486 (2014).
[19] S. Ostapchenko, Phys. Rev. D **89**, 074009 (2014).
[20] V. A. Khoze, A. D. Martin and M. G. Ryskin, Eur. Phys. J. C **74**, 2756 (2014).
[21] T. Pierog and K. Werner, Phys. Rev. Lett. **101**, 171101 (2008).
[22] D. S. Barton *et al.*, Phys. Rev. D **27**, 2580 (1983).
[23] N. Abgrall *et al.* (NA61 Collaboration), JINST **9**, P06005 (2014).
[24] A. Aab *et al.* (Pierre Auger Collaboration), Phys. Rev. D **90**, 012012 (2014); *ibid.*, **90**, 039904 (2014); *ibid.*, **92**, 019903 (2015).
[25] T. K. Gaisser, *Cosmic Rays and Particle Physics* (Cambridge University Press, Cambridge, England 1990).

EPJ Web of Conferences **120**, 04004 (2016) DOI: 10.1051/epjconf/201612004004
ISMD 2015

Measurement of the muon content in air showers at the Pierre Auger Observatory

Darko Veberič[1,a] for the Pierre Auger Collaboration[2,3]

[1] *Institute of Nuclear Physics, Karlsruhe Institute of Technology (KIT), Karlsruhe, Germany*
[2] *Observatorio Pierre Auger, Av. San Martín Norte 304, 5613 Malargüe, Argentina*
[3] *Full author list: http://www.auger.org/archive/authors_2015_10.html*

Abstract. The muon content of extensive air showers produced by ultra-high energy cosmic rays is an observable sensitive to the composition of primary particles and to the properties of hadronic interactions governing the evolution of air-shower cascades. We present different methods for estimation of the number of muons at the ground and the muon production depth. These methods use measurements of the longitudinal, lateral, and temporal distribution of particles in air showers recorded by the detectors of the Pierre Auger Observatory. The results, obtained at about 140 TeV center-of-mass energy for proton primaries, are compared to the predictions of LHC-tuned hadronic-interaction models used in simulations with different primary masses. The models exhibit a deficit in the predicted muon content. The combination of these results with other independent mass composition analyses, such as those involving the depth of shower maximum observable $X_{\max}$, provide additional constraints on hadronic-interaction models for energies beyond the reach of the LHC.

1 Introduction

Within a scheme of the generalized Heitler model of hadronic air showers, muons are produced by decays of charged mesons [1]. In the evolution of the cascade, energies of charged pions are reduced to a few dozens of GeV [2], at which point the decay length becomes shorter than the interaction length, and, therefore, the decay of pions into muons starts to dominate. Muon production in air showers depends on intricate details of the hadronic interactions; nevertheless, in the simple Heitler scheme, the number of produced muons N_μ approximately follows a power law

$$N_\mu = A\left(\frac{E/A}{\epsilon_c}\right)^\beta, \tag{1}$$

where E and A are the primary cosmic-ray energy and mass, respectively, ϵ_c is the critical energy at which charged pions decay, and the value for β, which is approximately 0.9 for a fixed value of A (for fixed composition as well), is obtained from the multiplicity and fraction of charged pions.

Measurements of the muon content of extensive air showers produced by ultra-high energy cosmic rays thus give excellent insight into the physics of hadronic interactions and multi-particle productions

[a] e-mail: darko.veberic@kit.edu

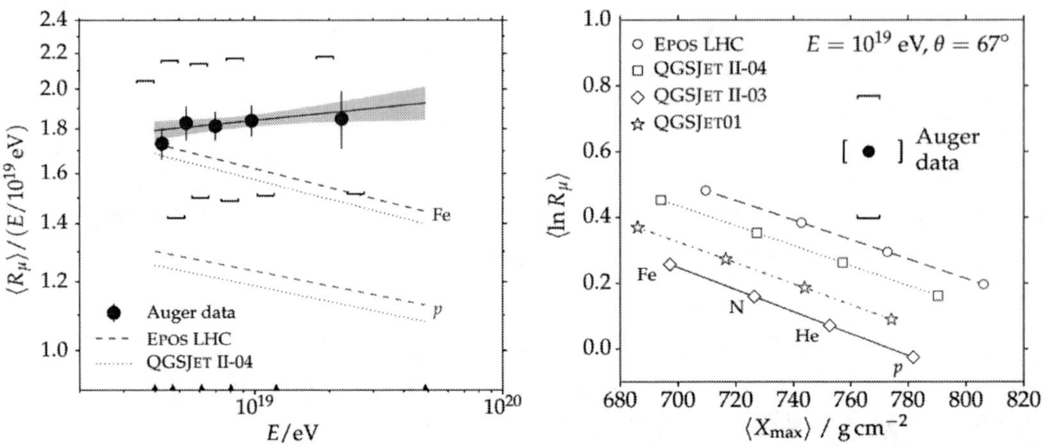

Figure 1. *Left:* Energy-normalized relative muon number $\langle R_\mu \rangle/(E/10^{19}\,\text{eV})$ for various interaction models and two primaries, p and Fe, as obtained from inclined events [5]. Note that the Auger data (black points) indicate a heavy composition and exhibit a different slope (black line) than the model predictions (dashed and dotted lines). *Right:* $\langle \ln R_\mu \rangle$ vs. $\langle X_{\text{max}} \rangle$ for various interaction models (lines) and primary masses (markers) and the Auger data (black point with systematic-uncertainty brackets) from events with energies around $E = 10^{19}$ eV [5].

that take place in such high-energy environments. When the primary energy and/or mass are obtained from independent observables, muon measurements can be used for direct tests of existing hadronic-interaction models and as such, these tests deliver the best constraints on model parameters. Furthermore, the dependence of muon numbers on the primary mass can be used as a sensitive estimator of cosmic-ray composition.

At the Pierre Auger Observatory we are performing extensive muon studies, and the planned upgrade of the detectors [3] is primarily aimed at further increasing the separability of the muonic and electromagnetic components of the showers. The two main parts of the Observatory [4] are the Surface Detector (SD), consisting of a ~3000 km^2 array of water-Cherenkov stations with an almost 100% duty cycle, and the Fluorescence Detector (FD) which is comprised of 27 telescopes operating on a ~15% duty cycle.

Due to the limited space available in the following Sections, we only briefly outline the main features of muon studies based on data from the Pierre Auger Observatory.

2 Hybrid observations of inclined events

Since the electromagnetic component of a shower is absorbed faster in the atmosphere than the muonic part, highly inclined showers deliver an almost pure muon "beam", which is suitable for direct studies. Here we use hybrid events, which were recorded by both the SD and FD, and with zenith angles in the range $62° < \theta < 80°$. While the whole procedure is more complicated [5] and involves construction of an unbiased estimator, it can be schematically described as a method where, for each event, the muon density at the ground ρ_μ^{rec}, which is reconstructed from measured SD signals, is separated into a footprint shape ρ_μ^{map} and a relative muon content R_μ as

$$\rho_\mu^{\text{rec}}(\vec{r}; \theta, \phi) = R_\mu \, \rho_\mu^{\text{map}}(\vec{r}; \theta, \phi) \tag{2}$$

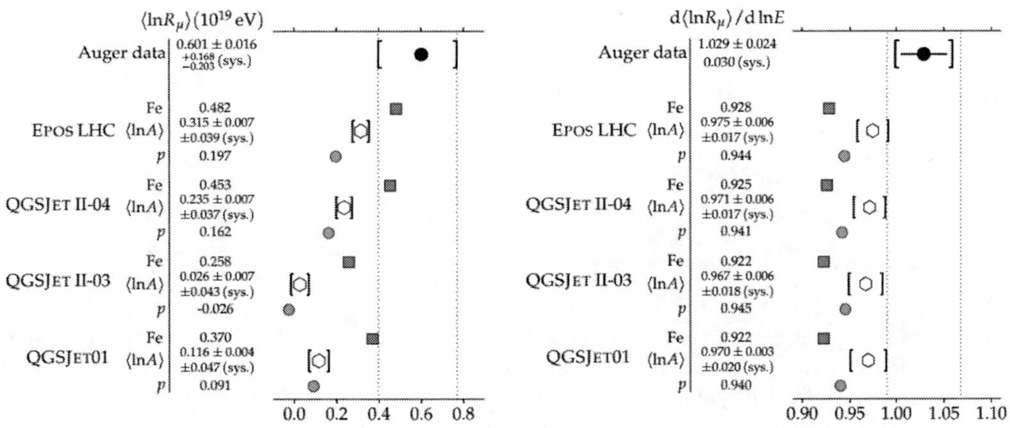

Figure 2. *Left:* Comparison of the mean logarithmic muon content $\langle R_\mu \rangle$ at $E = 10^{19}$ eV obtained from the Auger data (black point) and various model predictions for pure proton (red circles) and iron (blue squares) primary simulations and a mixed composition (empty hexagons) [5]. *Right:* Similar comparison for the logarithmic gain $\mathrm{d}\langle \ln R_\mu \rangle/\mathrm{d}\ln E$ [5].

where the template ρ_μ^{map} is derived from simulations of protons with energy $E = 10^{19}$ eV and with one selected hadronic-interaction model. The template takes into account most of the asymmetric and elongated shape of the footprint at the ground and depends only weakly on the cosmic-ray energy and mass.

As illustrated in Eq. (1), we expect a power-law dependence of the mean muon content on energy, $\langle R_\mu \rangle = a(E/10^{19}\,\mathrm{eV})^b$. A fit to 174 high-quality events gives $a = 1.84 \pm 0.03 \pm 0.32$(syst.) and $b = 1.03 \pm 0.02 \pm 0.03$(syst.). In Fig. 1-left, the energy-normalized $\langle R_\mu \rangle/(E/10^{19}\,\mathrm{eV})$ is shown for simulations with different primary masses and interaction models, and we can see that the proton and iron showers are well separated. While the large systematic uncertainty (square brackets) limits its power as a mass-sensitive estimator, R_μ still delivers valuable insights due to the discrepancy between its absolute scale in the data and the simulations. The diverging slopes between the data and simulated (pure) primaries also suggest a transition from lighter to heavier composition. The discrepancies with hadronic-interaction models become even more obvious when we combine muon results with those of our standard mass-sensitive method [6] involving the depth of shower maximum X_{max}. In Fig. 1-right, simulation results at $E = 10^{19}$ eV are shown in a $\langle \ln R_\mu \rangle$ vs. $\langle X_{\mathrm{max}} \rangle$ space together with the data point which lies far away from all model predictions for any primary mass.

The value obtained for $\langle \ln R_\mu \rangle$ at $E = 10^{19}$ eV and the logarithmic slope $\mathrm{d}\langle \ln R_\mu \rangle/\mathrm{d}\ln E$ can be compared to the corresponding results from simulations. In Fig. 2, the results from Auger data (black circles) are compared to simulations performed with various hadronic-interaction models for samples of purely proton (red circles) and purely iron (blue squares) primaries and a mixed composition sample where $\langle \ln A \rangle$ matches that obtained from conversions of $\langle X_{\mathrm{max}} \rangle$ (hexagons). In all cases, the models fail to match the measurements and would require substantial increases of muon production. Furthermore, the large value of the measured slope favors a changing composition.

3 Muon production depth

Under certain conditions (large zenith angles, large distances to the shower core), we can assume that the signal recorded in the SD stations is produced mostly by traversing muons. Through simple

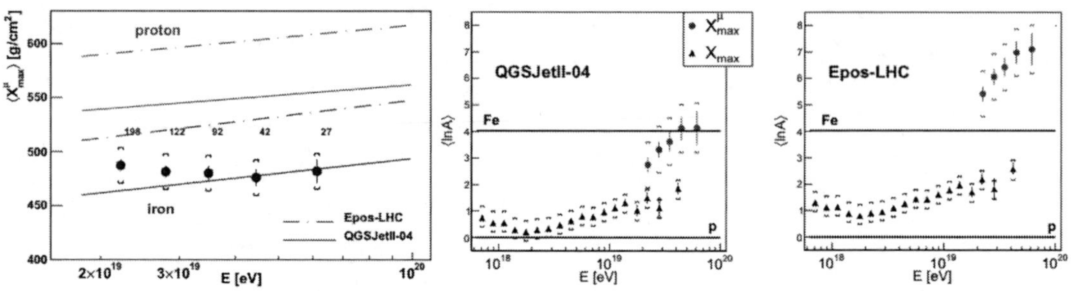

Figure 3. *Left:* $\langle X^{\mu}_{\max}\rangle$ obtained from muon production depth distributions using Auger data (black points) and simulations (lines) for proton and iron primaries [7]. *Center and right:* $\langle X^{\mu}_{\max}\rangle$ converted into $\langle \ln A\rangle$ values using the two interaction models (red points) shown together with $\langle \ln A\rangle$ values obtained from conversions of $\langle X_{\max}\rangle$ measurements (black triangles) [7].

geometrical considerations, the muon production distance z can be reconstructed from the time delay t_{g} between the passage of the shower front and the occurrence of the muon signal as

$$z \simeq \frac{1}{2}\left(\frac{r^2}{ct_{\mathrm{g}}} - ct_{\mathrm{g}}\right) + \Delta - \langle z_\pi\rangle \tag{3}$$

where r is the distance of the detector to the shower axis, Δ is the perpendicular distance of the SD station to the position of the shower plane at the time of muon occurrence, and $\langle z_\pi\rangle$ is the average distance parent mesons travel before decaying (see [7] for details). For a shower with a zenith angle θ, the production distance z is then converted to the equivalent atmospheric depth with

$$X^{\mu}(z) = \int_z^{\infty} \rho(z'\cos\theta)\,\mathrm{d}z' = \frac{1}{\cos\theta}\int_{z\cos\theta}^{\infty}\rho(h)\,\mathrm{d}h \tag{4}$$

where $\rho(h)$ is the atmospheric density at height h. In this way, signals detected by SD stations are converted into a muon production depth distribution for each event. The maximum, $X^{\mu}_{\max}$, of a muon production depth distribution is then obtained with a fit to a Gaisser-Hillas function.

In Fig. 3-left, the mean of the obtained $X^{\mu}_{\max}$ values from 481 events is shown in several energy bins, together with simulation predictions from two hadronic-interaction models and two primaries [7]. Again, the data indicates a change in composition as the energy increases and shows a flatter trend than for pure proton or for pure iron. The $\langle X^{\mu}_{\max}\rangle$ values can be converted to a mean logarithmic mass $\langle \ln A\rangle$ for each hadronic-interaction model and the rest of Fig. 3 shows this conversion for the two models together with the standard conversions of $\langle X_{\max}\rangle$. For both models, results indicate a transition to heavier primaries, but the conversion with the Epos-LHC model predicts implausibly heavy elements as primary cosmic rays.

4 Top-down reconstruction of hybrid events

In contrast to the inclined-shower case (as described in Sect. 2), the estimation of the muon content is more complicated for events with zenith angles $\theta < 60°$ due to the admixture of the electromagnetic component. Nevertheless, for 411 high-quality hybrid events in the energy range $18.8 < \lg(E/\mathrm{eV}) < 19.2$, a so-called "top-down" procedure can be performed [8] where the longitudinal profile, as measured with the FD, is first reproduced by running many simulations and finding

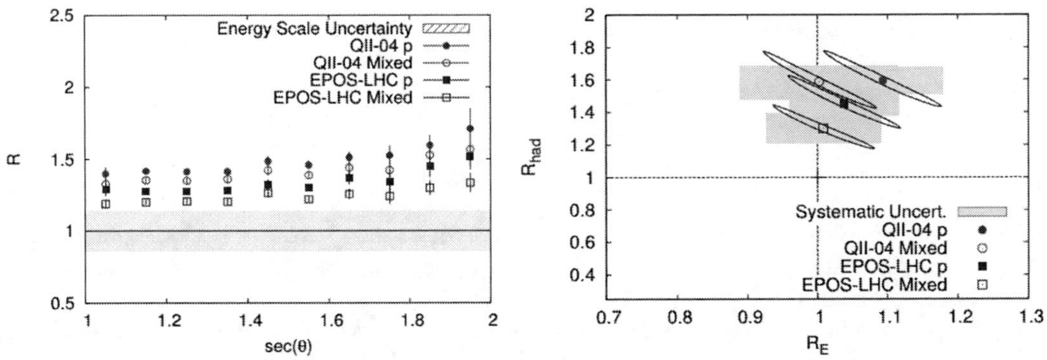

Figure 4. *Left:* The average ratio R of the SD signal at 1000 m in Auger data and the simulations that match the longitudinal profile measured by the FD [8]. Note the increase of R with the increasing zenith angle θ. *Right:* The 1σ contours of statistical uncertainty for fits of $R_{\rm had}$ and R_E rescalings together with gray bands for systematic uncertainties [8]. For both plots, shower simulations were performed with several hadronic-interaction models, for pure proton and mixed compositions.

the best match. Such simulated events thus have an $X_{\rm max}$ and energy compatible with the measured counterpart. In Fig. 4-left, the ratio of measured and simulated ground signals at 1000 m from the core, $R = S(1000)/S_{\rm MC}(1000)$, is shown as a function of $\sec\theta$. The increase of this ratio towards the large zenith angles, where showers have less of the electromagnetic component, may indicate that the measured data at the ground contain more muons than in simulations. Independently of the models and simulated composition, the ratio R is not compatible with 1 within the FD energy-scale uncertainty and actually increases with zenith angle where the muon component of the shower starts to dominate. This indicates that, indeed, it is the muon production that is not reproduced well by the interaction models.

To identify the main contribution to the ground-signal discrepancy, the different components of the shower are, in a second step, self-consistently modified in simulations, allowing rescaling of the hadronic interactions $R_{\rm had}$ and, independently, energy rescaling R_E. Note that $R_{\rm had}$ rescales not only the nominal muon component at the ground but also a part of the electromagnetic component originating from decays of muons and the part of the electromagnetic component from hadronic jets produced close to the detectors. In Fig. 4-right, the results of this rescaling analysis show that in order to reproduce the measured data, almost no energy rescaling R_E is needed and that most of the discrepancy can be explained by muon (or "hadronic") deficiency in the models. Note that *mixed* in both plots again refers to a mixed composition which reproduces the $\langle X_{\rm max}\rangle$ measurement at this energy.

5 Temporal structure of signals

While all the methods mentioned above were using only total integrated signal and ignored the detailed temporal structure of the SD signal at the ground, application of different filtering techniques on recorded traces can produce a certain degree of separation between the electromagnetic and muonic contributions [9].

Estimating the electromagnetic contribution by a smoothing method and a multi-variate method involving many time-structure related observables shows in either case that relative to the proton simulations, the muon proportion of the signal is up to a third larger in the measured data. Nevertheless, these methods are less powerful due to larger sensitivity to energy-scale systematics and are biased by the aging effects observed in SD response to shower particles from the data that has been gathered in nearly 12 years of the Observatory's operation.

6 Summary

In these proceedings, we only briefly described efforts within the Pierre Auger Collaboration related to the analysis of muon production in extensive air showers and the influence of muons on energy calibration and composition estimation for primary cosmic-ray energies around $E = 10^{19}$ eV [5, 7–9]. The numbers of muons in measured data relative to the air-shower simulations have been reported for small and large zenith angles, and in all cases indicate a non-negligible dearth in predictions by hadronic-interaction models. These discrepancies cannot easily be reconciled with measurements and even when they can be, then only marginally so.

On the other hand, there is no evidence of increased event-to-event fluctuations in ground signal for showers with fixed X_{max}, which indicates that the observed muon deficit in models cannot be attributed to some exotic scenario (like micro black-holes) taking place in the early stages of shower development that would only increase muon numbers for a certain subset of events [8].

Nevertheless, an important effect can be expected from the hadronic-interaction models when they are tuned [10] to reproduce the rho-meson spectra from collider experiments [11] more accurately, especially in the extreme forward region, where more neutral rho-mesons can be found in the data [12]. Due to their decays into charged pion mesons, the neutral rho-mesons can namely keep more energy in the hadronic part of a shower and therefore increase the muon production.

Acknowledgements

The author would like to thank the organizers of the ISMD 2015 symposium and the conveners of the Astroparticle Physics session for the invitation. The successful installation, commissioning, and operation of the Pierre Auger Observatory would not have been possible without the strong commitment and effort of all members of the Collaboration, and especially the technical and administrative staff in Malargüe, Province of Mendoza in Argentina, where the Observatory is located.

References

[1] J. Matthews, Astropart. Phys. **22**, 387–397 (2005).
[2] K.-H. Kampert and M. Unger, Astropart. Phys. **35**, 660–678 (2012).
[3] R.R. Engel for the Pierre Auger Collaboration, PoS(ICRC2015)686; arXiv:1509.03732.
[4] A. Aab et al. [Pierre Auger Collaboration], Nucl. Instrum. Meth. A **798**, 172–213 (2015).
[5] A. Aab et al. [Pierre Auger Collaboration], Phys. Rev. D **91**, 032003+059901 (2015).
[6] A. Aab et al. [Pierre Auger Collaboration], Phys. Rev. D **90**, 122005 (2014).
[7] A. Aab et al. [Pierre Auger Collaboration], Phys. Rev. D **90**, 012012+039904 (2014).
[8] G.R. Farrar for the Pierre Auger Collaboration, Proc. 33rd ICRC (2013); arXiv:1307.5059.
[9] B. Kégl for the Pierre Auger Collaboration, Proc. 33rd ICRC (2013); arXiv:1307.5059.
[10] F. Riehn, R. Engel, A. Fedynitch, T.K. Gaisser, and T. Stanev, PoS(ICRC2015)558; arXiv:1510.00568.
[11] H.-J. Drescher, Phys. Rev. D **77**, 056003 (2008).
[12] A.E. Hervé for the NA61/Shine Collaboration, PoS(ICRC2015)330; arXiv:1509.06586.

EPJ Web of Conferences **120**, 04005 (2016) DOI: 10.1051/epjconf/201612004005
ISMD 2015

Measurements of the proton-air cross section with high energy cosmic ray experiments

Rasha Abbasi[1,a]

[1] *Department of Physics and Astronomy at the University of Utah*

Abstract. Detecting Ultra High Energy Cosmic Rays (UHECRs) enables us to measure the proton-air inelastic cross section $\sigma^{\text{inel}}_{\text{p-air}}$ at energies that we are unable to access with particle accelerators. The proton-proton cross section $\sigma_{\text{p-p}}$ is subsequently inferred from the proton-air cross section at these energies. UHECR experiments have been reporting on the proton-air inelastic cross section starting with the Fly's Eye in 1984 at $\sqrt{s}$ = 30 TeV and ending with the most recent result of the Telescope Array experiment at $\sqrt{s}$ = 95 TeV in 2015. In this proceeding, I will summarize the most recent experimental results on the $\sigma^{\text{inel}}_{\text{p-air}}$ measurements from the UHECR experiments.

1 Introduction

Measuring the proton-air inelastic cross section $\sigma^{\text{inel}}_{\text{p-air}}$ from cosmic rays at ultra high energies allows us to achieve knowledge of a fundamental particle property that we are unable to attain with measurements at current accelerators. Such a fundamental measurement is important to both constraining the high energy hardronic models and to verifying fundamental physics assumptions. Hadronic models are in better agreement at lower energies, below 10^{15} eV, where they are tuned to measurements on multi-particle production provided by particle accelerators. However, above 10^{15} eV models rely solely on theoretical expectations [1]. Studying the energy dependence of the proton-air cross section is important in constraining these models.

The value of the proton-air cross section is determined from detecting Ultra High Energy Cosmic Ray (UHECR) showers. Optimally, by observing the slant depth of the first point of interaction between the cosmic ray particle and air, X_1. The slant depth referred to here is the amount of material penetrated by the cosmic ray particle before it arrives to the point in question (denoted by X and in units of g/cm^2). Since the observation of the first point of interaction is not possible with UHECR detectors, the inelastic proton air cross section is inferred from the measurement of the slant depth at which the cosmic ray shower maximum occurs, X_{max}.

The latest results reported on the proton-air cross section are from the UHECR detectors, the Telescope Array (TA) and the Auger experiments. In this proceeding, an overview of the data collection, selection, and methods used in both measurements of the $\sigma^{\text{inel}}_{\text{p-air}}$ is presented. The derived values of the $\sigma^{\text{tot}}_{\text{p-p}}$ is also summarized.

[a]e-mail: rasha@cosmic.utah.edu

2 Air Shower Data Samples

The most recent experimental results on the $\sigma^{\mathrm{inel}}_{\mathrm{p-air}}$ using UHECRs are from the Telescope Array and the Auger experiments. The Telescope Array detector started operation in 2008, it is located in the northern hemisphere in the southwestern desert of Utah in the United State of America. TA is composed of three Fluorescence Detectors (FDs) bounding 507 scintillation Surfase Detectors (SD). The SD array covers 700 km^2. The Auger detector started operating in 2004. It is located in the southern hemisphere near the town of Malargüe in Argentina. It consists of four FDs overlooking 1660 water Cherenkov stations spread out over an area of 3000 km^2.

The Telescope Array measurement of the $\sigma^{\mathrm{inel}}_{\mathrm{p-air}}$ is based on data collected between May 2008 to May 2013. The data contains 439 events detected by the TA northernmost FD referred to as Middle Drum (MD) and the SD in hybrid mode. The FD and the SD parts of the detector trigger independently. A hybrid data set is formed by time matching the events within 2μs from the two detectors. The FD overlooking the sky above the SDs provides the longitudinal part of the shower and the SD reporting the footprint of the shower provides the shower core and particle density. This hybrid technique improves the geometrical reconstruction of the shower significantly, a requirement for accurate $\mathrm{X_{max}}$ determination. The $\sigma^{\mathrm{inel}}_{\mathrm{p-air}}$ for the TA experiment is reported in the energy range of $10^{18.3}$ to $10^{19.3}$ eV. The quality cuts applied to the data were optimized to achieve the best observed $\mathrm{X_{max}}$ resolution [2]. The achieved $\mathrm{X_{max}}$ resolution for this data set was $\sim$ 23 $\mathrm{g/cm^2}$.

The data used in the Auger experiment is that collected from December-2004 to December-2012 using hybrid events from four fluorescence detectors. The $\sigma^{\mathrm{inel}}_{\mathrm{p-air}}$ is reported at two energy ranges: for the energy bin of $10^{17.8}$ to 10^{18} eV using 18090 events and for the energy bin of 10^{18} to $10^{18.5}$ eV using 21270 events [3]. Here the quality cuts applied are also chosen to achieve the best $\mathrm{X_{max}}$ resolution [4]. The achieved $\mathrm{X_{max}}$ resolution was approximately < 25 $\mathrm{g/cm^2}$ [5].

Note that, in a broad sense, both experiments use the same technique (hybrid detection) to study the cosmic ray showers. However, the TA published $\mathrm{X_{max}}$ distribution, using their event selection/quality cuts, also referred to as the "Pattern recognition cuts" [2], has the effects of detector efficiency and resolution folded in. The interpretation of the data or systematic uncertainties is then concluded from folding the detector efficiency and resolution in the simulation. On the other hand, the Auger collaboration approach is to select events/quality cuts without any detector efficiency and resolution bias, using what is referred to as "fiducial volume cuts" [4]. This way, the interpretation of the data or the systematic uncertainties is then concluded from comparing the data directly to the high energy shower simulation, without folding any detector properties or reconstruction details. A direct comparison of the $\mathrm{X_{max}}$ distributions from both detectors is being done by a working group comprised of members from both experiments [5]. The comparison is still a work in progress at the time of writing of this proceeding. Using the high energy model QGSJETII.3 [6] both distributions were found to be consistent within the statistical and systematic uncertainties.

3 Analysis and Proton-air cross section measurements

The Telescope Array determines the proton-air cross section using the "K-factor method". Assuming that the tail of the $\mathrm{X_{max}}$ distribution contains the lightest particles or the most penetrating particles (proton). This method infers the attenuation length Λ_{m} and hence the cross section value from fitting the tail of the $\mathrm{X_{max}}$ distribution to the exponential function $\exp(\frac{-\mathrm{X_{max}}}{\Lambda_m})$. Λ_m here is proportional to the interaction length $\lambda_{\mathrm{p-air}}$ such that:

$$\Lambda_m = K\lambda_{\mathrm{p-air}} = K\frac{14.45 m_p}{\sigma^{\mathrm{inel}}_{\mathrm{p-air}}} \tag{1}$$

The value of Λ_m is calculated by fitting the tail of the observed X_{max} distribution between 790 and 1000 g/cm^2 using the unbinnd maximum likelihood method, as shown in Figure 1. The Λ_m value from the fit is found to be 50.47 ± 6.26 [Stat.] g/cm^2.

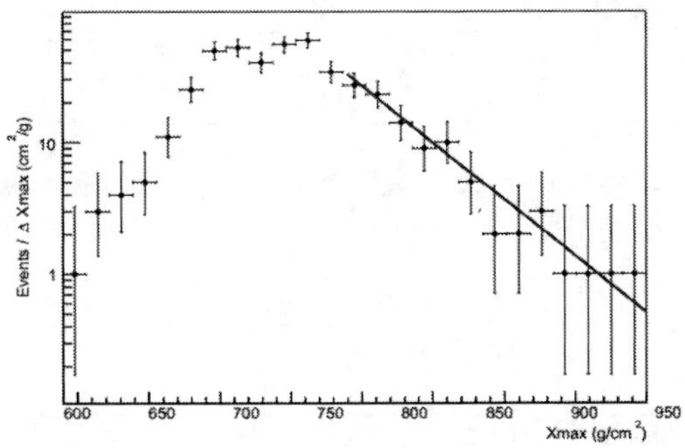

Figure 1. The number of events per X_{max} bin (ΔX_{max}) vs. X_{max} g/cm^2 for the Telescope Array data with the energy between $10^{18.3}$ and $10^{19.3}$ eV. The line is the exponential fit to the slope.

The high energy models used to determine the value of K are QGSJETII.4 [6], QGSJET01 [7], SIBYLL [8], and EPOS-LHC [9]. Several data sets of air showers in the energy range of $10^{18.3}$-$10^{19.3}$ eV were simulated using each of the above models. K is found to be fairly constant between the energies of $10^{18.3}$-$10^{19.3}$ eV, for each of the high energy models [10]. This finding justifies the use of a single average value over the range of interest. Table 1 summarizes each of the high energy models used, the corresponding value of K and consequently $\sigma^{\text{inel}}_{\text{p-air}}$.

The K-factor method has been used by several high energy experiments to deduce the value of $\sigma^{\text{inel}}_{\text{p-air}}$ from the X_{max} distribution. The first measurement of the proton air cross section using UHECR was performed by the Fly's Eye experiment, which used a calculated value of $K = 1.6$ [11]. Following the Fly's Eye result, the calculated values of K, which appeared in the literature, showed a continuous decrease as full Monte Carlo simulations came into use. By 2000, after the development of modern high energy hadronic models, the reported K-values still differed by approximately 7% [12]. Since then, as shown in Table 1, more complete hadronic shower simulations have converged on a smaller value of $K = 1.2$, with a model uncertainty of approximately 3%.

Several systematic checks were made including the model dependence (less than 3% uncertainty), detector bias, contamination by other primaries, energy dependence bias, and the error propagation of the systematic error in Λ_m. Table 2 summarizes the proton-air cross section systematics from the Telescope Array experiment [10]. The final value of the proton-air inelastic section reported by the Telescope Array collaboration is the average value of the $\sigma^{\text{inel}}_{\text{p-air}}$ obtained by the high energy models QGSJETII.4, QGSJET01, SIBYLL, and EPOS-LHC, and is found to be equal to 567.0 ± 70.5 [Stat.]$^{+29}_{-25}$ [Sys.] mb.

The Auger measurement of the proton-air cross section determines $\sigma^{\text{inel}}_{\text{p-air}}$ using, as referred to here by "the modified cross section method". This method starts similarly by inferring the proton-air cross section value from the exponential tail of the X_{max} distribution. This is done by fitting the tail of the distribution using the unbinned maximum likelihood fit to the exponential function $\exp(\frac{-X_{max}}{\Lambda_m})$. The value of Λ_m is determined for both energy bins of $10^{17.8}$-10^{18}eV and 10^{18}-$10^{18.5}$eV and is found to be 60.7 ± 2.1 [Stat.] g/cm^2 and 57.4 ± 1.8 [Stat.] g/cm^2 consecutively.

Table 1. The value of K obtained for each of the high energy models and the TA reported value of the $\sigma^{\text{inel}}_{\text{p-air}}$ in mb for that model.

Model	K	$\sigma^{\text{inel}}_{\text{p-air}}$ (mb)
QGSJETII.4	1.15± 0.01	550.3±68.5
QGSJET01	1.22±0.01	583.7±72.6
SIBYLL	1.18±0.01	564.6±70.2
EPOS-LHC	1.19±0.01	569.4±70.8

Table 2. The systematic source and uncertainties of the proton-air cross section estimated by the Telescope Array experiment.The total uncertainty was calculated in quadrature using the error for the 20% helium case.

Systematic source	Systematics (mb)
Model Dependence	(±17)
10% Helium	− 9
20% Helium	− 18
50% Helium	− 42
Gamma	+ 23
Total (20% Helium)	(+29,−25)

The cross section in the high energy model is then modified, such that the Λ_m obtained from the model matches that measured value by the detector. The cross-section is modified by multiplying the hadronic cross-section in the simulation by f [13]. Where f is:

$$f(E, f19) = 1 + (f19 - 1)\frac{\lg(E/10^{15})\text{ eV}}{\lg(10^{19}\text{ eV}/10^{15}\text{ eV})} \quad (2)$$

Here E is the shower energy and $f19$ is the value by which the shower at the energy 10^{19}eV is scaled.

After which, the model-modified-cross-section is calculated for the three high energy models: QGSJETII4, EPOS-LHC, and SIBYLL2.1. To calculate the systematic uncertainty of the measurement several checks were applied including: the model dependence (less than 3% uncertainty), propagating the uncertainties in Λ_m, the energy scale uncertainty, conversion of Λ_m to $\sigma^{\text{inel}}_{\text{p-air}}$ (mb), and contamination of other primaries. Table 3 summarizes the systematic uncertainties values for each of the energy bins [3].

The final values by Auger obtained for the proton air cross section for the energy bin of $10^{17.8}$-10^{18}eV is 457.5 ± 17.8[Stat.]$^{+19}_{-25}$[Sys.] and 485.8 ± 15.8[Stat.]$^{+19}_{-25}$[Sys.] for the energy bin of 10^{18}-$10^{18.5}$eV [3].

The results for the proton-air cross section for the high energy cosmic ray detectors is summarized in Figures 2. In addition, the predictions of all four of the high energy models (QGSJETII.4, QGSJET01, SIBYLL, EPOS-LHC) cross section are also shown.

4 Proton-proton cross section

The σ_{p-p} is calculated from the measured proton-air cross section σ^{inel}_{p-air}, using both Glauber Formalism [21] and the relation:

$$\sigma^{inel}_{p-air} = \sigma^{total}_{p-air} - \sigma^{el}_{p-air} - \sigma^{qel}_{p-air} \quad (3)$$

 DOI: 10.1051/epjconf/201612004005

Table 3. The systematic source and uncertainties in mb of the proton-air cross section estimated by the Auger experiment.

Systematic source	$10^{17.8} - 10^{18}$ eV	$10^{18} - 10^{18.5}$ eV
Model Dependence	10	10
Λ_m systematic uncertainty	13.5	14.1
Energy Scale uncertainty	2.1	1.3
conversion of Λ_m to $\sigma^{\text{inel}}_{\text{p-air}}$	7	7
Gamma	+ 4.7	+ 4.2
25% Helium	− 17.2	−15.8
Total (20% Helium)	(+19,−25)	(+19,−25)

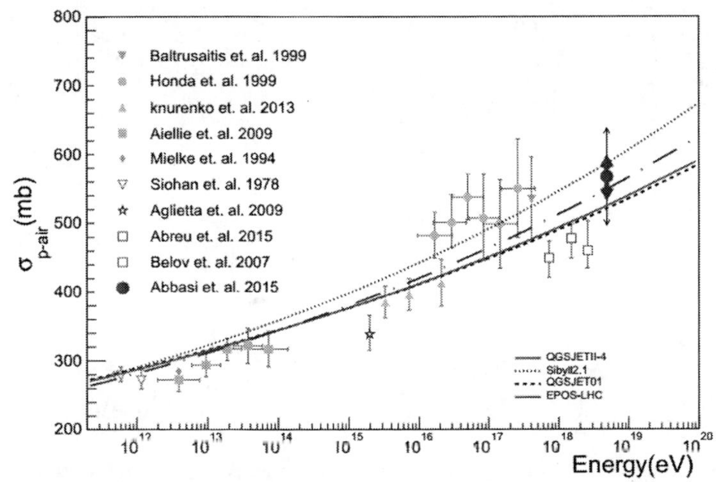

Figure 2. The proton-air cross section vs. energy for the high energy cosmic ray experiments ([10], [11], [14], [15], [16], [17], [18], [19], [20]). In addition, the high energy models (QGSJETII.4, QGSJET01, SIBYLL, EPOS-LHC) cross section predictions are also shown by solid line, fine dashed line, dotted line, and dashed line consecutively.

Where $\sigma^{\text{total}}_{p-air}$ is the total cross section, $\sigma^{\text{el}}_{p-air}$ is the elastic cross section, and $\sigma^{\text{qel}}_{p-air}$ is the quasi elastic cross section. The quasi-elastic cross section corresponds to scattering processes in which nuclear excitation occurs without particle production.

The relation between the $\sigma^{\text{inel}}_{p-air}$ and the $\sigma_{\text{p-p}}$ is highly dependent on the forward scattering elastic slope B.

$$B = \frac{d}{dt}\left[\ln \frac{d\sigma^{\text{el}}_{\text{p-p}}}{dt}\right]_{t=0} \tag{4}$$

There have been many attempts at predicting the relationship between B and $\sigma_{\text{p-p}}$. However many of these models either failed to describe the elastic scattering data, or the elastic slope energy dependence from the Tevatron, or failed to meet the unitarity condition ([22], [23], [24], [25]). But finally, a model developed by Block, Halzen, and Stanev (BHS) [26] to be consistent with unitarity while using a QCD inspired fit to the pp and p̄p data from the Tevatron seems to agree well with measurements at higher energies.

Figure 3 shows the $\sigma_{\text{p-p}}^{total}$ value of the Telescope Array result using the BHS fit [10]. Also shown are previous results from cosmic ray experiments like Fly's Eye [11], Akeno [15], HiRes [18], and Auger [13], together with accelerator pp and p̄p cross section measurement [27], and the recent result from the LHC, by TOTEM at $\sqrt{s}$ = 7 TeV [28]. The dotted curve is the QCD inspired fit of the total p-p cross section vs. the center of mass energy $\sqrt{s}$ (GeV) [29]. The result from the UHECR experiments and the reported recent result by the LHC at $\sqrt{s}$ = 7 TeV [28] are all in agreement with the BHS fit.

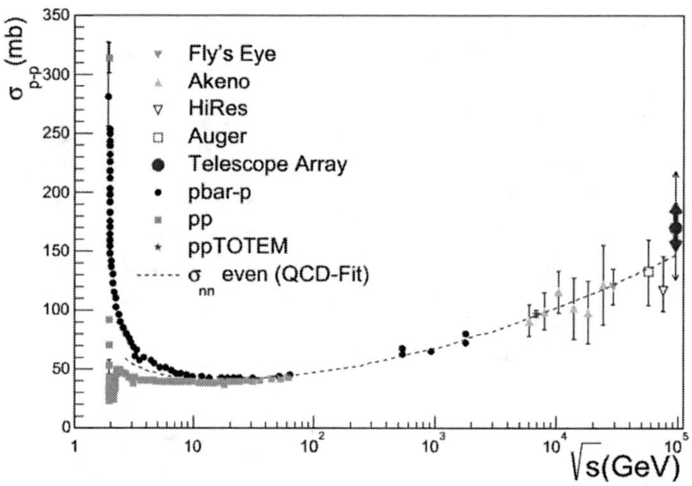

Figure 3. The proton-proton cross section vs. the center of mass energy for the high energy cosmic ray experiments ([10], [11], [13], [15], [18]). The dashed curve is the QCD inspired fit by BHS [29]. The p̄p and the pp data are shown in smaller darker circles and square symbols consecutively [27]. The recent result from LHC is also shown by the star marker [28].

5 Conclusion and Outlook

In this proceceeding we presented an overview of the most recent results on the proton-air cross section measurements from UHECR detectors. The data and methods used to deduce the proton air cross section values from the measured X_{max} distribution was also discussed. At the time of writing of this proceeding the most recent hadronic interaction models used in the conversion of Λ_m to the $\sigma_{\text{p-air}}^{\text{inel}}$ by the high energy cosmic ray experiments are QGSJETII4, Sibyll2.1, QGSJET01, and EPOS-LHC.

Results of the several systematic checks made during the analysis were summarized. The final value of $\sigma_{\text{p-air}}^{\text{inel}}$ was found by the Telescope Array experiment to be equal to 567.0±70.5[Stat.]$^{+29}_{-25}$[Sys.] mb at $\sqrt{s} = 95^{+5}_{-8}$ TeV [10]. The Auger experiment reported $\sigma_{\text{p-air}}^{\text{inel}}$ to be 457.5 ± 17.8[Stat.]$^{+19}_{-25}$[Sys.] at $\sqrt{s}$ = 38.7 ± 2.5 TeV and 485.8 ± 15.8[Stat.]$^{+19}_{-25}$[Sys.] at $\sqrt{s}$ = 55.5 ± 3.6 [3]. The value of $\sigma_{\text{p-p}}$ is also determined from $\sigma_{\text{p-air}}^{\text{inel}}$.

For the Telescope Array detector, while the events used in the published analysis were collected with the MD-SD part of the detector (only one out of three FD's), future cross section results, using events from all three FDs will likely be produced. In addition, for both the Telescope Array and the Auger detector an update with future versions of the high energy cosmic ray shower models is expected. With higher statistical power and updated models the proton air cross section measurement would in principle be able to better constrain the cross section values and predictions at high energies.

References

[1] R. Ulrich, R. Engel, M. Unger, Phys.Rev. **D83**, 054026 (2011), 1010.4310
[2] R. Abbasi, M. Abe, T. Abu-Zayyad, M. Allen, R. Anderson et al., Astropart.Phys. **64**, 49 (2014), 1408.1726
[3] *The Pierre Auger Observatory: Contributions to the 34th International Cosmic Ray Conference (ICRC 2015)* (2015), 1509.03732, http://inspirehep.net/record/1393211/files/arXiv:1509.03732.pdf
[4] J. Abraham et al. (Pierre Auger), Phys. Rev. Lett. **104**, 091101 (2010), 1002.0699
[5] *Report of the Working Group on the Composition of Ultra High Energy Cosmic Rays* (2015), 1503.07540
[6] S. Ostapchenko, Nucl.Phys.Proc.Suppl. **151**, 143 (2006), hep-ph/0412332
[7] N. Kalmykov, S. Ostapchenko, A. Pavlov, Nucl.Phys.Proc.Suppl. **52B**, 17 (1997)
[8] E.J. Ahn, R. Engel, T.K. Gaisser, P. Lipari, T. Stanev, Phys.Rev. **D80**, 094003 (2009), 0906.4113
[9] T. Pierog, I. Karpenko, J. Katzy, E. Yatsenko, K. Werner (2013), 1306.0121
[10] R.U. Abbasi et al. (Telescope Array), Phys. Rev. **D92**, 032007 (2015), 1505.01860
[11] R. Baltrusaitis, G. Cassiday, J. Elbert, P. Gerhardy, S. Ko et al., Phys.Rev.Lett. **52**, 1380 (1984)
[12] C. Pryke, Astropart.Phys. **14**, 319 (2001), astro-ph/0003442
[13] P. Abreu et al. (Pierre Auger Collaboration), Phys.Rev.Lett. **109**, 062002 (2012), 1208.1520
[14] F. Siohan, R. Ellsworth, A. Ito, J. Macfall, R. Streitmatter et al., J.Phys. **G4**, 1169 (1978)
[15] M. Honda, M. Nagano, S. Tonwar, K. Kasahara, T. Hara et al., Phys.Rev.Lett. **70**, 525 (1993)
[16] S. Knurenkoa, A. Sabourovb et al., EPJ Web of Conferences **53**, 07006 (2013)
[17] H. Mielke, M. Foeller, J. Engler, J. Knapp, J.Phys. **G20**, 637 (1994)
[18] K. Belov (HiRes Collaboration), Nucl.Phys.Proc.Suppl. **151**, 197 (2006)
[19] G. Aielli et al. (ARGO-YBJ Collaboration), Phys.Rev. **D80**, 092004 (2009), 0904.4198
[20] M. Aglietta et al. (EAS-TOP Collaboration), Phys. Rev. D **79**, 032004 (2009)
[21] R. Glauber, G. Matthiae, Nucl.Phys. **B21**, 135 (1970)
[22] R. Engel, T. Gaisser, P. Lipari, T. Stanev, Phys.Rev. **D58**, 014019 (1998), hep-ph/9802384
[23] J. Dias de Deus, P. Kroll, Acta Phys. Polon. **B9**, 157 (1978)
[24] A. Buras, J. Dias de Deus, Nucl.Phys. **B71**, 481 (1974)
[25] A. Donnachie, P. Landshoff, Phys.Lett. **B296**, 227 (1992), hep-ph/9209205
[26] M. Block, Phys.Rev. **D84**, 091501 (2011), 1109.2940
[27] C. Avila et al., Phys.Lett. B **445**, 419–422 (1999)
[28] G. Antchev, P. Aspell, I. Atanassov, V. Avati, J. Baechler et al., Europhys.Lett. **96**, 21002 (2011), 1110.1395
[29] M. Block, F. Halzen, Phys.Rev. **D72**, 036006 (2005), hep-ph/0506031

EPJ Web of Conferences **120**, 05001 (2016) DOI: 10.1051/epjconf/201612005001
ISMD 2015

Measurement of parton shower observables with OPAL

N. Fischer[1], S. Gieseke[2], S. Kluth[3,a], S. Plätzer[4], P. Skands[1], and OPAL collaboration

[1] *School of Physics and Astronomy, Monash University, Melbourne, Australia*
[2] *Institute for Theoretical Physics, Karlsruhe Institute of Technology, Karlsruhe, Germany*
[3] *Max-Planck-Institute for Physics, Munich, Germany*
[4] *Institute for Particle Physics Phenomenology, Durham University, Durham, United Kingdom*

Abstract. A study of QCD coherence is presented based on a sample of about 397,000 e+e- hadronic annihilation events collected at $\sqrt{s}$ = 91 GeV with the OPAL detector at LEP. The study is based on four recently proposed observables that are sensitive to coherence effects in the perturbative regime. The measurement of these observables is presented, along with a comparison with the predictions of different parton shower models. The models include both conventional parton shower models and dipole antenna models. Different ordering variables are used to investigate their influence on the predictions.

1 Introduction

In the evolution of a hard hadronic interaction to a final state observable in experiment the concept of the parton shower plays an important role. The parton shower connects the few energetic partons coming out of a hard scattering with the many soft partons thought to enter the hadronisation phase where observable hadrons are created. The main ingredient of the parton shower is that a hard parton transforms into a jet by repeatedly radiating gluons, which themselves can radiate, or create a quark-antiquark-pair, thus creating an avalanche or shower of partons. This process is regulated by colour coherence, i.e. the destructive interference of multiple colour-connected soft gluon emissions. The iterative nature of this process together with a simple probabilistic interpretation of each vertex or parton branching in the leading logarithmic approximation is the basis for many successful implementations in event based Monte Carlo simulation programs. A recent review about this subject is e.g. [1].

For many areas of particle physics the availability of universal numerical predictions for many different observables via Monte Carlo event generators is essential. For example at the LHC predictions for jet based observables like the distribution of the transverse momentum values of the most energetic jets in each event is usually calculated with an event generator program. Such a program contains calculations for the hard scattering of partons from the two colliding protons, a parton shower algorithm to model the evolution the energetic partons to jets of many soft partons, and a hadronisation model to describe the transition of partons to hadrons. The availability of next-to-leading order (NLO) calculations for complex final states in the event generators with consistent matching to the parton shower models has improved the accuracy of such predictions significantly. In turn, the theoretical accuracy

[a]e-mail: skluth@mpp.mpg.de

of the parton shower models has become a topic of interest for theorists in order to provide the best possible predictions.

It is difficult to construct decisive tests with the aim to distinguish between different parton shower model implementations in terms of their ability to describe data correctly. In [2] new observables based on e^+e^- annihilation into hadrons are proposed and studied to investigate differences between parton shower models. These new observables were shown to have sensitivity to explore differences between parton shower models and therefore a measurement with data from the OPAL experiment at LEP was performed [3].

2 Theory background

2.1 4-jet topology observables

The new observables are based on selecting 4-jet final states from e^+e^- annihilation to hadronic final states at a centre-of-mass energy of $\sqrt{s} = 91$ GeV corresponding to the Z^0 peak. The jets are reconstructed using the Durham iterative jet clustering algorithm [4]. The phase-space distance y_{ij} between any two particles i, j with energies $E_{i,j}$ and angle θ_{ij} between them is defined by $y_{ij} = 2\min(E_i, E_j)^2(1 - \cos\theta_{ij})/s$. The pair with the smallest value of y_{ij} is replaced by a pseudo-particle obtained by adding the 4-vectors; this procedure is repeated until two pseudo-particles or jets remain. The value of y_{ij} found when going from a 4-jet to a 3-jet configuration is required to be $y_{43} > 0.0045$. In the 4-jet configuration the jets are ordered with decreasing energy values counting from 1 to 4.

The four observables used in the analysis can now be explained. For the first three observables the angles between the jets are subjected to the requirements $\theta_{12} > 2\pi/3$, $\theta_{13} > 2\pi/3$ and $\theta_{23} < \pi/6$.

θ_{14}: The observable θ_{14} is defined as the angle between jets 1 and 4, as shown in figure 1 a) [5].

θ^*: With the additional requirement $\theta_{24} < \pi/2$ the observable $\theta^* = \theta_{24} - \theta_{23}$ is defined, see also figure 1 b) [6].

$C_2^{(1/5)}$: The so-called 2-point energy correlation double ratio [7].

ρ: For the observable ρ instead of restricting the angles between the jets the requirement $y_{43} > 0.5 \cdot y_{32}$ for the values of y_{ij} for going from a 4-jet to a 3-jet and for going from a 3-jet to a 2-jet configuration is made. In the final 2-jet configuration the invariant masses in both jets are calculated and denoted M_H for the larger and M_L for the smaller value. The ratio of these hemisphere masses $\rho = (M_L/M_H)^2$ defines the fourth observable [5]. Configurations corresponding to a small or large value of ρ are shown in figures 1 c) and d).

All observables are constructed to be sensitive to coherence effects and other properties of the parton shower models.

2.2 Parton shower models

We give a brief overview over the implementations of parton shower models in modern Monte Carlo generator programs, which will be used to produce predictions for comparison with measurements of the new observables explained above. In total six different parton shower implementations are studied, where three are variants of the Herwig++ program [8] and three are variants of the Pythia8 program [9] using the VINCIA plugin [10].

From the Herwig++ program we consider three different implementations of the parton shower:

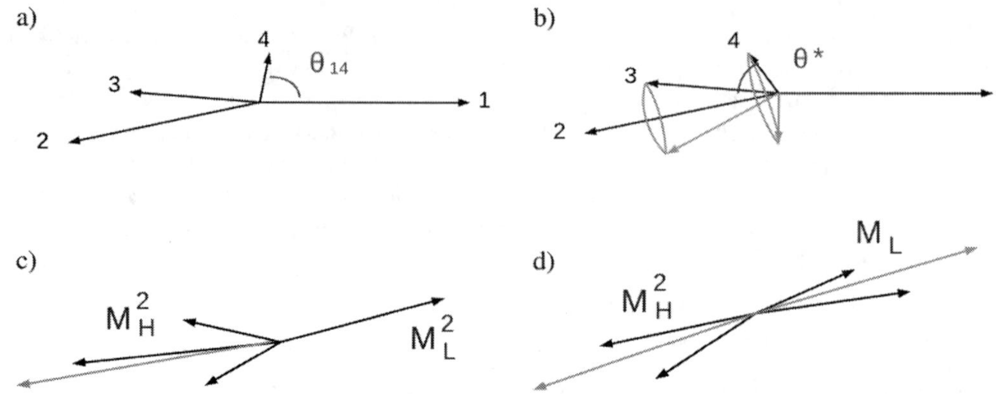

Figure 1. Figure a) shows the definition of θ_{14} as the angle between the most and least energetic jets in the 4-jet event with kinematic constraints. Figure b) displays the observable θ^* as the difference between angles θ_{23} and θ_{24}. Figure c) represents a configuration where three lower energy jets recoil against one energetic jet leading to small values of ρ while figure d) gives a configuration with two pairs of close jets recoiling against each other leading to $\rho \approx 1$.

$\tilde{q}^2$ ordering: The evolution of the parton shower is based on a DGLAP kernel with ordering variable $\tilde{q}^2$ proportional to the product of energy and angle.

$p^2_{\perp,dip}$ ordering: The partons shower model is based on Catani-Seymour (CS) dipoles and uses as ordering variable the relative transverse momentum of the splitting pair $p^2_{\perp,dip}$.

q^2_{dip} ordering: The same parton shower model based on CS dipoles is used but with the invariant mass or virtuality of the splitting pair q^2_{dip} as ordering variable. This shower model is expected to take colour coherence effects less well into account.

The VINCIA plugin for Pythia8 implements a parton shower based on antenna functions with local recoils within the antennae while the regular Pythia8 parton shower model is based on DGLAP splitting kernels. From the Pythia8 program and the VINCIA plugin we have the following implementations of the parton shower:

VINCIA $p^2_{\perp,ant}$ ordering: The shower model uses $p^2_{\perp,ant}$ without matrix element corrections. The shower model takes colour coherence into account through the antenna function formalism. This option is the preferred choice for the VINCIA shower [11].

VINCIA m^2_{ant} ordering: In this variation of the VINCIA parton shower as above the ordering variable is defined as the antenna mass m^2_{ant}.

Pythia8 $p^2_{\perp,evol}$ ordering: The regular Pythia8 shower model based on DGLAP splitting kernels is ordered in relative transverse momentum $p^2_{\perp,evol}$ of the parton pair, applies a LO matrix element correction and uses vetos on emission angles to be colour coherent.

The six different Monte Carlo models are tuned to the same set of observables measured in e^+e^- annihilation at LEP [2]. In this way differences between the models due to different quality of description of the data are removed.

3 Data analysis

We analyse data taken with the OPAL experiment at the LEP e^+e^- collider at centre-of-mass (cms) energies near the Z^0 peak $\sqrt{s} \approx 91$ GeV. The momenta of charged particle tracks are measured with the so-called jet chamber while the energies of all particles are measured from clusters of hit blocks of the lead glass electro-magnetic calorimeter; for details please see [3] and references therein. The tracking and calorimeter detectors provided almost complete coverage of the solid angle and in addition the event selection ensures well contained events. After the standard OPAL selection of hadronic events those with special 4-jet topology are selected as explained above in section 2.1.

In order to perform corrections for acceptance of the event selection and for resolution effects samples of simulated events are used. The simulations are based on Monte Carlo generators for e^+e^- annihilation into hadrons as used by the OPAL collaboration in its active period [12, 13] combined with a detailed simulation of the OPAL experiment [14]. These generators used parameter sets determined by the OPAL collaboration from tuning to OPAL data. The simulated events were processed with the same reconstruction and analysis software as the data. In simulated events we refer to results derived from the observed data as "detector-level" and to results derived from the output of the generators after hadronisation and decays of short-lived particles as "hadron-level"[1]. The experimental correction uses iterative bayesian unfolding [15] based on a response matrix derived from comparing event-by-event results at the detector- and hadron-level derived from the simulations.

It is observed that the distributions of the four observables at the detector level are well reproduced by the simulated data. As systematic variations of the analysis selection cuts are varied, the response matrix is calculated based on a different generator and the unfolding method is changed. The systematic variations are dominated by the effect of calculating the response matrix using a different generator and are larger than the statistical errors.

As a cross check the predictions of the simulations samples based on old Monte Carlo generators used for the experimental corrections are compared with the corrected data and good consistency within the combined statistical and experimental uncertainties is observed.

4 Comparison of parton shower models with data

The new parton shower models as implemented in Herwig++, Pythia8/VINCIA or Pythia8 are run with large statistics ($5 \cdot 10^6$ events) and are compared to the corrected data at the hadron-level.

Figure 2 shows in both panels the same corrected distribution of θ_{14} measured by OPAL with statistical and experimental uncertainties. In figure 2 a) the predictions of the Herwig++ variants and on figure 2 b) the predictions of the Pythia8/VINCIA variants are shown. In all cases the description of the data by the predictions is adequate with the exception of a narrow region $\theta_{14} \approx 0.7\pi$ where Herwig++ $p^2_{\perp,dip}$ deviates by about three standard deviations.

Figure 3 shows on both panels the asymmetry N_{left}/N_{right} of the distribution of ρ, which is defined by $N_{left}/N_{right} = \sum_{x_i<x_0} n_i / \sum_{x_i>x_0} n_i$ with bin centers x_i, bin contents n_i and dividing point x_0. The asymmetry is shown for two dividing points ρ_0. We observe comparatively large differences between the Herwig++ parton shower model predictions. We also find that the Herwig++ $\tilde{q}^2$ model provides among the Herwig++ variants the best description of the data while the Herwig++ q^2_{dip} model disagrees by up to four standard deviations with the data. The Pythia8/VINCIA model predictions lie closer together and also provide a reasonable description of the data.

In [3] comparisons to distributions and asymmetries of all four observables can be found together with tables of the data, their uncertainties and correlations.

[1]Hadron-level is often also referred to as particle-level.

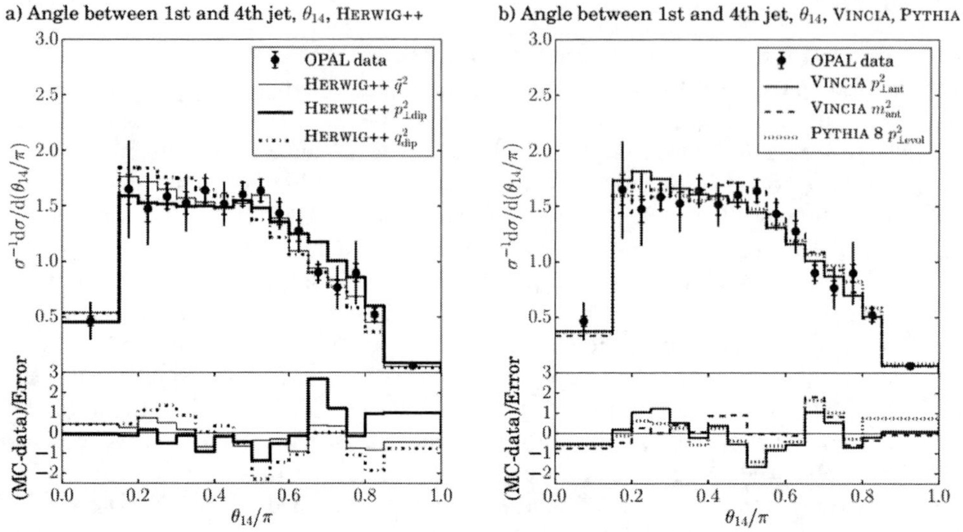

Figure 2. Figure a) shows as points with error bars the corrected distribution of θ_{14} with total errors. The statistical errors are indicated by the small horizontal bars. The lines represent the three variants of Herwig++ models as indicated. Figure b) shows the same data with the lines representing the three Pythia8 model predictions as indicated.

5 Conclusions and outlook

In general we find that all models give a fair description of the data measured by OPAL. We also observe discrepancies of up to four standard deviations for some predictions in some restricted regions of the asymmetries of the observables. These observations confirm that the new observables can have more sensitivity w.r.t. parton shower model differences compared with traditional observables. The Herwig++ q^2_{dip} model gives the least satisfactory description of the data; this is expected since this model on purpose takes coherence effects less well into account.

In this study the Herwig++ and Pythia8 variants use the cluster or string model for hadronisation, respectively, which limits the strength of our conclusions. It would be valuable to repeat the comparison with the data with all parton shower models coupled to the same hadronisation model.

References

[1] S. Höche, *Introduction to parton-shower event generators*, in *Theoretical Advanced Study Institute in Elementary Particle Physics: Journeys Through the Precision Frontier: Amplitudes for Colliders (TASI 2014) Boulder, Colorado, June 2-27, 2014* (2014), 1411.4085

[2] N. Fischer, S. Gieseke, S. Plätzer, P. Skands, Eur. Phys. J. **C74**, 2831 (2014), 1402.3186

[3] N. Fischer, S. Gieseke, S. Kluth, S. Plätzer, P. Skands (OPAL), Eur. Phys. J. **C75**, 571 (2015), 1505.01636

[4] S. Catani et al., Phys. Lett. **B**, 432 (1991)

[5] J. Alcaraz Maestre et al. (SM MC Working Group, SM and NLO MULTILEG Working Group), *The SM and NLO Multileg and SM MC Working Groups: Summary Report*, in *Proceedings, 7th Les Houches Workshop on Physics at TeV Colliders* (2012), pp. 1–220, 1203.6803

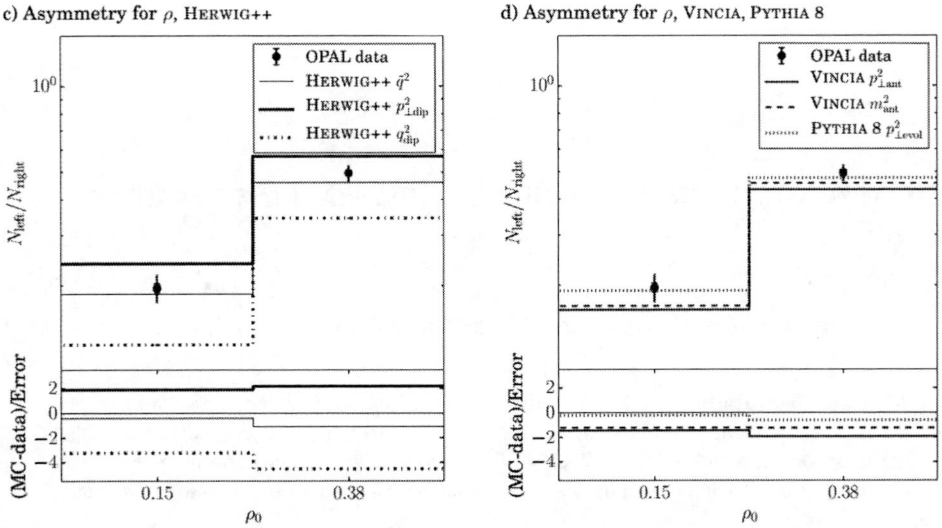

Figure 3. Figure c) shows as points with error bars the corrected asymmetry distribution (see text) of ρ with total uncertainties. The small horizontal bars give the statistical uncertainties. The lines display the predictions by the three Herwig++ models as indicated. Figure d) shows the same data compared with the three Pythia8 models as lines as indicated on the figure.

[6] S. Platzer, S. Gieseke, J. High Energy Phys. **01**, 024 (2011), 0909.5593
[7] A.J. Larkoski, G.P. Salam, J. Thaler, J. High Energy Phys. **06**, 108 (2013), 1305.0007
[8] S. Gieseke, P. Stephens, B. Webber, J. High Energy Phys. **12**, 045 (2003), hep-ph/0310083
[9] T. Sjöstrand, S. Mrenna, P. Skands, Comput. Phys. Commun. **178**, 852 (2008), 0710.3820
[10] W.T. Giele, D.A. Kosower, P.Z. Skands, Phys. Rev. **D**, 014026 (2008), 0707.3652
[11] L. Hartgring, E. Laenen, P. Skands, J. High Energy Phys. **10**, 127 (2013), 1303.4974
[12] T. Sjöstrand, Comput. Phys. Commun. **82**, 74 (1994)
[13] G. Marchesini et al., Comput. Phys. Commun. **67**, 465 (1992)
[14] J. Allison et al., Nucl. Instrum. Methods **A**, 47 (1992)
[15] G. D'Agostini, Nucl. Instrum. Methods **A**, 487 (1995)

EPJ Web of Conferences **120**, 05002 (2016) DOI: 10.1051/epjconf/201612005002
ISMD 2015

Recent developments in Monte-Carlo Event Generators

Marek Schönherr[1,a]

[1] *Institut für Theoretische Physik, Universität Zürich, 8057 Zürich, Switzerland*

Abstract. With Run II of the LHC having started, the need for high precision theory predictions whose uncertainty matches that of the data to be taken necessitated a range of new developments in Monte-Carlo Event Generators. This talk will give an overview of the progress in recent years in the field and what can and cannot be expected from these newly written tools.

1 Introduction

Modern Monte-Carlo Event Generators like PYTHIA8 [1], HERWIG++ [2, 3] and SHERPA [4] are instrumental in most physics analyses and measurements at the LHC. The current state-of-the art in usage at the experiments are either next-to-leading order to parton shower matched calculations (NLOPS) or multijet merged ones at leading order accuracy. Examples for their widespread use are shown in Fig. 1. In many instances the PYTHIA8 and HERWIG++ generators (or their older predecessors) receive input from parton level tools computing the hard core production matrix elements either at NLO for processes with few final state particles (e.g. MADGRAPH5_AMC@NLO [5] or POWHEGBOX [6]), or at LO for multileg processes (e.g. ALPGEN [7] or MADGRAPH5_AMC@NLO). The following contribution highlights a few important improvements thereupon effected in recent years.

2 Parton shower developments

The first avenue improvements in event generators have been accomplished in recent years are parton showers. Being instrumental for the description of many relevant observables parton showers are a main ingredient of all event generator frameworks and thus their continuing advancement is crucial to a better description of collider observables.

On the one hand side subleading colour information has been propagated into the algorithms otherwise operating in the leading colour limit. In the first such advancement it was a pure necessity to achieve a process independent NLO matching and was consequently only introduced in the first emission [10]. Later implementations trace subleading colour information in different limits through multiple, if not all, emissions of the parton shower evolution [11, 12]. Generally, the impact of such improvements is small, as shown in Fig. 2 (left), although also highly sensitive observables exist [13].

Other works build around gaining a higher degree of analytical control over the parton showers' resummation properties [14]. Through the accompanying scrutiny also their predictive power and

[a]e-mail: marek.schoenherr@physik.uzh.ch

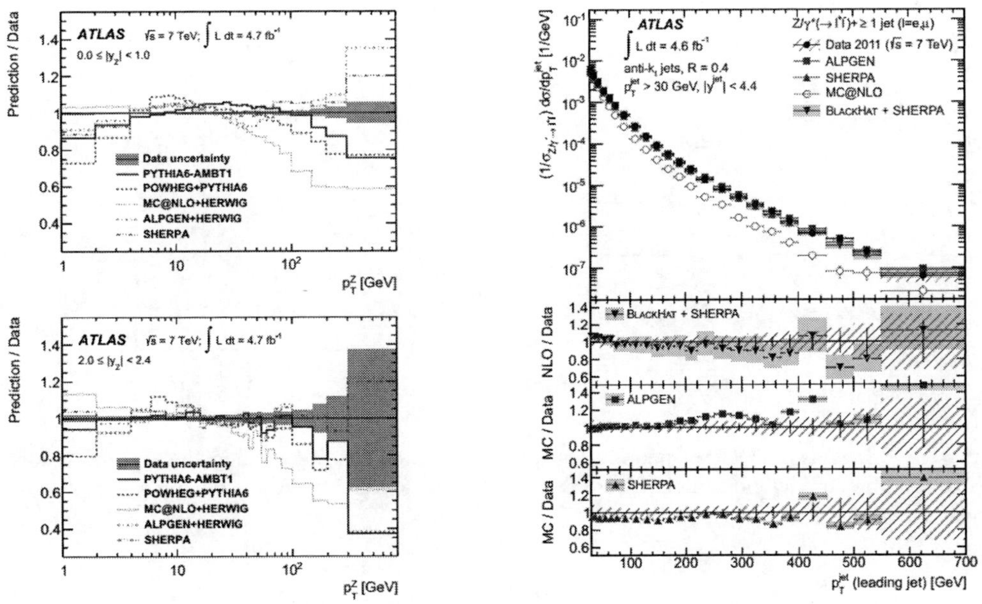

Figure 1. Left: Transverse momentum of the reconstructed Z boson in the central and the forward region, as measured by the ATLAS detector. Figure taken from [8]. Right: Transverse momentum of the leading jet in Z boson production in association with jets, as measured by the ATLAS detector. Figure taken from [9].

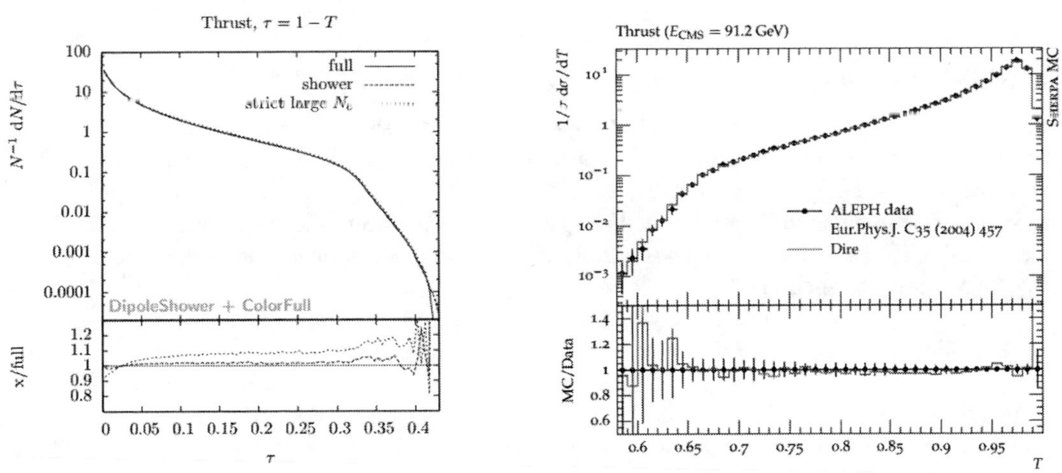

Figure 2. Left: Subleading colour effects in parton shower evolution in thrust in e^+e^--collisions at LEP. Figure taken from [11]. Right: Thrust in e^+e^--collisions at LEP as calculated by a new dipole shower implementation DIRE. Figure taken from [14].

ability to describe data has been improved. Fig. 2 (right) details the results of the newly written DIRE parton shower as compared to ALEPH data.

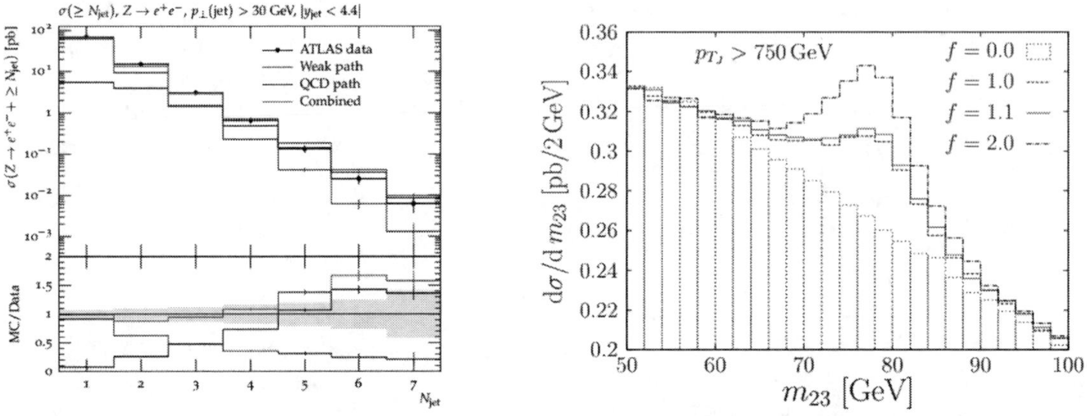

Figure 3. Left: Interplay of QCD evolution on top of W production and EW evolution on top of jet production in describing W plus mulitjet production. Figure taken from [15]. Right: Effects of adding EW evolution on subjet invariant masses. Figure taken from [16].

The third stream of development centres around incorporating electroweak effects into parton showers [15–17]. The emission of W and Z bosons, although rare, can be an important ingredient, especially in the highly boosted regime. Fig. 3 such effects for various observables. Such soft-collinear approximations to higher-order electroweak corrections complement the approximate NLO electroweak corrections of [18] and the recently achieved automation of NLO electroweak corrections [19–21].

3 NLOPs matching

Known under the names of MC@NLO [22] and POWHEG [23, 24], methods for matching NLO computations to parton showers are around for over ten years now. Recent years have seen small theoretical improvements on both schemes that lead to their application to a wider range of processes [10, 25–27] with a more complicated internal structure. The range of showers the respective matching schemes are available for has increased likewise [2, 3, 28, 29]. An systematically different matching method, UNLOPs, was developed in [30].

Similarly, CKKW [33] method of scale setting and Sudakov factor inclusion has been elevated to be applicable to NLO QCD computations in [34], leading to an improvement of NLOPs matched computations incorporating jets in the final state already at Born level. In colour singlet production in association with one additional jet the inclusion of a proper process dependent finite term can restore NLO accuracy for inclusive singlet production as well [35]. This formed the basis for the development of a NNLOPs matching method for colour singlet production [31, 36]. An exemplary result is shown in Fig. 4 (left). Another NNLOPs matching scheme basing basing on MC@NLO and UNLOPs matching was developed for the same process class in [32, 37]. Fig. 4 (right) details the results for this scheme named UN^{2}LOPs.

4 Multijet merging

Multijet merging aims at consistently combining calculations for the production of a certain experimental signature, like lepton pairs, Higgs bosons or top quark pairs, in association with any number

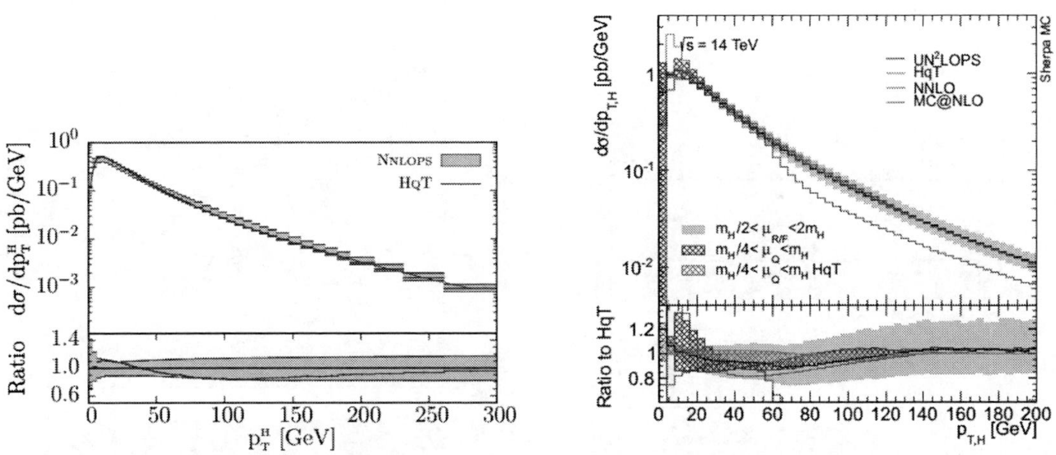

Figure 4. Left: Transverse momentum of the Higgs boson described at NNLOPS in the MiNLO approach. Figure taken from [31]. Right: Transverse momentum of the Higgs boson described at NNLOPS in the UN²LOPS approach. Figure taken from [32].

of jets. As many observables do not clearly separate between different jet multiplicities but instead receive substantial contributions by e.g. one, two and three jet final states, such multijet merging schemes are the best way to calculate these observables with the highest accuracies.

At the NLO, this was pioneered in [40]. Modern implementations for hadron colliders first appeared as MEPS@NLO [41–43] and were applied to a wide range of processes [13, 39, 44–47]. Other implementations using other methods to calculated the matched processes for each jet multiplicity have been established in [38] and [30]. Fig. 5 details results of all three mentioned methods.

5 Conclusions

Monte-Carlo Event Generators are in good shape for Run II of the LHC. Tremendous progress in terms of the achieved accuracy in calculating the hard scattering process has been achieved. They can thus be used as for precise theoretical predictions including an evaluation of the theoretical uncertainty. Developments for the non-perturbative component of high-energy collisions, however, remain sparse. In that regime, playing a role in every hadron collider event, still phenomenologically motivated models with a large number of to-be-tuned parameters are instrumental in all generators. Thus, for precision calculations one should still try to minimise the influence of that regime on the considered observables.

MS acknowledges funding by the Swiss National Science Foundation (SNF) under contract PP00P2-128552.

References

[1] T. Sjöstrand, S. Ask, J.R. Christiansen, R. Corke, N. Desai, P. Ilten, S. Mrenna, S. Prestel, C.O. Rasmussen, P.Z. Skands, Comput. Phys. Commun. **191**, 159 (2015), 1410.3012

[2] J. Bellm et al. (2013), 1310.6877

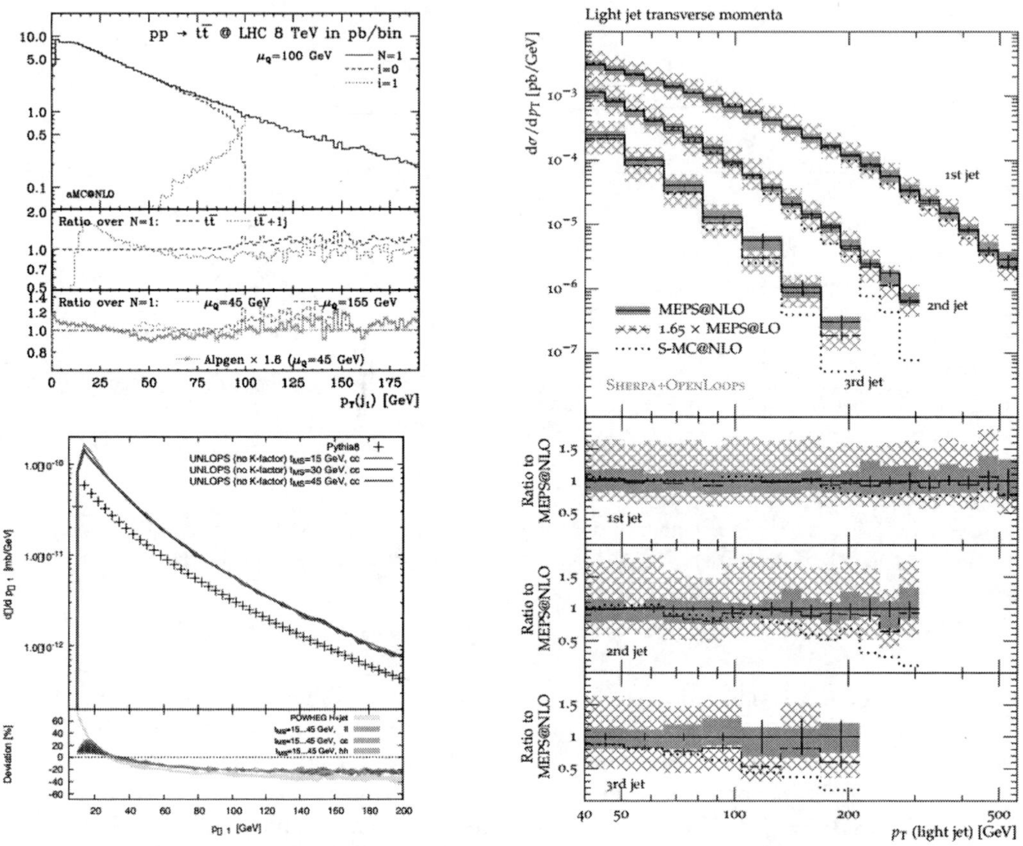

Figure 5. Left top: Transverse momentum of the leading jet in top pair production in association with jets described through a FxFx combination. Figure taken from [38]. Left bottom: Transverse momentum of the leading jet in Higgs production in association with jets described through UNLOPs merging. Figure taken from [30]. Right: Transverse momentum of the three leading jets in top pair production in association with jets described through MEPs@NLO merging. Figure taken from [39].

[3] J. Bellm et al. (2015), 1512.01178

[4] T. Gleisberg, S. Höche, F. Krauss, M. Schönherr, S. Schumann, F. Siegert, J. Winter, JHEP **02**, 007 (2009), 0811.4622

[5] J. Alwall, R. Frederix, S. Frixione, V. Hirschi, F. Maltoni, O. Mattelaer, H.S. Shao, T. Stelzer, P. Torrielli, M. Zaro, JHEP **07**, 079 (2014), 1405.0301

[6] S. Alioli, P. Nason, C. Oleari, E. Re, JHEP **06**, 043 (2010), 1002.2581

[7] M.L. Mangano, M. Moretti, F. Piccinini, R. Pittau, A.D. Polosa, JHEP **07**, 001 (2003), hep-ph/0206293

[8] G. Aad et al. (ATLAS), JHEP **09**, 145 (2014), 1406.3660

[9] G. Aad et al. (ATLAS), JHEP **07**, 032 (2013), 1304.7098

[10] S. Höche, F. Krauss, M. Schönherr, F. Siegert, JHEP **09**, 049 (2012), 1111.1220

[11] S. Plätzer, M. Sjödahl, JHEP **07**, 042 (2012), 1201.0260

[12] Z. Nagy, D.E. Soper, JHEP **06**, 097 (2014), 1401.6364

[13] S. Höche, J. Huang, G. Luisoni, M. Schönherr, J. Winter, Phys. Rev. **D88**, 014040 (2013), 1306.2703
[14] S. Höche, S. Prestel, Eur. Phys. J. **C75**, 461 (2015), 1506.05057
[15] J.R. Christiansen, T. Sjöstrand, JHEP **04**, 115 (2014), 1401.5238
[16] F. Krauss, P. Petrov, M. Schönherr, M. Spannowsky, Phys. Rev. **D89**, 114006 (2014), 1403.4788
[17] J.R. Christiansen, S. Prestel (2015), 1510.01517
[18] S. Gieseke, T. Kasprzik, J.H. Kühn, Eur. Phys. J. **C74**, 2988 (2014), 1401.3964
[19] S. Kallweit, J.M. Lindert, P. Maierhöfer, S. Pozzorini, M. Schönherr, JHEP **04**, 012 (2015), 1412.5157
[20] S. Frixione, V. Hirschi, D. Pagani, H.S. Shao, M. Zaro, JHEP **06**, 184 (2015), 1504.03446
[21] S. Kallweit, J.M. Lindert, S. Pozzorini, M. Schönherr, P. Maierhöfer (2015), 1511.08692
[22] S. Frixione, B.R. Webber, JHEP **06**, 029 (2002), hep-ph/0204244
[23] P. Nason, JHEP **11**, 040 (2004), hep-ph/0409146
[24] S. Frixione, P. Nason, G. Ridolfi, JHEP **09**, 126 (2007), 0707.3088
[25] S. Höche, F. Krauss, M. Schönherr, F. Siegert, Phys. Rev. Lett. **110**, 052001 (2013), 1201.5882
[26] S. Höche, M. Schönherr, Phys. Rev. **D86**, 094042 (2012), 1208.2815
[27] F. Cascioli, P. Maierhöfer, N. Moretti, S. Pozzorini, F. Siegert, Phys. Lett. **B734**, 210 (2014), 1309.5912
[28] S. Höche, F. Krauss, M. Schönherr, F. Siegert, JHEP **04**, 024 (2011), 1008.5399
[29] M. Czakon, H.B. Hartanto, M. Kraus, M. Worek, JHEP **06**, 033 (2015), 1502.00925
[30] L. Lönnblad, S. Prestel, JHEP **03**, 166 (2013), 1211.7278
[31] K. Hamilton, P. Nason, E. Re, G. Zanderighi, JHEP **10**, 222 (2013), 1309.0017
[32] S. Höche, Y. Li, S. Prestel, Phys. Rev. **D90**, 054011 (2014), 1407.3773
[33] S. Catani, F. Krauss, R. Kuhn, B.R. Webber, JHEP **11**, 063 (2001), hep-ph/0109231
[34] K. Hamilton, P. Nason, G. Zanderighi, JHEP **10**, 155 (2012), 1206.3572
[35] K. Hamilton, P. Nason, C. Oleari, G. Zanderighi, JHEP **05**, 082 (2013), 1212.4504
[36] A. Karlberg, E. Re, G. Zanderighi, JHEP **09**, 134 (2014), 1407.2940
[37] S. Höche, Y. Li, S. Prestel, Phys. Rev. **D91**, 074015 (2015), 1405.3607
[38] R. Frederix, S. Frixione, JHEP **12**, 061 (2012), 1209.6215
[39] S. Höche, F. Krauss, P. Maierhöfer, S. Pozzorini, M. Schönherr, F. Siegert, Phys. Lett. **B748**, 74 (2015), 1402.6293
[40] N. Lavesson, L. Lonnblad, JHEP **12**, 070 (2008), 0811.2912
[41] S. Höche, F. Krauss, M. Schönherr, F. Siegert, JHEP **04**, 027 (2013), 1207.5030
[42] T. Gehrmann, S. Höche, F. Krauss, M. Schönherr, F. Siegert, JHEP **01**, 144 (2013), 1207.5031
[43] S. Höche, F. Krauss, M. Schönherr, F. Siegert, JHEP **08**, 123 (2011), 1009.1127
[44] F. Cascioli, S. Höche, F. Krauss, P. Maierhöfer, S. Pozzorini, F. Siegert, JHEP **01**, 046 (2014), 1309.0500
[45] S. Höche, F. Krauss, M. Schönherr, Phys. Rev. **D90**, 014012 (2014), 1401.7971
[46] S. Höche, F. Krauss, S. Pozzorini, M. Schönherr, J.M. Thompson, K.C. Zapp, Phys. Rev. **D89**, 093015 (2014), 1403.7516
[47] M. Buschmann, D. Goncalves, S. Kuttimalai, M. Schönherr, F. Krauss, T. Plehn, JHEP **02**, 038 (2015), 1410.5806

Measurements of the production of a vector boson in association with jets in the ATLAS and CMS detectors

Vieri Candelise[1,a]

[1]*University of Trieste and INFN Trieste, Via Valerio 2, 34128, Trieste, Italy*

Abstract. The production of a vector boson (V = W, Z) in association with jets is can be used for a precise test of perturbative QCD and is a dominant background process in many searches for physics beyond the Standard Model. The differential cross section measurements from CMS and ATLAS collaborations would be presented, and several theoretical predictions at leading order (LO) and next-to leading order (NLO) are compared to data, after the deconvolution of detector effects, with proton-proton collisions at a center of mass energy of 7, 8 and 13 TeV at LHC.

1 Introduction

The associated production of vector bosons and jets (V+jets) in proton-proton collisions at LHC is of prime interest in the physics program of the ATLAS and CMS experiments [1,2]. This type of processes is important for several physics aspects. First of all in perturbative QCD (pQCD) studies, where direct measurement of the quark content of the proton and parton density functions (PDF) can be made by measuring the production cross sections. Studying the differential cross sections is also essential to test the latest pQCD calculations at the highest precision available in the perturbative expansion, making these measurements an unique tool for the test the Monte Carlo event generators. In Standard Model physics, the W, Z plus jets production is an essential measurement in order to precisely estimate the background component of the Higgs boson and top pairs production. Furthermore, many Beyond the Standard Model searches (such as supersymmetry, new Higgs bosons, 4th generation of heavy quarks) have a similar topology with respect to the V+jets production, and thus the control on their backgrounds might be highly improved by means of these measurements.

2 Associated production of a W boson and jets

The associated production of a W boson and one or more jets has been measured in proton-proton collisions at the center of mass energy of 7 TeV both in ATLAS and CMS. The differential spectra are deconvolved for detector effects and compared to the theoretical predictions at LO and NLO precision.

2.1 W+jets at 7 TeV in ATLAS

The W+jets measurement in ATLAS [3] has been performed using an integrated luminosity of 4.6 fb^{-1}. Electrons and muons from the W boson decay must pass isolation criteria and are required to have a transverse momentum (p_T) of at least 25 GeV and pseudorapidity (η) less than 2.4. The reconstructed missing transverse energy is used to account for the presence of the neutrino, and the associated jets are clustered using the anti-K_T algorithm [4] with R=0.4. Jets must also have p_T>25 GeV and rapidity (y) within the range of 4.4. The background coming from the production of top quark pairs is suppressed by requiring a b quark tag veto, and the multi-jet QCD background is estimated in data. Measured cross sections are unfolded to particle level using a Bayesian algorithm based on the Monte Carlo information. The differential cross sections are presented as a function of the leading jet

[a] Corresponding author: vieri@CERN.ch

momentum (Fig.1). Data are compared to the theoretical prediction of BlackHat+Sherpa (NLO) [5], ALPGEN (LO) [6], and Sherpa v.2 [7]. It can be seen that the LO calculations describe data in the hard part of the spectrum within the uncertainties, while a 10% overestimation is seen in the 200-400 GeV range. NLO calculations show good agreement in the comparison with unfolded data.

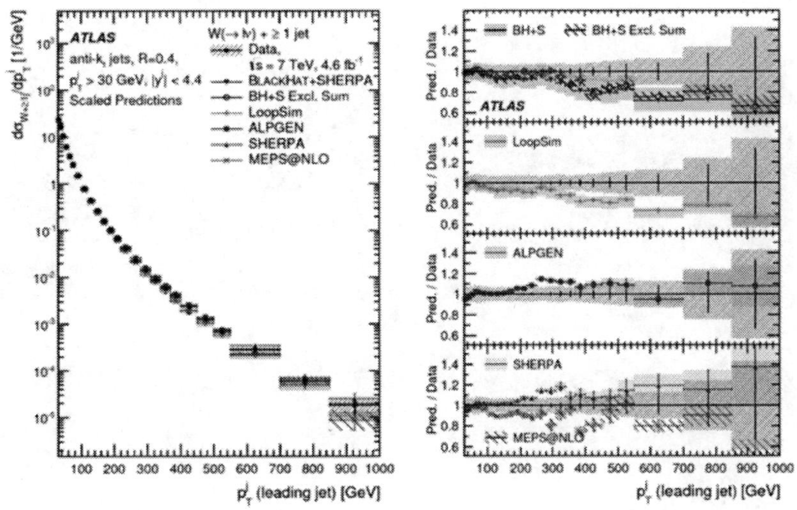

Figure 1. Differential W+jets cross sections as a function of the leading jet transverse momentum compared to the theoretical predictions of BlackHat+Sherpa at NLO, Sherpa (LO) and MadGraph+Pythia6 (LO) measured in ATLAS.

2.2 W+jets at 7 TeV in CMS

The W+jets cross section in CMS [8] has been measured using an integrated luminosity of 5 fb^{-1}. Electrons and muons from the W boson decay must be isolated and are required to have $p_T>25$ GeV and $|\eta|<2.1$. Associated jets are clustered using the anti-K_T algorithm with R=0.5. Jets must also have $p_T>25$ GeV and have $|\eta|<2.4$. A cut on the missing transverse mass (M_T) of the neutrino and the muon is applied, requiring $M_T>50$ GeV. The background coming from the production of top quark pairs is taken into account requiring a b quark tag veto, and the multi-jet QCD background is estimated again with a data-driven technique. Measured cross sections are unfolded to particle level using the SVD algorithm based on the Monte Carlo simulation. The differential cross sections are presented as a function of the leading jet momentum and as a function of the scalar sum of all the jets in the event, namely H_T (Fig.2). Data are compared to the theoretical expectations of MadGraph5 [9]+Pythia6 [10] (LO), BlackHat+Sherpa (NLO) and Sherpa (LO). It can be seen that the NLO predictions of BlackHat +Sherpa are not describing properly the spectra of the H_T distribution. It is shown that the LO MadGraph+Pythia6 prediction is over-estimating data in the full range of p_T of the leading jet and H_T.

3 Associated production of a Z boson and jets

The associated production of a Z boson and at least one jet measurement in proton-proton collisions at the center of mass energy of 8 TeV in CMS is presented. The differential cross sections are unfolded for detector effects and compared to the theoretical predictions at the LO and NLO precision.

DOI: 10.1051/epjconf/201612005003

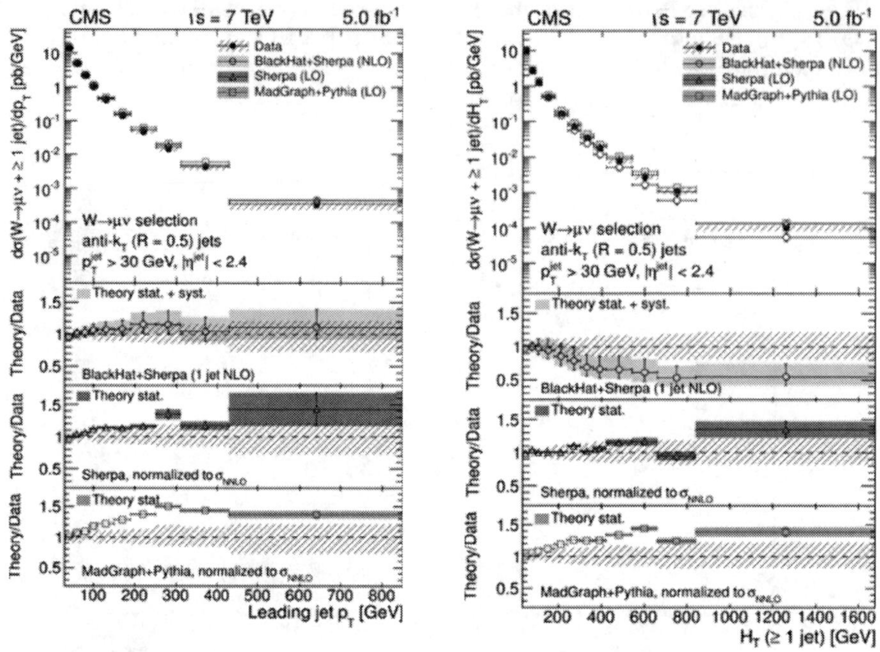

Figure 2. Differential W+jets cross sections as a function of the leading jet transverse momentum (left) and H_T (right) compared to the theoretical predictions of BlackHat+Sherpa at NLO, Sherpa (LO) and MadGraph+Pythia6 (LO) measured in CMS.

3.1 Z+jets differential and double differential cross sections at 8 TeV in CMS

The Z+jets differential cross section measurement in CMS [11] has been performed using an integrated luminosity of 5 fb^{-1}. Electron and muon pairs from the Z boson decay must be isolated and must have p_T>20 GeV and $|\eta|$<2.4. The dilepton pairs are selected if their invariant mass lies within the mass range of [71;111] GeV. Jets are clustered using the anti-K_T algorithm with R=0.5 and must pass a p_T>25 GeV threshold, and have $|\eta|$<2.4. Measured cross sections are unfolded to particle level using the Bayesian unfolding algorithm exploiting the simulation information. The differential cross sections are presented as a function of the jet multiplicity (Fig.3). Data are compared to the theoretical expectations of MadGraph5+Pythia6 (LO) and Sherpa v.2 (NLO up to the second jet). Good agreement is seen in the unfolded cross sections. The double differential cross sections has also been measured as a function of the leading jet momentum and several ranges of rapidity, up to $|y|$ <4.7 [6].

3.2 Z+jets differential cross sections at 13 TeV in ATLAS

The Z+jets differential cross sections, measured at a center of mass energy of 13 TeV, have been studied with the ATLAS detector [12] with an integrated luminosity of 85 pb^{-1}. Dielectron (dimuon) pairs coming from the Z boson decay must be isolated and must have p_T>25 GeV and $|\eta|$<2.47(2.4). The dilepton invariant mass is taken within the mass range of [66;116] GeV. Jets are clustered using anti-K_T algorithm with R=0.4 and must also have a transverse momentum of at least 30 GeV and $|y|$ <2.5. Measured cross sections are corrected to particle level using a correction factor calculated with the Sherpa event generator. The differential cross sections are presented as a function of the jet multiplicity (Fig.4). Data are compared to the theoretical predictions of MadGraph_aMC@NLO +Pythia8 [13] (up to four partons at LO) and Sherpa v.2 (NLO up to the second jet). A good agreement is seen in the measured cross sections.

DOI: 10.1051/epjconf/201612005003

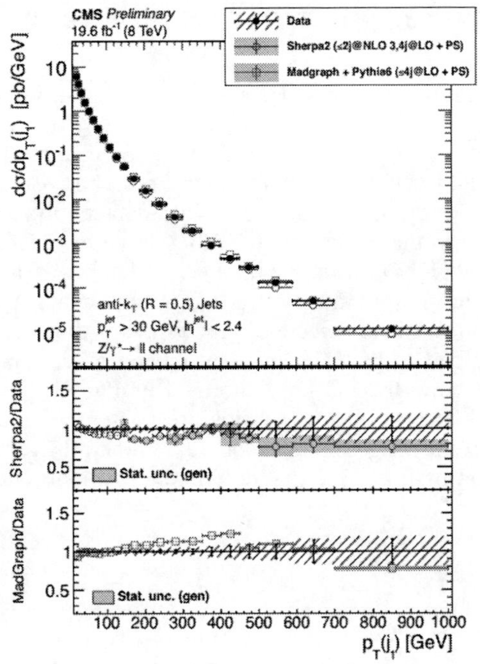

Figure 3. Differential Z+jets cross sections as a function of the leading jet transverse momentum compared to the theoretical predictions of Sherpa v2 (NLO up to the second jet) and MadGraph+Pythia6 (LO) measured in CMS.

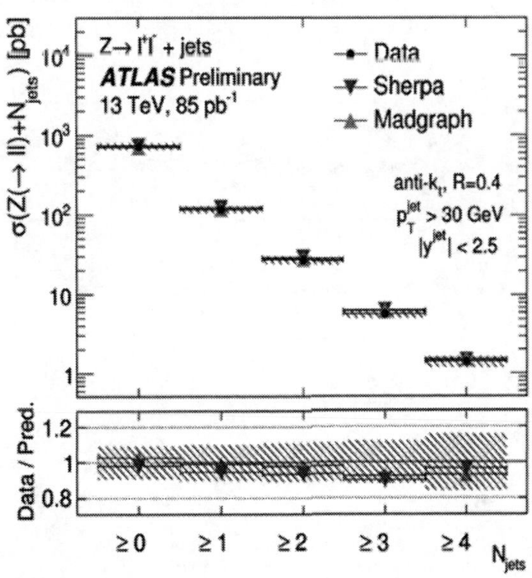

Figure 4. The 13 TeV differential Z+jets cross sections as a function of the jet multiplicity compared to the theoretical predictions of Sherpa v2 (NLO up to the second jet) and MadGraph_aMC@NLO+Pythia8 (LO) measured with the ATLAS detector.

4 Differential W+jets and Z+jets cross-section ratio in ATLAS

The differential cross section ratio between W+jets and Z+jets is an important tool to test the pQCD predictions, exploiting the advantage of many important systematics cancelation in the ratio, thus increasing the precision in the final results. Furthermore, the precise modeling of the jet recoil against the vector boson is an important measurement for many searches for physics beyond the Standard Model. Events are collected using 4.6 fb^{-1} of the 7 TeV data taking of the ATLAS detector [14]. Isolated leptons are selected from the boson decay if they have p_T>25 GeV and $|\eta|$<2.47. Associated jets are clustered using the anti-K_T algorithm with R=0.4. Jets must have p_T>25 GeV and $|y|$<4.4. Additional requirements on the missing transverse energy and mass are imposed for the W boson reconstruction, and the dilepton pairs are selected in the range of the Z boson mass according to the Z +jets selection. Data are unfolded to particle level by means of the Bayesian unfolding procedure. Unfolded data are then compared to the LO predictions of ALPGEN, Sherpa and the NLO prediction of BlackHat+Sherpa. Differential cross section ratios are presented as a function of the leading jet transverse momentum (Fig. 5), showing general good agreement all over the spectrum, but with some tensions that can be seen comparing data to the Sherpa prediction around 300 GeV, where the generator overestimates data.

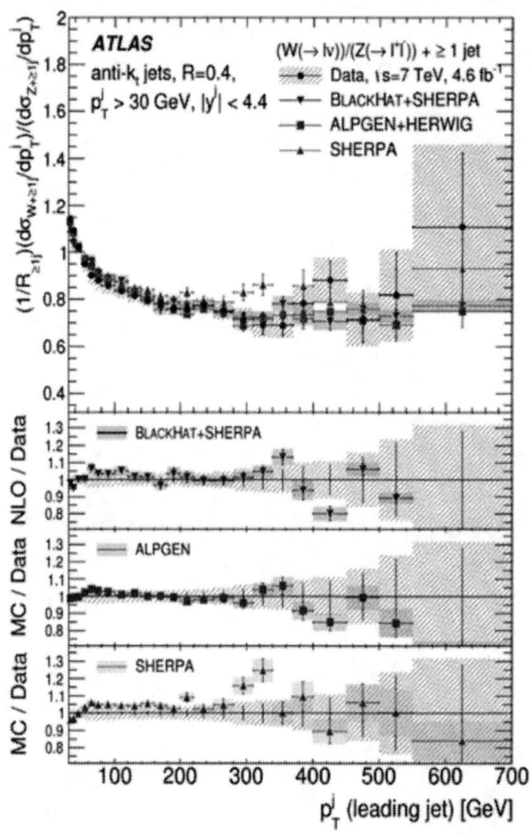

Figure 5. Differential W+jets/Z+jets cross sections ratio as a function of the leading jet transverse momentum compared to the theoretical predictions of Sherpa (LO), ALPGEN (LO) and BlackHat+Sherpa (NLO) measured with the ATLAS detector.

5 Associated production of a Z boson and b jets in CMS

The associated production of a Z boson and jets originated from the hadronization of a b quark is a unique subprocess of the V+jets events. The Z+b final state is a dominant background for the Higgs production in the HZ mode and it appears in the final state of many beyond the Standard Model models processes. The Z+b final state is also essential to understand the b quark production mechanism in high energy proton collisions, allowing to test the so called 4-flavour scheme (4FS), with massive b quarks in the final state, and the 5-flavour scheme (5FS), with massless b quarks in the initial state. The Z+b differential cross sections have been measured by CMS with 20 fb^{-1} in 8 TeV collisions [15]. The b-jets are identified with a b-tagging technique that exploits the characteristics of the secondary vertex reconstruction, and must pass a threshold in p_T of 30 GeV and pseudorapidity less than 2.4. The Z boson decay products are selected using the same Z+jets criteria. The dominant top-induced background is evaluated in data, and the true fraction of bottom, charm and light flavours is extracted by fitting the secondary vertex mass distribution. Data ara unfolded using the SVD algorithm and the results are compared to the theoretical calculations in the 4- and 5-flavour schemes in pQCD made with MadGraph+Pythia6, and the 5-flavour scheme prediction of Powheg [16] (NLO for the leading jet). Differential cross section are presented as a function of the leading b-jet pt and as a function of the invariant mass of the bb system (Fig. 6). An overall normalization discrepancy of about 20% is seen in the 4-flavour scheme for the b-jet spectrum, while the NLO prediction of Powheg correctly describes the data. The 5-flavour scheme appears to overestimate data in the soft Z momentum region. Good agreement is seen in the bb invariant mass distribution comparing data with all the predictions.

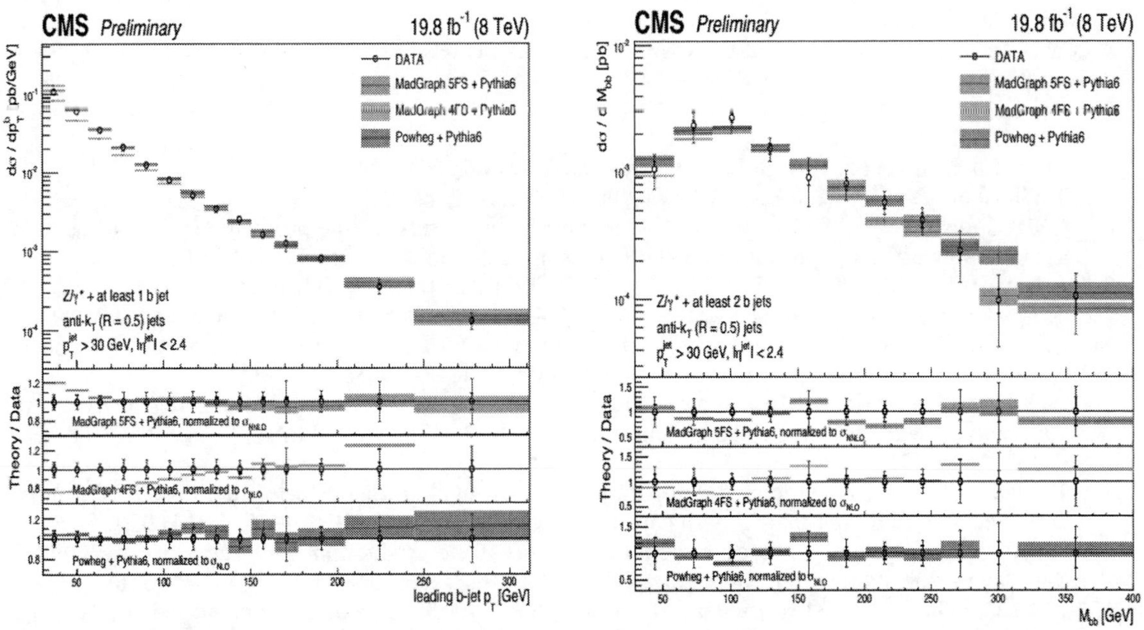

Figure 6. Differential Z+b cross sections as a function of the leading b-jet transverse momentum (left) and as a function of the bb invariant mass compared to the theoretical predictions of MadGraph 4FS and 5FS (LO) and Powheg (NLO for the leading jet) with the CMS detector.

6 Conclusions

Differential cross sections for the associated production of a W or Z boson and jets have been measured in the ATLAS and CMS experiments at a center of mass energy of 7, 8 and 13 TeV in proton-proton collisions at LHC. The W boson plus jets has been measured in ATLAS and CMS at 7 TeV and compared to several LO and NLO predictions; the Z+jets has been measured by CMS at 8 TeV, showing also the double-differential cross sections as a function of the momentum and rapidity of the leading jet, and in ATLAS at 13 TeV. The ratio between W and Z bosons plus with at least one associated jet has been measured has been measured as a function of the leading jet momentum and compared to the NLO calculations. The Z+b differential cross sections has been measured by CMS as a function of the most energetic b-jet momentum and as a function of the invariant mass of the leading and sub-leading b jet. The results have been compared to the LO predictions in the 4FS and 5FS and the NLO predictions for the first jet. All the differential cross sections measured in data presented have been unfolded to particle level in order to be compared to the theoretical predictions in perturbative QCD, allowing a test of the theoretical interpretation with collision data.

References

1. ATLAS Collaboration, "The ATLAS Experiment at the CERN LHC" , JINST 3 (2008) S08003
2. CMS Collaboration, "The CMS experiment at the CERN LHC", JINST 3 (2008) S08004
3. ATLAS Collaboration, "Measurements of the W production cross sections in association with jets with the ATLAS detector ", Eur.Phys.J. C75 (2015) 2, 82.
4. M. Cacciari, G. P. Salam, and G. Soyez, "The anti-k(t) jet clustering algorithm", JHEP600 0804 (2008) 063
5. Berger, C.F. et al., "Vector Boson + Jets with BlackHat and Sherpa" Nucl.Phys.Proc.Suppl. 205-206 (2010) 92-97
6. M. L. Mangano et al., J. High Energy Phys. 0307, 001 (2003)
7. T. Gleisberg et al., J. High Energy Phys. 0902, 007 (2009)
8. CMS Collaboration, "Differential cross section measurements for the production of a W boson in association with jets in proton-proton collisions at $\sqrt{s} = 7$ TeV", PLB 741 (2015) 12
9. J. Alwall et al., "MadGraph 5: going beyond", JHEP 06 (2011) 128, JHEP06(2011)128
10. T. Sjostrand, S. Mrenna, and P. Skands, "PYTHIA 6.4 physics and manual", JHEP 05 (2006) 026
11. CMS Collaboration, "Measurement of the differential production cross section of Z bosons in association with jets in pp collisions at $\sqrt{s} = 8$ TeV", CMS-PAS-SMP-13-007
12. ATLAS Collaboration, "Measurement of the Production Cross Sections of a Z boson Boson in Association with Jets in collisions at $\sqrt{s} = 13$ TeV with the ATLAS Detector", ATLAS-CONF-2015-051
13. J. Alwall et al., "The automated computation of tree-level and next-to-leading order differential cross sections, and their matching to parton shower simulations", JHEP 1407 (2014) 079
14. ATLAS Collaboration, "A measurement of the ratio of the production cross sections for W and Z bosons in association with jets with the ATLAS detector", Eur. Phys. J. C (2014) 74: 3168
15. CMS Collaboration, "Measurement of the associated production of a Z boson and b jets in pp collisions at 8 TeV", CMS-PAS-SMP-14-010
16. S. Frixione, P. Nason, and C. Oleari, "Matching NLO QCD computations with parton shower simulations: the POWHEG method", JHEP 11 (2007) 070.

EPJ Web of Conferences **120**, 05004 (2016)
ISMD 2015 DOI: 10.1051/epjconf/201612005004

Resummed Results for Hadron Collider Observables

Heather McAslan[1,a]

[1] *University of Sussex, Brighton, UK*

Abstract. Event shapes are invaluable QCD tools for theoretical calculations and experimental measurements. We revise the definition of these observables in e^+e^- annihilation and in hadron collisions, and give a review of the state-of-the-art results for their resummation. Then we detail how recent work on the resummation of event shapes in electron-positron annihilation can provide us with the tools to extend resummation of generic hadronic event shapes to NNLL accuracy. We match our findings to fixed-order results at NNLO accuracy, showing the sizeable effects of resummation in the relevant regions of phase space.

1 Event Shapes

Event shapes are measures of the hadronic energy flow in a system of particles. These observables are particularly useful for tuning Monte Carlo generators and extraction of the strong coupling α_s.

1.1 Event Shapes' Definitions in e^+e^-

Event shapes were originally defined in e^+e^- annihilation events. The canonical event shape, and a useful observable for encoding the physical geometries of different event types, is the thrust.

$$\tau \equiv 1 - \max_{\vec{n}} \frac{\sum_i |\vec{p}_i \cdot \vec{n}|}{Q} \,, \tag{1}$$

where the vector $\vec{n}$ maximised by the sum defines the thrust axis, $\vec{n}_T$ and Q is simply the total centre of mass energy. The thrust axis divides the system into two hemispheres, $\mathcal{H}(1)$ and $\mathcal{H}(2)$, as shown in Fig. 1. The value of τ is equal to zero for events with two back-to-back jets, and equal to one for a spherically spread-out event, as seen in Figs. 1 and 2.
Using these hemispheres we can define jet broadening

$$B_T \equiv B_L + B_R \,, \tag{2}$$

where

$$B_L \equiv \sum_{i \in \mathcal{H}^{(1)}} \frac{|\vec{p}_i \times \vec{n}_T|}{2Q} \,, \quad B_R \equiv \sum_{i \in \mathcal{H}^{(2)}} \frac{|\vec{p}_i \times \vec{n}_T|}{2Q} \,. \tag{3}$$

[a] e-mail: h.mcaslan@sussex.ac.uk

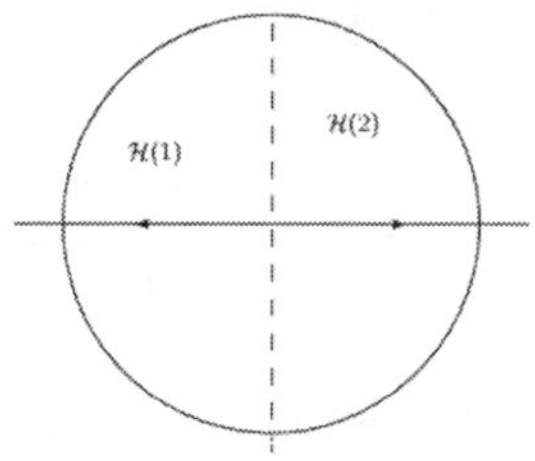

Figure 1. Pencil-like geometry of an event: $\tau = 0$

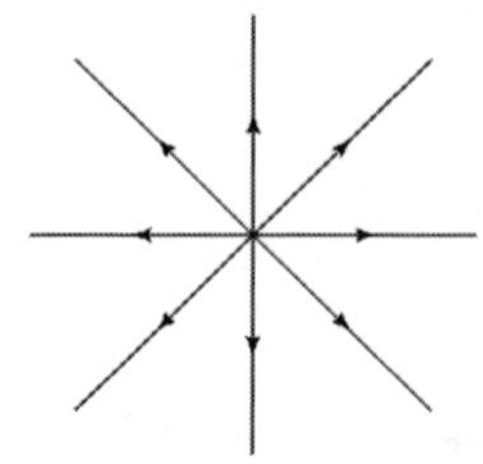

Figure 2. Spherical geometry of event: $\tau = 1$

1.2 Event Shapes' Definitions in Hadronic Collisions

Hadronic event shapes are defined analogously to those in e^+e^-. [1] However, they measure in the transverse plane, for boost invariance, and require a different choice of normalisation. Hadronic event shapes can be conveniently normalised by the sum of the transverse momentum of the final state particles. Transverse thrust is defined in analogy to thrust

$$\tau_\perp \equiv \max_{\vec{n_T}} \left(\frac{\sum_i |\vec{p}_{\perp i} \cdot \vec{n_T}|}{\sum_i |\vec{p}_{\perp i}|} \right), \tag{4}$$

where the maximum $\vec{n_T}$ defines the transverse thrust axis.
As before, we can define a jet broadening for hadronic collisions:

$$B_T \equiv \frac{1}{2P_\perp^2} \left(\sum_{i \in C^{(1)}} p_{\perp i} \sqrt{(\Delta\eta_{1,i})^2 + (\Delta\phi_{1,i})^2} + \sum_{i \in C^{(2)}} p_{\perp i} \sqrt{(\Delta\eta_{2,i})^2 + (\Delta\phi_{2,i})^2} \right), \tag{5}$$

where the normalisation $P_\perp^2 = \sum_i |\vec{p}_{\perp i}|$. Here $C^{(1)}$ and $C^{(2)}$ are the hemispheres defined by the transverse thrust axis, but excluding any particle i that is not within the measured 'central region'. The issue of constraining emissions in a central region of phase space (i.e. with small η) is discussed in the next section.

2 Globalness Issues and Solutions

Experimentally, measurements of QCD observables are made in the central region of the detector geometry where detector information is most extensive. This is implemented in event analysis by placing a cut on events outwith a central rapidity, $|\eta| < \eta_C$. When carrying out theoretical predictions, however, we want to avoid cutting on regions of phase space. Restricting the phase space of emissions causes large non-global logarithms to arise, which pose difficulties to resummation. The crucial interplay between theory and experiment calls for some compromise in the calculation/measurement of observables outside of the central region.

Techniques exist to suppress the effects of non-global effects. [2] One can define a new global observable which is split into a central piece and a piece sensitive to radiation away from the central region. For example, global thrust is defined

$$\tau_{\perp,\mathcal{E}} \equiv 1 - T_{\perp,\mathcal{E}} \equiv \tau_\perp + \mathcal{E}_{\bar{C}}, \tag{6}$$

where

$$\mathcal{E}_{\bar{C}} = \frac{1}{Q_{\perp,C}} \sum_{i \notin C} p_{\perp i} e^{-|\eta_i - \eta_C|}, \tag{7}$$

and

$$\eta_C = \frac{1}{Q_{\perp,C}} \sum_{i \in C} p_{\perp i} \eta_i \,, \tag{8}$$

and $\tau_\perp$ is the hadronic transverse thrust as given in Eq. (4). The normalisation $Q_{\perp,C}$ is the sum of transverse momenta of particles within the central region. The sum in Eqs (7) and (8) spans whichever objects we work with, be that individual particles or anti-k_t jets, for example. This parameterisation allows for a hadronic global thrust definition, sensitive to radiation in all regions of phase space. Global broadening, and other global event shapes, can be constructed in the same way.

There has been recent progress in techniques for tackling non-global resummation head-on. See for example [3] [4].)

3 Resummation

The infra-red and collinear safety of event shapes' definitions ensures that their cross-sections are finite in every order of perturbation theory. However, when the value of the event shape is small ($v \approx 0$ where e.g. $v \equiv \tau$, for thrust) the implicit constraint on QCD radiation being soft and collinear leads to large logarithms appearing at every order in α_s. These large logarithms are remnants of the cancellation of real and virtual singularities. The effective coupling becomes $\alpha_s \ln(1/v) \approx 1$ and of course the perturbative expansion is no longer reliable. The series is rearranged in terms of the dominant logarithms and the cumulative distribution is resummed via the following expression [5]

$$\Sigma(v) = \frac{1}{\sigma_0} \int_0^v dv' \frac{d\sigma(v')}{dv'} = e^{Lg_1(\lambda)+g_2(\lambda)} \mathcal{F}_{NLL}(\lambda), \qquad \lambda = \alpha_s(Q)\beta_0 \ln(1/v) \,, \tag{9}$$

where the exponential function is a Sudakov form factor, containing contributions arising from virtual corrections and unresolved real emissions. The $\mathcal{F}(\lambda)$ contains contributions from real resolved emissions. The $\{g_i(\lambda)\}$ are ordered in decreasing logarithmic-dominance, with $g_1(\lambda)$ resumming the leading logarithms - or terms of form $\alpha_s^n \ln^{n+1}(1/v)$ - $g_2(\lambda)$ resumming the next-to-leading logarithms - terms of form $\alpha_s^n \ln^n(1/v)$ - and so on. Eq. (9) is accurate to next-to-leading logarithmic order (NLL). Physically, truncating at NLL order corresponds to requiring that all radiation be soft and collinear, and widely separated in rapidity. To achieve NNLL accuracy one would also need to include $(\alpha_s/\pi)g_3(\lambda)$ in the exponent, and any NNLL contributions to the $\mathcal{F}$-function.

4 Resummation Tools

4.1 CAESAR

Resummation of event shapes (as well as other observables) to NLL accuracy has been carried out in the semi-numerical method of CAESAR, via the automated resummation of a generic observable. [5] CAESAR's generic observable is defined in the Born system, with an extra soft and collinear gluon (Fig. (3))

$$V(\{\tilde{p}\}, k) = d_\ell \left(\frac{k_t^{(\ell)}}{Q} \right)^{a_\ell} e^{-b_\ell \eta^{(\ell)}} g_\ell(\phi^{(\ell)}) \tag{10}$$

(Thrust can be recovered by setting $a_\ell = b_\ell = 1$.)
$\{\tilde{p}\}$ are the hard partons (or 'legs' ℓ) involved in the event, recoiled against the additional emission. Referring to fig. (3) they are the quark-antiquark dipole, $\{\tilde{p}\} = \{\tilde{p}_1, \tilde{p}_2\}$. k_t is the transverse momentum of the soft and collinear emission k, with respect to the dipole, with some normalisation Q. η is its

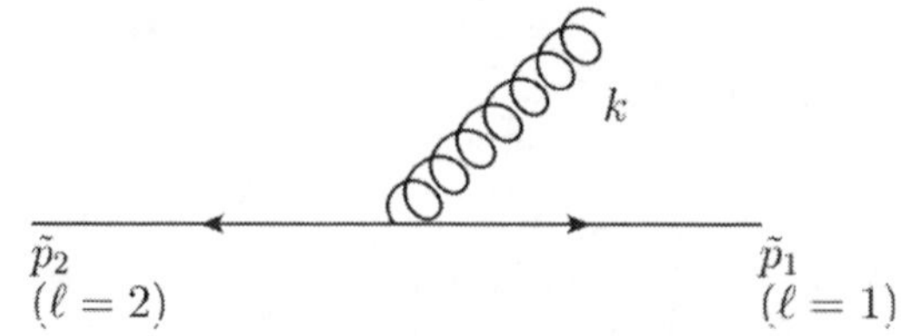

Figure 3. A system of a back-to-back quark-antiquark pair, and an additional gluon, k.

rapidity and ϕ its azimuth, also with respect to the dipole. The (a, b, d, g) parameterise each different observable.

CAESAR requires that the observable being resummed is continuously global, thereby prohibiting non-global logarithms. This implies that the scaling of the observable is the same everywhere, i.e. that $a = a_1 = a_2$. The observables must also satisfy recursive infrared and collinear (rIRC) safety. rIRC safety stipulates that the observable's scaling properties should be the same in the presence of any number of extra soft and/or collinear emissions, i.e. in the presence of a series of soft and collinear emisisons $(k_1, k_2, k_3, \ldots)$.

4.2 Resummation of Event Shapes to NNLL

CAESAR's NLL results are the current state-of-the-art for many event shapes. Recently there have been results appearing at NNLL using Soft Collinear Effective Theory, for example the transverse thrust [6], the beam thrust [7] and N-jettiness [8].

4.3 ARES (Automated Resummer of Event Shapes)

Using the formalism of CAESAR we have recently developed a resummation method (and accompanying code, soon to be made public) for event shapes in e^+e^- annihilation at NNLL order. [9] To achieve NNLL accuracy, we relax the kinematic assumptions made at NLL one by one: allowing one emission to roam in phase space, i.e. for a single hard-collinear or soft-wide angle emission to occur; allowing two emissions to be close in rapidity. Every other emission must obey the stronger NLL constraints. This reults in the calculation remaining purely NNLL accurate - we do no want to introduce subleading contributions that are not fully under our control. Each of these scenarios produces a correction to the real-emission function $\mathcal{F}$ in Eq. (9), giving $\mathcal{F}_{NNLL} = \mathcal{F}_{NLL} + \delta\mathcal{F}_i$. As mentioned in Sec. 3, we also include the function $g_3(\lambda)$ in the Sudakov factor's exponent. Thus the cumulative resummed distribution of an event shape v at NNLL is

$$\Sigma(v) = \frac{1}{\sigma_0}\int_0^v dv' \frac{d\sigma(v')}{dv'} = e^{Lg_1(\lambda)+g_2(\lambda)+\frac{\alpha_s(Q)}{\pi}g_3(\lambda)}\left[\mathcal{F}_{\text{NLL}}(\lambda) + \frac{\alpha_s(Q)}{\pi}\delta\mathcal{F}_{\text{NNLL}}(\lambda)\right] \quad (11)$$

Figures (4) and (5) show the effect of adding the ARES resummation to an NNLO fixed order Monte Carlo cross-section for thrust and broadening in e^+e^- (we use the generator `EERAD3` [10]). In the low τ and B_T regions the resummed results are dominant, as expected. It is clear that the fixed order result on its own does not provide a good prediction of the event shape cross-section in this region, and resummation is necessary. For larger values of the observable our resummation tends to the fixed order result, which is adequate in this region. There is full agreement to $O(\alpha_s^3)$ between the results from ARES and observables for which an analytic resummation was already available (thrust

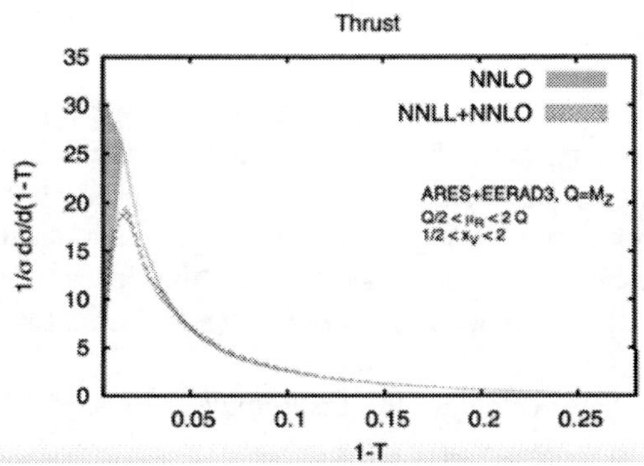

Figure 4. Effect of the resummation on the thrust distribution.

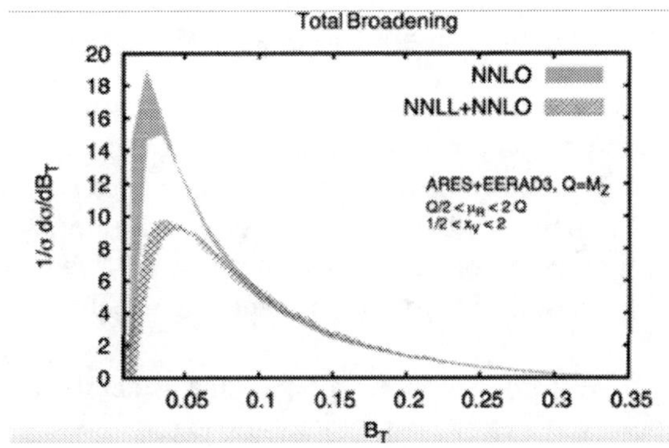

Figure 5. Effect of the resummation on the broadening distribution.

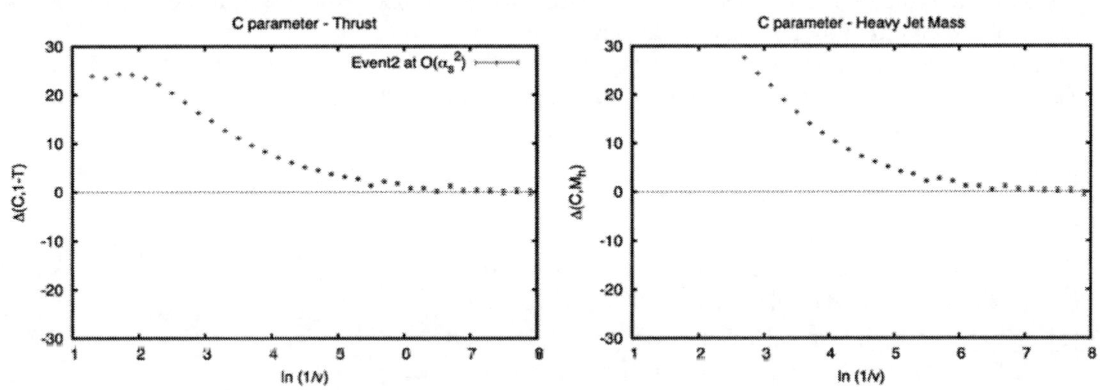

Figure 6. The difference between the expansion of our results and the full result must go to zero as $\ln(1/v) \to$ large positive values.

[13], total and wide broadenings[16], heavy jet mass[17]). To check our new results we compare our resummation to fixed order by expanding it in powers of the strong coupling. Figure (6) shows the difference between the fixed-order generator `Event2` [18] and the expansion of our resummation to order in α_s^2. To obtain stable distributions the difference of observables with the same k_t-scaling are plotted,

$$\Delta(v_1, v_2) = \left(\frac{1}{\sigma_0} \frac{d\sigma^{\mathrm{NLO}}}{d \ln \frac{1}{v_1}} - \frac{1}{\sigma_0} \frac{d\sigma^{\mathrm{NNLL}}|_{\mathrm{expanded}}}{d \ln \frac{1}{v_1}} \right) - \{v_1 \to v_2\} \,. \tag{12}$$

As expected, this difference tends to zero for low-v where the resummation is necessary to correctly capture the behaviour of the cross-section. This same trend can be seen for all of the observables we have resummed [9].

4.4 Hadronic Event Shapes in ARES

Having built the tools to carry out the resummation of a generic event shape in e^+e^- annihilation, we turn our attention to the resummation of hadronic event shapes at NNLL accuracy. A few adjustments are required in dealing with the extra beam information, however the resummation can be viewed as a combination of the same observable in e^+e^- annihilation and a beam part. For example, the resummation of the transverse thrust will be equivalent to the thrust in e^+e^- plus the beam thrust. Similarly, the hadronic jet broadening is equivalent to the jet broadening in e^+e^- plus the beam thrust.

We have developed a novel and general framework to calculate the resummed cross-section of observables to NNLL order, utilising the code to resum event-shape variables in $e^+e^- \to$ 2jets. We are currently working to extend this framework to allow the resummation of hadronic event shapes to NNLL accuracy.

Acknowledgements

I would like to thank the organisers of ISMD2015 for the chance to present this work, and for an enjoyable week full of discussion at Wildbad Kreuth.
This work was carried out in collaboration with Andrea Banfi, Pier Francesco Monni and Giulia Zanderighi. I would like to thank them for their support and input whilst I was preparing this talk.

References

[1] A. Banfi, G. P. Salam, G. Zanderighi, JHEP **0408**, 62 (2004)
[2] A. Banfi, G. P. Salam, G. Zanderighi, JHEP **1006**, 038 (2010)
[3] Y. Hatta and T. Ueda, Nucl. Phys. B **874**, 808820 (2013)
[4] A. J. Larkoski, I. Moult, D. Neill, JHEP **1509**, 143 (2015)
[5] A. Banfi, G. P. Salam, G. Zanderighi, JHEP **0503**, 073 (2005)
[6] T. Becher and X. G. i Tormo, JHEP **1505**, 71 (2015)
[7] I. W Stewart, F. J. Tackmann and W. J. Waalewijn, Phys. Rev. Lett. **106** (2011)
[8] I. W Stewart, F. J. Tackmann and W. J. Waalewijn, Phys. Rev. Lett. **105** (2010)
[9] A. Banfi, H. McAslan, P. F. Monni, G. Zanderighi, JHEP **1505**, 102 (2015)
[10] A. Gehrmann-De Ridder, T. Gehrmann, E. W. N. Glover and G. Heinrich, Comput. Phys. Commun. **185** 3331 (2014)
[11] A. H. Hoang, D. W. Kolodrubetz, V. Mateu and I. W. Stewart, Phys. Rev. D **91**, 094017 (2015)
[12] S. Alioli, C. W. Bauer, C. J. Berggren, A. Hornig, F. J. Tackmann, C. K. Vermilion, J. R. Walsh and S. Zuberi, JHEP **1309** 120 (2013)
[13] T. Becher and M. D. Schwartz, JHEP **0807**, 034 (2008)
[14] P. F. Monni, T. Gehrmann and G. Luisoni, JHEP **1108**, 010 (2011)
[15] A. Banfi, P. F. Monni, G. P. Salam and G. Zanderighi, Phys. Rev. Lett. **109**, 202001 (2012)
[16] T. Becher and G. Bell, JHEP **1211**, 126 (2012)
[17] Y. T. Chien and M. D. Schwartz, JHEP **1008**, 058 (2010)
[18] S. Catani and M. H. Seymour, Nucl. Phys. B **485**, 291 (1997) [Erratum-ibid. B **510**, 503 (1998)]

EPJ Web of Conferences **120**, 05005 (2016) DOI: 10.1051/epjconf/201612005005
ISMD 2015

Studies of jet cross-sections and production properties with the ATLAS and CMS detectors

Nuno Anjos[1,a,b] on behalf of the ATLAS and CMS Collaborations

[1] *Institut de Física d'Altes Energies (IFAE), The Barcelona Institute of Science and Technology, Campus UAB, 08193 Bellaterra (Barcelona) Spain.*

Abstract. Several characteristics of jet production in pp collisions have been measured by the ATLAS and CMS collaborations at the LHC. Measurements of event shapes and multi-jet production probe the dynamics of QCD in the soft regime and can constrain parton shower and hadronisation models. Measurements of multi-jet systems with a veto on additional jets probe QCD radiation effects. Double-differential cross-sections for three- and four-jet final states are measured at different centre-of-mass energies of pp collisions and are compared to expectations based on NLO QCD calculations. The distribution of the jet charge has been measured in di-jet events and compared to predictions from different hadronisation models and tunes. Jet-jet energy correlations are sensitive to the strong coupling constant. These measurements constitute precision tests of QCD in a new energy regime.

1 Introduction

Jets are narrow collimated clusters of stable particles (mainly hadrons) produced in high-energy particle interactions. In the theory of strong interactions, quantum chromodynamics (QCD), jets arise from the fragmentation of partons (quarks and gluons). In a first (perturbative) phase, partons with high-transverse momentum (p_{T}) are produced in a hard-scattering process. Partons evolve from hard to soft scales by a process of emission of additional gluons and production of quark-antiquark pairs, known as the parton shower (PS). This phase can be described by perturbative QCD (pQCD) approximations. In the second (non-perturbative) phase, jet evolution is an interplay between the hadronisation process and the underlying event (UE). The hadronisation process governs the transition from partons to hadrons. The UE represents initial-state radiation, multiple parton interactions and colour-reconnection effects. Non-perturbative (NP) effects are most pronounced at low p_{T}, and are accounted for through the use of phenomenological models that are tuned to the data.

ATLAS [1] and CMS [2] are multi-purpose experiments at the Large Hadron Collider (LHC), the proton-proton (pp) collider at CERN. This paper presents several studies [3–11] of jet production

[a] e-mail: nuno.anjos@cern.ch

[b] Work supported by the Beatriu de Pinós program managed by Agència de Gestió d'Ajuts Universitaris i de Recerca with the support of the Secretaria d'Universitats i Recerca of the Departament d'Economia i Coneixement of the Generalitat de Catalunya, and the Cofund program of the Marie Curie Actions of the 7th R&D Framework Program of the European Union. Work partially supported by MINECO under grants SEV-2012-0234, FPA2013-48308, and FPA2012-38713, which include FEDER funds from the European Union.

performed by ATLAS and CMS with data taken at centre-of-mass energies ($\sqrt{s}$) of 7 and 8 TeV during Run 1 of the LHC, from 2010 until 2012. The ATLAS and CMS detectors cover a solid angle of almost 4π. Closest to the interaction point are precise tracking devices, surrounded by non-compensating calorimeters and muon chambers. All analyses presented here use single jet calorimeter based triggers to select events online. The ATLAS analysis in Sec. 3 also uses minimum bias triggers. As inputs to offline jet reconstruction, ATLAS uses topoclusters [12] while CMS uses Particle Flow (PF) objects [13]. The topoclustering algorithm groups cells from the finely segmented calorimeters in three-dimensional topological clusters according to an iterative procedure designed to suppress noise. The PF algorithm utilizes the best energy measurements of each particle candidate from the most suitable combination of detector components. Both experiments use the anti-k_T [14] algorithm as implemented in the FastJet [15] package to reconstruct jets. The values of the jet distance parameter R used by ATLAS are 0.4 and 0.6, while CMS uses 0.5 and 0.7. Both experiments apply a factorized jet calibration procedure with corrections for pile-up and jet energy scale.

Both experiments apply unfolding procedures to correct data for detector effects and allow comparisons with theory. Theoretical predictions are obtained from Monte Carlo (MC) event generators. The Pythia 6 [16], Pythia 8 [17], Herwig 6 [18] and Herwig++ [19] MCs complement leading-order (LO) $2 \rightarrow 2$ matrix elements (ME) with parton showers to simulate higher-order processes. MadGraph [20] supplies an improved LO ME calculation with $2 \rightarrow 4$ outgoing partons that can be matched to one of the above PS MCs. Alpgen provides LO MEs for $2 \rightarrow n$ multi-parton final states, with $n \leq 6$, interfaced with Herwig 6 for the PS. The Powheg framework [21–23] provides a full next-to-leading-order (NLO) di-jet calculation accounting for the emission of an additional third hard parton, and can also be matched to a PS MC.

2 Hadronic Event Shapes

Event shapes (ES) are observables that measure the energy flow in an event. They are combinations of the momenta of final state particles in a number related to the event geometry. In terms of hadronic jet production, ES are an indirect probe of multi-jet topologies. They are theoretically defined in an infrared- and collinear-safe manner and can be computed in pQCD. A large set of ES sensitive to QCD dynamics from soft (hadronisation) to hard (multi-jet radiation) scales has been proposed [24, 25] for studies in pp collisions. CMS has studied ES [3] using a dataset with an integrated luminosity of 5 fb^{-1} recorded at $\sqrt{s}$ = 7 TeV during 2011. The ES studied by CMS are the transverse thrust ($\tau_\perp$), jet broadening (B_{tot}), jet mass (both total, ρ_{tot}, and in the transverse plane, $\rho^{\mathrm{T}}_{\mathrm{tot}}$) and the third-jet resolution parameter (T_{23}). Events are selected with at least one $R = 0.5$ jet with transverse momentum $p_{\mathrm{T}} > 110$ GeV and absolute pseudorapidity $|\eta| < 2.4$. The ES distributions are studied in five leading jet p_{T} ranges and compared to predictions from standard and improved LO MCs, combined with several PS and UE tunes: D6T [26], Perugia-P0 [27], Z2 [28] and tune4C [29]. Fig. 1 (left) shows the $\tau_\perp$ distribution for one range of leading jet p_{T}. The MCs agrees with data to within 20%. ES more sensitive to the longitudinal energy flow or hard parton emission, such as ρ_{tot} in Fig. 1 (right), exhibit larger disagreement between data and MC. MadGraph consistently reproduces all distributions within uncertainties.

3 Di-jet Azimuthal Decorrelation

At leading order (LO) in pQCD, two final-state partons balanced in p_{T} are produced in a back-to-back topology. The azimuthal angular separation between the two p_{T} leading jets, $\Delta\phi = |\phi_1 - \phi_2|$, equals π. The production of a third high-p_{T} jet, however, leads to a decorrelation in azimuthal angle. The smallest achievable value of $\Delta\phi = 2\pi/3$ is reached with a symmetric star-shaped three-jet configuration.

DOI: 10.1051/epjconf/201612005005

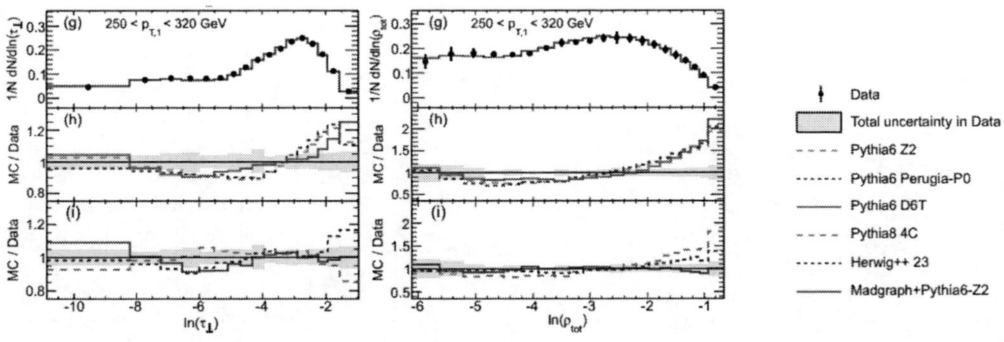

Figure 1. The transverse thrust (left) and jet mass (right) distributions in the 250 to 320 GeV leading jet p_T range from 5 fb^{-1} of CMS data recorded at $\sqrt{s} = 7$ TeV, compared to MadGraph predictions. The prediction/data ratios for the MC models listed in the caption are also shown [3].

If more than three jets are found, $\Delta\phi$ decreases and approaches the limit of the jet size. Therefore, the measurement of the di-jet azimuthal decorrelation is a way to study multi-jet production and test NLO predictions without actually measuring jets beyond the leading two. CMS measured the di-jet azimuthal decorrelation [4] using 19.7 fb^{-1} of data recorded at $\sqrt{s} = 8$ TeV in 2012. Events with at least two $R = 0.7$ jets with rapidity $|y| < 2.5$ and (sub)leading jet p_T >(100) 200 GeV are selected. To reduce the background from $t\bar{t}$ and heavy vector boson production to hard-QCD jet events, events for which the ratio of missing transverse momentum E_T^{miss} to total transverse energy $\sum E_T$ is larger than 0.1 are rejected. The di-jet cross-section is measured as a function of $\Delta\phi$ and normalized to the total di-jet cross-section, reducing many experimental and theoretical uncertainties. The measured differential cross-section is shown in Fig. 2 (left) in slices of the leading jet p_T up to 2.2 TeV. It is compared with LO and NLO MC predictions. MadGraph agrees with data throughout the entire $\Delta\phi$ range, Pythia 8 also provides a reasonable description of the data, while the other MCs overestimate the data in specific regions.

Although fixed order computations provide a good description of jet production, for certain observables and in some regions of the phase space higher-order contributions are important and must be resummed to all orders. This is the case for di-jet events with large rapidity separation between the two jets, Δy, and/or when a veto is applied on additional jets with $p_T > Q_0$ produced in the rapidity interval between them; Q_0 is called a veto scale. There are two available resummation approaches, BFKL and DGLAP. BFKL resums the higher orders in terms of $\ln(1/x)$ while DGLAP does it in terms of $\ln(Q^2)$. The DGLAP inspired procedure is implemented in Powheg interfaced with Herwig 6 or Pythia 8. The predictions based on BFKL are generated with Hej [30, 31], alone or interfaced with Ariadne [32]. Hej resums leading-logarithms relevant in the Mueller-Navelet limit. Ariadne provides parton showers based on the colour-dipole cascade model. ATLAS has studied jet production in events with large-y separated di-jets and a veto on a third in-between jet [5], using 36.1 pb^{-1} and 4.5 fb^{-1} of data collected at $\sqrt{s} = 7$ TeV during 2010 and 2011 respectively. Events with at least two $R = 0.6$ jets, with (sub)leading jet p_T >(50) 60 GeV are selected. For 2010 data, the full rapidity range of $|y| < 4.9$ was used. For 2011 data, $|\Delta y| \leq 1$ and $|y| < 2.4$ are required on the two leading jets. Events are classified into two categories: the inclusive and gap di-jet events. An event contributes to the gap sample if there is no jet with $p_T > Q_0$ in the rapidity interval between the two leading jets. The veto scale Q_0 was set to $p_T > 20$ (30) GeV for data collected in 2010 (2011). The gap fraction, the ratio of the di-jet cross-sections measured using the gap and the inclusive samples, and the mean multiplicity

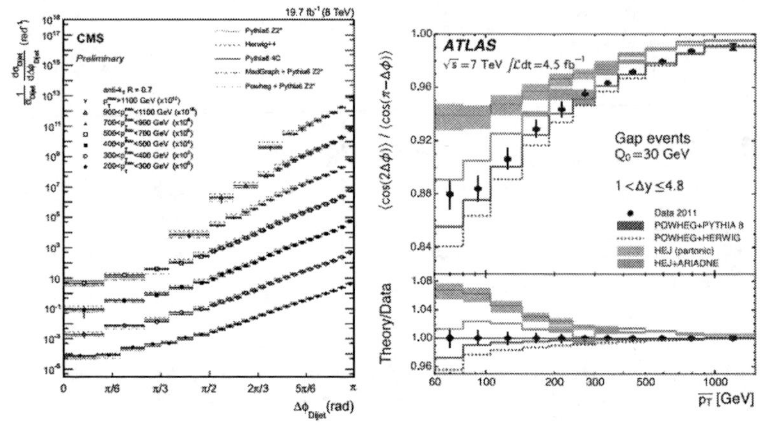

Figure 2. Normalised CMS di-jet cross-section, differential in $\Delta\phi$ for seven leading p_T regions, compared to MC predictions and scaled for presentation purposes (left) [4]. Ratio of the first two angular moments measured in the ATLAS di-jet gap sample, compared to MC predictions (right) [5].

of jets with $p_T > Q_0$ in the gap are measured. Powheg+Pythia 8 and Hej+Ariadne provide the best agreement with data. The observable with best discrimination is the ratio of the first two angular moments, $\langle\cos(2\Delta\phi)\rangle/\langle\cos(\pi - \Delta\phi)\rangle$. It is shown in Fig. 2 (right) for the gap sample. The data is below the Hej+Ariadne prediction and slightly above the Powheg+Pythia 8 prediction.

4 Multi-jet Topologies

Multi-jet topologies test both higher order effects in pQCD and NP dynamics. Topological variables are sensitive to QCD colour factors, gluon spin structure and hadronisation.

Both ATLAS and CMS have measured the double differential tri-jet cross-section with data recorded at $\sqrt{s} = 7$ TeV in 2011. The cross-sections are presented as a function of the three-jet mass and in bins of an angular variable. The CMS measurement [6] uses a data set of 5 fb^{-1}. Events where at least three $R = 0.7$ jets are found with $|y| \leq 3$ and $p_T > 100$ GeV are selected. The angular variable is the maximum jet rapidity, $|y|_{max} = \max(|y_1|, |y_2|, |y_3|)$. Two bins of $|y|_{max}$ are defined, $|y|_{max} \leq 1$ and $1 < |y|_{max} \leq 2$, with events outside this range being discarded. The ATLAS measurement [7] uses a data set of 4.5 fb^{-1}. Events where at least three jets are found with $|y| \leq 3$ and with $p_T > 150$, 100 and 50 GeV, respectively, are selected. The angular variable is the sum of the jet separations, $|Y^*| = |\Delta y_{1,2}| + |\Delta y_{2,3}| + |\Delta y_{3,4}|$, and five bins of $|Y^*|$ were defined. Fig. 3 shows the measured tri-jet cross-section as a function of the three-jet mass in bins of the angular variable for CMS (left) and ATLAS (right). The ATLAS results on display use jets with $R = 0.4$, but results were also obtained for $R = 0.6$. Also shown is the comparison with the NLOJet++ [33] prediction with the CT10 [34] parton distribution function (PDF) set and corrected for NP effects. Good agreement between the data and the theoretical predictions was found by both experiments.

ATLAS measured four-jet differential cross-setions [8] using 20.3 fb^{-1} of data recorded at $\sqrt{s} =$ 8 TeV in 2012. Events with at least four $R = 0.4$ jets within $|y| < 2.8$, leading jet $p_T > 100$ GeV, three subleading jets with $p_T > 64$ GeV and minimal angular distance between any pair of four leading jets larger than 0.65 are selected. Cross-sections are measured as a function of several variables including the four-jet mass, as shown in Fig. 4 (left). Results are compared to predictions from several LO and NLO MCs.

CMS studied distributions of topological observables in inclusive three- and four-jet events [9] using 5.1 fb^{-1} of data recorded at $\sqrt{s} = 7$ TeV in 2011. All $R = 0.5$ jets are required to have $p_T > 50$ GeV and $|y| < 2.5$. The leading jet p_T must be in the 190–300 GeV range or above 500 GeV

 DOI: 10.1051/epjconf/201612005005

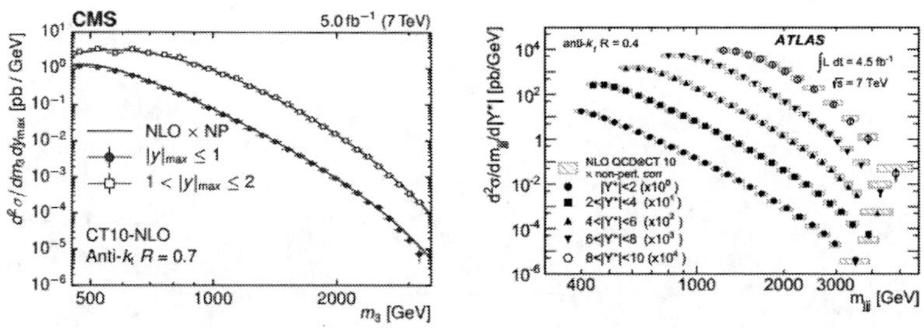

Figure 3. The tri-jet cross-section as a function of the three-jet mass $m_3 = m_{jjj} = \sqrt{p_1^2 + p_2^2 + p_3^2}$ and in bins of the angular variable, measured by CMS (left) [6] and ATLAS (right) [7] compared with NLO predictions corrected for NP effects.

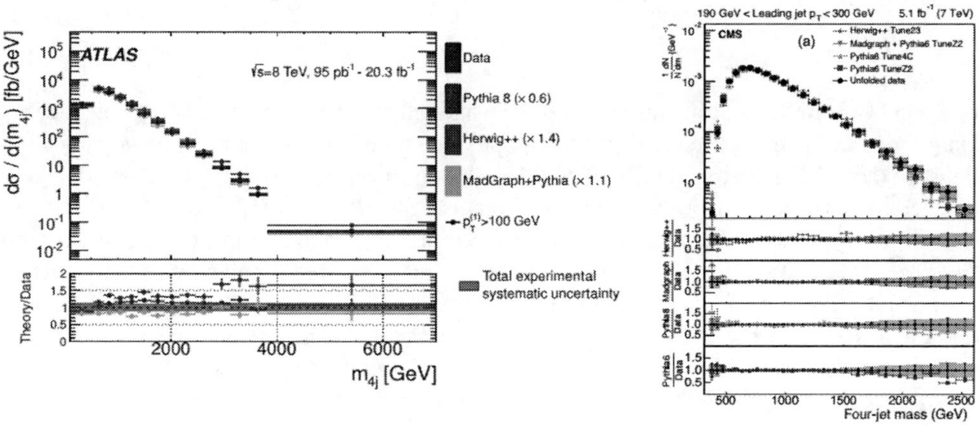

Figure 4. Four jet differential cross-section a function of $m_{4j} = \sqrt{p_1^2 + p_2^2 + p_3^2 + p_4^2}$ measured by ATLAS at $\sqrt{s} = 8$ TeV (left) [8]. Distribution of m_{4j} in events with leading jet in the 190–300 GeV range measured by CMS at $\sqrt{s} = 7$ TeV (right) [9]. Both are compared to different theoretical predictions.

to satisfy trigger criteria. Events with at least three (four) jets satisfying the selection are classified as three(four)-jet events. Distributions are obtained as a function of multiple variables, including dedicated angular variables and the three- and four-jet masses, the last one shown in Fig. 4 (right). Results are compared to predictions from several LO MCs.

Both ATLAS and CMS find good overall agreement between data and theory, with discrepancies in specific regions. The LO MCs can disagree with the data due to the missing higher multiplicities, while the NLO MCs can have large uncertainties, specially at low momenta.

5 Jet Charge

Due the confining nature of the strong force, the electric charge of quarks is not directly observable. It is however possible to define a jet charge observable as the momentum-weighted sum of the charges of tracks associated to a jet. This observable is sensitive to the charge of the initiating quark or gluon and can provide constraints on models of jet formation. ATLAS measured the jet charge [10] using 20.3 fb^{-1} of data recorded at $\sqrt{s} = 8$ TeV in 2012. Events are required to have at least two $R = 0.4$ jets

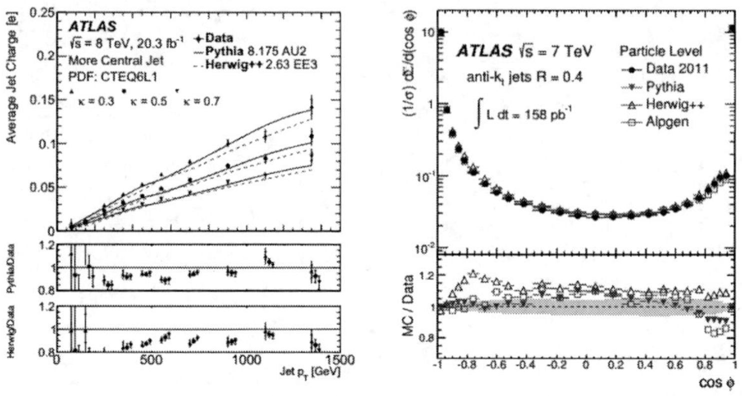

Figure 5. The average jet charge in units of positron charge measured by ATLAS as a function of jet p_T, for $\kappa = 0.3, 0.5, 0.7$ and for the most central jets, using the CT10 PDF set. (left) [10]. The TEEC distribution measured by ATLAS data with comparisons to MC expectations.(right) [11].

with $p_T > 50$ GeV in the central region, ($|y| < 2.1$), and only the two leading jets are considered for the jet charge measurement. To select di-jet topologies, the two leading jets must have $p_{T,1}/p_{T,2} < 1.5$. The jet with the smallest (largest) $|\eta|$ is classified as the most central (most forward) jet. The jet charge Q_J of a jet J is calculated using a transverse-momentum weighting, $Q_J = (1/p^{\kappa}_{T,J}) \sum_i q_i p^{\kappa}_{T,i}$, where i runs over all tracks associated to jet J, $p_{T,i}$ is the transverse momentum of track i with associated charge q_i, $p_{T,J}$ is the transverse momentum of the jet and κ is a free regularisation parameter (values of 0.3, 0.5 and 0.7 are used) that controls the sensitivity to soft radiation. The average of the jet charge distribution is shown in Fig. 5 (left) compared to several MCs. The calculations are generally about 10% below the data, possibly due to fragmentation modeling. The jet charge increases with jet p_T, which is expected to happen due to the increase in up-flavour jets.

6 Transverse Energy-energy Correlations

The transverse energy-energy correlation (TEEC) [35] is an event shape proposed for the study of hadronic collisions. that exhibits a quadratic dependence on the strong coupling constant. The TEEC distribution is obtained by calculating the cosines of the angles in the transverse plane between all possible pairs of jets in each event, ϕ_{ij}. Every pair (i, j) represents an entry in the distribution, which is then weighted with the normalised product of the transverse energies. ATLAS measured the TEEC [11] using 158 pb^{-1} of data recorded at $\sqrt{s} = 7$ TeV in 2011. The event selection requires at least two $R = 0.4$ jets with $p_T > 50$ GeV and $|\eta| < 2.5$, as well as $p_{T,1} + p_{T,2} > 500$ GeV. The comparison of the measured TEEC spectrum to the MC predictions is shown in Fig. 5 (right). The description of the TEEC is good in the back-to-back region $\cos\phi \simeq -1$ for both Pythia 6 and Alpgen+Herwig. Differences up to 10% are observed in the central part, while the region of small angles shows differences as large as about 15%. The description by Herwig++ is poorer. The $\alpha_s(m_Z^2)$ value is extracted by performing a χ^2 fit of theoretical predictions to the TEEC data. The theoretical predictions are provided by NLO pQCD with NP corrections. This yields $\alpha_s = 0.1173 \pm 0.0010$ (exp.) $^{+0.0063}_{-0.0020}$ (scale) $\pm$ 0.0017 (PDF) $\pm$ 0.0002 (NP corrections), where the uncertainties are experimental, from the scale choice, the choice of PDFs and the NP corrections, respectively.

7 Conclusions

Multi-jet production in pp collisions was extensively studied by the ATLAS and CMS collaborations at centre-of-mass energies of 7 and 8 TeV during Run 1 of the LHC . Studies presented include hadronic event shapes, TEEC, jet charge in di-jet events, di-jet azimuthal decorrelation with or without a jet veto, double-differential cross-sections for three- and four-jet final states, and multi-jet topological distributions. Measurements were compared with NLO predictions corrected for NP effects, and ME generators at LO and NLO. Overall, good agreement between data and theory was found, and the TEEC were used to measure the strong coupling constant.

References

[1] ATLAS Collaboration, JINST **3**, S08003 (2008)
[2] CMS Collaboration, JINST **3**, S08004 (2008)
[3] CMS Collaboration, JHEP **10**, 087 (2014).
[4] CMS Collaboration, CMS-PAS-SMP-14-015, https://cdsweb.cern.ch/record/2019401.
[5] ATLAS Collaboration, Eur. Phys. J. **C 74**, 3117 (2014).
[6] CMS Collaboration, Eur. Phys. J. **C 75**, 186 (2015).
[7] ATLAS Collaboration, Eur. Phys. J. **C 75**, 228 (2015).
[8] ATLAS Collaboration, arXiv:1509.07335 [hep-ex].
[9] CMS Collaboration, Eur. Phys. J. **C 75**, 302 (2015).
[10] ATLAS Collaboration, arXiv:1509.05190 [hep-ex].
[11] ATLAS Collaboration, Phys. Lett. **B 750**, 427 (2015).
[12] W. Lampl et al, ATL-LARG-PUB-2008-002, http://cdsweb.cern.ch/record/1099735.
[13] CMS Collaboration, CMS-PAS-PFT-09-001, https://cdsweb.cern.ch/record/1194487.
[14] M. Cacciari, G. P. Salam, and G. Soyez, JHEP **04** 063 (2008).
[15] M. Cacciari, G. P. Salam, and G. Soyez, Eur. Phys. J. **C 72**, 1896 (2012).
[16] T. Sjöstrand, S. Mrenna and P. Z. Skands, JHEP **05**, 026 (2006).
[17] T. Sjöstrand, S. Mrenna and P. Z. Skands, Comput. Phys. Commun. **178**, 852 (2008).
[18] G. Corcella et al., JHEP **0101**, 010 (2001).
[19] M. Bähr et al., Eur. Phys. J. **C 58**, 639 (2008).
[20] J. Alwall et al., JHEP **06**, 128 (2011).
[21] S. Frixione, P. Nason, and C. Oleari, JHEP **11**, 070 (2007).
[22] S. Alioli, P. Nason, C. Oleari, and E. Re, JHEP **06** 043, (2010).
[23] P. Nason, JHEP **0411**, 040 (2004).
[24] A. Banfi, G. P. Salam and G. Zanderighi, JHEP **08**, 062 (2004).
[25] A. Banfi, G. P. Salam and G. Zanderighi, JHEP **06**, 038 (2010).
[26] R. Field, Acta Phys. Polon. **B 39**, 2611 (2008).
[27] P. Z. Skands, Phys. Rev. **D 82**, 074018 (2010).
[28] R. Field, Acta Phys. Polon. **B 42**, 2631 (2011).
[29] R. Corke and T. Sjöstrand, JHEP **03**, 032 (2011).
[30] J. R. Andersen and J. M. Smillie, Nucl.Phys.Proc.Suppl. **205-206**, 205–210 (2010).
[31] J. R. Andersen and J. M. Smillie, JHEP **1106**, 010 (2011).
[32] L. Lonnblad, Comput.Phys.Commun. **71**, 15–31 (1992).
[33] Z. Nagy, Phys. Rev. **D 68** 094002 (2003).
[34] H.-L. Lai et al., Phys. Rev. **D 82**, 074024 (2010).
[35] A. Ali, E. Pietarinen and W. J. Stirling, Phys. Lett. **B 141**, 447 (1984).

EPJ Web of Conferences **120**, 06001 (2016) DOI: 10.1051/epjconf/201612006001
ISMD 2015

A viscous blast-wave model for high energy heavy-ion collisions

Amaresh Jaiswal[1,a] and Volker Koch[2]

[1] *GSI, Helmholtzzentrum für Schwerionenforschung, Planckstrasse 1, D-64291 Darmstadt, Germany*
[2] *Lawrence Berkeley National Laboratory, Nuclear Science Division, MS 70R0319, Berkeley, California 94720, USA*

Abstract. Employing a viscosity-based survival scale for initial geometrical perturbations formed in relativistic heavy-ion collisions, we model the radial flow velocity at freeze-out. Subsequently, we use the Cooper-Frye freeze-out prescription, with viscous corrections to the distribution function, to extract the transverse momentum dependence of particle yields and flow harmonics. We fit the model parameters for central collisions, by fitting the spectra of identified particles at the Large Hadron Collider (LHC), and estimate them for other centralities using simple hydrodynamic relations. We use the results of Monte Carlo Glauber model for initial eccentricities. We demonstrate that this improved viscous blast-wave model leads to good agreement with transverse momentum distribution of elliptic and triangular flow for all centralities and estimate the shear viscosity to entropy density ratio $\eta/s \simeq 0.24$ at the LHC.

1 Introduction

The quark-gluon plasma (QGP) formed in relativistic heavy-ion collisions exhibit strong collective behaviour and hence can be studied within the framework of relativistic hydrodynamics. The hydrodynamical modelling of heavy-ion collisions suggests that the QGP behaves like a nearly perfect fluid having an extremely small shear viscosity to entropy density ratio η/s [1–3]. Apart from hydrodynamics, the collective behaviour of QGP can also be studied within the so-called blast-wave model. Schnedermann *et. al.* used the blast-wave model for the first time to fit the transverse momentum spectra using a simple functional form for the phase-space density distribution at kinetic freeze-out [4]. They approximated hydrodynamical results with only two parameters: a kinetic temperature, and a radial flow strength. Subsequently, there were several improvements incorporated in this model to account for the anisotropies in the transverse flow profile [5, 6]. Within the blast-wave model, Teaney made the first attempt to estimate the effect of shear viscosity on spectra and elliptic flow [7]. However, the viscous effects were included only in the freeze-out prescription and the model parameters had to be fit separately for central and non-central collisions.

Here we generalize the blast-wave model to include viscous effects by using a viscosity-based survival scale for initial geometrical anisotropies, formed in relativistic heavy-ion collisions, to parametrize the radial flow velocity. Including η/s, the present model has five parameters which has to be fitted only for one centrality. We obtain obtain relations to account for the centrality dependence of model parameters, so that one only need to fit them for central collisions, rendering the model

[a] e-mail: jaiswal.amaresh@gmail.com

with increased predictive power. We also consider the first-order viscous corrections to the distribution function in the Cooper-Frye freeze-out prescription for particle production, [7]. In essence, we provide a model which incorporates the key features of viscous hydrodynamic evolution without requiring to do the actual evolution. We employ this viscous blast-wave model to obtain the transverse momentum dependence of particle yields and flow harmonics for LHC. We fix the model parameters by fitting the transverse momentum distribution of identified particle spectra. Subsequently, we demonstrate that this leads to reasonably good agreement with transverse momentum dependence of elliptic and triangular flow for various centralities. Within the present model, we estimate the shear viscosity to entropy density ratio $\eta/s \simeq 0.24$ at the LHC.

2 Blast wave model

The blast-wave model has been used extensively to fit experimental data and it provides good description of spectra and elliptic flow observed in relativistic heavy-ion collisions [5–9]. The most important feature of the blast-wave model is the Hubble like parametrization of transverse velocity which is found to be in agreement with hydro results [10, 11]. The model assumes boost invariance [12], as well as rotational invariance. We work in the Milne co-ordinate system where,

$$\tau = \sqrt{t^2 - z^2}, \qquad \eta_s = \tanh^{-1}(z/t), \qquad r = \sqrt{x^2 + y^2}, \qquad \varphi = \mathrm{atan2}(y, x). \tag{1}$$

Here the metric tensor for this co-ordinate system is $g_{\mu\nu} = \mathrm{diag}(1, -\tau^2, -1, -r^2)$. The hydrodynamic fields are parametrized as [7]

$$T = T_f\,\Theta(R - r), \qquad u^r = u_0\,\frac{r}{R}\,\Theta(R - r), \qquad u^\varphi = u^{\eta_s} = 0, \qquad u^\tau = \sqrt{1 + (u^r)^2}, \tag{2}$$

where R is the transverse radius of the fireball at freeze-out. The condition $u^\mu u_\mu = 1$ leads to the expression for u^τ.

Using the Cooper-Frye prescription for particle production, the hadron spectra can be obtained as [13]

$$\frac{dN}{d^2p_T dy} = \frac{1}{(2\pi)^3}\int p_\mu d\Sigma^\mu f(x, p), \tag{3}$$

where $d\Sigma_\mu$ is the oriented freeze-out hyper-surface and $f(x, p)$ is the distribution function of the particles. The distribution function can be split as equilibrium and non-equilibrium parts, $f = f_0 + \delta f$ where the equilibrium distribution function is given by

$$f_0 = \frac{1}{\exp(u_\mu p^\mu/T) + a}, \tag{4}$$

where $a = +1$ for baryons and $a = -1$ mesons. For a system close to equilibrium, i.e., $\delta f \ll f_0$, we employ the Grad's 14-moment approximation for the non-equilibrium part [14]

$$\delta f = \frac{f_0\tilde{f}_0}{2(\epsilon + P)T^2}\, p^\alpha p^\beta \pi_{\alpha\beta} \quad \Rightarrow \quad \delta f_1 = \frac{f_0\tilde{f}_0}{T^3}\left(\frac{\eta}{s}\right) p^\alpha p^\beta \nabla_{\langle\alpha}u_{\beta\rangle}. \tag{5}$$

where $\tilde{f}_0 = 1 - af_0$ is the quantum factor, $\pi_{\alpha\beta}$ is the shear stress tensor, η is the coefficient of shear viscosity, $s = (\epsilon + P)/T$ is the entropy density and the angular brackets denote traceless symmetric projection orthogonal to the fluid four-velocity [15]. In the second equality, we have approximated the shear stress tensor with its first-order relativistic Navier-Stokes expression, $\pi_{\alpha\beta} = 2\eta\nabla_{\langle\alpha}u_{\beta\rangle}$. In the case of blast-wave model, the form of $p^\alpha p^\beta \nabla_{\langle\alpha}u_{\beta\rangle}$ is calculated in Refs. [7, 16].

The anisotropic flow is defined as

$$v_n(p_T) \equiv \frac{\displaystyle\int_{-\pi}^{\pi} d\phi \, \cos[n(\phi - \Psi_n)] \, \frac{dN}{dy \, p_t \, dp_T \, d\phi}}{\displaystyle\int_{-\pi}^{\pi} d\phi \, \frac{dN}{dy \, p_t \, dp_T \, d\phi}}, \tag{6}$$

where Ψ_n is the event-plane angle for the n-th harmonic. Up to first order in viscosity [7],

$$v_n(p_T) = v_n^{(0)}(p_T) \left(1 - \frac{\displaystyle\int d\phi \, \frac{dN^{(1)}}{dy \, p_t \, dp_T \, d\phi}}{\displaystyle\int d\phi \, \frac{dN^{(0)}}{dy \, p_t \, dp_T \, d\phi}} \right) + \frac{\displaystyle\int d\phi \, \cos[n(\phi - \Psi_n)] \, \frac{dN^{(1)}}{dy \, p_t \, dp_T \, d\phi}}{\displaystyle\int d\phi \, \frac{dN^{(0)}}{dy \, p_t \, dp_T \, d\phi}}, \tag{7}$$

where the superscript '(0)' and '(1)' denote quantities calculated using the ideal distribution function, Eq. (4), and first-order viscous correction, Eq. (5), respectively.

3 Viscous blast-wave model

In terms of the Fourier expansion for a single-particle distribution, the definition of the participant anisotropies, ε_n, is

$$f(\varphi) = \frac{1}{2\pi} \left[1 + 2 \sum_{n=1}^{\infty} \varepsilon_n \, \cos[n(\varphi - \psi_n)] \right]. \tag{8}$$

Here ψ_n is the angle between the x axis and the major axis of the participant distribution. The initial geometrical anisotropies, ε_n, eventually converts to anisotropies in the radial fluid velocity,

$$u^r = u_0 \, \frac{r}{R} \left[1 + 2 \sum_{n=1}^{\infty} u_n \, \cos[n(\varphi - \psi_n)] \right]. \tag{9}$$

Next, we determine the conversion efficiency of the initial geometrical eccentricity to final anisotropy in the radial fluid velocity, u_n/ε_n.

In Ref. [17], a viscosity-based survival scale for anisotropic structures formed by point like perturbations was proposed using the dispersion relation for sound in a viscous medium,

$$\omega = c_s k + i k^2 \frac{1}{T} \left(\frac{2}{3} \frac{\eta}{s} \right), \tag{10}$$

In the above equation, c_s is the speed of sound in the medium and η is the coefficients of shear viscosity. Using a Fourier ansatz, $\exp(i\omega t - ikx)$, we see that the amplitudes of the stress tensor harmonics, with momentum k, are attenuated by a factor

$$\delta T^{\mu\nu}(t, k) = \exp \left[- \left(\frac{2}{3} \frac{\eta}{s} \right) \frac{k^2 t}{T} \right] \delta T^{\mu\nu}(0, k), \tag{11}$$

where we have ignored the oscillatory pre-factor. The presence of momentum squared in the exponent results in enhanced effect of viscosity for the higher harmonics.

We note that each harmonics is essentially a damped oscillator with wave-vector k whose amplitude is progressively suppressed due to viscous effects. They form standing waves on the fireball circumference throughout the evolution, i.e.,

$$2\pi R = n\frac{2\pi}{k}, \tag{12}$$

where R is the transverse radius of the fireball at freeze-out. Therefore, at the freeze-out time t_f, the wave amplitude reaction is

$$\frac{\delta T^{\mu\nu}|_{t=t_f}}{\delta T^{\mu\nu}|_{t=0}} = \exp\left[-n^2\left(\frac{2}{3}\frac{\eta}{s}\right)\frac{t_f}{R^2 T_f}\right], \tag{13}$$

where T_f is the freeze-out temperature.

The final radial flow velocity is due to initial geometrical perturbations and in absence of viscosity, the conversion efficiency remains the same for all harmonics. For a viscous medium, the conversion efficiency must be proportional to the wave amplitude reaction,

$$\frac{u_n}{\varepsilon_n} = \alpha_0 \exp\left[-n^2\left(\frac{2}{3}\frac{\eta}{s}\right)\frac{t_f}{R^2 T_f}\right], \tag{14}$$

where α_0 is the constant of proportionality. In Ref. [18], Shuryak and Zahed proposed that the r.h.s. of Eq. (14) should be equal to v_n/ε_n where v_n is the final p_T-integrated flow harmonics. This assumption misses several key features of hydrodynamic evolution [19–21]. We therefore emphasize that the acoustic damping should be applicable to the hydrodynamic variables u_n rather then to the final state observables v_n [16].

4 Initial conditions and numerical results

In order to reduce the number of free parameters of the blast-wave model, we set-up initial conditions of the collisions. We consider two identical nuclei with mass number A colliding at relativistic energies with impact parameter b. The radius r_0 of the circle, drawn in the transverse plane, which divides the boundary of the overlap zone equally in four parts is given by [16]

$$r_0 = \frac{1}{2}\left(b^2 - 2\,bR_0\sqrt{2+\frac{b}{R_0}} + 4R_0^2\right)^{1/2}, \tag{15}$$

where $R_0 = 1.25A^{1/3}$ fm is the radius of each colliding nuclei. This is done in order to approximate the boundary of the overlap zone as the second harmonics of initial geometrical fluctuations ε_2, on r_0. Note that ε_2 is the most prominent eccentricity for non-central collisions.

The transverse expansion of the fireball is obtained by using the radial velocity, Eq. (2),

$$u^r \equiv \frac{dr}{d\tau} = u_0\frac{r}{R} \quad\Rightarrow\quad \int_{r_0}^{R}\frac{dr}{r} = \int_0^{\tau_f}\frac{u_0}{R}\,d\tau \quad\Rightarrow\quad R = r_0\exp\left(\frac{u_0\,\tau_f}{R}\right). \tag{16}$$

which can be solved for R given the freeze-out time τ_0 and the isotropic expansion velocity u_0. Next we determine the freeze-out times for non-central collisions when it is fixed for the central one. To this end, we employ the analytical result of Bjorken expansion $\epsilon \propto \tau^{-4/3}$. Assuming the initial thermalization time and the freeze-out energy density to be same for all collisions, we get

$$\tau_f = \tau_{f0}\left(\frac{\epsilon_i}{\epsilon_{i0}}\right)^{3/4}. \tag{17}$$

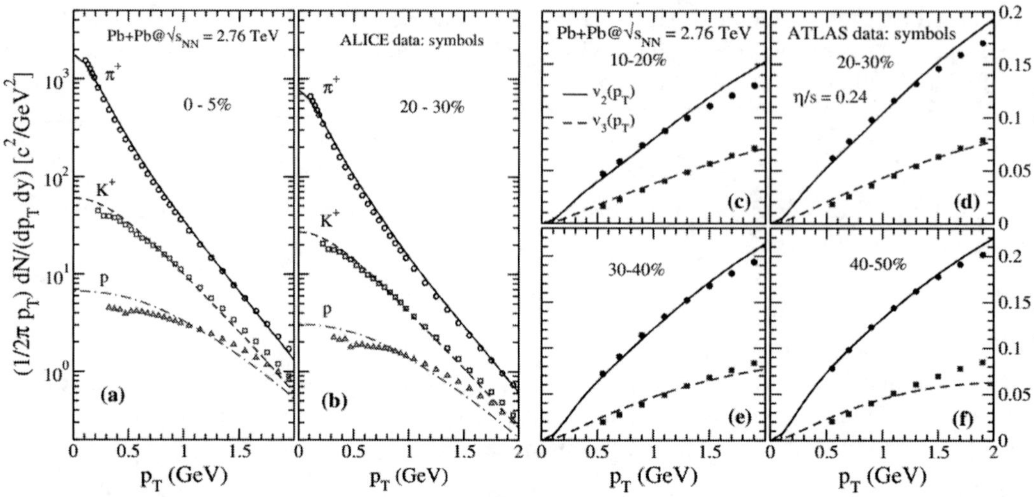

Figure 1. Transverse momentum distribution of particle multiplicities in Pb+Pb collisions at $\sqrt{s_{NN}}$ = 2.76 TeV. We show results for π^+, K^+, and p in two centrality ranges, (a): $0-5\%$ and (b): $20-30\%$. The symbols represent ALICE data [22] at midrapidity and the lines correspond to viscous blast-wave calculations. (c)–(f): Transverse momentum dependence of the anisotropic flow coefficients $v_n(p_T)$ of charged hadrons, for $n = 2$ and 3, calculated at various centralities in Pb+Pb collisions at $\sqrt{s_{NN}}$ = 2.76 TeV in the viscous blast-wave model (lines) with $\eta/s = 0.24$ as compared to the ATLAS data [23] (symbols).

Here τ_{f0} is the freeze-out time for most central collisions, which is fixed by fitting the corresponding transverse momentum spectra. The ratio ϵ_i/ϵ_{i0} is the initial central energy density for all centrality, scaled by its corresponding value in most central collisions, and can be determined using an initial condition model. Therefore the parameters that needs to be fixed within the viscous-blast wave model, in order to fit the spectra, are the freeze-out temperature T_f, the freeze-out time for central collision τ_{f0} and the maximum radial flow velocity u_0. An interplay between α_0 in Eq. (14) and η/s is important to reproduce the flow harmonics.

In Fig. 1, we show our results for transverse momentum distribution of identified particles and $v_n(p_T)$ for Pb+Pb collisions at $\sqrt{s_{NN}}$ = 2.76 TeV and compare it with experimental data measured by the ALICE [22] and ATLAS [23] collaborations. For ϵ_i/ϵ_{i0} we use the results from the Glauber model for various centralities. Figure 1 shows the transverse momentum distribution of pions, kaons, and protons spectra for (a): $0-5\%$ and (b): $20-30\%$ centrality. We observe that, for a freeze-out temperature of 120 MeV, the spectra for π^+ and K^+ from the viscous blast-wave model are in good overall agreement with the ALICE data. On the other hand, the freeze-out temperature for protons is considered to be 135 MeV in order to fit the experimental data. The freeze-out time $\tau_{f0} = 8$ fm was used to fit $0-5\%$ most central collision data. Figure 1 (c)–(f) shows our results for the $v_n(p_T)$, for various centralities, in comparison with the ATLAS data [23]. We find fairly good agreement with the experimental data for elliptic flow, $v_2(p_T)$, and triangular flow, $v_3(p_T)$, by choosing $\alpha_0 = 0.4$ and $\eta/s = 0.24$. We have used the root-mean square values of the eccentricities obtained in the MC-Glauber model for initial eccentricity ε_n [24].

5 Conclusions

We have generalized the blast-wave model to incorporate viscous effects by using a viscosity-based survival scale for initial geometrical anisotropies formed in relativistic heavy-ion collisions. This viscous damping formula is introduced in the parametrization of the radial flow velocity. Including η/s, the viscous blast wave model presented here involved five parameters which has to be fitted for only one centrality. This model therefore incorporates the key features of viscous hydrodynamic evolution without requiring to do the actual evolution. This model was used to obtain the transverse momentum distribution of particle yields and anisotropic flow at LHC. The model parameters were fixed by fitting the transverse momentum spectra of identified particles. We showed that, for various centralities, a fairly good agreement was achieved for transverse momentum distribution of elliptic and triangular flow. Within the present model, we estimated the ratio of shear viscosity to entropy density $\eta/s \simeq 0.24$ at the LHC.

Acknowledgements:- A.J. was supported by the Frankfurt Institute for Advanced Studies (FIAS), Germany. The work of V.K. was supported by the ExtreMe Matter Institute (EMMI) and by the Office of Nuclear Physics in the US Department of Energy's Office of Science under Contract No. DE-AC02-05CH11231.

References

[1] P. Romatschke and U. Romatschke, Phys. Rev. Lett. **99**, 172301 (2007).
[2] H. Song and U. W. Heinz, Phys. Rev. C **77**, 064901 (2008).
[3] R. S. Bhalerao, A. Jaiswal and S. Pal, Phys. Rev. C **92**, 014903 (2015).
[4] E. Schnedermann, J. Sollfrank and U. W. Heinz, Phys. Rev. C **48**, 2462 (1993).
[5] P. Huovinen, P. F. Kolb, U. W. Heinz, P. V. Ruuskanen and S. A. Voloshin, Phys. Lett. B **503**, 58 (2001).
[6] C. Adler *et al.* [STAR Collaboration], Phys. Rev. Lett. **87**, 182301 (2001).
[7] D. Teaney, Phys. Rev. C **68**, 034913 (2003).
[8] Z. Tang, Y. Xu, L. Ruan, G. van Buren, F. Wang and Z. Xu, Phys. Rev. C **79**, 051901 (2009).
[9] X. Sun, H. Masui, A. M. Poskanzer and A. Schmah, Phys. Rev. C **91**, 024903 (2015).
[10] D. Teaney, J. Lauret and E. V. Shuryak, nucl-th/0110037.
[11] M. Habich, J. L. Nagle and P. Romatschke, Eur. Phys. J. C **75**, 15 (2015).
[12] J. D. Bjorken, Phys. Rev. D **27**, 140 (1983).
[13] F. Cooper and G. Frye, Phys. Rev. D **10**, 186 (1974).
[14] H. Grad, Comm. Pure Appl. Math. **2**, 331 (1949).
[15] R. S. Bhalerao, A. Jaiswal, S. Pal and V. Sreekanth, Phys. Rev. C **89**, 054903 (2014).
[16] A. Jaiswal and V. Koch, arXiv:1508.05878 [nucl-th].
[17] P. Staig and E. Shuryak, Phys. Rev. C **84**, 034908 (2011).
[18] E. Shuryak and I. Zahed, Phys. Rev. C **88**, 044915 (2013).
[19] R. A. Lacey, Y. Gu, X. Gong, D. Reynolds, N. N. Ajitanand, J. M. Alexander, A. Mwai and A. Taranenko, arXiv:1301.0165.
[20] R. A. Lacey, A. Taranenko, J. Jia, D. Reynolds, N. N. Ajitanand, J. M. Alexander, Y. Gu and A. Mwai, Phys. Rev. Lett. **112**, 082302 (2014).
[21] R. A. Lacey, D. Reynolds, A. Taranenko, N. N. Ajitanand, J. M. Alexander, F. H. Liu, Y. Gu and A. Mwai, arXiv:1311.1728 [nucl-ex].
[22] B. Abelev *et al.* [ALICE Collaboration], Phys. Rev. C **88**, 044910 (2013).
[23] G. Aad *et al.* [ATLAS Collaboration], Phys. Rev. C **86**, 014907 (2012).
[24] E. Retinskaya, M. Luzum and J. Y. Ollitrault, Phys. Rev. C **89**, 014902 (2014).

EPJ Web of Conferences **120**, 06002 (2016) DOI: 10.1051/epjconf/201612006002
ISMD 2015

Overview of quarkonium production in heavy-ion collisions at LHC

Byungsik Hong[1,a] on behalf of the ALICE, ATLAS and CMS Collaborations

[1] *Department of Physics, Korea University, Seoul 02841, Republic of Korea*

Abstract. Quarkonium has been regarded as one of the golden probes to identify the phase transition from confined hadronic matter to the deconfined quark-gluon plasma (QGP) in heavy-ion collisions. Recent data on the yields and momentum distributions of J/ψ and Υ families in pp, pPb, and PbPb collisions at the Large Hadron Collider (LHC) are reviewed. The possible implications related to the propagation of quarkonia in the deconfined hot, dense matter and the modified parton distribution function (PDF) in cold nuclei are also discussed.

1 Introduction

Several observables have been proposed to identify and study the detailed characteristics of the phase transition from the confined hadronic matter to the deconfined QGP. The production of various $c\bar{c}$ and $b\bar{b}$ bound states is one of them. The heavy quarkonia are generated in early stage of the collision due to their large masses and expected to be sequentially resolved by color Debye screening at different temperatures, depending on their binding energies, during evolution [1]. Therefore, the data on the yields and momentum spectra of quarkonia may provide invaluable information on the medium temperature and the characteristics of quantum chromodynamics (QCD) in high density and temperature environment [2].

Another production mechanism is the random recombination (or regeneration) of heavy quarks and their antiquarks [3]. Contrary to the color Debye screening effect, the regeneration process enhances the quarkonium yields, especially, at the low-transverse momentum (p_T) range. Recently, various transport models have incorporated with not only color Debye screening, but also the regeneration effects to reproduce the experimental data [4, 5]. As a result, the quarkonium data are expected to constrain probabilities of the two competing processes.

More recently, the quarkonium production in asymmetric pPb collisions became available at LHC. The quarkonium data in pPb are useful to recognize and quantify the initial-state effects, such as nuclear modification of PDF, energy loss in cold nuclear matter, and p_T broadening due to multiple scattering [6]. These initial-state nuclear effects need to be taken into account to isolate any genuin signals caused by the high-density QCD medium in heavy-ion collisions. In the meanwhile the quarkonium data in pp are important for heavy-ion physics as they serve as important references. Furthermore, the production mechanism of quarkonia in pp is yet to be understood.

[a] e-mail: bhong@korea.ac.kr

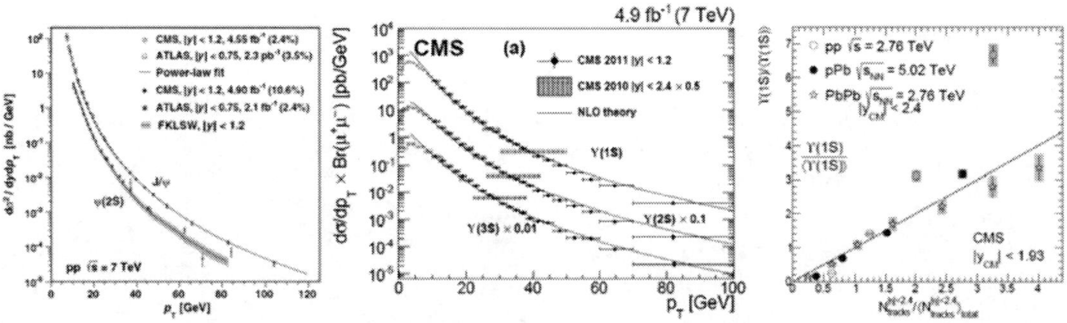

Figure 1. (Left) Differential p_T cross sections from CMS ($|y| < 1.2$) [7] and ATLAS ($|y| < 0.75$) [8, 9] for J/ψ and $\psi(2S)$ in pp at 7 TeV. The error bars show the statistical and systematic uncertainties added in quadrature. The uncertainties from integrated luminosities and branching fractions are given by the percentages in the legend. The curve is a fit to the J/ψ cross section from CMS by using a power-law function, and the band labeled by FKLSW shows a cross-section calculation from NRQCD+NLO for $\psi(2S)$ [11]. (Center) Differential p_T cross sections times dimuon branching fractions for $\Upsilon(nS)$ within $|y| < 1.2$ in pp at 7 TeV from CMS [12]. The $\Upsilon(2S)$ and $\Upsilon(3S)$ data are scaled by 0.1 and 0.01, respectively, for clearer display. The error bars show the total uncertainty, excluding the systematic uncertainty in the integrated luminosity, and the horizontal bars show the bin widths. Previous CMS measurements for $|y| < 2.4$ are shown by the cross-hatched areas [13]. The solid lines are the NRQCD+NLO calculations [14]. (Right) The $\Upsilon(1S)$ cross sections as a function of charged-track multiplicity measured in $|\eta| < 2.4$ in pp and PbPb at 2.76 TeV and pPb at 5.02 TeV [15]. Cross sections and $N_{tracks}^{|\eta|<2.4}$ values are normalized by their corresponding activity-integrated values. The dotted line is a linear function with a slope of unity. The error bars indicate the statistical uncertainties, and the boxes represent the point-to-point systematic uncertainties.

This paper summarizes some of recent experimental data on the J/ψ and Υ productions in pp, pPb, and PbPb collisions at LHC. The implications of the data for the production mechanism, the initial- and final-state effects are also discussed.

2 Quarkonium production in pp

The left panel of Fig. 1 shows the double-differential cross sections of promptly produced J/ψ and $\psi(2S)$ in pp at $\sqrt{s} = 7$ TeV from ATLAS and CMS [7–9]. The curve represents a power-law fit function to the J/ψ cross section [10], whereas the band labeled by FKLSW represents the global fit to the $\psi(2S)$ cross section by the perturbative QCD (pQCD) at the next-to-leading order (NLO) level [11]. The experimental $\psi(2S)$ cross section are well described by the NLO pQCD calculations.

The $\Upsilon(nS)$ differential cross sections, multiplied by dimuon branching fractions, are plotted as a function of p_T in the middle panel of Fig. 1 over the rapidity range $|y| < 1.2$ in pp at 7 TeV [12]. They are compared with the earlier CMS data [13], scaled by 0.5 to account for the smaller rapidity coverage in the latest data. Previously, the NLO analysis of non-relativistic QCD (NRQCD) successfully described the $\Upsilon(nS)$ spectra for $p_T < 50$ GeV. The extended calculations to $p_T = 100$ GeV [14], shown by solid lines in the middle panel of Fig. 1, agree with the recent CMS data. The new data with an extended p_T reach should be useful in evaluating long-distance matrix element (LDME) and studying the p_T-dependent corrections in NRQCD and other models.

One of the most interesting features in the quarkonium production in pp is shown in the right panel of Fig. 1, which compares the $\Upsilon(1S)$ cross section vs. charged-track multiplicity measured in $|\eta| < 2.4$,

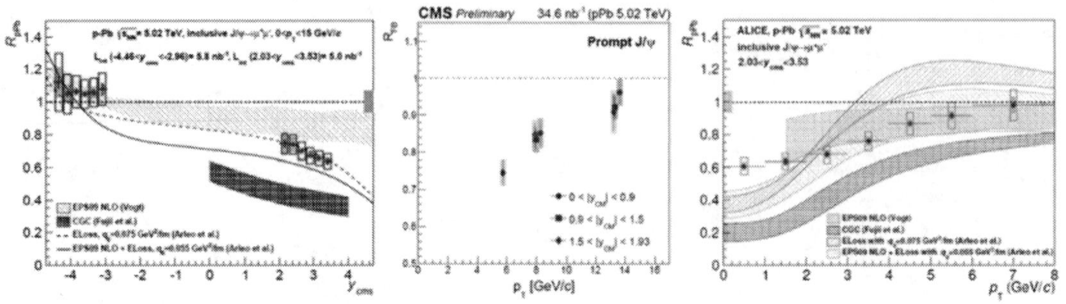

Figure 2. (Left) Nuclear modification factor R_{pPb} of inclusive J/ψ as a function of the center-of-mass rapidity in pPb at 5.02 TeV from ALICE [18]. The error bars represent the statistical uncertainties. The open and the shaded boxes correspond to the uncorrelated and the partially correlated, respectively, systematic uncertainties. The full boxes around $R_{pPb} = 1$ show the the correlated uncertainties. Various model calculations are also shown. (see the text for details.) (Center) Forward-to-backward yield ratio R_{FB} of prompt J/ψ in three different rapidity ranges as a function of p_T in pPb at 5.02 TeV from CMS [22]. (Right) R_{pPb} of inclusive J/ψ as a function of p_T at forward rapidities in pPb at 5.02 TeV from ALICE [23]. Various model calculations are also displayed. (see the text for details.)

$N_{tracks}^{|\eta|<2.4}$, in pp, pPb, and PbPb [15]. Note that the cross sections and $N_{tracks}^{|\eta|<2.4}$ values are normalized by their corresponding activity-integrated values. The self-normalized cross section ratios increase with event multiplicity: the pPb and PbPb data increase almost linearly with $N_{tracks}^{|\eta|<2.4}$, but the pp data show much stronger positive correlation between the Υ yield and the underlying event activity. A similar trend can be also observed in the J/ψ production at 7 TeV [16]. A possible interpretation would be the multiple parton-parton interactions in a single pp collision [17].

3 Quarkonium production in pPb

Figure 2 shows the center-of-mass rapidity and p_T dependences of the J/ψ yields in pPb at 5.02 TeV. The left panel of Fig. 2 from ALICE shows that the inclusive J/ψ yield in pPb is suppressed with respect to the one in binary-scaled pp collisions at the forward (p-going) rapidities, whereas it is unchanged at the backward (Pb-going) rapidities [18]. The data are compared with the calculations by a NLO color-evaporation-model (CEM) using the EPS09 nPDF parameterization [19] and a coherent parton-energy-loss model with and without EPS09 shadowing effect [20]. The calculation in the color-glass condensate (CGC) framework [21] is also displayed. Within the uncertainties, the models based on shadowing and the coherent energy loss are able to describe the data, but the CGC-based calculation overestimates the suppression. Note that none of the above mentioned models include a suppression due to the break-up of the $c\bar{c}$ pairs.

CMS analyzed the prompt J/ψ and non-prompt J/ψ from B decay, separately, using the precise measurement of the decay length. For example, the middle panel of Fig. 2 shows the forward-to-backward yield ratios R_{FB} of prompt J/ψ in three different rapidity bins as a functions of p_T [22]. The R_{FB} distributions decrease monotonically towards smaller p_T without any strong rapidity dependence. The prompt (and also non-prompt, not shown here) J/ψ yields at forward rapidities are smaller than those at backward rapidities, which is consistent with the ALICE data.

The right panel of Fig. 2 displays the R_{pPb} distribution of inclusive J/ψ as a function of p_T at forward rapidity region where the suppression is largest [23]. The data are compared to the calculations

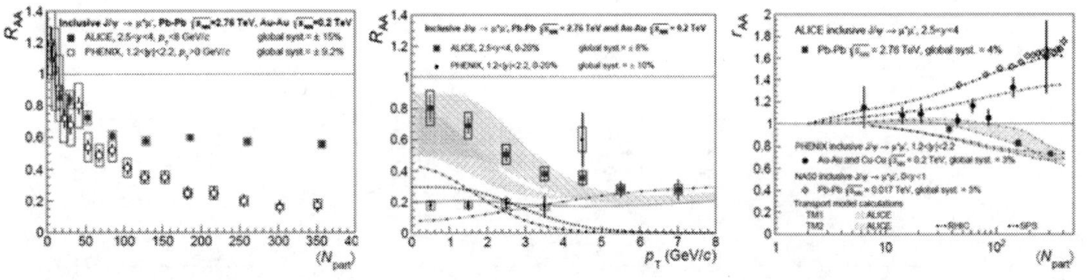

Figure 3. (Left) Nuclear modification factor R_{AA} of inclusive J/ψ as a function of the number of participant nucleons N_{part} in PbPb at 2.76 TeV from ALICE [25], compared to the PHENIX data in AuAu at 0.2 TeV [26]. (Center) R_{AA} of inclusive J/ψ as a function of p_T for $2.5 < y < 4$ in central (0 - 20%) PbPb at 2.76 TeV from ALICE [25], compared to the PHENIX data in AuAu at 0.2 TeV [26]. The calculations from transport models (TM1 [28] and TM2 [5]) are shown for comparison. (Right) Mean p_T variable r_{AA} of inclusive J/ψ as a function of N_{part} from ALICE at 2.76 TeV [25], PHENIX at 0.2 TeV [26, 29, 30], and NA50 at 0.017 TeV [31]. The lines and bands are the calculations from TM1 and TM2.

from various models. The pure energy loss scenario predicts a much stronger p_T dependence at forward rapidities. A calculation based on NLO CEM with EPS09 reproduces the p_T dependence within uncertainties [24], but a discrepancy still exist between the data and both calculations at the low-p_T region. Note that the theoretical uncertainties arise from the uncertainties on EPS09, the charm-quark mass, and the renormalization and factorization scales for the cross section calculation.

4 Quarkonium production in PbPb

The centrality dependence of nuclear modification factor R_{AA} for inclusive J/ψ in PbPb at 2.76 TeV measured by ALICE is shown in the left panel of Fig. 3 [25]. The J/ψ is measured at $p_T < 8$ GeV in forward rapidity of $2.5 < y < 4$ as a function of the number of participant nucleons N_{part}. The suppression is significantly smaller than that observed at lower energy by PHENIX in a similar kinematic range [26]. The suppression is also smaller than that measured at higher p_T (> 6.5 TeV) at 2.76 TeV from CMS [27]. For $N_{part} > 70$, corresponding to ~50% most central PbPb collision, the J/ψ R_{AA} from ALICE is almost constant, which is distinct from the lower energy data.

The p_T dependence of R_{AA} for inclusive J/ψ is shown in the middle panel of Fig. 3 for central (0 - 20%) PbPb. R_{AA} for inclusive J/ψ shows very little suppression at low p_T at LHC, and decreases towards higher p_T. On contrary the low-energy PHENIX data show no significant p_T dependence [26]. The calculations from the two transport models (TM1 [28] and TM2 [5]) are compared to the ALICE data. (The main difference between the two models is the rate equation, controlling the J/ψ dissociation and regeneration processes.) In addition, the primordial component from the initial hard parton scattering and the regenerated component from coalescence of c and $\bar{c}$ are displayed, separately. The primordial components are constant or slightly increase with p_T, but the regenerated components are concentrated at the low-p_T region for both models.

The centrality dependences of the p_T-width variable $r_{AA} = \langle p_T^2 \rangle_{AA} / \langle p_T^2 \rangle_{pp}$ for inclusive J/ψ are compared each other at different energies from ALICE at 2.76 TeV [25], PHENIX at 0.2 TeV [26, 29, 30], and NA50 at 0.017 TeV [31] in the right panel of Fig. 3. A very different N_{part} dependence is observed: the increase at the lowest energy was attributed to the extra p_T kick due to multiple

 DOI: 10.1051/epjconf/201612006002

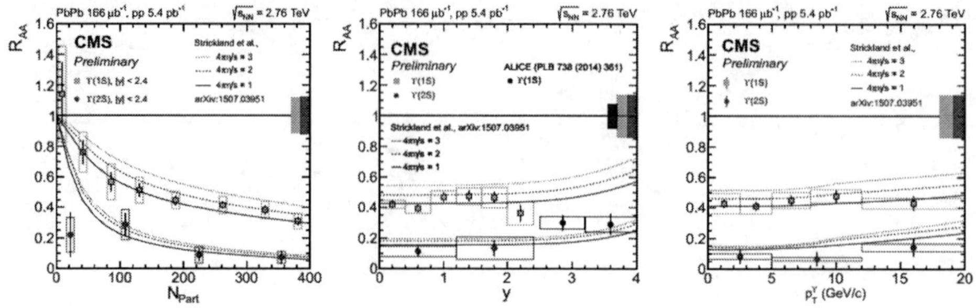

Figure 4. (Left) Nuclear modification factor R_{AA} of $\Upsilon(1S)$ and $\Upsilon(2S)$ in minimum bias (0 - 100%) PbPb at 2.76 TeV from CMS as functions of the number of participant nucleons N_{part} (left), rapidity (center), and p_T (right) [33]. The error bars and boxes represent the statistical and systematic, respectively, uncertainties. The boxes at unity show the fully-correlated global uncertainties. The various lines represent the calculations from the anisotropic hydrodynamics with different assumption on the shear-viscosity-to-entropy-density ratio $4\pi\eta/s$ [35]. For the rapidity distribution, the ALICE data for $\Upsilon(1S)$ in $2.5 < y < 4$ [36] are also plotted for comparison.

scatterings of initial partons (the so-called Cronin effect) [32], and the decrease at the highest energy could be related to the regeneration and the thermalization of charm quarks.

The R_{AA} distributions of $\Upsilon(1S)$ and $\Upsilon(2S)$ in minimum bias (0 - 100%) PbPb at 2.76 TeV from CMS [33] are shown in Fig. 4 as functions of N_{part}, rapidity, and p_T. The strong centrality dependence, previously observed by CMS with larger uncertainties [34], is confirmed with higher precision. The improvement was achieved by ~30% more statistics for $\Upsilon(1S)$ in PbPb and fully analyzed pp reference events recorded in 2013. On the other hand, the R_{AA} distributions are constant within uncertainties as functions of p_T and y, and no pronounced dependence on the kinematic variables can be observed. The R_{AA} values, integrated over p_T and y, are 0.425 ± 0.029(stat) ± 0.070(syst) for $\Upsilon(1S)$ and 0.116 ± 0.028(stat) ± 0.022(syst) for $\Upsilon(2S)$. $\Upsilon(3S)$ cannot be identified in PbPb, and only upper limit of 0.14 is given at 95% confidence level.

The $\Upsilon(nS)$ data are compared with the calculations from various models. For example, Fig. 4 shows the comparison with anisotropic hydrodynamics model with different assumptions on the shear-viscosity-to-entropy-density ratio $4\pi\eta/s$ [35]. The centrality dependence of $\Upsilon(1S)$ yield agrees better with the model calculation, when smaller shear viscosity ($1 \lesssim 4\pi\eta/s \lesssim 2$) is employed. For the rapidity dependence, the model describes the $\Upsilon(1S)$ data in $|y| < 2.4$ covered by CMS, but fails to reproduce the ALICE data at forward rapidities in $2.5 < y < 4$ [36]. In the meanwhile, the transport model [4] can also reproduce the centrality dependences of the $\Upsilon(1S)$ and $\Upsilon(2S)$ yields only when the regeneration components are included. In particular, the $\Upsilon(2S)$ yield from semi-central to central PbPb collisions consists of mostly the regenerated particles according to the transport model.

5 Summary

The J/ψ, $\psi(2S)$, and $\Upsilon(nS)$ mesons have been measured in pp, pPb, and PbPb collisions by the LHC experiments. The new pp data at 7 TeV extends the p_T reach up to 100 GeV, which helps to understand the production mechanism. The pPb data at 5.02 TeV show less yields in the forward p-going direction than in the backward Pb-going direction at low p_T, which hints a modification of the parton distribution function in Pb nucleus. The J/ψ yield in central PbPb requires the regeneration component. The width of the p_T distribution of J/ψ in PbPb decreases with centrality at LHC energy, which can be attributed to the regeneration component. The nuclear modification factors of $\Upsilon(1S)$ and $\Upsilon(2S)$ do not show any pronounced dependence on the kinematic variables like p_T and y.

References

[1] T. Matsui and H. Satz, Phys. Lett. B **178**, 416 (1986)
[2] A. Mocsy and P. Petreczky, Phys. Rev. D **77**, 014501 (2008)
[3] A. Andronic, P. Braun-Munzinger, K. Redlich, and J. Stachel, J. Phys. G **38**, 124081 (2011)
[4] A. Emerick, X. Zhao, and R. Rapp, Eur. Phys. J. A **48**, 72 (2012)
[5] K. Zhou, N. Xu, Z. Xu, and P. Zhuang, Phys. Rev. C **89**, 054911 (2014)
[6] R. Vogt, Phys. Rev. C **81**, 044903 (2010)
[7] CMS Collaboration, Phys. Rev. Lett. **114**, 191802 (2015)
[8] ATLAS Collaboration, Nucl. Phys. B **850**, 387 (2011)
[9] ATLAS Collaboration, J. High Energy Phys. **09**, 079 (2014)
[10] HERA-B Collaboration, Eur. Phys. J. C **49**, 545 (2007)
[11] P. Faccioli, V. Knünz, C. Lourenco, J. Seixas, and H. Wöhri, Phys. Lett. B **736**, 98 (2014)
[12] CMS Collaboration, arXiv:1501.07750
[13] CMS Collaboration, Phys. Lett. B **727**, 101 (2013)
[14] B. Gong, L.-P. Wan, J.-X.Wang, and H.-F. Zhang, Phys. Rev. Lett. **112**, 032001 (2014)
[15] CMS Collaboration, J. High Energy Phys. **04**, 103 (2014)
[16] ALICE Collaboration, Phys. Lett. B **712**, 165 (2012)
[17] H. Abramowicz, *et al.*, arXiv:1306.5413
[18] ALICE Collaboration, J. High Energy Phys. **02**, 073 (2014)
[19] J. L. Albacete, *et al.*, Int. J. Mod. Phys. E **22**, 1330007 (2013)
[20] F. Arleo and S. Peigne, J. High Energy Phys. **03**, 122 (2013)
[21] H. Fujii and K. Watanabe, Nucl. Phys. A **915**, 1 (2013)
[22] CMS Collaboration, "Measurement of prompt and non-prompt J/ψ production in pPb collisions at $\sqrt{s_{NN}} = 5.02$ TeV TeV", CMS PAS HIN-14-009
[23] ALICE Collaboration, J. High Energy Phys. **06**, 055 (2015)
[24] F. Arleo, R. Kolevatov, S. Peigne and M. Rustamova, J. High Energy Phys. **05**, 155 (2013)
[25] ALICE Collaboration, arXiv:1506.08804
[26] PHENIX Collaboration, Phys. Rev. C **84**, 054912 (2011)
[27] CMS Collaboration, "Prompt and non-prompt J/ψ R_{AA} with 150 μb^{-1} integrated PbPb luminosity at $\sqrt{s_{NN}} = 2.76$ TeV", CMS PAS HIN-12-014
[28] X. Zhao and R. Rapp, Nucl. Phys. A **859**, 114 (2011)
[29] PHENIX Collaboration, Phys. Rev. Lett. **98**, 232002 (2007)
[30] PHENIX Collaboration, Phys. Rev. Lett. **101**, 122301 (2008)
[31] NA50 Collaboration, Phys. Lett. B **499**, 85 (2001)
[32] J. Cronin, H. J. Frisch, M. Shochet, J. Boymond, and R. Mermod *et al.*, Phys. Rev. D **11**, 3105 (1975)
[33] CMS Collaboration, "Suppression of $\Upsilon(1S)$, $\Upsilon(2S)$ and $\Upsilon(3S)$ in PbPb collisions at $\sqrt{s_{NN}} = 2.76$ TeV TeV", CMS PAS HIN-15-001
[34] CMS Collaboration, Phys. Rev. Lett. **109**, 222301 (2012)
[35] B. Krouppa, R. Ryblewski, and M. Strickland, arXiv:1507.03951
[36] ALICE Collaboration, Phys. Lett. B **738**, 361 (2014)

EPJ Web of Conferences **120**, 06003 (2016) DOI: 10.1051/epjconf/201612006003
ISMD 2015

Bulk and shear viscosities for the Gribov-Zwanziger plasma

Wojciech Florkowski[1,a]

[1] *Institute of Nuclear Physics, Polish Academy of Sciences*
ul. Radzikowskiego 152, 31-342 Kraków, Poland

Abstract. The concept of the Gribov-Zwanziger plasma is introduced and used to calculate the bulk and shear viscosities of the system of gluons. The kinetic coefficients are obtained in two different ways which are shown to yield equivalent results.

1 Introduction

In this contribution I report on the recent results obtained in collaboration with Radoslaw Ryblewski, Nan Su, and Konrad Tywoniuk [1, 2] describing the form of the bulk and shear viscosity coefficients for the Gribov-Zwanziger (GZ) plasma. The concept of the GZ plasma follows from the use of the dispersion relation

$$E(\boldsymbol{k}) = \sqrt{\boldsymbol{k}^2 + \frac{\gamma_G^4}{\boldsymbol{k}^2}} \tag{1}$$

for the gluons [3]. Here $\boldsymbol{k}$ is the three-momentum of gluons and E denotes their energy. The parameter γ_G appears during the quantisation of the Yang-Mills theory. Equation (1) has been derived for the first time by Gribov in his famous paper [3]. It takes into account the expected behavior of gluons in the infrared, namely, a large energy cost connected with the excitation of soft modes and the corresponding reduction of the physical phase space [3–7]. The use of (1) suggests that interacting massless gluons are equivalent to non-interacting gluons having this very special form of the dispersion relation.

Thermodynamic properties of the system with the particles obeying Eq. (1) have been worked out in Refs. [8, 9]. The main results follow directly from the use of (1) in the Bose-Einstein distribution function

$$f_{GZ} = \frac{1}{\exp(E(\boldsymbol{k})/T) - 1}, \tag{2}$$

where T is the system's temperature. This approach provides good agreement with the lattice results on thermodynamic variables [10]. Encouraged by the success connected with the use of the dispersion relation (1) in equilibrium, in Refs. [1, 2] we have made an attempt to construct a simple kinetic theory that incorporates (1) and may describe processes out of equilibrium. In particular, in [1] we have concentrated on the calculation of the bulk viscosity which attracts more and more attention in the last years [11–15].

[a] e-mail: wojciech.florkowski@ifj.edu.pl

2 Lorentz covariance and boost-invariance

2.1 Reimplementation of Lorentz covariance

At first it should be emphasised that the original dispersion relation (1) is derived in the Coulomb gauge which explicitly breaks Lorentz invariance. In order to regain a covariant framework, that is required for description of relativistic fluids, one has to make certain assumptions about the Lorentz transformation properties of the quantities that appear in Eq. (1). The strategy to recover the covariant description is not straightforward and we discuss this problem in greater detail in Ref. [2].

We choose the procedure where the Gribov dispersion relation (1) is transformed into the following covariant expression [1, 2]

$$E(k \cdot u) = \sqrt{(k \cdot u)^2 + \frac{\gamma_G^4}{(k \cdot u)^2}}. \tag{3}$$

Here u is the four-velocity of the fluid element. We assume that the Coulomb gauge as well as the in-medium value of the Gribov parameter γ_G are fixed in the local rest frame where $u^\mu = (1, 0, 0, 0)$. We introduce $k^0 \equiv |\boldsymbol{k}|$, which is the magnitude of the three-vector $\boldsymbol{k} \equiv (k_x, k_y, k_\parallel)$, and $k_\perp = \sqrt{k_x^2 + k_y^2}$ such that the resulting four-vector $k^\mu = (k^0, \boldsymbol{k})$ has standard Lorentz transformation properties with $k^2 = k \cdot k = 0$.

2.2 Boost-invariance

One of the simplest ways to obtain the formulas for the kinetic coefficients is to analyse a (0+1)-dimensional, boost-invariant and transversally homogeneous system. We stress that in contrast to the problem of Lorentz covariance, which is the fundamental issue connected with the use of the Coulomb gauge, imposing boost-invariance is just a technical method to facilitate our manipulations.

The boost-invariant systems are characterised by the flow vector $u^\mu = (t/\tau, 0, 0, z/\tau)$ [16]. In this case, all scalar functions of space and time depend only on the longitudinal proper time $\tau = \sqrt{t^2 - z^2}$. One may furthermore introduce the boost-invariant variables $v = k^0 t - k_\parallel z$ and $w = k_\parallel t - k^0 z$ [17]. In this case the energy of the particle is given by the formula

$$E(\tau, w, k_\perp) = \sqrt{\frac{w^2}{\tau^2} + k_\perp^2 + \frac{\gamma_G^4}{\frac{w^2}{\tau^2} + k_\perp^2}}. \tag{4}$$

The phase space integration measure in the three-momentum space can be written correspondingly as

$$\int dK(\ldots) = \frac{g_0}{(2\pi)^3} \int_{-\infty}^{\infty} \frac{dw}{\tau} \int d^2 k_\perp (\ldots), \tag{5}$$

where g_0 is the number of internal degrees of freedom. We note that the phase space distribution function f, which is also a Lorentz scalar, may depend in our case only on τ, w and $k_\perp$, namely $f = f(\tau, w, k_\perp)$ [17].

3 Kinetic equation

3.1 Relaxation time approximation

The arguments presented in [1, 2] suggest that we can use the standard kinetic equation in the relaxation-time approximation (RTA) of the form [18–20]

$$\frac{\partial f(\tau, w, k_\perp)}{\partial \tau} = \frac{f_{GZ}(\tau, w, k_\perp) - f(\tau, w, k_\perp)}{\tau_{rel}(\tau)}, \tag{6}$$

where $\tau_{\rm rel}$ is the relaxation time and the (effective) temperature T appearing in the equilibrium distribution function $f_{\rm GZ}$ is determined from the equation

$$\int \mathrm{d}K\, E(\tau, w, k_\perp)\, f_{\rm GZ}(\tau, w, k_\perp) = \int \mathrm{d}K\, E(\tau, w, k_\perp)\, f(\tau, w, k_\perp)\,. \tag{7}$$

Equation (7) is known as the Landau matching condition for the energy. The formal solution of Eq. (6) is [21–23]

$$f(\tau, w, k_\perp) = f_0(w, k_\perp) D(\tau, \tau_0) + \int_{\tau_0}^{\tau} \frac{\mathrm{d}\tau'}{\tau_{\rm rel}(\tau')} D(\tau, \tau') f_{\rm GZ}(\tau', w, k_\perp)\,, \tag{8}$$

where the damping function $D(\tau_2, \tau_1)$ has the form

$$D(\tau_2, \tau_1) = \exp\left[-\int_{\tau_1}^{\tau_2} \frac{\mathrm{d}\tau}{\tau_{\rm rel}(\tau)} \right]. \tag{9}$$

Inserting the solution (8) into the Landau matching condition (7) allows us to find the time dependence of the system's temperature, $T(\tau)$. This, in turn, when used back in (8), allows us to find the time dependence of various system's characteristics such as the energy density and transverse and longitudinal pressures.

3.2 Linear response method

Besides the exact treatment of the kinetic equation that has been sketched above, we may use the linear respond method. In this case we seek the solution of Eq. (6) in the form

$$f \approx f_{\rm GZ} + \delta f + \cdots\,, \tag{10}$$

where $\delta f = -\tau_{\rm rel} \mathrm{d} f_{\rm GZ}/\mathrm{d}\tau$. This leads directly to the following result

$$\delta f = -\frac{E\,\tau_{\rm rel}}{T\tau} \left\{ \frac{w^2}{E^2 \tau^2} \left[1 - \frac{\gamma_{\rm G}^4}{\left(\frac{w^2}{\tau^2} + k_\perp^2\right)^2} \right] + c_s^2 \right\} f_{\rm GZ}\,(1 + f_{\rm GZ})\,. \tag{11}$$

The speed of sound squared, c_s^2, that appears in (11) can be obtained from the equation of state. The correction to the distribution function (11) can be used further to obtain the shear tensor $\pi^{\mu\nu}$ and the bulk pressure Π. These quantities define directly the bulk and shear viscosities if the Navier-Stokes limit is considered. Straightforward manipulations lead to the following expression for the bulk viscosity coefficient [1]

$$\zeta = \frac{g_0 \gamma_{\rm G}^5}{3\pi^2} \frac{\tau_{\rm rel}}{T} \int_0^\infty \mathrm{d}y \left[c_s^2 - \frac{1}{3}\frac{y^4 - 1}{y^4 + 1} \right] f_{\rm GZ}(1 + f_{\rm GZ}), \tag{12}$$

where $f_{\rm GZ} = \{\exp[\gamma_{\rm G} \sqrt{y^2 + y^{-2}}/T] - 1\}^{-1}$. In the similar way we obtain the shear viscosity coefficient [1, 2]

$$\eta = \frac{g_0 \gamma_{\rm G}^5}{30\pi^2} \frac{\tau_{\rm rel}}{T} \int_0^\infty \mathrm{d}y \frac{\left(y^4 - 1\right)^2}{y^4 + 1} f_{\rm GZ}(1 + f_{\rm GZ})\,. \tag{13}$$

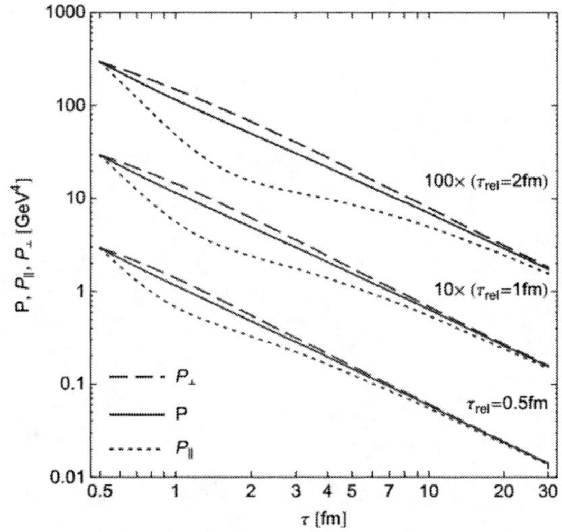

Figure 1. Time dependence of pressure for three different values of the relaxation times. The initial condition corresponds to the equilibrium state with the initial temperature $T_0 = T(\tau_0) = 600$ MeV. The initial time is $\tau_0 = 0.5$ fm/c.

4 Numerical results

Let us now turn back to the discussion of the exact solutions of the kinetic equation. Knowing the time dependence of the effective temperature T we find the longitudinal and transverse pressures from the equations

$$P_{\parallel} = \int dK \, \frac{w^2}{\tau^2 E(\tau, w, k_\perp)} \left[1 - \frac{\gamma_{\rm G}^4}{(w^2/\tau^2 + k_\perp^2)^2} \right] f, \tag{14}$$

$$P_{\perp} = \int dK \, \frac{k_\perp^2}{2\, E(\tau, w, k_\perp)} \left[1 - \frac{\gamma_{\rm G}^4}{(w^2/\tau^2 + k_\perp^2)^2} \right] f, \tag{15}$$

and the total pressure is obtained as $P = \frac{1}{3}(P_{\parallel} + 2P_{\perp})$. We note that the parallel pressure acts in the direction of the beam axis. The transverse pressure acts in the transverse direction to the beam and is the same for all such directions. The numerical results for the time evolution of pressures is shown in Fig. 1. Knowing $P_{\parallel}$ and $P_{\perp}$ we find the shear and bulk viscous pressures (in the general case the shear tensor $\pi^{\mu\nu}$ has 5 independent components, they are reduced to one independent variable π for (0+1)D systems)

$$\pi = \frac{2}{3}\left(P_{\perp} - P_{\parallel}\right), \tag{16}$$

$$\Pi = P - P_{\rm GZ}. \tag{17}$$

The last two equations allow us to define the effective shear and bulk viscosities

$$\pi = \frac{4}{3}\frac{\eta_{\rm eff}}{\tau}, \tag{18}$$

$$\Pi = -\frac{\zeta_{\rm eff}}{\tau}. \tag{19}$$

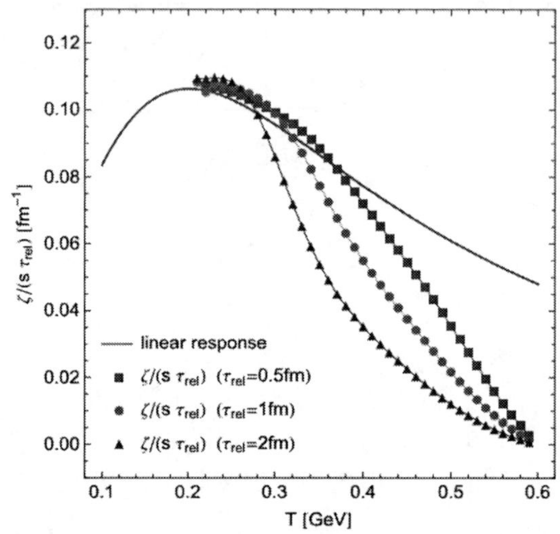

Figure 2. Temperature dependence of the bulk viscosity scaled by the relaxation time and entropy density. The triangles, circles and squares show the effective bulk viscosity for three different values of the relaxation time (defined in the figure), whereas the solid line shows the result of the linear-response formula.

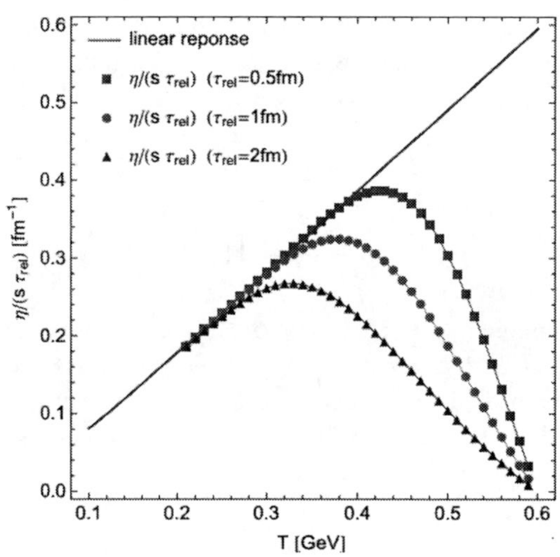

Figure 3. The same as Fig. 2 but for the shear viscosity.

The effective coefficients should agree with the standard coefficients when the system is close to equilibrium. This is demonstrated in Figs. 2 and 3.

5 Conclusions

The Gribov-Zwanziger quantisation of the Yang-Mills theory leads to the dispersion relation that can be successfully used in the studies of gluon thermodynamics and kinetic theory. In this contribution we have presented the results for the bulk and shear viscosities of the Gribov-Zwanziger plasma. These results may be useful for future phenomenological applications for ultrarelativistic heavy-ion collisions.

Acknowledgements: I thank my collaborators, Radoslaw Ryblewski, Nan Su, and Konrad Tywoniuk for very fruitful collaboration. This research has been supported in part by Polish National Science Center grants No. DEC-2012/05/B/ST2/02528 and No. DEC-2012/06/A/ST2/00390.

References

[1] W. Florkowski, R. Ryblewski, N. Su, K. Tywoniuk (2015), 1504.03176
[2] W. Florkowski, R. Ryblewski, N. Su, K. Tywoniuk (2015), 1509.01242
[3] V.N. Gribov, Nucl. Phys. **B139**, 1 (1978)
[4] D. Zwanziger, Nucl. Phys. **B323**, 513 (1989)
[5] R.P. Feynman, Nucl. Phys. **B188**, 479 (1981)
[6] Y.L. Dokshitzer, D.E. Kharzeev, Ann. Rev. Nucl. Part. Sci. **54**, 487 (2004), hep-ph/0404216
[7] N. Su, K. Tywoniuk, Phys. Rev. Lett. **114**, 161601 (2015), 1409.3203
[8] D. Zwanziger, Phys. Rev. Lett. **94**, 182301 (2005), hep-ph/0407103
[9] K. Fukushima, N. Su, Phys.Rev. **D88**, 076008 (2013), 1304.8004
[10] S. Borsanyi, G. Endrodi, Z. Fodor, S.D. Katz, K.K. Szabo, JHEP **07**, 056 (2012), 1204.6184
[11] P.B. Arnold, C. Dogan, G.D. Moore, Phys. Rev. **D74**, 085021 (2006), hep-ph/0608012
[12] H.B. Meyer, Phys. Rev. Lett. **100**, 162001 (2008), 0710.3717
[13] D. Kharzeev, K. Tuchin, JHEP **09**, 093 (2008), 0705.4280
[14] G. Torrieri, I. Mishustin, Phys. Rev. **C78**, 021901 (2008), 0805.0442
[15] G.D. Moore, O. Saremi, JHEP **09**, 015 (2008), 0805.4201
[16] J.D. Bjorken, Phys. Rev. **D27**, 140 (1983)
[17] A. Bialas, W. Czyz, A. Dyrek, W. Florkowski, Nucl. Phys. **B296**, 611 (1988)
[18] P.L. Bhatnagar, E.P. Gross, M. Krook, Phys. Rev. **94**, 511 (1954)
[19] G. Baym, Phys. Lett. **B138**, 18 (1984)
[20] G. Baym, Nucl. Phys. **A418**, 525C (1984)
[21] W. Florkowski, R. Ryblewski, M. Strickland, Nucl.Phys. **A916**, 249 (2013), 1304.0665
[22] W. Florkowski, R. Ryblewski, M. Strickland, Phys.Rev. **C88**, 024903 (2013), 1305.7234
[23] W. Florkowski, E. Maksymiuk, R. Ryblewski, M. Strickland, Phys. Rev. **C89**, 054908 (2014), 1402.7348

EPJ Web of Conferences **120**, 06004 (2016)
ISMD 2015 DOI: 10.1051/epjconf/201612006004

LHCb results from proton ion collisions

Laure Massacrier on behalf of the LHCb Collaboration[1,2,a]

[1] *LAL, Univ. Paris-Sud, CNRS/IN2P3, Université Paris-Saclay, Orsay, France*
[2] *IPNO, Univ. Paris-Sud, CNRS/IN2P3, Université Paris-Saclay, Orsay, France*

Abstract. Proton-lead and lead-proton data taking during 2013 has allowed LHCb to expand its physics program to heavy ion physics. Results include the first forward measurement of Z production in proton-lead collisions as well as a measurement of the nuclear modification factor and forward-backward production of prompt and displaced J/ψ, $\psi(2S)$ and Υ. Angular particle correlations have also been measured for events of varying charged particle activity.

1 Introduction

Ultra-relativistic heavy-ion collisions are used to study the nuclear matter at high temperature and pressure where the formation of the Quark Gluon Plasma (QGP), a state of matter which consists of asymptotically free quarks and gluons, occurs. Open heavy flavours and quarkonia are produced at the early stages of the collision and might interact with the deconfined medium, making them ideal probes of the QGP. It was indeed predicted that in hot nuclear matter, charmonia are suppressed due to color screening of the heavy quarks potential [1]. Quarkonia can also be suppressed in absence of the QGP formation by Cold Nuclear Matter (CNM) effects. Proton-Nucleus collisions (pA or Ap), which are interesting by themselves, are therefore essential to interpret Nucleus-Nucleus data in order to disentangle QGP effects from CNM effects in those collisions. The main CNM effects affecting quarkonium production include initial-state nuclear effects on the parton densities (shadowing) [2], the initial-state parton energy loss and final-state energy loss (coherent energy loss) [3], the final-state absorption of the pre-resonant heavy quark pair by the spectator nucleons (nuclear absorption) albeit small at LHC [4] and the final-state interaction of the quarkonium with the produced medium (comovers) [5].

Proton-nucleus collisions are also useful for the determination of nuclear parton distribution functions (nPDF) [6]. The measurement of the Z electroweak boson in pA and Ap collisions in the LHCb acceptance permits to probe low x_A values[1] (2×10^{-4} - 3×10^{-3}) in the forward case, and high x_A value (0.2-1.0) in the backward case for an energy scale $Q^2 = M_Z^2$. Such a measurement (the first performed at the LHC in pA collisions) should offer large constraining power to nPDF fits especially at small x_A. Finally, two-particle angular correlations are interesting tools to probe collective effects in the dense environment of high-energy collisions. Thanks to the high particle density and multiplicities reached in pA collisions, of similar size to that of non central nucleus-nucleus collisions (AA), LHCb can

[a] e-mail: massacri@lal.in2p3.fr
[1] x_A is the momentum fraction of a certain parton inside a given nucleon bound in the nucleus A.

investigate at forward rapidity the long-range correlation on the near side (ie the so-called ridge [7]) which was already observed in proton-proton (pp), proton-lead (pPb) and lead-lead (PbPb) collisions at mid-rapidity.

2 The LHCb detector and data taking

The LHCb detector is a single arm spectrometer in the forward region. It is fully instrumented in its angular acceptance and covers the pseudo-rapidity range $2 < \eta < 5$. Initially designed for b-physics, the LHCb detector is becoming a general purpose detector by carrying out an extensive heavy-ion program (including pA, AA data taking in the collider mode and a rich fixed target program). More details on the LHCb detector apparatus can be found in [8, 9]. In 2013, LHCb collected pPb and Pbp data at a center-of-mass (CMS) energy of $\sqrt{s_{NN}} = 5$ TeV. In such asymmetric collisions, the nucleon-nucleon center-of-mass system is shifted by 0.47 unit of rapidity in the direction of the proton beam. In the forward (backward) configuration pPb (Pbp), the proton (lead) beam traverses LHCb from the vertex locator to the muon system, respectively. The LHCb acceptance is $1.5 < \mathrm{y_{CMS}} < 4.0$ in the forward configuration and $-5.0 < \mathrm{y_{CMS}} < -2.5$ in the backward configuration, leading to a common rapidy range coverage for both configurations of $2.5 <| \mathrm{y_{CMS}} |< 4.0$. Data collected in the forward configuration amounts to $L_{int} = 1.1$ nb^{-1} while in the backward configuration it amounts to $L_{int} = 0.5$ nb^{-1}.

3 J/ψ, $\psi(2S)$, Υ production in pPb/Pbp and CNM studies

J/ψ, ψ(2S) and Υ are studied in the dimuon final state with a dimuon transverse momentum restricted to $p_T < 14$ GeV/c for charmonia and to $p_T < 15$ GeV/c for bottomonia. Prompt J/ψ and ψ(2S) can be disentangled from J/ψ and ψ(2S) coming from b-hadron decays thanks to the excellent vertexing capability of LHCb. The yields of prompt charmonia and charmonia from b-hadron decays are obtained, in each kinematic bin, from a simultaneous fit of the dimuon invariant mass and pseudo-proper time distributions (see Fig. 1 for the ψ(2S) signal). An unbinned extended maximum likelihood fit to the dimuon invariant mass of the Υ candidates was performed to determine the signal yields of Υ(1S), Υ(2S) and Υ(3S). While the three Υ resonances are observed in the forward configuration, only Υ(1S) gives a significant signal in the backward configuration. Nuclear effects are usually quantified by the nuclear modification factor R_{pA} and the forward to backward ratio R_{FB}. R_{pA} is defined as the production cross-section of a given particle in pA collisions divided by its production cross section in pp collisions at the same center-of-masse energy, and scaled by the atomic mass number A of the nuclei, while R_{FB} is the ratio of the production cross-section of a given particle in pA over its production cross-section in Ap configuration, measured in the same absolute center-of-mass rapidity range:

$$R_{\mathrm{pA}} = \frac{1}{A}\frac{\mathrm{d}^2\sigma_{\mathrm{pA}}(\mathrm{y},\mathrm{p_T})/\mathrm{d}\sigma\mathrm{dy}}{\mathrm{d}^2\sigma_{\mathrm{pp}}(\mathrm{y},\mathrm{p_T})/\mathrm{d}\sigma\mathrm{dy}}, \qquad R_{\mathrm{FB}} = \frac{\sigma_{\mathrm{pA}}(+\mid y\mid, p_{\mathrm{T}})}{\sigma_{\mathrm{pA}}(-\mid y\mid, p_{\mathrm{T}})}. \tag{1}$$

R_{FB} has the advantage to not rely on the pp reference cross-section and that part of the experimental systematic uncertainties and theoretical scale uncertainties cancel.
To determine the nuclear modification factor R_{pPb} of the J/ψ, ψ(2S) and Υ(1S), their pp reference production cross-section at $\sqrt{s_{NN}} = 5$ TeV are required. Since no data at this energy were available so far[2], the reference cross-sections for J/ψ and Υ(1S) are obtained by a power-law fit to existing

[2]A first data dating period in pp collisions at $\sqrt{s} = 5$ TeV has just been completed end of November 2015 by the LHC.

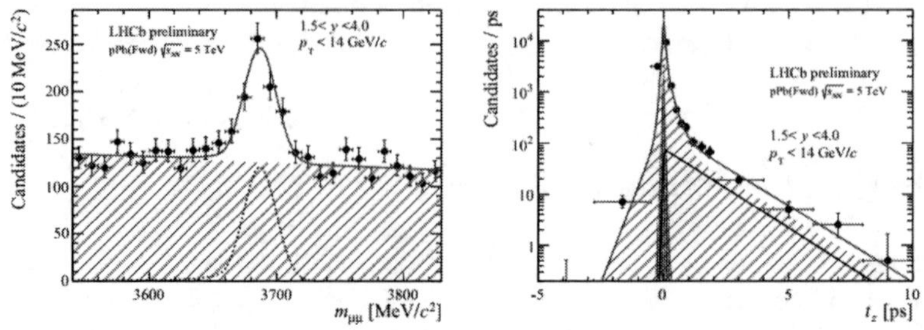

Figure 1. Projections of the ψ(2S) fit results to the dimuon invariant mass (left) and pseudo-proper time t_z (right) in pPb forward configuration ($1.5 < y_{CMS} < 4.0$). For the mass projection, the total fitted function is the red line, the signal distribution is the blue dotted line and the combinatorial background is represented by the green hatched area. For the t_z projection, the same representation is adopted and in addition the ψ(2S) from b-hadron signal is given by the black line.

LHCb measurements at 2.76, 7 and 8 TeV[10, 11]. To get the ψ(2S) reference pp cross-section at $\sqrt{s} = 5$ TeV, the following assumption has been made:

$$\frac{\sigma_{\rm pp}^{J/\psi}(5\ {\rm TeV})}{\sigma_{\rm pp}^{\psi(2S)}(5\ {\rm TeV})} \approx \frac{\sigma_{\rm pp}^{J/\psi}(7\ {\rm TeV})}{\sigma_{\rm pp}^{\psi(2S)}(7\ {\rm TeV})}, \tag{2}$$

assuming that the systematic uncertainty on this hypothesis is negligible compared to the statistical uncertainty on the ψ(2S) measurement in pA collisions. The ψ(2S) and J/ψ production cross-sections at $\sqrt{s} = 7$ TeV are taken from the LHCb measurements in Refs [12, 13]. Under this assumption, the ψ(2S) nuclear modification factor can be derived from the J/ψ nuclear modification factor[14], with the following equation:

$$R_{\rm pPb}^{\psi(2S)} = \frac{\sigma_{\rm pPb}^{\psi(2S)}(5\ {\rm TeV})}{\sigma_{\rm pPb}^{J/\psi}(5\ {\rm TeV})}\frac{\sigma_{\rm pp}^{J/\psi}(5\ {\rm TeV})}{\sigma_{\rm pp}^{\psi(2S)}(5\ {\rm TeV})} \times R_{\rm pPb}^{J/\psi} \approx \frac{\sigma_{\rm pPb}^{\psi(2S)}(5\ {\rm TeV})}{\sigma_{\rm pPb}^{J/\psi}(5\ {\rm TeV})}\frac{\sigma_{\rm pp}^{J/\psi}(7\ {\rm TeV})}{\sigma_{\rm pp}^{\psi(2S)}(7\ {\rm TeV})} \times R_{\rm pPb}^{J/\psi} \tag{3}$$

Figure 2 left shows the $R_{\rm pPb}$ of prompt J/ψ and ψ(2S) in the rapidity ranges $-4.0 < y < -2.5$ and $2.5 < y < 4.0$ [14, 15]. The results are compared to several theoretical calculations of parton shadowing and of coherent energy loss with or without shadowing[2–4, 16]. A strong suppression of prompt J/ψ is observed at the forward rapidity which is compatible with most of the theoretical predictions. The comparison of prompt J/ψ and ψ(2S) $R_{\rm pPb}$ suggests that ψ(2S) are more suppressed than J/ψ, especially in the backward region. Models describing the J/ψ data might still be able to describe the forward ψ(2S) suppression but are not able to reproduce the suppression in the backward region. This intriguing result could be a first indication that another CNM effect is at play. Recently, a theoretical calculation based on comovers scenario [5] tries to explain this behaviour. Figure 2 right shows the $R_{\rm pPb}$ of J/ψ and ψ(2S) from b-hadrons in the rapidity ranges $-4.0 < y < -2.5$ and $2.5 < y < 4.0$. J/ψ from b are slightly suppressed in the forward region and both models including the shadowing effect can describe the data. This is the first indication of the b-hadron production in pPb collisions. In the backward region, J/ψ from b are slightly less suppressed than prompt J/ψ as expected from the models, however models are in worse agreement with the J/ψ from b data. Given the large experimental uncertainties on the ψ(2S) from b measurement, no conclusions can be made on the comparison of the ψ(2S) from b suppression with respect to the J/ψ from b suppression. Figure

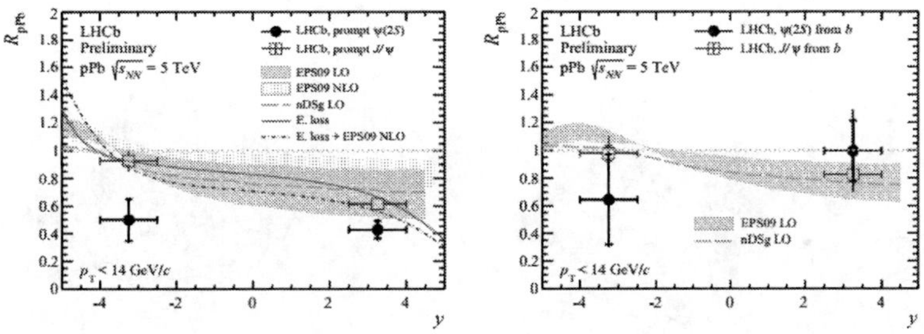

Figure 2. Nuclear modification factor R_{pPb} as a function of rapidity for prompt J/ψ and ψ(2S) (left) and for J/ψ and ψ(2S) from b-hadrons (right). Results are compared with theoretical models from Refs. [4, 16] (yellow dashed line and brown band), from Ref. [2] (blue band) and from Ref. [3] (green solid and blue dash-dotted lines). Only the models from [4, 16] are available for ψ(2S) from b.

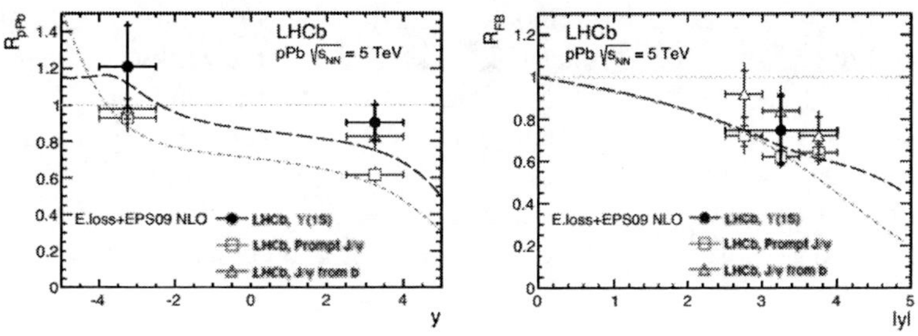

Figure 3. R_{pA} (left) and R_{FB} (right) as a function of rapidity for prompt J/ψ (red), J/ψ from b (blue) and Υ(1S) (black). The results are compared with theoretical predictions including energy loss and nuclear shadowing[3].

3 left shows the measurement of R_{pPb} for Υ(1S) as a function of rapidity [17] compared with the R_{pPb} measurement of prompt J/ψ and J/ψ from b. Υ(1S) is suppressed in the forward region while there is an indication for an enhancement of Υ(1S) production with respect to pp in the backward region, which could be attributed to anti-shadowing. The R_{pPb} measurement of $\Upsilon(1S)$ agrees within uncertainties (albeit large in the backward region) with the R_{pPb} measurement of J/ψ from b, reflecting the fact that CNM effects affect the b-hadrons production. Υ(1S) data agree with coherent energy loss model including nuclear shadowing as parametrized with EPS09 [3]. A comparison with more models is available in Ref. [17]. Figure 3 right shows the forward-backward production ratio R_{FB} as a function of rapidity for Υ(1S), prompt J/ψ and J/ψ from b. A smaller forward-backward asymmetry is observed for J/ψ from b with respect to prompt J/ψ. The forward backward asymmetry measurement of Υ(1S) and prompt J/ψ agree with theoretical calculation of coherent energy loss with nuclear shadowing parametrized with EPS09[3].

4 Z production in pPb and Pbp

Z boson candidates have been reconstructed in the dimuon decay channel, in the fiducial region defined by $2.0 < \eta(\mu^{\pm}) < 4.5$, $p_T(\mu^{\pm}) > 20$ GeV/c and $60 < m_{\mu^+\mu^-} < 120$ GeV/c^2. A clean signal of 11 forward candidates in the forward region and 4 candidates in the backward region has been observed[18].

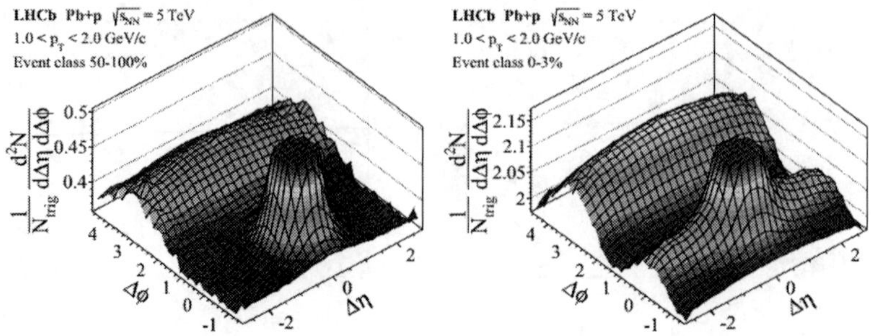

Figure 4. Two-particle correlation functions in the backward Pbp configuration, for low event-activity class (left) and high event-acticity class (right). The analysed pairs of prompt charged particles are selected in the p_T range $1 < p_T < 2$ GeV/c.

The cross section for Z production is found to be $\sigma_{Z\to\mu^+\mu^-} = 13.5^{+5.4}_{-4.0}$(stat.) ± 1.2(syst.) nb in the forward direction and $\sigma_{Z\to\mu^+\mu^-} = 10.7^{+8.4}_{-5.1}$(stat.) ± 1.0(syst.) nb in the backward direction. The result in the forward region agrees with predictions from NNLO calculations using FEWZ [19] and the MSTW08 PDF set [20] with and without nuclear effects. In the backward region, data are higher than theoretical calculations. The forward-backward ratio has also been measured in the rapidity range 2.5 $<| y_{CMS} |<$ 4.0. It is found to be lower than expectations with a 2.2σ deviation from one. The present measurements have a limited statistical precision, preventing to conclude on the presence of nuclear effects.

5 Multi-particle correlations in pPb and Pbp

Two-particle angular correlations between prompt charged particles [21] in pPb collisions have been measured in the forward (backward) region in the rapidity ranges $1.5 < y < 4.4$ ($2.5 < -y < 5.4$), respectively. They were studied in event-activity classes, which are defined as fractions of the VELO hit-multiplicity distributions in a minimum bias sample. In events with high event-activity a long-range correlation on the near side, the so-called ridge, is observed both in the forward and backward configurations, for $1 < p_T < 2$ GeV/c, while it is not observed in the 50-100% event-activity class (see Fig. 4 for the backward case). It was also observed that when probing identical absolute activity ranges in the pPb and Pbp configurations, the observed long-range correlations are compatible with each other (albeit there is a shift of one unit in rapidity coverage between both configurations).

6 Summary and prospects for heavy ion studies

LHCb sucessfully participated in the pPb data taking in 2013. The production of prompt J/ψ, $\psi(2S)$, J/ψ and ψ(2S) from b and Υ has been measured in pPb collisions at $\sqrt{s_{NN}}$ = 5 TeV. J/ψ from b and Υ(1S) are less affected by cold nuclear matter effects than prompt J/ψ which exhibits a stronger suppression. Prompt ψ(2S) seems even more suppressed than prompt J/ψ, especially in the backward region. Models of coherent energy loss with and without nuclear shadowing effects give a good description of the data, except for the production of ψ(2S) in the backward region. The prompt ψ(2S) data suggest that another CNM effect could be at play in the backward region. LHCb also performed the first measurement of Z production in pPb collisions at the LHC and the current analysis with limited statistical precision, will highly benefit from the larger statistics data sample which will be collected during run 2. Thanks to two-particle correlation analysis, LHCb observed for the first time the ridge in the forward region, in pPb collisions, thus complementing measurement at mid-rapidity from other experiments. The LHCb dectector will also collect for the first time PbPb data at $\sqrt{s_{NN}} = 5$ TeV at the end of 2015. LHCb will thus lead a rich physics program covering heavy flavour physics, electroweak physics, soft QCD physics and Quark Gluon Plasma studies. It is expected to collect $L_{int} \approx 50\text{-}80\ \mu b^{-1}$ at the end of 2015. In addition, LHCb is in the unique position at the LHC to perform fixed target physics. Thanks to its System for Measuring Overlap with Gas (SMOG), initially used as luminosity monitor, noble gases can be injected inside the inner tracker of LHCb. With the proton and lead beams delivered by the LHC, and the various species of gas which can be injected in LHCb, a large number of fixed target configurations can be studied. LHCb already successfully collected proton-helium, proton-neon, proton-argon, lead-neon data and a lead-argon data taking is foreseen in parallel of the 2015 PbPb data taking. The highest center-of-mass energy which can be achieved in proton-gas collisions is $\sqrt{s_{NN}} = 115$ GeV while in lead-gas collision it is $\sqrt{s_{NN}} = 72$ GeV. Through its fixed target program, LHCb can bridge the gap from SPS to LHC with a single experiment. LHCb is definitely becoming a truly general purpose detector in the forward region.

References

[1] T. Matsui, H. Satz, Phys. Lett. **B178**, 416 (1986)
[2] J.L. Albacete et al., Int. J. Mod. Phys. **E22**, 1330007 (2013), 1301.3395
[3] F. Arleo, S. Peigne, JHEP **03**, 122 (2013), 1212.0434
[4] E.G. Ferreiro et al., Phys. Rev. **C88**, 047901 (2013), 1305.4569
[5] E.G. Ferreiro, Phys. Lett. **B749**, 98 (2015), 1411.0549
[6] C.A. Salgado et al., J. Phys. **G39**, 015010 (2012), 1105.3919
[7] V. Khachatryan et al. (CMS Collaboration), JHEP **09**, 091 (2010), 1009.4122
[8] A.A. Alves, Jr. et al. (LHCb Collaboration), JINST **3**, S08005 (2008)
[9] R. Aaij et al. (LHCb Collaboration), Int. J. Mod. Phys. **A30**, 1530022 (2015), 1412.6352
[10] (LHCb and ALICE Collaborations), CERN-LHCb-CONF-2013-013, ALICE-PUBLIC-2013-002 (2013)
[11] (LHCb and ALICE Collaborations), CERN-LHCb-CONF-2014-003, ALICE-PUBLIC-2014-002 (2014)
[12] R. Aaij et al. (LHCb Collaboration), Eur. Phys. J. **C71**, 1645 (2011), 1103.0423
[13] R. Aaij et al. (LHCb Collaboration), Eur. Phys. J. **C72**, 2100 (2012), 1204.1258
[14] R. Aaij et al. (LHCb Collaboration), JHEP **02**, 072 (2014), 1308.6729
[15] M. Zhao et al. (LHCb Collaboration), CERN-LHCb-CONF-2015-005 (2015)
[16] Z. Conesa del Valle et al. (2014), [Nucl. Phys.A926,236(2014)], 1402.1747
[17] R. Aaij et al. (LHCb Collaboration), JHEP **07**, 094 (2014), 1405.5152
[18] R. Aaij et al. (LHCb Collaboration), JHEP **09**, 030 (2014), 1406.2885
[19] R. Gavin et al., Comput. Phys. Commun. **182**, 2388 (2011), 1011.3540
[20] A.D. Martin et al., Eur. Phys. J. **C63**, 189 (2009), 0901.0002
[21] R. Aaij et al. (LHCb Collaboration) (2015), 1512.00439

De-Confinement in small systems: Clustering of color sources in high multiplicity $\bar{p}p$ collisions at $\sqrt{s}$= 1.8 TeV

L. J. Gutay[1,a]

[1] *Department of Physics and Astronomy, Purdue University, West Lafayette, IN-47907, USA*

Abstract. It is shown that de-confinement can be achieved in high multiplicity non jet $\bar{p}p$ collisions at $\sqrt{s}$= 1.8 TeV Fermi National Accelerator Laboratory(FNAL- E735) experiment. In this paper we have analyzed the transverse momentum spectrum in the frame-work of the clustering of color sources. This frame-work naturally predicts the reduction in the charged particle multiplicity with respect to the value expected from the number of independent strings. Results are presented for both thermodynamic and transport properties. The initial temperature and energy density are obtained from the data via the color reduction factor $F(\xi)$ and the associated string density parameter ξ. The results for he trace anomaly Δ and shear viscosity to entropy density ratio(η/s) are presented. These results confirm our earlier observation that the de-confined state of matter was created in high multiplicity events in $\bar{p}p$ collisions at $\sqrt{s}$=1.8 TeV.

1 Introduction

The observation of high total multiplicity, high transverse energy, non-jet, isotropic events led Van Hove [1] and Bjorken [2] to conclude that high energy density events are produced in high energy $\bar{p}p$ collisions [3]. These events have a far greater cross section than the jet production. In these events the transverse energy is proportional to the number of low transverse momentum particles. This basic correspondence can be explored over a wide range of the charged particle pseudorapidity density $\langle dN_c/d\eta\rangle$ in $\bar{p}p$ collisions at center of mass energy $\sqrt{s}$ = 1.8 TeV. In our earlier work, published in 2002, the evidence of hadronic de-confinement in $\bar{p}p$ collisions at $\sqrt{s}$ = 1.8 TeV was presented [4].

The objective of this work is to further analyze the published E735 data on the transverse momentum spectra of charged particles in the framework of Color String Percolation Model(CSPM). CSPM has been successfully applied to heavy ion data for thermodynamics and transport coefficients [5–9].

2 Color Strings and Percolation

Multiparticle production at high energies is currently described in terms of color strings stretched between the projectile and target. Hadronizing these strings produce the observed hadrons. The individual strings act as independent color sources of particles through the creation of $q\bar{q}$ pairs from the sea. At low energies only valence quarks of nucleons form strings that then hadronize. The number of strings grows with the energy and with the number of nucleons of participating nuclei. Color strings

[a] e-mail: gutay@purdue.edu

may be viewed as small discs in the transverse space filled with the color field created by colliding partons. Particles are produced by the Schwinger mechanisms [10]. With growing energy and size of the colliding nuclei the number of strings grow and start to overlap and interact to form clusters [11, 12].

We assume that a cluster of n strings behaves as a single string with an energy-momentum that corresponds to the sum of energy-momenta of the individual strings and with a higher color field, corresponding to the vectorial sum of the color field of each individual string [11, 12]. One can obtain the multiplicity μ and the mean transverse momentum squared $\langle p_t^2 \rangle$ of the particles produced by a cluster of n strings [12]

$$\mu_n = \sqrt{\frac{nS_n}{S_1}}\mu_0; \qquad \langle p_t^2 \rangle = \sqrt{\frac{nS_1}{S_n}}\langle p_t^2 \rangle_1, \tag{1}$$

where μ_0 and $\langle p_t^2 \rangle_1$ are the mean multiplicity and $\langle p_t^2 \rangle$ of particles produced from a single string with a transverse area $S_1 = \pi r_0^2$. In the limit of high string density, one obtains [11, 12]

$$\langle \frac{nS_1}{S_n} \rangle = \frac{\xi}{1 - e^{-\xi}} \equiv \frac{1}{F(\xi)^2}, \tag{2}$$

where $F(\xi)$ is the color suppression factor. $F(\xi)$ is related to the ξ

$$F(\xi) = \sqrt{\frac{1 - e^{-\xi}}{\xi}}. \tag{3}$$

The net effect due to $F(\xi)$ is the reduction in hadron multiplicity and increase in the average transverse momentum of particles. The CSPM model calculation for hadron multiplicities and momentum spectra for pp and AA collisions at RHIC and LHC energies were found to be in excellent agreement with the experiments [8, 9].

3 E735 Experiment

Experiment E735 was run during the 1988-1989 Tevatron running period, primarily collecting data triggered to enrich high multiplicity events in $\bar{p}p$ collisions at $\sqrt{s} = 1.8$ TeV [13]. The multiplicity dependence of the transverse momentum p_t spectra was measured. The invariant p_t spectra was fitted with a power law $A/(p_0 + p_t)^n$ [14]. The power law function is the QCD inspired function which goes asymptotically to p_t^{-n} at large p_t and approximates an exponential at small p_t.

4 Color Suppression Factor and Temperature

The suppression factor is determined by comparing the charged particle spectra for $0.1 < p_t < 1.5$ GeV/c for minimum bias events from pp collisions and high multiplicity transverse momentum spectra from $\bar{p}$p. To evaluate the initial value of ξ from data a parameterization of pp events at $\sqrt{s} = 200$ GeV is used to compute the p_t distribution

$$dN_c/dp_t^2 = a/(p_0 + p_t)^n, \tag{4}$$

where a is the normalization factor. $p_0 = 1.98$ and $n = 12.88$ are parameters used to fit the data [5]. A modified parameterization is used for high multiplicity events in $\bar{p}$p collisions at $\sqrt{s} = 1.8$ TeV to take into account the interactions of the strings [12].

$$dN_c/dp_t^2 = \frac{a'}{\left(p_0\sqrt{F(\xi_{\bar{p}p})_{HM}/F(\xi_{pp})} + p_t\right)^n}, \tag{5}$$

DOI: 10.1051/epjconf/201612006005

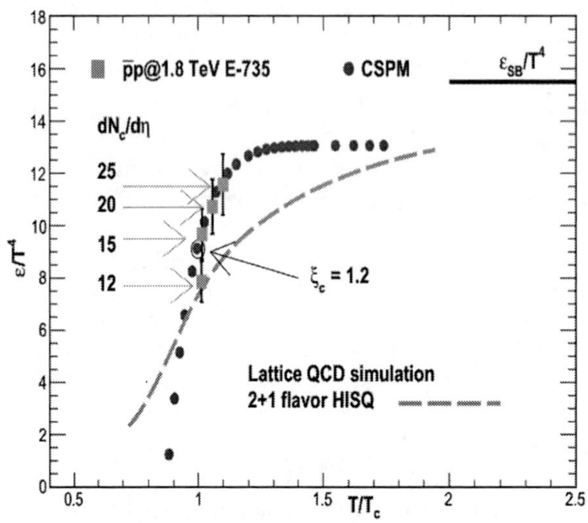

Figure 1. ε/T^4 as a function of T/T_c. Solid green squares are from $\bar{p}p$ collisions at $\sqrt{s}$ = 1.8 TeV. The lattice QCD calculation is shown as dotted green line [22]. CSPM for Au+Au at $\sqrt{s_{NN}}$ = 200 GeV are red solid circles [7].

where $F(\xi_{\bar{p}p})_{HM}$ is the color suppression factor in high multiplicity events in $\bar{p}p$ collisions at $\sqrt{s}$ = 1.8 TeV. In pp collisions $F(\xi_{pp}) \sim 1$ at $\sqrt{s}$ = 200 GeV due to the low string overlap probability [5].

The connection between the measured $F(\xi)$ and the temperature $T(\xi)$ involves the Schwinger mechanism (SM) for particle production. The Schwinger distribution for massless particles is expressed in terms of p_t^2 [10, 15]

$$dn/dp_t^2 \sim e^{-\pi p_t^2/x^2}, \tag{6}$$

where the average value of the string tension is $\langle x^2 \rangle$. The tension of the macroscopic cluster fluctuates around its mean value because the chromo-electric field is not constant. Quadratic fluctuations lead to a Gaussian distribution of the string tension for the cluster, which gives rise to the thermal distribution [5, 16, 17]

$$dn/dp_t^2 \sim e^{(-p_t\sqrt{\frac{2\pi}{\langle x^2 \rangle}})}, \tag{7}$$

with $\langle x^2 \rangle = \pi \langle p_t^2 \rangle_1 / F(\xi)$. The initial temperature is expressed as

$$T(\xi) = \sqrt{\frac{\langle p_t^2 \rangle_1}{2F(\xi)}}, \tag{8}$$

where $\langle p_t^2 \rangle_1$ is the average transverse momentum squared of particles produced from a single string and is evaluated using Eq. (8). At the percolation transition ξ_c = 1.2 the temperature T_c = 167 ± 2.6 MeV as obtained from the statistical model analysis of data, which is the universal chemical freeze-out temperature and is a good measure of the phase transition temperature [18, 19]. This determines $\sqrt{\langle p_t^2 \rangle_1}$ = 207 ± 3.3 MeV [5] which calibrates the temperature.

 DOI: 10.1051/epjconf/201612006005

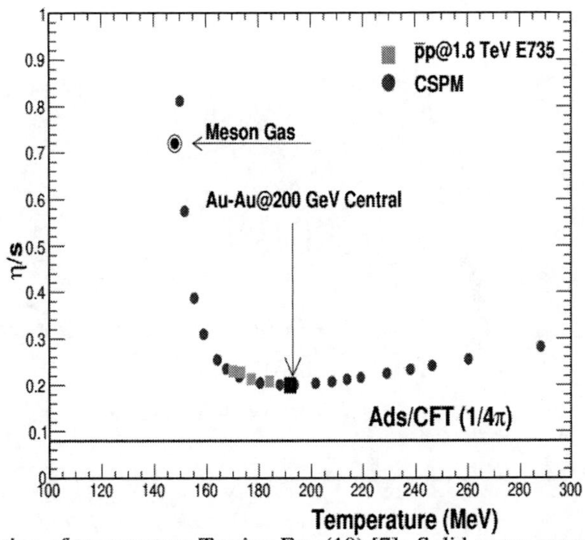

Figure 2. η/s as a function of temperature T using Eq. (10) [7]. Solid green squares are from $\bar{p}p$ collisions at $\sqrt{s}$ = 1.8 TeV. CSPM results for Au+Au at $\sqrt{s_{NN}}$ = 200 GeV are shown as solid red circles [7]. The meson gas value for $\eta/s \sim 0.7$ is shown as solid black circle at T ~ 150 MeV [23]. The lower bound shown is given by the AdS/CFT [24].

5 Energy Density

The QGP according to CSPM is born in local thermal equilibrium because the temperature is determined at the string level. After the initial temperature $T > T_c$ the CSPM is described as a perfect fluid and may expand according to Bjorken boost invariant 1D hydrodynamics [20]

$$\varepsilon = \frac{3}{2} \frac{\frac{dN_c}{dy}\langle m_t \rangle}{S_\perp \tau_{pro}}, \tag{9}$$

where ε is the energy density, $\langle m_t \rangle$ the average transverse mass, $S_\perp$ the transverse overlap area of the colliding nucleons, and τ_{pro} the proper time. Above the critical temperature only massless particles are present in CSPM. In Schwinger model τ_{pro} is the formation time for gluon production and is given by $\tau_{pro} = \frac{2.405\hbar}{\langle m_t \rangle}$ [10]. With the determination of T and ε one can compare the energy density expressed as ε/T^4 with the available lattice QCD results [21, 22]. Figure 1 shows a plot of ε/T^4 as a function of T/T_c. The error on ε/T^4 is 8-11% mainly due to the uncertainty in temperature. The LQCD results are from the HotQCD Collaboration [22]. It is observed that for highest multiplicity $\langle dN_c/d\eta \rangle \sim 27$, ε/T^4 from $\bar{p}p$ collisions is close to the value obtained for Au+Au at 200 GeV. This result strongly supports the formation of the QGP above the phase transition temperature in high multiplicity events in $\bar{p}p$ interactions at $\sqrt{s}$ = 1.8 TeV. This emphasizes that high string densities determines the initial energy density and temperature.

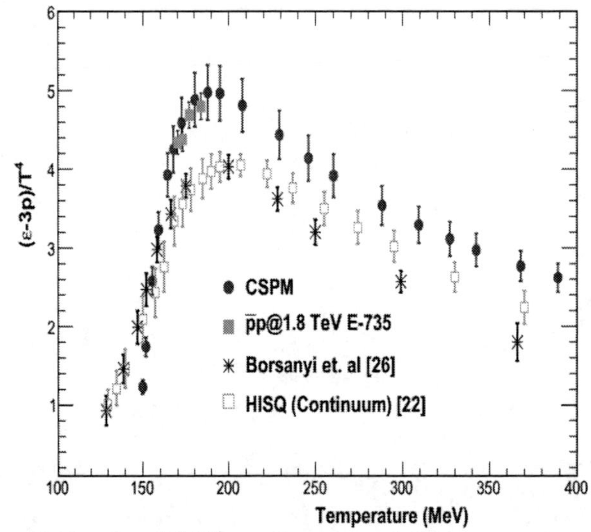

Figure 3. The trace anomaly $\Delta = (\varepsilon - 3p)/T^4$ vs temperature. Solid green squares are from $\bar{p}p$ collisions at $\sqrt{s}$ = 1.8 TeV. Green open squares are from HotQCD Collaboration [22]. Black stars are from Borsanyi et al. [26].

6 Shear viscosity to entropy density ratio η/s

Shear viscosity to entropy density ratio η/s was obtained in the framework of kinetic theory and the string percolation [7] and is given by

$$\frac{\eta}{s} = \frac{TL}{5(1 - e^{-\xi})}, \tag{10}$$

where T is the temperature and L is the longitudinal extension of the source ~ 1 fm. Fig. 2 shows η/s as a function of the temperature [7]. The error on η/s is driven by the error on temperature which is ~ 2.5-3.5%. The lower bound shown in Fig. 2 is given by the AdS/CFT conjecture [24]. It is seen that for $\langle dN_c/d\eta \rangle \sim 25$ the η/s is equal to the value from the most central collisions in Au+Au at $\sqrt{s_{NN}}$ = 200 GeV.

7 Trace anomaly Δ

The trace anomaly Δ is the expectation value of the trace of the energy-momentum tensor, $\langle \Theta^{\mu}_{\mu} \rangle = (\varepsilon - 3p)$, which measures the deviation from conformal behavior and thus identifies the interaction still present in the medium [25]. We find that the reciprocal of η/s is in numerical quantitative agreement with the dimensionless ratio $(\varepsilon - 3p)/T^4$ from LQCD over a wide range of temperatures [6, 8]. Fig. 3 shows Δ for $\bar{p}p$ along with the CSPM calculation and LQCD simulations. The minimum in $\eta/s \sim 0.20$ determines the peak of the interaction measure ~ 5 in agreement with the recent HotQCD values [22]. The maximum in Δ corresponds to the minimum in η/s. This is true for the highest multiplicity events in $\bar{p}p$ collisions.

8 Summary

The $\bar{p}p$ collisions at $\sqrt{s}$ = 1.8 TeV from the E735 experiment have been re-analyzed in using the clustering of color sources data based phenomenology. The results for temperature T, energy density ε and shear viscosity to entropy density ratio η/s have been extracted for high multiplicity events. The trace anomaly is also obtained and compared with the LQCD results. These results strongly argue that even in small systems at high energy and high multiplicity events QGP formation is possible as seen in $\bar{p}p$ collisions at $\sqrt{s}$ = 1.8 TeV.

Acknowledgment

This research was supported by the Office of Nuclear Physics within the U.S. Department of Energy Office of Science under Grant No. DE-FG02-88ER40412. Author is grateful to A. S. Hirsch, R. P. Scharenberg and B. K. Srivastava for useful and simulating discussions.

References

[1] L. Van Hove, Phys. Lett. B **118**, 138 (1982).
[2] J. D. Bjorken, Fermilab Pub. 82/44-THY (1982).
[3] L. McLerran, Rev. Mod. Phys. **58**, 1021 (1986).
[4] T. Alexopoulos et al. (E735 Collaboration), Phys. Lett. B **528**, 43 (2002).
[5] R. P. Scharenberg, B. K. Srivastava and A. S. Hirsch, Eur. Phys. J. C **71**, 1510 (2011).
[6] R. P. Scharenberg, PoS (CPOD 2013), **017** (2013).
[7] J. Dias de Deus, A. S. Hirsch, C. Pajares, R. P. Scharenberg and B. K. Srivastava, Eur. Phys. J. C **72**, 2123 (2012).
[8] B. K. Srivastava, Nucl. Phys. A **926**, 142 (2014).
[9] M. A. Braun, J. Dias de Deus, A. S. Hirsch, C. Pajares, R. P. Scharenberg and B. K. Srivastava, Phys. Rep. **599**, 1 (2015).
[10] J. Schwinger, Phys. Rev. **128**, 2425 (1962).
[11] M. A. Braun and C. Pajares, Eur. Phys. J. C **16**, 349 (2000).
[12] M. A. Braun, F. del Moral and C. Pajares, Phys. Rev. C **65**, 024907 (2002).
[13] T. Alexopoulos et al. (E735 Collaboration), Phys. Rev. D **48**, 984 (1993).
[14] C. Albajar et al. (UA1 Collaboration), Nucl. Phys. B **335**, 261 (1990).
[15] C. Y. Wong , *Introduction to high energy heavy ion collisions* (World Scintific, 1994).
[16] A. Bialas, Phys. Lett. B **466**, 301 (1999).
[17] J. Dias de Deus and C. Pajares, Phys. Lett. **642**, 455 (2006).
[18] P. Braun-Munzinger, J. Stachel and C. Wetterich, Phys. Lett. B **596**, 61 (2004).
[19] F. Becattini, P. Castorina, A. Milov and H. Satz, Eur. Phys. J. C **66**, 377 (2010).
[20] J. D. Bjorken, Phys. Rev. D **27**, 140 (1983).
[21] A. Bazavov et al., Phys. Rev. D **80**, 014504 (2009).
[22] A. Bazavov et al., Phys. Rev. D **90**, 094503 (2014).
[23] M. Prakash, M. Prakash, R. Venugopalan and G. Wleke, Phys. Rep. **227**, 321 (1993).
[24] P. K. Kovtun, D. T. Son and A. O. Starinets, Phys. Rev. Lett. **94**, 111601 (2005).
[25] M. Cheng et al., Phys. Rev. D **81**, 054504 (2010).
[26] S. Borsanyi et al., Phys. Lett. **730**, 99 (2014).

EPJ Web of Conferences **120**, 06006 (2016) DOI: 10.1051/epjconf/201612006006
ISMD 2015

p-p minimum-bias dijets and nonjet quadrupole in relation to conjectured collectivity (flows) in high-energy nuclear collisions

Thomas A. Trainor[1]

[1] *CENPA 354290 University of Washington, Seattle, Washington, USA*

Abstract. Recent observations of ridge-like structure in p-p and p-A angular correlations at the RHIC and LHC have been interpreted to imply collective motion in smaller collision systems. It is argued that if correlation structures accepted as manifestations of flow in A-A collisions appear in smaller systems collectivity (flow) must extend to the smaller systems. But the argument could be reversed to conclude that such structures appearing in A-A collisions may not imply hydrodynamic flow. I present spectrum, correlation and fluctuation data from RHIC p-p and Au-Au collisions and p-p, p-Pb and Pb-Pb results from the LHC described accurately by a two-component (soft+dijet) model of hadron production. I also present evidence for a significant p-p nonjet (NJ) quadrupole (v_2) component with n_{ch} systematics directly related to A-A NJ quadrupole systematics. The combination suggests that soft, dijet and NJ quadrupole components are distinct phenomena in all cases, inconsistent with hadron production from a common bulk medium exhibiting collective motion (flow).

1 Introduction

Certain analysis techniques applied to LHC data for smaller collision systems have lead to claims for "collectivity" (flows) even in p-p collisions at higher energies. The original flow concept has been extended to a universal property of all high-energy nuclear collisions, not just a subset corresponding to the highest particle and energy densities. That conceptual trend is ironic in that accumulating evidence from alternative analysis techniques argues against any hydrodynamic phenomenon in high-energy collisions, a conclusion buttressed by the observation that certain phenomena associated with flows in central A-A collisions also appear in p-p collisions with *negligible* particle and energy densities. Given the limitation on article length I summarize here only a few results most relevant to recent LHC claims.

2 p-p 2D angular correlations vs n_{ch} – the three-component model

Figure 1 shows a 2D model fit to data from high-multiplicity (n_{ch} index $n = 6$) 200 GeV p-p collisions (one of seven multiplicity classes). The standard 2D fit model includes soft +

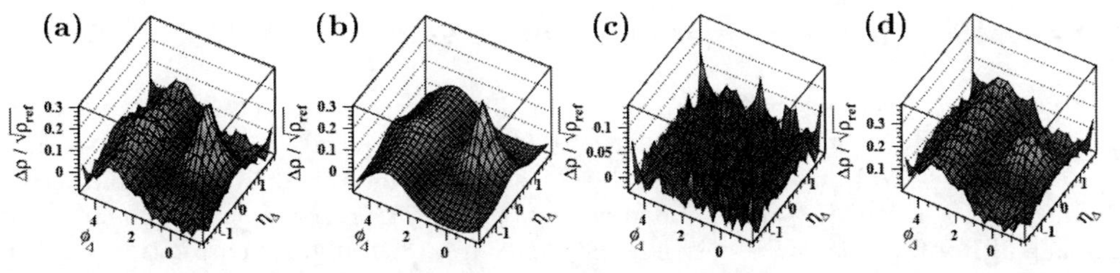

Figure 1. Example 2D model fit to $n = 6$ angular correlations from 200 GeV p-p collisions: (a) data, (b) fit model, (c) fit residuals, (d) dijet and NJ quadrupole components of the data.

dijet + NJ quadrupole elements [1]. The fit residuals (c) indicate that the 2D model fits are typically excellent. Panel (d) shows dijet plus NJ quadrupole data components in isolation.

Figure 2 shows model-fit results for the principal model elements vs soft multiplicity density $\bar{\rho}_s = n_s/\Delta\eta$ within acceptance $\Delta\eta = 2$. Those results are consistent with previous studies of minimum-bias (MB) p-p collisions [2, 3]. In terms of correlated hadron-pair number (e.g. $n_{ch}A_{\text{soft}}$) the soft (projectile dissociation) pair number scales $\propto \bar{\rho}_s$, the dijet pair number scales $\propto \bar{\rho}_s^2$ and the NJ-quadrupole pair number scales $\propto \bar{\rho}_s^3$. Since multiplicity $n_{ch} = n_s + n_h$ varies ten-fold the range of dijet production is 100-fold and the NJ quadrupole varies 1000-fold for this data sample. The dijet trend confirms results from a p_t spectrum analysis leading to a *two*-component model (TCM) where the hard-component (dijet) hadron yield n_h is related to the soft-component (projectile-dissociation) yield n_s as $n_h \approx 0.01n_s^2$ within $\Delta\eta = 1$ [4].

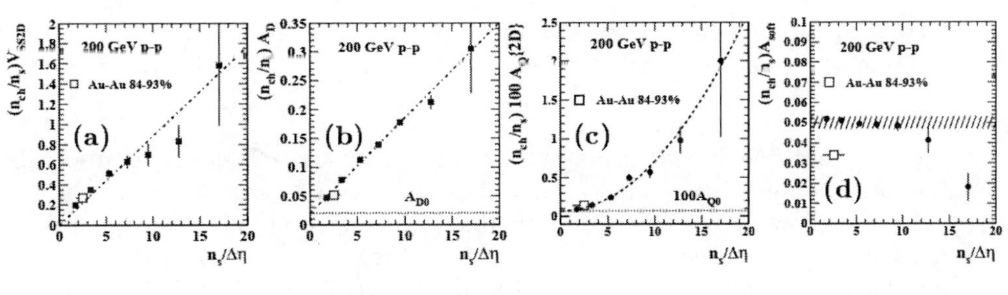

Figure 2. Fit parameters vs soft-multiplicity density $\bar{\rho}_s = n_s/\Delta\eta$: (a) SS 2D peak volume, (b) AS 1D peak amplitude, (c) NJ quadrupole amplitude, (d) soft-component (projectile-nucleon) amplitude.

The dijet trend for spectra and 2D angular correlations is consistent with recent LHC measurements of ensemble-mean p_t [5] and p_t fluctuations [6] vs n_{ch} where p-p collisions at LHC energies are found to be dominated by dijet production described accurately by the same TCM (see Sec. 4). Dijet correlation trends in Au-Au collisions at the RHIC are consistent with spectrum hard components [7, 8] and with QCD (via event-wise reconstructed jets) [9].

The new *third* model element for p-p collisions (NJ quadrupole) is found to be very significant for larger multiplicities. Evidence from Au-Au collisions suggests that the NJ quadrupole component in A-A collisions is carried by a small fraction of total hadrons ($<$5%) [10].

3 The CMS "ridge" – nonjet quadrupole vs minimum-bias dijets

One argument for collectivity (flows) in small collision systems proceeds from identification of a same-side (SS) "ridge" in 2D angular correlations from 7 TeV *p-p* collisions with certain cuts applied [11]. Note that several "ridges" have been identified in RHIC and LHC data including a "soft ridge" [12] and a ridge associated with trigger-associated combinatoric jet analysis [13] (both probably jet-related). Evidence from 2D angular correlations as in Sec. 2 suggests that the CMS ridge is associated with the NJ quadrupole component [14].

Figure 3 (a) repeats results from the $n = 6$ multiplicity class of 200 GeV *p-p* collisions in Fig. 1 (d) that can be compared with high-multiplicity events from 7 TeV *p-p* collisions in panel (b). The structures are quite similar, including a large negative curvature near $\phi_\Delta = \pi$ and nearly zero curvature at the origin. The critical issue is the *net* azimuth curvature near $\phi_\Delta = 0$ in the interval $|\eta_\Delta| > 1$ that is determined by a superposition of the away-side (AS) dipole (positive curvature) and NJ quadrupole (negative curvature). A "ridge" appears when the net SS curvature becomes negative (corresponding to a maximum). With increasing n_{ch}, collision energy and p_t cut the NJ quadrupole amplitude increases relative to the AS dipole amplitude. At some point a change in sign of the curvature may occur and a "ridge" appears.

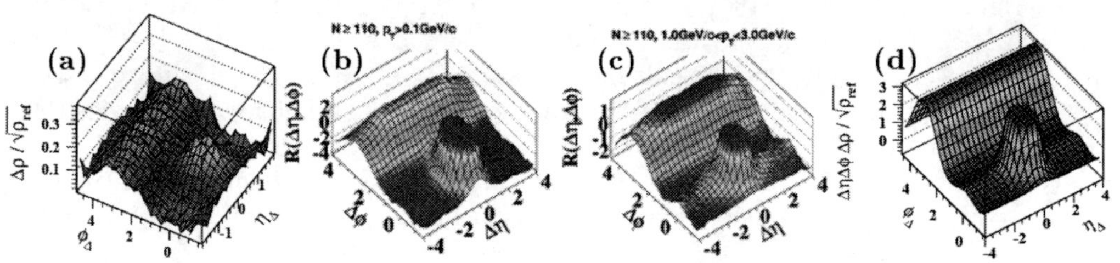

Figure 3. (a) 200 GeV *p-p* dijets + NJ quadrupole, (b) CMS 7 TeV *p-p* data for $N_{\rm trk} > 110$ (c) CMS data with additional p_t cut (d) extrapolation from 200 GeV data trends for CMS conditions.

Figure 3 (c) shows the appearance of a SS "ridge" when a p_t cut is applied to CMS high-multiplicity data, the primary evidence for claims of a novel ridge phenomenon interpreted by some to signal flow in *p-p* collisions. In Ref. [14] dijet and NJ quadrupole trends for 62 and 200 GeV Au-Au collisions were extrapolated first to *N-N* collisions and then to 7 TeV. The prediction shown in panel (d) agrees quantitatively with the CMS result in panel (c). One motivation for the *p-p* correlation study in Sec. 2 was to confirm the extrapolation to *N-N* collisions in Ref. [14] and that has been achieved. It is notable that as the SS curvature changes sign from positive to negative the negative AS curvature *doubles for the same conditions*, confirming the role of the NJ quadrupole with its two maxima at 0 and π.

4 p_t fluctuations at the RHIC and LHC – minimum-bias dijets

Figure 4 (a) shows p_t fluctuations (ALICE) as measured by relative r.m.s. measure $\sqrt{C}/\bar{p}_t$ vs multiplicity density $\bar{\rho}_0 = n_{ch}/\Delta\eta$. The ALICE result is interpreted to suggest "collectivity" (flow) in *p-p* collisions and no significant energy dependence over a large interval. The choice of fluctuation measure is motivated as a p_t proxy for relative temperature fluctuations in the form $\delta T/T_0$ in the context of a QCD phase boundary. But C is simply related to a conventional variance-based fluctuation measure B as $C = B/\overline{n_{ch}(n_{ch}-1)}$, with conditional variance difference $B \equiv \overline{(P_t - n_{ch}\bar{p}_t)^2} - \bar{n}_{ch}\sigma^2_{p_t}$ (the second term is a central-limit reference).

Figure 4 (b) shows ALICE data from panel (a) transformed to measure B (points) per the relations above. The dashed curve is (soft + hard) TCM representation $B = b_s\bar{\rho}_s + b_h\bar{\rho}_s^2$. The second term is the contribution from MB dijet production that dominates p_t fluctuations. Corresponding results for other collision energies inferred from known ensemble-mean $\bar{p}_t$ systematics [5] indicate that B (and p_t fluctuations) are strongly energy dependent as expected for dijets [6]. The ALICE choice of p_t fluctuation measure (a *ratio of ratios*) has as one consequence the near cancellation of the dominant dijet contribution, suggesting as a consequence that small collision systems may be substantially thermalized with no significant correlation structure and that event-wise temperature as a state variable may be relevant.

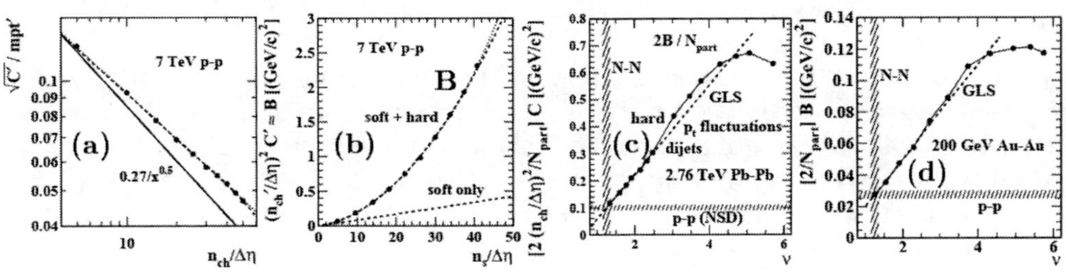

Figure 4. (a) p-p p_t fluctuations from ALICE, (b) same data converted to variance difference B, (c) per-participant form $2B/N_{part}$ for 2.76 TeV Pb-Pb, (d) STAR data for 200 GeV Au-Au collisions.

Figure 4 (c) shows ALICE p_t fluctuation data for 2.76 TeV Pb-Pb collisions transformed to the *per-particpant* statistical measure $2B/N_{part}$. The TCM for transparent A-A collisions is given by the GLS curve (dashed line, Glauber linear superposition) with $\nu = 2N_{bin}/N_{part}$. The hard-component trend $\propto \nu$ is a signature for dijet production that appears to dominate p_t fluctuations up to central collisions. Figure 4 (d) shows STAR 200 GeV Au-Au fluctuation data in the same format (obtained almost ten years ago) that exhibit the same trend [15, 16]. The five-fold increase in overall amplitude from RHIC to LHC is expected from the energy dependence of dijet production. The STAR study derived the underlying p_t 2D angular correlations corresponding to measured p_t fluctuations (via inversion of fluctuation scale dependence), and the dominant jet-related correlation structure is undeniable.

5 The nonjet quadrupole – flow or nonflow?

Figure 5 (a) shows the per-binary-collision trend of the SS 2D jet-peak amplitude vs Au-Au centrality. For the more-peripheral half of total cross section σ_0 Au-Au collisions exhibit transparency – jets remain unmodified. For the more-central half jet formation is modified resulting in a *larger* yield of jet fragments, the change occurring at a *sharp transition* (ST) (hatched band) [1]. One could speculate that above the ST a flowing bulk medium (QGP?) may cause jet modification, in which case a corresponding change should occur to elliptic flow measure v_2. In panel (b) NJ quadrupole measure $A_Q = \bar{\rho}_0 v_2^2$ follows the same trend from peripheral to central Au-Au with no significant deviation, no response to a "bulk medium" [17].

Figure 5 (c) shows $v_2(p_t)$ vs p_t for three hadron species in a conventional plotting format. The horizontal displacement of data trends at lower p_t referred to as "mass ordering" is

 DOI: 10.1051/epjconf/201612006006

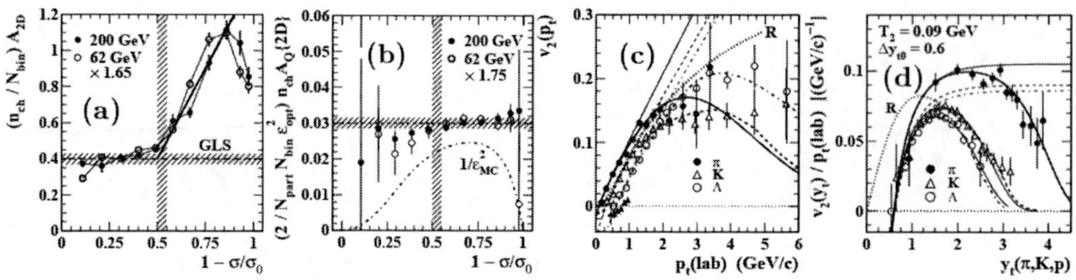

Figure 5. (a) 200 GeV Au-Au jet correlations showing sharp transition, (b) Au-Au NJ quadrupole (c) $v_2(p_t)$ vs p_t, (d) $v_2(p_t)/p_t$ vs transverse rapidity y_t revealing common source boost at $y_t = 0.6$.

interpreted to confirm a hydrodynamic flow mechanism (i.e. elliptic flow). However, hydro theory itself relating to the particle-source boost distribution can only describe a ***tiny fraction*** of that plot space near the origin. Panel (d) shows the result of a simple transformation to $v_2/p_t(lab)$ vs transverse rapidity $y_t \equiv \ln[(p_t + m_t)/m_h]$ (logarithmic measure of speed or boost). The common zero intercept at $y_t = 0.6$ reveals the hadron-source boost distribution consistent with a ***single value***. Conventional hydro theories based on a Hubble-expanding bulk medium include a broad source-boost distribution (as for theory curve R). Thus, $v_2(p_t)$ data for identified hadrons demonstrate that the NJ quadrupole corresponds to an expanding cylindrical shell with ***fixed source boost for all collisions systems*** [18, 19]. The combination of results in Fig. 5 plus other evidence against a hydrodynamic mechanism argues against a flow interpretation for the NJ quadrupole in any high-energy collision system [20]. The possibility emerges that the NJ quadrupole represents an alternative (nonflow) QCD mechanism [21].

6 Summary

The two-component (soft + hard) model (TCM) provides a remarkably accurate description of hadron production over a broad range of collision systems – *p-p* and *p-A* or *d-A* vs n_{ch} and *A-A* vs centrality measure ν at RHIC and LHC collision energies. The dijet (hard) component of the TCM is quantitatively consistent with spectrum hard components and with data from event-wise reconstructed jets. Evidence from p_t spectra, ensemble-mean p_t, p_t fluctuations and 2D angular correlations shows that minimum-bias (MB) dijets dominate high-energy collisions. However, alternative spectrum, fluctuation and correlation measures can act to suppress evidence for large dijet contributions, for instance in the form of statistical "ratios of ratios" that cancel the hard components of TCM trends or spectrum ratios (e.g. R_{AA}) that suppress spectrum hard components at smaller p_t *where they achieve their maximum values*.

A third component, the azimuth quadrupole (conventionally identified with elliptic flow), emerges as a significant element only in angular correlations. A nonjet (NJ) quadrupole contribution can be identified unambiguously via 2D model fits, with a substantial amplitude even in *p-p* collisions as demonstrated in the present study. Careful examination of NJ quadrupole trends on *p-p* n_{ch}, *A-A* centrality and on p_t for identified hadron species reveals systematic trends incompatible with the conventional hydro interpretation. For instance, the NJ quadrupole vs centrality trend in Au-Au collisions shows no correspondence with the large change in jet formation ("jet quenching") in more-central collisions attributed to formation of a dense medium. The v_2 vs p_t trends for identified hadrons argue against a Hubble-expanding dense bulk medium, support instead an expanding thin cylindrical shell with source boost independent of collision centrality *for those few hadrons carrying the NJ quadrupole.*

The CMS "ridge," seen as a novel phenomenon suggesting "collectivity" (flow) in small collision systems at LHC energies, is simply explained as an interplay of the NJ quadrupole with the away-side dijet peak that controls the same-side net azimuth curvature. The LHC phenomenon can be predicted quantitatively from quadrupole and dijet trends observed already in RHIC collisions. One could argue that appearance of an NJ quadrupole component in p-p collisions does support a flow interpretation there, as recently claimed. But the opposite is more likely: the appearance of a quadrupole component in small systems where the particle density is negligible argues against a hydro interpretation, instead is consistent with mounting evidence against a hydrodynamic description in any high-energy collision system.

The unanticipated abundance of resolved low-energy (MB) jets or minijets in more-central Au-Au collisions at the RHIC became apparent in STAR angular-correlation data more than ten years ago, already casting doubt on claims for local thermalization in such collisions and therefore for a flowing bulk medium or QGP. Evidence against thermalization and hydrodynamic flows has mounted steadily since then. Statistical and correlation measures and spectrum-analysis techniques that reveal the persistent large dijet contribution and evidence against flows have been developed and published throughout that period. It is remarkable how little influence those results have had on the larger community over the ensuing ten years.

References

[1] G. Agakishiev, *et al.* (STAR Collaboration), Phys. Rev. C **86**, 064902 (2012).
[2] R. J. Porter and T. A. Trainor (STAR Collaboration), J. Phys. Conf. Ser. **27**, 98 (2005).
[3] R. J. Porter and T. A. Trainor (STAR Collaboration), PoS C **FRNC2006**, 004 (2006).
[4] J. Adams *et al.* (STAR Collaboration), Phys. Rev. D **74**, 032006 (2006).
[5] T. A. Trainor, Phys. Rev. C **90**, no. 2, 024909 (2014)
[6] T. A. Trainor, Phys. Rev. C **92**, no. 2, 024915 (2015).
[7] T. A. Trainor, Int. J. Mod. Phys. E **17**, 1499 (2008).
[8] T. A. Trainor and D. T. Kettler, Phys. Rev. C **83**, 034903 (2011).
[9] T. A. Trainor, Phys. Rev. C **80**, 044901 (2009).
[10] T. A. Trainor, Phys. Rev. C **78**, 064908 (2008).
[11] CMS Collaboration, JHEP **1009**, 091 (2010).
[12] S. Gavin, L. McLerran and G. Moschelli, Phys. Rev. C **79**, 051902 (2009).
[13] B. I. Abelev *et al.* (STAR Collaboration), Phys. Rev. C **80**, 064912 (2009).
[14] T. A. Trainor and D. T. Kettler, Phys. Rev. C **84**, 024910 (2011).
[15] J. Adams *et al.* (STAR Collaboration), J. Phys. G **32**, L37 (2006).
[16] J. Adams *et al.* (STAR Collaboration), J. Phys. G **33**, 451 (2007).
[17] D. T. Kettler (STAR collaboration), Eur. Phys. J. C **62**, 175 (2009).
[18] D. Kettler (STAR Collaboration), J. Phys. Conf. Ser. **270**, 012058 (2011).
[19] D. T. Kettler, D. J. Prindle and T. A. Trainor, Phys. Rev. C **91**, 064910 (2015).
[20] T. A. Trainor, J. Phys. G **37**, 085004 (2010).
[21] T. A. Trainor, Mod. Phys. Lett. A **23**, 569 (2008).

EPJ Web of Conferences **120**, 07001 (2016) DOI: 10.1051/epjconf/201612007001
ISMD 2015

Deconfinement and color screening in 2+1 flavor QCD

Péter Petreczky[1,a] and Johannes Weber[2,b]

[1] *Brookhaven National Laboratory, Physics Department, Upton NY 11973-5000, USA*
[2] *Technische Universität München, Physik Department T30f, James-Franck-Str. 1, 85748 Garching, Germany*

Abstract. We discuss the deconfinement an color screening in 2+1 flavor QCD in terms of the free energy of static quarks.

1 Introduction

QCD is known to exhibit a confined phase at low temperatures and a deconfined phase at high temperatures. In the confined phase, quarks and gluons are bound inside of hadrons and the iso-vector chiral symmetry in the light quark sector is broken spontaneously. This spontaneously broken chiral symmetry is restored in the deconfined phase.

In $SU(N_c)$ pure gauge theory without dynamical quarks, the deconfinement transition is a phase transition, with the Polyakov loop being the order parameter. This phase transition is associated with the $Z(N_c)$ center symmetry. The logarithm of the Polyakov loop corresponds to the negative free energy of a static quark in temperature units [1]. The Polyakov loop is not invariant under $Z(N_c)$ transformation and therefore strictly vanishes for temperatures below the phase transition temperature, i.e. the free energy of a static quark is infinite. On the other hand, it assumes a finite expectation value above the transition temperature due to color screening. Thus, the deconfinement transition is closely related to the onset of color screening [1].

The center symmetry is broken explicitly by the presence of sea quarks in full QCD. As two color charges are separated beyond the string breaking distance, the flux tube breaks apart and a light quark-antiquark pair is created from the sea. Hence, the Polyakov loop takes a small, but positive expectation value in the confined phase of QCD, which is related to the binding energy of static-light mesons. Therefore, it is not clear to what extent the Polyakov loop is suitable for the discussion of the deconfinement transition in QCD, and therefore, the deconfinement in QCD is often discussed in terms of fluctuations of conserved charges (see [2] and references therein).

In this contribution we discuss the deconfinement transition and color screening in QCD in terms of the Polyakov loop or equivalently the free energy $F_Q(T)$ of a static quark. We will also consider related quantities like the singlet free energy $F_1(r,T)$ and the color-averaged free energy $F_{\rm avg}(r,T)$ of a static quark anti-quark pair in full QCD.

[a] e-mail: petreczk@quark.phy.bnl.gov
[b] e-mail: johannes.weber@tum.de

DOI: 10.1051/epjconf/201612007001

2 Lattice setup

We study the Polyakov loop and related correlators in full QCD for a large range of temperatures and four different temporal extents $N_\tau = \{12, 10, 8, 6\}$ of the lattice. The temperature T is related to the gauge coupling β through the lattice spacing $a(\beta)$ and is defined as $T(\beta, N_\tau) = 1/(a(\beta)N_\tau)$. Taking the continuum limit at fixed temperature requires simultaneous variation of both β and N_τ, which is why numerical studies of QCD at finite temperature are quite demanding. In order to avoid finite volume effects, we use lattices with aspect ratio $N_\sigma/N_\tau = 4$, where N_σ is the extent of each spatial direction. We have physical strange quark mass and nearly physical light quark masses, $m_l = m_s/20$, corresponding to a pion mass of $m_\pi = 161\,\text{MeV}$ in the continuum limit. The gauge configurations were generated with tree-level improved Symanzik gauge action and highly improved staggered quarks (HISQ) by HotQCD collaboration [8, 9]. We use the publicly available MILC code [3] in combination with libraries provided by the USCQD consortium [4].

2.1 Observables

The Polyakov loop $L(\beta, N_\tau, \mathbf{x})$ is defined in lattice QCD as a normalized trace of a temporal Wilson line $W(\beta, N_\tau, \mathbf{x})$ wrapping around the time direction once,

$$L(\beta, N_\tau, \mathbf{x}) = \frac{1}{N_c} \operatorname{Tr} W(\beta, N_\tau, \mathbf{x}), \quad W(\beta, N_\tau, \mathbf{x}) = \prod_{x_0=1}^{N_\tau} U_0(x_0, \mathbf{x}). \tag{1}$$

We obtain the expectation values of Polyakov loop, Polyakov loop correlator and color-singlet Wilson line correlator by taking the ensemble average (average over $\mathbf{x}$ included) as

$$L^{\text{bare}}(\beta, N_\tau) = \langle L(\beta, N_\tau, \mathbf{x})\rangle, \tag{2}$$

$$C_P^{\text{bare}}(\beta, N_\tau, r) = \langle L(\beta, N_\tau, \mathbf{x}) L^\dagger(\beta, N_\tau, \mathbf{x}+\mathbf{r})\rangle, \tag{3}$$

$$C_S^{\text{bare}}(\beta, N_\tau, r) = \langle \frac{1}{N_c} \operatorname{Tr} \left[W(\beta, N_\tau, \mathbf{x}) W^\dagger(\beta, N_\tau, \mathbf{x}+\mathbf{r})\right]\rangle. \tag{4}$$

Since the Wilson line $W(\beta, N_\tau, \mathbf{x})$ itself is not gauge invariant, $C_S^{\text{bare}}(\beta, N_\tau, r)$ vanishes unless a gauge-fixing procedure is used. We fix the gauge fields to Coulomb gauge [5, 6]. All three bare observables diverge in the continuum limit and require multiplicative renormalization with a factor $\exp\left[-c_Q(\beta)\right]$ for each matrix $W(\mathbf{x})$. The renormalized observables are related to free energies for finite N_τ and in the continuum limit as

$$L^{\text{ren}}(T) = e^{-F_Q(T)/T} \quad , \quad C_P^{\text{ren}}(T, r) = e^{-F_{\text{avg}}(r,T)/T} \quad \text{and} \quad C_S^{\text{ren}}(T, r) = e^{-F_S(r,T)/T}. \tag{5}$$

In the following discussion, we make use of free energies in units of the temperature, e.g. $f_Q(T, N_\tau) = F_Q(T, N_\tau)/T$, which vary more slowly with the temperature than the primary observables and are renormalized by adding $N_\tau c_Q(\beta)$ (or the respective combinations for the other free energies). We also usually omit the indication ^{ren} for renormalized quantities.

2.2 Renormalization and scale setting with $Q\bar{Q}$ procedure

We obtain the renormalization constant $c_Q(\beta)$ by normalizing the static $Q\bar{Q}$ potential at zero temperature $V(r)$ to a prescribed value. Namely, the static potential at zero temperature for different lattice spacings a is fixed to be $0.954/r_0$ or $0.2065/r_1$ at distances $r = r_0$ [8] or $r = r_1$ [9], respectively. The scales r_0 and r_1 are defined as

$$r^2 \frac{dV(r)}{dr}\bigg|_{r=r_i} = C_i, \quad i = 0, 1, \tag{6}$$

where $C_0 = 1.65$ and $C_1 = 1.0$. In physical units we have $r_0 = 0.4688(41)\,\mathrm{fm}$ and $r_1 = 0.3106\,\mathrm{fm}$ [7]. Either distance r_i defines the lattice spacing $a(\beta)$ in physical units as a function of the gauge coupling $\beta = 10/g^2$. The renormalization procedure ($Q\bar{Q}$ procedure) is limited by availability of lattice data at zero temperature ($\beta \leq 7.825$, lattice spacing $a \gtrsim 0.04\,\mathrm{fm}$). Thus, continuum extrapolation including the finest lattices ($N_\tau = 12$ is restricted to $T \lesssim 407\,\mathrm{MeV}$. The renormalization constants $c_Q(\beta)$ have been obtained using the results on $V(r)$ from [8, 9]. We have to interpolate the renormalization constant in β, since we require intermediate β values for renormalization at fixed temperature. We interpolate $c_Q(\beta)$ using R statistical package [10] with error propagation via bootstrap method. Further details of the interpolation are deferred to a publication in preparation [12].

3 Deconfinement observables from the Polyakov loop

3.1 Free energy $F_Q(T)$

We use bare Polyakov loop data from HotQCD given in [8, 9] and interpolate $f_Q^{\mathrm{bare}}(\beta, N_\tau)$ for each N_τ in β using R statistical package [10] with error propagation via bootstrap method. Further details of the interpolation are deferred to a publication in preparation [12]. Interpolation is necessary to obtain the Polyakov loop for a set of common temperatures for all N_τ. We add the errors of $N_\tau c_Q(\beta)$ in quadrature, since they are statistically independent from the errors of the bare Polyakov loop.

Above $T \gtrsim 200\,\mathrm{MeV}$, cutoff effects are similar to the size of uncertainties of the interpolation, which are dominated by the uncertainty of $N_\tau c_Q(\beta)$. We extrapolate $F_Q(T, N_\tau)$ to the continuum limit with point-wise extrapolations for each temperature and with a global fit using a polynomial Ansatz,

$$F_Q(T, N_\tau) = \sum_{i_0=0}^{N_0} a_{i_0} T^{i_0} + \frac{1}{N_\tau^2} \sum_{i_2=0}^{N_2} a_{i_2} T^{i_2}. \tag{7}$$

The Ansatz is specified in terms of the orders N_0 and N_2 of two temperature polynomials, which parametrize the temperature dependence of the continuum limit and of cutoff effects. In both cases, the extrapolations are performed with and without the $N_\tau = 6$ data. We finally include estimates of the uncertainties of the continuum extrapolation using different global and point-wise fits. Further details of the interpolation are deferred to a publication in preparation [12].

Our continuum result is shown as a black band in figure 1. This result has been discussed already in a comparison between conventional approach to the renormalization of the Polyakov loop and the renormalization of the Polyakov with gradient flow [11]. It has been found that the continuum limit of $F_Q(T, N_\tau)$ at finite flow time is consistent up to an additive constant with $F_Q(T)$ using $Q\bar{Q}$ procedure for temperatures up to $T \lesssim 400\,\mathrm{MeV}$. We show comparison of results for $F_Q(T, 12)$ using either $Q\bar{Q}$ procedure or flow time $f_t = f_0$ in figure 1. Details of the gradient flow scheme and the choice for f_0 are discussed in [11]. Both renormalization

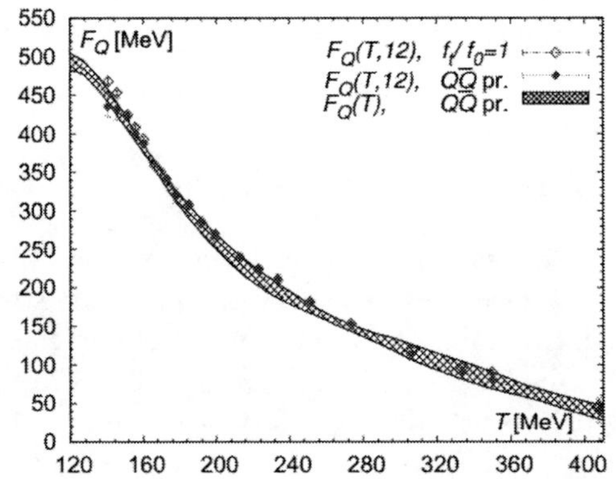

Figure 1. Continuum limit $F_Q(T)$ using $Q\bar{Q}$ procedure as black 1σ band. Error bars show $F_Q(T,12)$ in $Q\bar{Q}$ procedure (filled symbols) and at flow time $f_t = f_0$ (open symbols). The flow time result is taken from [11].

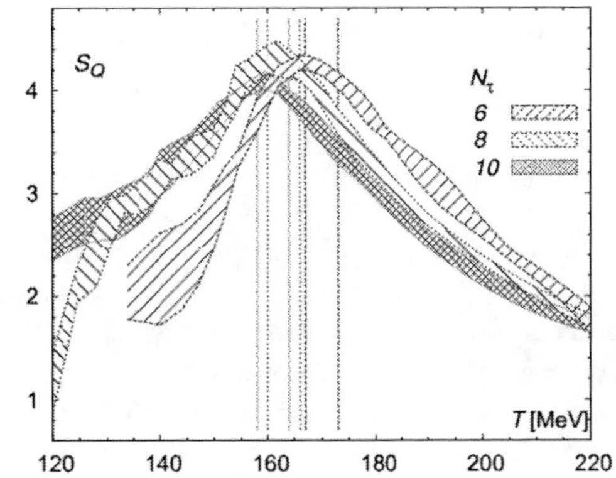

Figure 2. $S_Q(T, N_\tau)$ using $Q\bar{Q}$ procedure has a peak close to $T_c(N_\tau)$, which is indicated by the three pairs of vertical lines.

schemes agree within errors for $N_\tau = 12$ with exception of a few low temperature points which have relatively poor statistics. For coarser lattices, flow-time dependent cutoff effects are observed for low temperatures, but are taken care of by continuum extrapolation.

3.2 Entropy $S_Q(T)$

The temperature derivative of the free energy is related to the entropy of a static quark as

$$S_Q(T, N_\tau) = -\frac{\partial F_Q(T, N_\tau)}{\partial T} = \log L(T, N_\tau) + \frac{L'(T, N_\tau)}{L(T, N_\tau)}, \quad L'(T, N_\tau) = \frac{\partial L(T, N_\tau)}{\partial T}. \quad (8)$$

Though the Polyakov loop in full QCD does not vanish below the crossover region, the presence of a peak in the entropy of a static quark is indicative of rapid change of the properties of the medium and the onset of deconfinement. The calculation of the peak position in S_Q and the comparison of its value with the pseudo-critical temperature T_c of the chiral crossover is important for improving our understanding of the chiral and deconfinement aspects of the finite temperature QCD crossover.

We calculate the temperature derivative of the free energy numerically. We use spline and polynomial interpolations using R statistical package [10] with error propagation via bootstrap method for the renormalization constant $c_Q(\beta)$ and for the bare free energy $f_Q^{\text{bare}}(\beta, N_\tau)$. We estimate the systematical error of our interpolation from the difference between both results. Further details of the interpolation are deferred to a publication in preparation [12]. Since our current data quality for $N_\tau = 12$ is insufficient for precise calculation of the derivative, we show only results for $N_\tau = \{10, 8, 6\}$ in figure 2.

The position of the peak agrees quite well with the pseudo-critical temperature T_c obtained from chiral susceptibilities in [8]. We obtain $T_{\text{peak}} = 159(3), 162(4)$ and $166(4)$ MeV for $N_\tau = \{10, 8, 6\}$, which compares well with $T_c = 161.9(1.4), 164.1(1.8)$ and $171.3(2.1)$ MeV [8]. Below $T \lesssim 170$ MeV, $N_\tau = 6$ is not in the $1/N_\tau^2$ scaling regime for continuum extrapolation of $S_Q(T, N_\tau)$. Hence, we cannot present a robust continuum extrapolation of the entropy for temperatures in the crossover region. Above $T \gtrsim 250\,\text{MeV}$, cutoff effects in $S_Q(T, N_\tau)$ are as large as the statistical uncertainties.

4 Color screening observables from the static quark correlators

Due to color screening, static quark-antiquark free energies approach the same constant value at sufficiently large distances. This constant equals to twice the free energy of an isolated static quark because the color charges are not correlated anymore. In other words the correlators defined above will approach L^2 at large distances. On the other hand, thermal effects are no more than a perturbation to correlators at sufficiently short distances. At distances $r \lesssim 1/T$ we expect to see the interplay between vacuum physics and medium effects, while at still larger distance we should see the onset of color screening. It is convenient to subtract the known asymptotic constant and obtain correlators that vanish for sufficiently large distances. By subtracting the bare Polyakov loop squared from the correlation functions the correlated noise at large distances is greatly reduced. The renormalized correlators are obtained by adding back the asymptotic constant in terms of the renormalized instead of the bare Polyakov loops.

4.1 Color-singlet free energy $F_S(r, T)$

Applying the previously discussed approach to the color-singlet Wilson line correlator $C_S(r, T, N_\tau)$, we compute $f_S^{\text{sub}}(r, T, N_\tau) = f_S(r, \beta, N_\tau) - 2f_Q^{\text{bare}}(\beta, N_\tau)$, the subtracted color singlet free energy. As for the Poylakov loop, interpolation in the temperature is required for continuum extrapolation at fixed temperature. Moreover, the available lattice distances $r/a(\beta)$ correspond to different physical lengths for different β. We correct for leading cutoff effects of the tree-level gauge action at short distances by using the improved distance r_I (see e.g. [13]) instead of bare distance r_b and interpolate in the distance. We perform both point-wise continuum extrapolation for each temperature and distance and with a global fit using a polynomial Ansatz similar to eq. (7) in both the temperature and the distance. Finally we add the continuum limit of the renormalized asymptotic value, $2f_Q^{\text{ren}}(T)$, and obtain the renormalized singlet free energy. The continuum limit of this result is shown together with the static energy at zero temperature, $V(r)$, in figure 3. Both quantities are numerically similar up to $rT \lesssim 0.45$. This indicates a regime of almost vacuum-like physics with medium effects suppressed for small rT. These two regimes can be made even more explicit by defining an effective coupling constant $\alpha_{Q\bar{Q}}(r, T)$ as

$$\alpha_{Q\bar{Q}}(r, T, N_\tau) = -\frac{3}{4} r^2 \frac{\partial E(r, T, N_\tau)}{\partial r}, \quad E(r, T, N_\tau) = \{F_S(r, T, N_\tau), V(r)\}, \tag{9}$$

which has only a mild N_τ dependence. We show $\alpha_{Q\bar{Q}}(r, T, 6)$ together with $\alpha_{Q\bar{Q}}(r, 0)$ obtained from the zero temperature static energy $V(r)$ in figure 5. The maximum of $\alpha_{Q\bar{Q}}(r, T, N_\tau)$ defines a distance $r_{\max}(T)$ at which screening overcomes the string tension. Thus, for $r < r_{\max}(T)$ we see vacuum-like behavior, while for $r > r_{\max}(T)$ we see the onset of color screening.

4.2 Color-averaged free energy $F_{\text{avg}}(r, T)$

Applying the same machinery to the Polyakov loop correlator $C_P(r, T, N_\tau)$, we compute $f_{\text{avg}}^{\text{sub}}(r, T, N_\tau) = f_{\text{avg}}(r, \beta, N_\tau) - 2f_Q^{\text{bare}}(\beta, N_\tau)$, the subtracted color-averaged free energy.

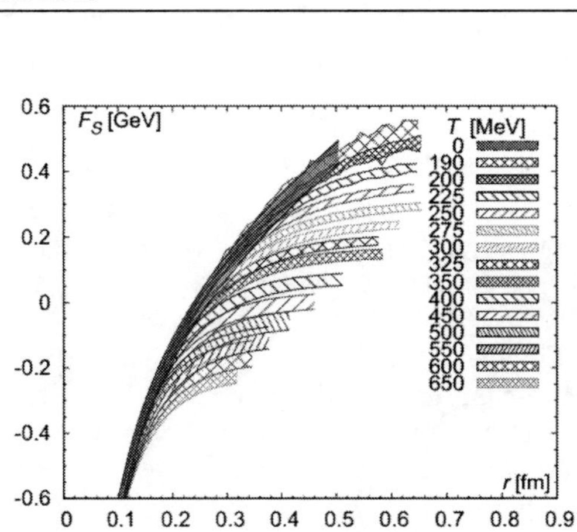

Figure 3. The singlet free energy $F_S(r,T)$ is numerically similar to the zero temperature static energy $V(r)$ up to $rT \lesssim 0.45$.

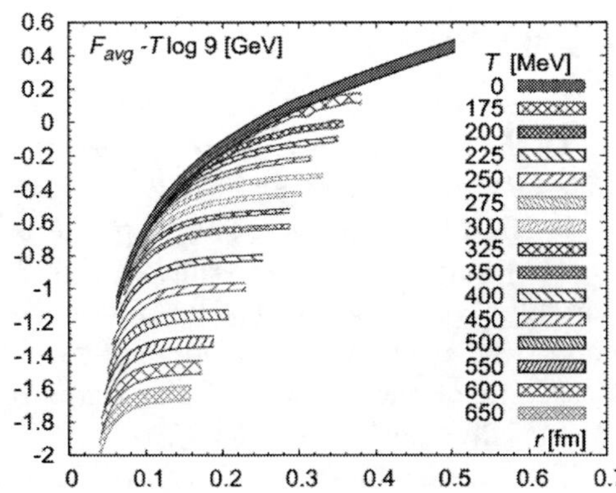

Figure 4. Once a trivial color factor $T\log 9$ is subtracted from the color-averaged free energy; $F_{\mathrm{avg}}(r,T)$ it is numerically similar to the zero temperature static energy $V(r)$ up to $rT \lesssim 0.15$.

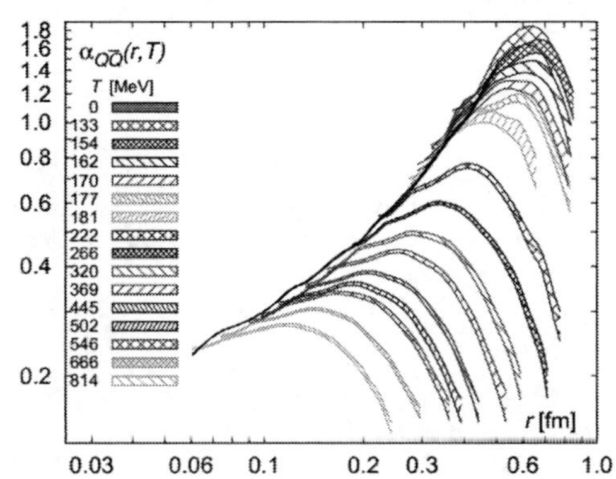

Figure 5. The effective coupling constant $\alpha_{Q\bar{Q}}(r,T)$ highlights the transition between confining and screening regimes (here $N_\tau = 6$). The quadratic rise of the black band for $T = 0$ is indicative of the string tension.

Due to combinatorics, it can be written in terms of the color-singlet free energy and the color-octet free energy [1]

$$\exp(-f_{\mathrm{avg}}(r,\beta,N_\tau)) = \frac{1}{9}\exp(-f_S(r,\beta,N_\tau)) + \frac{8}{9}\exp(-f_O(r,\beta,N_\tau)). \tag{10}$$

Hadronic states with $Q\bar{Q}$ pair in color-octet configuration require valence gluons and correspond to hybrid mesons, which are energetically disfavored at low temperatures. Furthermore, in octet channel we have repulsive interactions at short distances. Hence, it is expected that the color-averaged free energy is dominated by the singlet contribution for short enough distances: $F_{\mathrm{avg}}(r,\beta,N_\tau) \simeq F_S(r,T,N_\tau) + T\log 9$. We account for the trivial color factor $T\log 9$ and show the modified continuum limit of the color-averaged free energy together with the static energy at zero temperature, $V(r)$, in figure 4. It is no surprise that differences between $F_{\mathrm{avg}}(r,T)$ and $V(r)$ due to in-medium modification become significant for much smaller r due to the rising contribution from states in color-octet configuration.

[1] The decomposition of the $Q\bar{Q}$ free energy into singlet and octet contributions can be rigorously derived in the small distance limit using effective field theory approach [14].

5 Conclusions

We have studied the Polyakov loop, its correlator and a color-singlet Wilson line correlator in full QCD with 2+1 flavors of quarks almost at the physical point. Using these observables, we calculate free energies and study deconfinement and color screening in the thermal medium.

We find that the Polyakov loop is a useful order parameter for deconfinement in full QCD. For low temperatures, renormalization using either $Q\bar{Q}$ procedure or gradient flow yields numerically consistent results up to cutoff effects. The deconfinement crossover is clearly indicated by a peak in the entropy $S_Q(T)$. The cutoff dependence of the associated temperature T_{peak} and of the pseudo-critical temperature T_c defined in terms of chiral observables are numerically consistent.

The free energies of static $Q\bar{Q}$ pairs in the medium are consistent with the zero temperature static energy $V(r)$ for short distances, indicating vacuum-like physics as required by asymptotic freedom. For larger distances, the normalized color-averaged free energy $F_{\text{avg}} - T \log 9$ starts to deviate from $V(r)$ at $rT \approx 0.15$ due to the presence of the octet contribution, whereas the color-singlet free energy $F_S(r,T)$ stays close to $V(r)$ up to $rT \approx 0.45$. We attribute this deviation to the onset of color screening and deconfinement, which is highlighted more explicitly in an effective coupling constant $\alpha_{Q\bar{Q}}(r,T)$.

Acknowledgments

This work was supported by U.S. Department of Energy under Contract No. DE-SC0012704. We acknowledge the support by the DFG Cluster of Excellence "Origin and Structure of the Universe". The calculations have been carried out on the computing facilities of the Computational Center for Particle and Astrophysics (C2PAP).

References

[1] L. D. McLerran and B. Svetitsky, Phys. Rev. D **24**, 450 (1981).
[2] S. Mukherjee, P. Petreczky and S. Sharma, arXiv:1509.08887 [hep-lat].
[3] http://www.physics.utah.edu/~detar/milc/milc_qcd.html
[4] http://www.usqcd.org/software.html
[5] O. Kaczmarek, F. Karsch, P. Petreczky and F. Zantow, Phys. Lett. B **543**, 41 (2002)
[6] S. Digal, S. Fortunato and P. Petreczky, Phys. Rev. D **68**, 034008 (2003)
[7] A. Bazavov *et al.* [MILC Collaboration], PoS LATTICE **2010** (2010) 074
[8] A. Bazavov *et al.*, Phys. Rev. D **85** (2012) 054503
[9] A. Bazavov *et al.* [HotQCD Collaboration], Phys. Rev. D **90** (2014) 9, 094503
[10] R Development Core Team (2011), R: A Language and Environment for Statistical Computing. Vienna, Austria : the R Foundation for Statistical Computing. ISBN: 3-900051-07-0, http://www.R-project.org/.
[11] P. Petreczky and H.-P. Schadler, Phys. Rev. D **92** (2015) 9, 094517
[12] Preprint: A. Bazavov, N. Brambilla, H.T. Ding, P. Petrczky, H.-P. Schadler, A. Vairo, J. Weber, Polyakov loop in 2+1 flavor QCD from low to high temperatures, TUM-EFT 76/15
[13] A. Bazavov, N. Brambilla, X. Garcia i Tormo, P. Petreczky, J. Soto and A. Vairo, Phys. Rev. D **90**, 074038 (2014)
[14] N. Brambilla, J. Ghiglieri, P. Petreczky and A. Vairo, Phys. Rev. D **82**, 074019 (2010)

EPJ Web of Conferences **120**, 07002 (2016) DOI: 10.1051/epjconf/201612007002
ISMD 2015

Heavy quark masses from lattice QCD

Andrew T. Lytle[1,a]

[1] *SUPA, School of Physics and Astronomy*
University of Glasgow, Glasgow, G12 8QQ, UK

Abstract. Progress in quark mass determinations from lattice QCD is reviewed, focusing on results for charm and bottom mass. These are of particular interest for precision Higgs studies. Recent determinations have achieved percent-level uncertainties with controlled systematics. Future prospects for these calculations are also discussed.

1 Introduction

Quark masses are fundamental parameters of QCD, arising in the Standard Model (SM) from Yukawa interactions with the Higgs field. LHC measurements of the Higgs' couplings are consistent with SM predictions, but high-precision studies at future accelerators such as the ILC will measure couplings at the per mil level [1]. In order to test the SM at this level, and discriminate potential new physics effects, correspondingly precise determinations of quark mass are required.

Lattice calculations of quark masses have seen considerable progress in recent years, with charm and bottom uncertainties being quoted at the percent to few-percent level. This is due to a combination of new measurement techniques and increasingly realistic simulations. State-of-the-art simulations include dynamical u, d, s, and frequently c quarks, with pion masses at or near their physical values, and typically at several lattice spacings. This realism translates to increasingly accurate results, and with fewer systematic errors. In order to reliably determine quark masses at the sub-percent level, it is important to have multiple calculational techniques/strategies available, as well as independent determinations from different groups.

The outline of the rest of this article is as follows: Sec. 2 briefly discusses quark mass parameters in a general context, and how they are determined in lattice QCD simulations. Sec. 3 discusses recent progress in the charm mass determinations, focusing on a method using current-current correlators. Sec. 4 will look at strategies and results for bottom mass determinations, and Sec. 5 discusses the important role played by mass ratios. Sec. 6 presents a summary and discusses future prospects for these calculations.

2 Quark mass and LQCD

Quark masses are scheme and scale dependent quantities that, along with α_s, specify QCD at the Lagrangian level. These parameters must be determined from comparison of theory and experiment.

[a] e-mail: andrew.lytle@glasgow.ac.uk

In the absence of nonperturbative lattice simulations, one must focus on experimentally measurable observables that are 1) sensitive to quark masses and 2) can be reliably computed in perturbation theory. One set of observables satisfying these criteria are derived from the the R-ratio, for which the relevant perturbation series are known to N^3LO [2–4]. To be discussed in Sec. 3.1, an effective way to calculate m_q for heavy quarks via lattice simulations uses these same perturbative calculations, but substitutes experimental data with data from LQCD simulations.

Lattice QCD simulations are well suited for mass determinations, since the mass parameters are simulation inputs controlled by the "experimenter". By changing the input masses, one can measure the resultant change in physical observables such as meson masses. In a standard LQCD simulation, one tunes the input masses in order to reproduce a number of low energy observables – one for each quark in the theory. In this way one obtains (typically very precise) bare quark masses, but in the particular lattice regularization one happens to be using. In order to make contact with a continuum-regularized determination such as the $\overline{\text{MS}}$ scheme, one needs an additional calculation of the lattice to $\overline{\text{MS}}$ matching factor. This can be found using lattice perturbation theory or via non-perturbative renormalization (NPR) techniques. The ratios of bare quark masses in a given regularization are however immediately useful, as they are, up to lattice artfifacts, equal to renormalized mass ratios.

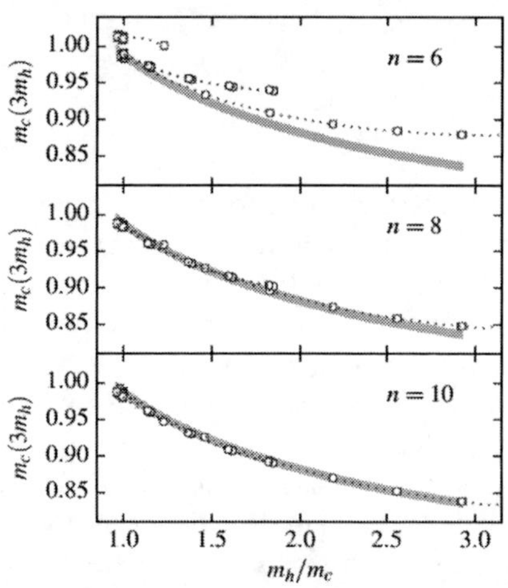

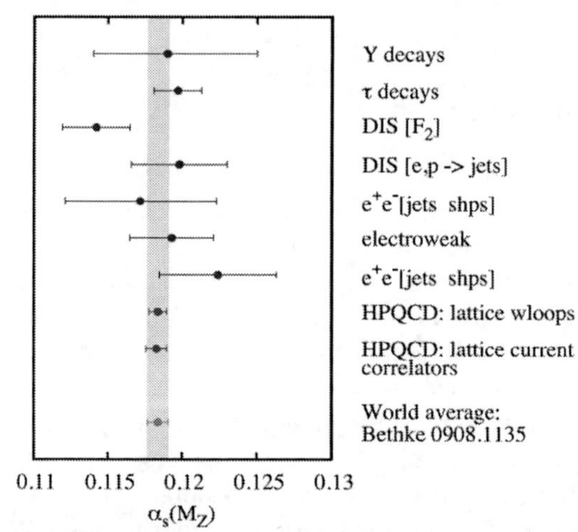

Figure 1. (Left) Data from [5] showing $m_c^{\overline{\text{MS}}}(\mu = 3m_h)$ extracted from lattice data and perturbation theory for moments $n = 6, 8, 10$ using Eq. (8). The green/blue/red data points correspond to lattice spacings of 0.12/0.09/0.06 fm. The gray band shows the evolution of the best-fit value for m_c using perturbation theory. (Right) Value of $\alpha_s^{\overline{\text{MS}}}(M_Z)$ from [5] compared with determinations based on various experimental inputs and a world average.

3 Charm quark mass

3.1 Current-current correlator method.

The current-current correlator method uses time moments of Euclidean-time twopoint functions,

$$G(t) = a^6 \sum_{\mathbf{x}} (am_{0h})^2 \langle J_5(t,\mathbf{x}) J_5(0,0) \rangle \,. \tag{1}$$

Here $J_5 \equiv \bar{\psi}_h \gamma_5 \psi_h$, a is the lattice spacing, and am_{0h} is the bare quark mass parameter in lattice units. In formalisms with sufficient chiral symmetry, the current J_5 is absolutely normalized. The correlator $G(t)$ is UV finite, so that

$$G(t)_{\rm cont} = G(t)_{\rm latt} + O(a^2) \qquad (t \neq 0)\,. \tag{2}$$

The correlators $G(t)_{\rm latt}$ are the same ones used in the computation of pseudoscalar masses and decay constants, in which case it is the large-t exponential tail of the correlator that is of interest. For the mass calculation it is the small-t short distance behavior that is extracted via time-moments of $G(t)$, defined as:

$$G_{n,{\rm latt}} = \sum_{t=0}^{T} (t/a)^n \, G(t)_{\rm latt}\,. \tag{3}$$

The time-moments G_n have also been computed to N^{3}LO in perturbation theory [2–4]. For $n \geq 4$,

$$G_{n,{\rm pert}} = \frac{g_n(\alpha_{\overline{\rm MS}}, \mu)}{(am_h(\mu))^{n-4}}\,. \tag{4}$$

Here g_n is a series in α_s and $m_h(\mu)$ is the $\overline{\rm MS}$ quark mass at the scale μ. The basic strategy to extract the quark mass is to compare $G_{n,{\rm cont}}$, the continuum extrapolated $G_{n,{\rm latt}}$ values, with the perturbative expressions $G_{n,{\rm pert}}$ in Eq. (4) (evaluated at a scale $\mu \sim m_h$), and from these determine best-fit values for $\alpha_{\overline{\rm MS}}(\mu)$ and $m_h(\mu)$. For example, computing the continuum limit of $G_{4,{\rm latt}}$ with physically tuned input charm masses m_{0c}, one can obtain $\alpha_{\overline{\rm MS}}(m_c)$, and then use this value in G_6 to obtain $m_c(m_c)$.

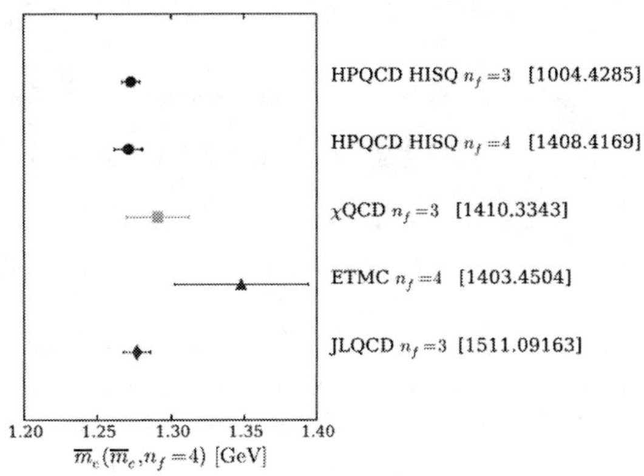

Figure 2. Comparison plot for determinations of $m_c^{\overline{\rm MS}}(m_c^{\overline{\rm MS}}, n_f = 4)$, computed from $n_f = 2{+}1$ and $n_f = 2{+}1{+}1$ simulations.

The HPQCD collaboration carried out an analysis in [5] using reduced moments, R_n, which are related simply to the time-moments as

$$R_4 = G_4/G_4^{(0)} \tag{5}$$

$$R_n = \frac{1}{m_{0c}} (G_n/G_n^{(0)})^{1/(n-4)} \qquad (n \geq 6)\,, \tag{6}$$

where $G_n^{(0)}$ are the tree-level results for the moments. Dividing by $G_n^{(0)}$ has the advantage of reducing lattice-spacing effects. In continuum perturbation theory,

$$R_4 = r_4(\alpha_{\overline{\rm MS}}, \mu) \tag{7}$$

$$R_n = \frac{1}{m_c(\mu)}\, r_n(\alpha_{\overline{\rm MS}}, \mu) \quad (n \geq 6)\,. \tag{8}$$

Here r_n are the perturbative expressions obtained from powers of $g_n/g_n^{(0)}$, with $g_n^{(0)}$ the lowest order perturbative result. For a given m_{0h} one computes the values R_n from Eq. (6) to get estimates of $m_c(3m_h) = R_n/r_n(3m_h)$, via Eq. (8) (here the scale μ is taken to be $3m_h$).

The running of $m_c(\mu)$ was calculated this way in [5] using $n_f = 2 + 1 + 1$ HISQ ensembles. The $n = 4, 6, 8, 10$ moments were obtained using three different lattice spacings $a \approx 0.12, 0.09, 0.06$ fm and for seven input masses from $m_h = m_c$ to $0.7m_b$. The extractions of $m_c(3m_h)$ from each of these data points are shown in Fig. 1 (left), along with the perturbative running. Fig. 1 (right) shows the corresponding estimate of $\alpha_s(M_Z)$ extracted from this data, compared with results based on other experimental inputs.

Estimates of $m_c(\mu)$ from time-moments are subject to a number of systematic uncertainties. The truncation of perturbation theory limits the precision, but fortunately the expansions of $r_n = 1 + \sum_j \alpha^j r_{nj}$ are known for $j = 1, 2, 3$ and $n \leq 10$. The lattice moments are sensitive to condensate terms not captured in the perturbative expansions. These effects are suppressed like $(\Lambda_{\text{QCD}}/2m_h)^4$, but grow with n. The lattice data also has cut-off effects, which grow like $\alpha_s (am_h)^2$ and decrease with increasing n; these trends are visible in Fig. 1.

Fitting the moments data for $n = 4, 6, 8, 10$ to Eqs. (7) and (8), HPQCD find

$$m_c^{\overline{\text{MS}}}(3\text{ GeV}, n_f = 4) = 0.9851(63)\text{ GeV} \tag{9}$$

$$\alpha_s^{\overline{\text{MS}}}(3\text{ GeV}, n_f = 4) = 0.2545(37)\,. \tag{10}$$

These are compatible with earlier $n_f = 2 + 1$ results [6]. The compatibility of $n_f = 2 + 1$ and $n_f = 2 + 1 + 1$ results suggests that the effect of charm quarks in the sea can be treated perturbatively to this level of precision.

The JLQCD collaboration has also utilized the current-current correlator method with $n_f = 2 + 1$ domain-wall fermions to determine m_c and α_s [7]. Their calculation uses three lattices spacings $a \approx 0.08, 0.055, 0.044$ fm, and focuses on R_6, R_8, and R_{10}, from which they find

$$m_c^{\overline{\text{MS}}}(3\text{ GeV}, n_f = 3) = 0.9948(71)\text{ GeV} \tag{11}$$

$$\alpha_s^{\overline{\text{MS}}}(3\text{ GeV}, n_f = 3) = 0.2514(95)\,. \tag{12}$$

3.2 Results comparison

In [8] the ETMC collaboration used lattice RI/MOM techniques to determine a mass renormalization factor $Z_m^{\text{RI}}(\mu, 1/a)$ connecting the bare mass to the RI-scheme mass, $m_c^{\text{RI}}(\mu) = Z_m^{\text{RI}}(\mu, 1/a)\, m_{c0}$, which is then converted to the $\overline{\text{MS}}$ scheme using continuum perturbation theory. Unlike the current-current correlator method, which uses a heavy input mass to set the scale μ, the RI/MOM calculation is extrapolated to the chiral limit, and ETMC have generated mass degenerate $n_f = 4$ ensembles for this purpose. The χQCD collaboration have also used RI/MOM methods for their $n_f = 2 + 1$ determination [9, 10].

A comparison of recent lattice results for $m_c^{\overline{\text{MS}}}$ is shown in Fig. 2.

4 Bottom mass

It is challenging to directly simulate the b quark in relativistic lattice simulations, since one would like $am_{b0} \ll 1$ in order for discretization effects to be under control. As an alternative, effective theories may be employed such as non-relativistic QCD (NRQCD) or heavy-quark effective theory (HQET). It has also recently become possible to approach the b mass using improved relativistic actions, making extrapolation methods viable.

4.1 Ratio method

The ETMC collaboration have used the *ratio method* [11] to extrapolate relativistic $n_f = 2 + 1 + 1$ simulation results around the charm mass to the bottom mass [12]. The method is based on the expectation from HQET that

$$\lim_{m_h^{\text{pole}} \to \infty} \frac{M_{hl}}{m_h^{\text{pole}}} = \text{constant}, \tag{13}$$

where M_{hl} is the mass of a heavy-light meson and m_h^{pole} is the heavy quark pole mass.

They calculate ratios of meson masses, $M_{hl}(m_h)/M_{hl}(m_h/\lambda)$, computed for a series of masses m_h around the charm mass, e.g.: $m_h^{(0)} = m_c$, $m_h^{(1)} = \lambda m_c$, ..., $m_h^{(n)} = \lambda^n m_c$. These ratios have the advantage that discretization effects proportional to $(am_h)^2$ are largely canceled. From these ratios they construct the function

$$y(m_h, \lambda) = \lambda^{-1} \frac{M_{hl}(m_h)}{M_{hl}(m_h/\lambda)} \frac{\rho(m_h/\lambda)}{\rho(m_h)}. \tag{14}$$

The functions $\rho(m_h)$ on the r.h.s. of Eq. (14) relate the pole mass to the $\overline{\text{MS}}$ mass and are known to N^{3}LO in perturbation theory. $y(m_h, \lambda)$ satisfies $\lim_{m_h \to \infty} y(m_h, \lambda) = 1$ on account of Eq. (13), and so its value can interpolated between the charm region and the static limit using a fit ansatz. Rewriting Eq. (14), the combination $\lambda\, y(m_h, \lambda) \frac{\rho(m_h)}{\rho(m_h/\lambda)}$ is a known function that evolves $M_{hl}(m_h/\lambda)$ to $M_{hl}(m_h)$. Choosing λ such that $M_{hl}(m_h^{(N)}) = M_{bl}^{\text{phys}}$ for some N, they determine the b mass from $m_b = \lambda^N m_c$.

4.2 NRQCD approach

The NRQCD Hamiltonian is written as an expansion in v^2, where v is a typical velocity of a b quark in the system of interest. NRQCD calculations should be carried out with $am_{b0} > 1$, which has the advantage that the b can be simulated using relatively coarse lattices. On the other hand it not as straightforward to extract continuum physics ($a \to 0$ limit) as compared to relativistic calculations.

The NRQCD current-correlator approach [13] is similar to the relativistic approach described in Sec. 3.1. Here however the currents need to be normalized,

$$J_\mu^{\text{NRQCD}} = Z_V J_\mu^{\text{cont}}. \tag{15}$$

Then the time moments are related to continuum perturbation theory,

$$G_n^{\text{NRQCD}} = Z_V^2 \frac{g_n(\alpha_{\overline{\text{MS}}}, \mu)}{(am_b(\mu))^{n-2}}. \tag{16}$$

The factors of Z_V can be canceled by constructing ratios of successive moments. Because the continuum limit cannot be approached directly one instead studies m_b as a function of the moment number. Compared to the charm case, condensate contributions which grow with moment number are more suppressed at the heavier quark mass. A "plateau" in m_b as a function of moment number implies that n is sufficiently large for discretization effects to be small. Such a plateau from [13] is shown in Fig. 3 (left).

Results at three lattice spacings and with two different light-quark masses for $n = 18$ are shown in Fig. 3 (right). A fit to this data, including systematic errors, and perturbatively evolved to m_b gives

$$m_b^{\overline{\text{MS}}}(m_b^{\overline{\text{MS}}}, n_f = 5) = 4.196(23)\ \text{GeV}. \tag{17}$$

This result is compared with others in Fig. 4. It is significant that the values in the figure are calculated using a range of techniques. In [6] results are extrapolated to m_b from below, using a

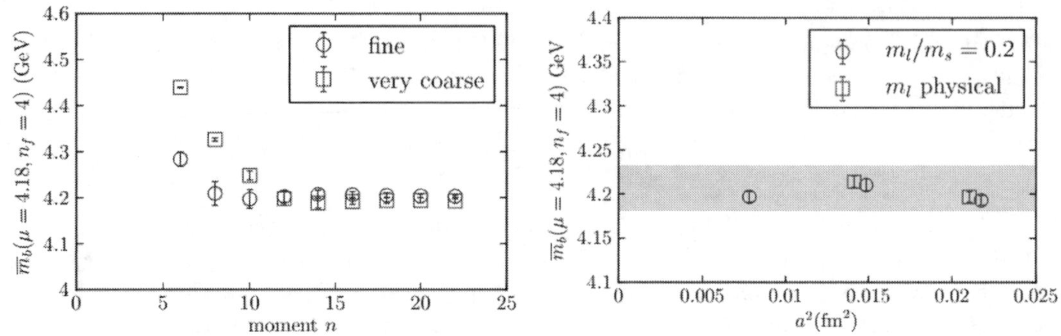

Figure 3. (Left) $m_b^{\overline{\rm MS}}$ extracted from the moments of NRQCD current-current correlators at two different lattice spacings from [13]. (Right) Results from the $n = 18$ moment as a function of lattice spacing and for two different light-quark masses. The gray band gives the continuum determination with the total error budget.

relativistic action as described in Sec. 3.1. This calculation is based on a different range of moment numbers, and uses a different action than [13]. The work of [14] uses the binding energy of Υ and B_s mesons, computed using NRQCD and lattice perturbation theory, to determine the heavy quark pole mass, which is then converted to the $\overline{\rm MS}$ mass with continuum perturbation theory.

5 Mass ratios

The bare mass parameters input to lattice QCD simulations are pure numbers (am_0) corresponding to masses expressed in units of the inverse lattice spacing. These are tuned by requiring that a number of observables, one for each input mass, achieve their physical values, resulting in a set of numbers $\{(am_{ud0}), (am_{s0}), (am_{c0})\}$. The bare lattice inputs are defined at the cutoff scale and depend on the details of the discretization. However, ratios of input masses are equal to the ratios of $\overline{\rm MS}$ masses, up to discretization effects that vanish in the continuum,

$$\frac{am1_0}{am2_0} = \frac{m1^{\overline{\rm MS}}(\mu)}{m2^{\overline{\rm MS}}(\mu)} + O(a^2). \qquad (18)$$

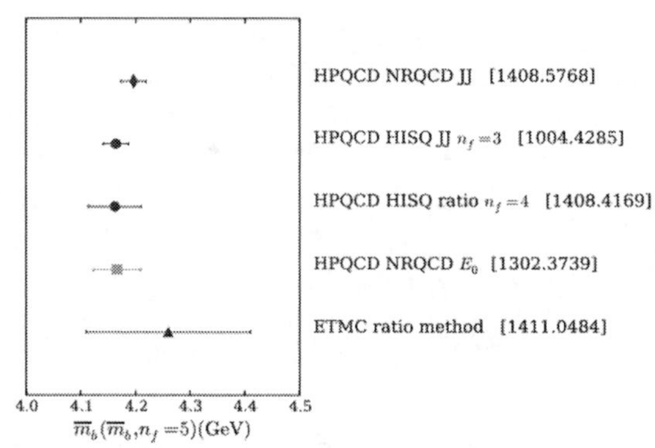

Figure 4. Comparison plot for determinations of $m_b^{\overline{\rm MS}}(m_b, n_f = 5)$, from $n_f = 2+1$ and $n_f = 2+1+1$ simulations.

Once the $\overline{\rm MS}$ mass is known for one quark in the theory, this can be converted to the $\overline{\rm MS}$ masses for the others using ratios of the input mass parameters.

In [5] HPQCD use this technique to obtain values for m_s and m_b using bare mass ratios and a precise determination of m_c. Taking the continuum limit of m_{c0}/m_{s0} they find

$$\frac{m_c(\mu, n_f)}{m_s(\mu, n_f)} = 11.652(65) \qquad (19)$$

Using their result for $m_c^{\overline{\mathrm{MS}}}(\mu)$ from the current-current correlator method discussed in Sec. 3.1, they obtain

$$m_s^{\overline{\mathrm{MS}}}(3\ \mathrm{GeV}, n_f = 3) = 84.7(7)\ \mathrm{MeV}\,, \tag{20}$$

which is the most precise estimate to date. Using input mass ratios involving a heavy mass m_{h0}/m_{c0}, and extrapolating to m_{b0}, they obtain

$$\frac{m_b(\mu, n_f)}{m_c(\mu, n_f)} = 4.528(54) \tag{21}$$

$$m_b^{\overline{\mathrm{MS}}}(m_b, n_f = 5) = 4.162(48)\ \mathrm{GeV} \tag{22}$$

6 Conclusion

Progress in lattice calculations of charm and bottom quark mass was reviewed. In order to achieve reliable (sub-)percent level uncertainties for these quantities, it is important that a variety of calculational strategies are used, and also that independent measurements are carried out by different groups.

At present the most precise quoted values for c mass come from the current-current correlator technique. Precision in the value of the charm mass can be cascaded to the other masses using bare quark-mass ratios, which are determined in the tuning of simulation parameters to the physical point. Calculations of the b mass either use an effective-theory framework for the b quark or extrapolate relativistic simulation results from the lower-mass region where discretization effects are under control. Extrapolation methods will continue to improve as ensembles with smaller lattice spacings become available.

References

[1] G. P. Lepage, P. B. Mackenzie and M. E. Peskin, arXiv:1404.0319 [hep-ph].
[2] K. G. Chetyrkin, J. H. Kuhn and C. Sturm, Eur. Phys. J. C **48** (2006) 107 [hep-ph/0604234].
[3] R. Boughezal, M. Czakon and T. Schutzmeier, Phys. Rev. D **74** (2006) 074006 [hep-ph/0605023].
[4] A. Maier, P. Maierhofer, P. Marquard and A. V. Smirnov, Nucl. Phys. B **824** (2010) 1 [arXiv:0907.2117 [hep-ph]].
[5] B. Chakraborty *et al.*, Phys. Rev. D **91** (2015) 5, 054508 [arXiv:1408.4169 [hep-lat]].
[6] C. McNeile, C. T. H. Davies, E. Follana, K. Hornbostel and G. P. Lepage, Phys. Rev. D **82** (2010) 034512 [arXiv:1004.4285 [hep-lat]].
[7] K. Nakayama, B. Fahy and S. Hashimoto, arXiv:1511.09163 [hep-lat].
[8] N. Carrasco *et al.* [European Twisted Mass Collaboration], Nucl. Phys. B **887** (2014) 19 [arXiv:1403.4504 [hep-lat]].
[9] Y. B. Yang *et al.*, Phys. Rev. D **92** (2015) 3, 034517 [arXiv:1410.3343 [hep-lat]].
[10] Z. Liu *et al.* [chiQCD Collaboration], Phys. Rev. D **90** (2014) 3, 034505 [arXiv:1312.7628 [hep-lat]].
[11] B. Blossier *et al.* [ETM Collaboration], JHEP **1004** (2010) 049 [arXiv:0909.3187 [hep-lat]].
[12] A. Bussone *et al.*, arXiv:1411.0484 [hep-lat].
[13] B. Colquhoun, R. J. Dowdall, C. T. H. Davies, K. Hornbostel and G. P. Lepage, Phys. Rev. D **91** (2015) 7, 074514 [arXiv:1408.5768 [hep-lat]].
[14] A. J. Lee *et al.* [HPQCD Collaboration], Phys. Rev. D **87** (2013) 7, 074018 [arXiv:1302.3739 [hep-lat]].

The structure of the proton: The CT14 QCD global analysis

S. Dulat[1,5], T.-J. Hou[2], J. Gao[3], M. Guzzi[4,a] J. Huston[5], P. Nadolsky[2], J. Pumplin[5], C. Schmidt[5], D. Stump[5], and C.-P. Yuan[5]

[1]*School of Physics Science and Technology, Xinjiang University, Urumqi, Xinjiang 830046 China.*
[2]*Department of Physics, Southern Methodist University, Dallas, TX 75275-0181, U.S.A.*
[3]*High Energy Physics Division, Argonne National Laboratory, Argonne, Illinois 60439, U.S.A.*
[4]*The University of Manchester, Oxford Road, M13 9PL Manchester, United Kingdom.*
[5]*Department of Physics and Astronomy, Michigan State University, East Lansing, MI 48824 U.S.A.*

Abstract. A brief description of the CT14 global analysis of quantum chromodynamics at next-to-next-to leading order (NNLO) from the CTEQ-TEA group is given. Predictions for standard candle processes at the LHC are illustrated together with the main features of the CT14 parton distribution functions.

1 Introduction

Understanding of the structure of the proton is crucial to a multitude of high-energy physics programs at the Large Hadron Collider (LHC). Interpretation of experimental measurements at hadron colliders relies on precise knowledge of fundamental QCD parameters and distribution of proton's constituent quarks and gluons: the parton distribution functions (PDFs). The LHC Run-1 culminated in the discovery of the Higgs boson [1, 2]. The LHC Run-2 will perform measurements of the Higgs boson properties and of various electroweak (EW) observables with extraordinary accuracy in new kinematic regimes. For both precision measurements and discovery of possible new physics, it is important to have the proper tools for the calculation of the relevant cross sections. These tools include both matrix element determinations at higher orders in perturbative QCD and EW theory, and precision PDFs.

The CT10 parton distribution functions were published at next-to-leading order (NLO) in 2010 [3], followed by the CT10 next-to-next-to leading order (NNLO) parton distribution functions in 2013 [4]. These PDF ensembles were determined using diverse experimental data from fixed-target experiments, HERA and the Tevatron collider, but without data from the LHC. The new CT14 global analysis [5] includes data from the LHC for the first time, as well as updated data from the Tevatron and from HERA experiments. Various CT14 PDF sets have been produced at the leading order (LO), NLO and NNLO and are available from LHAPDF [6].

Features of the CT14 analysis. The CT14 PDFs are determined from data on inclusive high-momentum transfer processes, for which perturbative QCD is expected to be reliable. For example, in the case of deep inelastic lepton scattering, only data with $Q > 2$ GeV and $W > 3.5$ GeV are used. Data in this region are expected to be relatively free of non-perturbative effects, such as higher twists or nuclear corrections.

[a]Presenter, e-mail: marco.guzzi@manchester.ac.uk

For the majority of processes in the CT14 fit, theoretical predictions are included at the NNLO level of accuracy. In particular, a NNLO treatment [7] of heavy-quark mass effects in neutral-current DIS is realized in the ACOT-χ scheme [8–10] and is essential for obtaining correct predictions for LHC electroweak cross sections [11, 12]. Measurements for charged-current DIS and inclusive jet production represent the only exception to this rule and are included at NLO only. In both cases, the complete NNLO contributions are not yet available, but it can be argued based on some investigations that the expected effect of missing NNLO effects is small relatively to current experimental errors (see Sec.II in Ref. [5]).

The new LHC measurements [13–15] of W/Z cross sections directly probe flavor separation of u and d (anti-)quarks in an x-range around 0.01 that was not directly assessed by the previously available experiments. The updated measurements of electron charge asymmetry from the DØ collaboration [16] included in the CT14 analysis probe the d quark PDF at $x > 0.1$. To better estimate variations in relevant PDF combinations, such as $d(x, Q)/u(x, Q)$ and $\bar{u}(x, Q)/\bar{d}(x, Q)$, the number of free PDF parameters is increased to 28, compared to 25 in CT10 NNLO.

As another important modification, CT14 employs a novel flexible parametrization for the PDFs, based on the use of Bernstein polynomials (reviewed in the Appendix of Ref. [5]). The shape of the Bernstein polynomials is such that a single polynomial is dominant in each given x range, reducing undesirable correlations among the PDF parameters that sometimes occurred in CT10. In the asymptotic limits of $x \to 0$ or $x \to 1$, the new parametrization forms allow for the possibility of arbitrary constant ratios of d/u or $\bar{d}/\bar{u}$, in contrast to the more constrained behavior assumed in CT10.

The CT14 PDF error sets is obtained using two techniques, the Hessian method [17] and Monte-Carlo sampling [18]. Lagrange multiplier studies [19] have also been used to verify the Hessian uncertainties, especially in regions not well constrained by data. This applies at NNLO and NLO; no error sets are provided at LO due to the difficulty of defining meaningful uncertainties at that order.

A central value of $\alpha_s(M_Z)$ of 0.118 has been assumed in the global fits at NLO and NNLO, but PDF sets at alternative values of $\alpha_s(m_Z)$ are also provided. CT14 prefers $\alpha_s(M_Z) = 0.115^{+0.006}_{-0.004}$ at NNLO (0.117 ± 0.005 at NLO) at 90 % confidence level (C.L.). These uncertainties from the global QCD fits are larger than those of the data from LEP and other experiments included into the world average [20]. Thus, the central PDF sets are obtained using the value of 0.118, which is consistent with the world average value and was recommended by the PDF4LHC group [21]. The CT14LO PDFs are provided by supplying two versions, one with a 1-loop $\alpha_s(M_Z)$ value of 0.130, and the other with a 2-loop $\alpha_s(M_Z)$ value of 0.118.

2 Results

The CT14 fit has a total of 2947 data points included from 33 experiments, producing $\chi^2 = 3252$ at the best fit (with $\chi^2/N_{pt} = 1.10$). The flavor composition of CT14 PDFs has changed somewhat compared to CT10 due to the inclusion of new LHC and Tevatron data sets and to the use of modified parametrization forms. The new PDFs are largely compatible with CT10 within the estimated PDF uncertainty. In Fig. 1 one observes that the CT14 NNLO PDFs have a softer strange-quark distribution at low x and a somewhat softer gluon at high x, compared to CT10 NNLO. The CT14 d-quark has increased by 5% at $x \approx 0.05$, after ATLAS and CMS W/Z production data sets at 7 TeV were included. At $x \gtrsim 0.1$, the update of the DØ charge asymmetry data set in the electron channel has reduced the magnitude of the d quark PDFs by a large amount, and has moderately increased the $u(x, Q)$ distribution. The d/u ratio in Fig. 2 has decreased at high x in comparison to CT10, as a consequence of replacing the 2008 DØ electron charge asymmetry (0.75 fb^{-1} [22]) measurement by the new 9.7fb^{-1} data set [16]. The d/u ratio approaches a constant value in the $x \to 1$ limit due to the

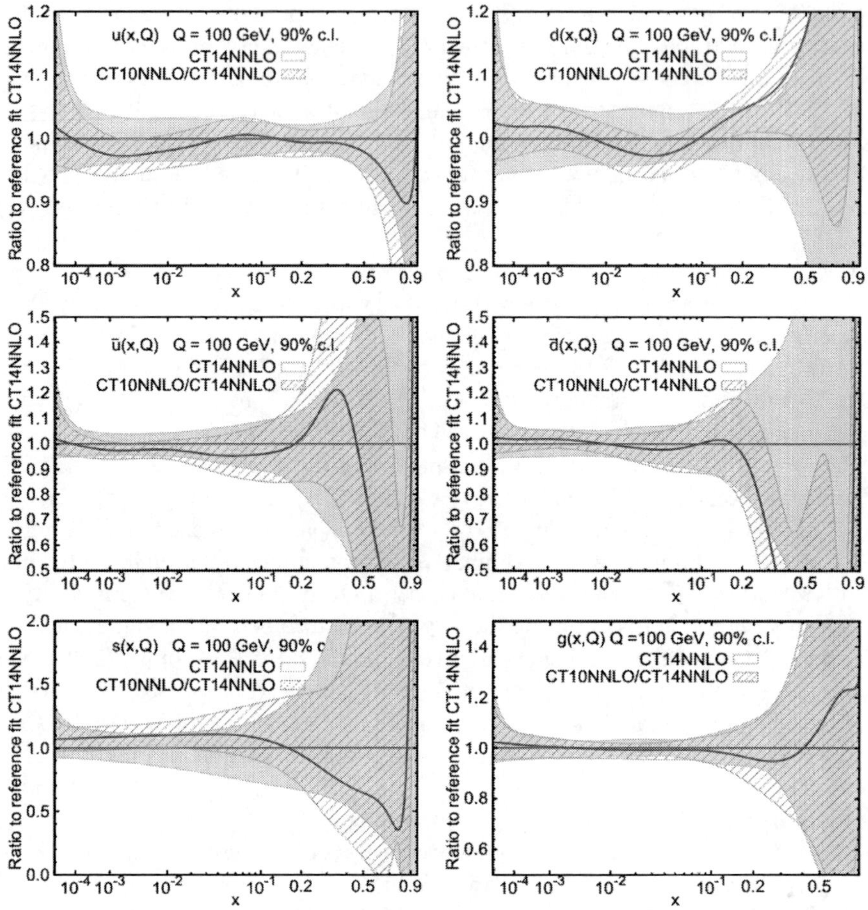

Figure 1. Comparison of 90% C.L. PDF uncertainties from CT14 NNLO (solid blue) and CT10 NNLO (red dashed) error sets. Both error bands are normalized to the respective central CT14 NNLO PDFs.

input physics assumption that both $d_{\rm val}$ and $u_{\rm val}$ behave as $(1-x)^{a_2}$ at $x \to 1$ with the same value of a_2 (reflecting expectations from spectator counting rules), but allowing for independent normalizations. The $\bar{u}/\bar{d}$ ratio has also changed as a consequence of the new data and the new parametrization form. The CTEQ-JLab analysis (CJ12) [23] has independently determined the ratio d/u at NLO, and it is compared to that of CT14 in Fig. 2.

***W/Z* production cross sections at the LHC.** Fig. 3 shows central predictions and 90% C.L. regions for (W^+, W^-), (Z,$W^\pm$) pairs of inclusive cross sections at the LHC 8 and 13 TeV. In each subfigure, two elliptical confidence regions are shown, obtained with either CT14 or CT10 NNLO PDFs. These can be used to read off PDF uncertainties and correlations for each pair of cross sections. The plots in Fig. 3 indicate that the PDF uncertainties, at the 90% C.L., are about 3.9%, 3.7%, and 3.7% for W^+, W^-, and Z boson production at the LHC 13 TeV, respectively. The central predictions at 8 TeV are in agreement with the recent CMS measurements [24]. They also show that the electroweak gauge boson cross sections are highly correlated with each other; in fact, much of the uncertainty is driven in this case by the small-x gluon [12].

 DOI: 10.1051/epjconf/201612007003

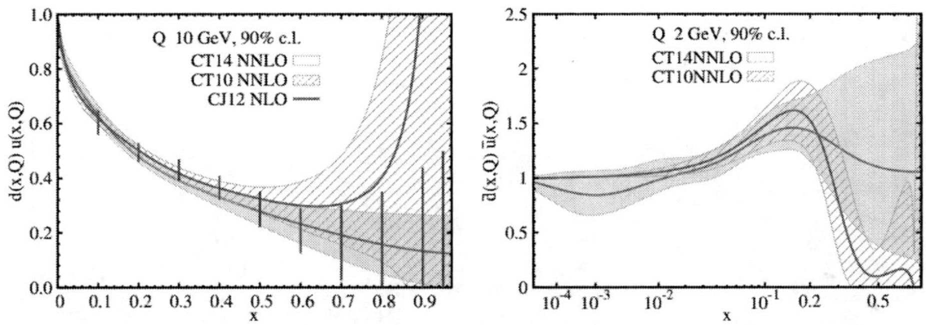

Figure 2. Comparison of 90% C.L. uncertainties on the ratios $d(x, Q)/u(x, Q)$ for CT14 NNLO (solid blue) and CT10 NNLO (red dashed), and CJ12NLO (green lines) error ensembles and $\bar{d}(x, Q)/\bar{u}(x, Q)$.

Differential cross sections for production of massive vector bosons set important constraints on the flavor composition of the proton, notably on the u and d quarks, anti-quarks and their ratios. In Fig. 4 CT14 NNLO theoretical predictions for lepton charge asymmetries of inclusive $W^{\pm}$ boson production are compared to data measured at the DØ Tevatron Run-2 [16] and ATLAS [13], CMS [14], and LHCb [15] experiments at the LHC 7 TeV. Theoretical predictions are computed using the program ResBos [25–28]. The black data points represent the unshifted central values of the data. The error bars indicate the total (statistical+systematic) experimental error, while the blue band is the CT14 PDF uncertainty evaluated at the 68% C.L. In total, constraints from the LHC and Tevatron W/Z differential cross sections and asymmetries lead to important changes in the quark sector PDFs. At $x \lesssim 0.02$, more realistic error bands for the u, $\bar{u}$, d, $\bar{d}$ PDFs are obtained upon including the ATLAS and CMS data sets. At $x > 0.1$, the high-luminosity DØ charge asymmetry and other compatible experiments predict a softer behavior of $d(x, Q)/u(x, Q)$ than in CT10W.

Compared to CT10, the new inputs and theoretical advancements of the CT14 fit resulted in a slight increase in the large-x gluon (of order 1%). The CT14 gg luminosities are compared to those of MMHT2014 [29] and NNPDF3.0 [30] PDFs at 13 TeV in Fig. 5. The parton luminosity is defined as in Ref. [31]. All central values and uncertainty bands agree very well among the three global PDFs, in the x range sensitive to Higgs production.

Conclusions

A short overview of the CT14 global analysis of proton's PDFs has been given. With rapid advancements in LHC measurements, the focus of the global analysis has shifted toward providing accurate predictions in the wide range of x and Q covered by the LHC data. This development requires long-term multi-prong effort in theoretical, experimental, and statistical areas. On the theory side, a more flexible parametrization has been introduced to better capture variations in the PDF dependence. A series of benchmark tests of NNLO cross sections, carried out in the run-up for the CT14 fit for all key fitted processes, have resulted in better agreement with most experiments and brought accuracy of most predictions to the truly NNLO level. The final CT14 PDFs are presented in the form of 1 central and 56 Hessian eigenvector sets at NLO and NNLO. The 90% C.L. PDF uncertainties for physical observables can be estimated from these sets using the symmetric [32] or asymmetric [3, 33] master formulae. These PDFs are determined for the central QCD coupling of $\alpha_s(M_Z) = 0.118$, consistent with the world-average α_s value.

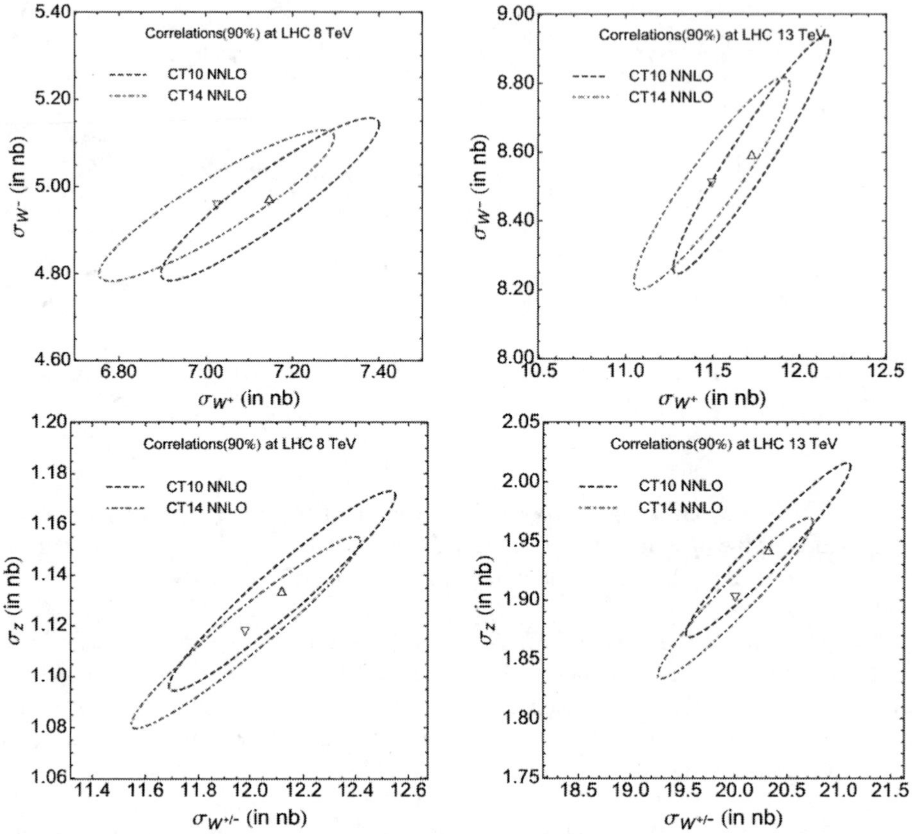

Figure 3. CT14 and CT10 NNLO 90% C.L. error ellipses for W^- and W^+ cross sections, at the LHC 8 and 13 TeV. and Z and $W^{\pm}$.

Acknowledgments.
M.G. would like to thank the organizers of the ISMD2015 conference for the kind invitation. This work was supported by the STFC (ST/L000431/1) and by the Lancaster-Manchester-Sheffield Consortium for Fundamental Physics under STFC grant ST/L000520/1; by the U.S. DOE Early Career Research Award DE-SC0003870; by the U.S. Department of Energy under Grant No. DE-FG02-96ER40969, DE-SC0013681, and DE-AC02-06CH11357; by the U.S. National Science Foundation under Grant No. PHY-0855561 and PHY-1417326; by Lightner-Sams Foundation; and by the National Natural Science Foundation of China under Grant No. 11165014 and 11465018.

References

[1] G. Aad et al. (ATLAS), Phys. Lett. **B716**, 1 (2012), 1207.7214
[2] S. Chatrchyan et al. (CMS), Phys. Lett. **B716**, 30 (2012), 1207.7235
[3] H.-L. Lai, M. Guzzi, J. Huston, Z. Li, P.M. Nadolsky, J. Pumplin, C.-P. Yuan, Phys. Rev. **D82**, 074024 (2010), 1007.2241

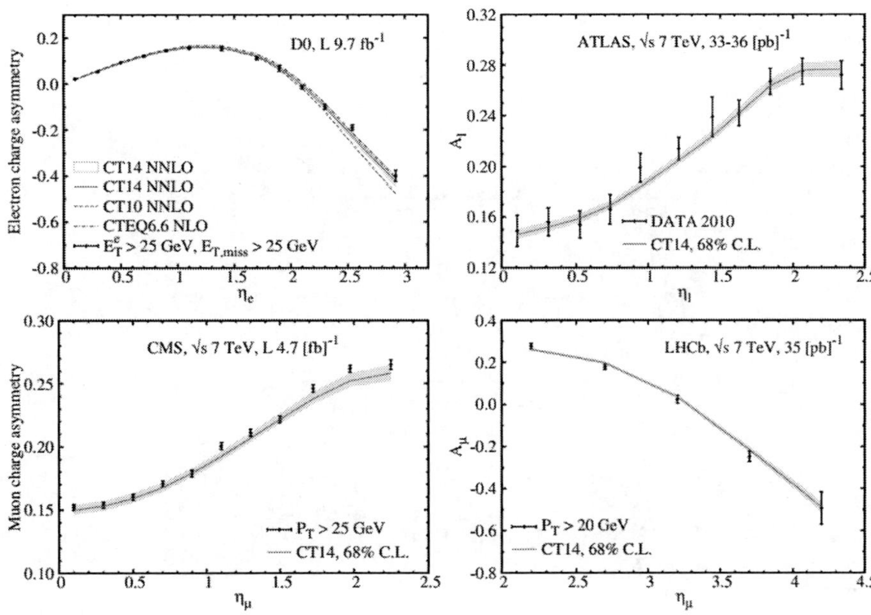

Figure 4. Charge asymmetries as a function of lepton pseudorapidity measured by the DØ, ATLAS, CMS, and LHCb Collaborations, compared to the 68% C.L. CT14 NNLO uncertainty band.

[4] J. Gao, M. Guzzi, J. Huston, H.L. Lai, Z. Li, P. Nadolsky, J. Pumplin, D. Stump, C.-P. Yuan, Phys. Rev. **D89**, 033009 (2014), 1302.6246

[5] S. Dulat, T.J. Hou, J. Gao, M. Guzzi, J. Huston, P. Nadolsky, J. Pumplin, C. Schmidt, D. Stump, C.-P. Yuan (2015), 1506.07443

[6] https://lhapdf.hepforge.org/

[7] M. Guzzi, P.M. Nadolsky, H.L. Lai, C.-P. Yuan, Phys. Rev. **D86**, 053005 (2012), 1108.5112

[8] M.A.G. Aivazis, J.C. Collins, F.I. Olness, W.-K. Tung, Phys. Rev. **D50**, 3102 (1994), hep-ph/9312319

[9] J.C. Collins, Phys. Rev. **D58**, 094002 (1998), hep-ph/9806259

[10] W.-K. Tung, S. Kretzer, C. Schmidt, J. Phys. **G28**, 983 (2002), hep-ph/0110247

[11] W.-K. Tung, H.L. Lai, A. Belyaev, J. Pumplin, D. Stump, C.-P. Yuan, JHEP **02**, 053 (2007), hep-ph/0611254

[12] P.M. Nadolsky, H.L. Lai, Q.H. Cao, J. Huston, J. Pumplin, D. Stump, W.-K. Tung, C.-P. Yuan, Phys. Rev. **D78**, 013004 (2008), 0802.0007

[13] G. Aad et al. (ATLAS), Phys. Rev. **D85**, 072004 (2012), 1109.5141

[14] S. Chatrchyan et al. (CMS), Phys. Rev. **D90**, 032004 (2014), 1312.6283

[15] R. Aaij et al. (LHCb), JHEP **06**, 058 (2012), 1204.1620

[16] V.M. Abazov et al. (D0), Phys. Rev. **D91**, 032007 (2015), [Erratum: Phys. Rev.D91,no.7,079901(2015)], 1412.2862

[17] J. Pumplin, D. Stump, R. Brock, D. Casey, J. Huston, J. Kalk, H.L. Lai, W.-K. Tung, Phys. Rev. **D65**, 014013 (2001), hep-ph/0101032

[18] G. Watt, R.S. Thorne, JHEP **08**, 052 (2012), 1205.4024

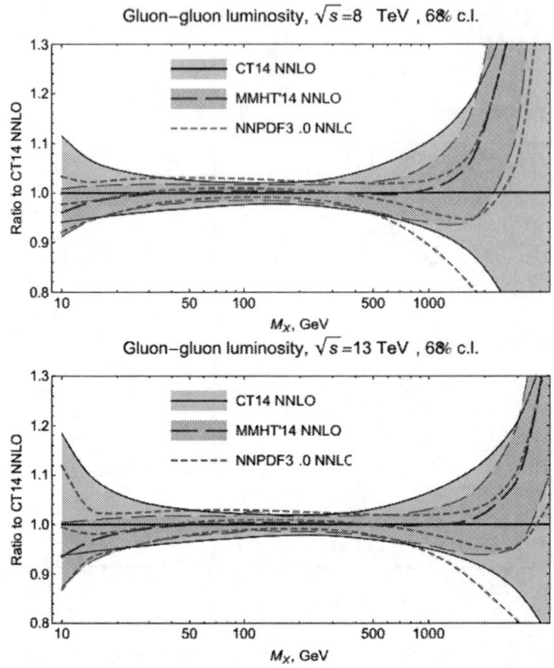

Figure 5. The *gg* PDF luminosities for CT14, MMHT2014 and NNPDF3.0 PDFs at the LHC with $\sqrt{s}$ = 8 and 13 TeV, with $\alpha_s(M_Z) = 0.118$.

[19] D. Stump, J. Pumplin, R. Brock, D. Casey, J. Huston, J. Kalk, H.L. Lai, W.-K. Tung, Phys. Rev. **D65**, 014012 (2001), hep-ph/0101051
[20] K.A. Olive et al. (Particle Data Group), Chin. Phys. **C38**, 090001 (2014)
[21] S. Alekhin et al. (2011), 1101.0536
[22] V.M. Abazov et al. (D0), Phys. Rev. Lett. **101**, 211801 (2008), 0807.3367
[23] J. Owens, A. Accardi, W. Melnitchouk, Phys.Rev. **D87**, 094012 (2013), 1212.1702
[24] S. Chatrchyan et al. (CMS), Phys. Rev. Lett. **112**, 191802 (2014), 1402.0923
[25] C. Balazs, J.w. Qiu, C.-P. Yuan, Phys. Lett. **B355**, 548 (1995), hep-ph/9505203
[26] C. Balazs, C.-P. Yuan, Phys. Rev. **D56**, 5558 (1997), hep-ph/9704258
[27] F. Landry, R. Brock, P.M. Nadolsky, C.-P. Yuan, Phys. Rev. **D67**, 073016 (2003), hep-ph/0212159
[28] M. Guzzi, P.M. Nadolsky, B. Wang, Phys. Rev. **D90**, 014030 (2014), 1309.1393
[29] L.A. Harland-Lang, A.D. Martin, P. Motylinski, R.S. Thorne, Eur. Phys. J. **C75**, 204 (2015), 1412.3989
[30] R.D. Ball et al. (NNPDF), JHEP **04**, 040 (2015), 1410.8849
[31] J.M. Campbell, J.W. Huston, W.J. Stirling, Rept. Prog. Phys. **70**, 89 (2007), hep-ph/0611148
[32] J. Pumplin, D.R. Stump, J. Huston, H.L. Lai, P.M. Nadolsky, W.-K. Tung, JHEP **07**, 012 (2002), hep-ph/0201195
[33] P.M. Nadolsky, Z. Sullivan, eConf **C010630**, P510 (2001), hep-ph/0110378

A low-energy determination of α_s at three loops

Antonio Vairo[1,a]

[1] *Physik-Department, Technische Universität München, James-Franck-Str. 1, 85748 Garching, Germany*

Abstract. We review one of the most accurate low-energy determinations of α_s. Comparing at short distances the QCD static energy at three loops and resummation of the next-to-next-to leading logarithms with its determination in 2+1-flavor lattice QCD, we obtain $\alpha_s(1.5\ \text{GeV}) = 0.336^{+0.012}_{-0.008}$, which corresponds to $\alpha_s(M_Z) = 0.1166^{+0.0012}_{-0.0008}$. We discuss future perspectives.

1 α_s in 2015

For many years the average of the strong coupling constant, α_s, provided by the Particle Data Group (PDG) (the latest printed edition is [1]) has shown a rather stable central value and a steady decrease in the associated error, see figure 1. This satisfactory situation has however been challenged in the last years (see for example the summary tables in [2–4] and the recent collection [5]): precise determinations pointing often towards a lower value of α_s have appeared, traditionally larger and precise determinations from τ decay have become smaller and less precise [6], accurate determinations from lattice QCD have been questioned. The 2016 PDG average will reflect this changed situation and, for the first time in twenty years, the α_s world average will see an increase in the error [7].

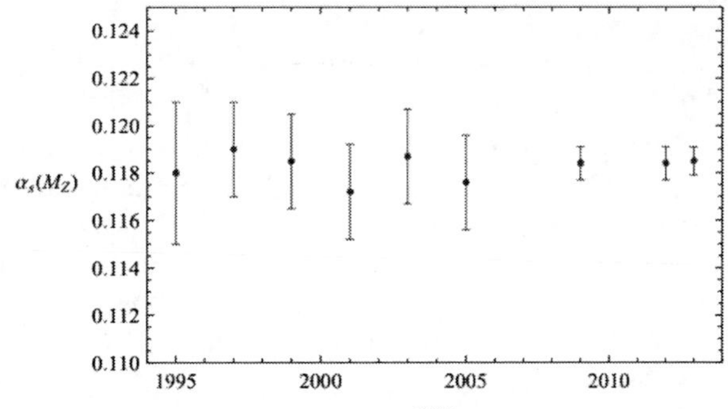

Figure 1. List of PDG averages for $\alpha_s(M_Z)$ from 1995 to 2013, source [8].

[a] e-mail: antonio.vairo@ph.tum.de

This change of scenario makes it even more important to look for accurate determinations of α_s. The extraction of α_s that follows from comparing perturbative QCD with short distance lattice computations of the QCD static energy is one of these determinations. Perturbatively this quantity is known at three loops and next-to-next-to-next-to leading logarithmic (NNNLL) accuracy. On the lattice side, the static Wilson loop is one of the most accurately known quantities that has been computed since the establishment of lattice QCD. Nowadays computations are with quarks at or very close to their physical masses so that the lattice spacing can be matched to a physical scale, eventually leading to a determination of the physical (unquenched) α_s. The strong coupling constant extracted in this way is typically at a low energy scale because the lattice cannot explore too short distances. In fact the method provides one of the most precise low-energy determinations of α_s. In this sense the determination of α_s from the QCD static energy not only provides a competitive but also a complementary determination with respect to high-energy determinations. Obviously there is also a value *per se* in having a low-energy determination of α_s, for it adds to our understanding of low-energy QCD and non-trivially constraints our modeling of it. Here we will review our latest determination of α_s from the QCD static energy [3], which improves our previous determination [9]. An earlier quenched analysis can be found in [10]. Finally, as we will discuss briefly in the outlook, although the extraction of α_s from the static Wilson loop provides already today an accurate value, there is room for improvement in several directions.

2 QCD static energy in perturbation theory

The QCD static energy, $E_0(r)$, is defined (in Minkowski spacetime) as

$$E_0(r) = \lim_{T\to\infty} \frac{i}{T} \ln \left\langle \mathrm{Tr\,P} \exp\left\{ ig \oint_{r\times T} dz^\mu A_\mu \right\} \right\rangle, \qquad (1)$$

where the integral is over a rectangle of spatial length r, the distance between the static quark and the static antiquark, and time length T; $\langle \dots \rangle$ stands for the path integral over the gauge fields A_μ and the light quark fields, P is the path-ordering operator of the color matrices and g is the SU(3) gauge coupling ($\alpha_s = g^2/(4\pi)$). The above definition of $E_0(r)$ is valid at any distance r. In the short range, $r\Lambda_{\rm QCD} \ll 1$ for which $\alpha_s(1/r) \ll 1$, $E_0(r)$ may be computed in perturbative QCD and expressed as a series in α_s (computed at a typical scale of order $1/r$):

$$E_0(r) = \Lambda_s - \frac{4\alpha_s}{3r}(1 + \dots), \qquad (2)$$

where Λ_s is a constant that accounts for the normalization of the static energy and the dots stand for higher-order terms. The expansion of $E_0(r)$ in powers of α_s is known up to three loops. At three loops, a contribution proportional to $\ln\alpha_s$ appears for the first time. This three-loop logarithm has been computed in [11, 12]. The complete three-loop contribution can be derived from [13, 14]. The leading logarithms have been resummed to all orders in [15], providing, among others, also the four-loop contribution proportional to $\ln^2\alpha_s$. The four-loop contribution proportional to $\ln\alpha_s$ has been computed in [16]. Next-to-leading logarithms have been resummed to all orders in [17], which represents the status of the art of the QCD static energy in perturbation theory. A concise but complete summary with all relevant formulas can be found in [18]. $E_0(r)$ is, therefore, one of the best known quantities in perturbative QCD.

The appearance of $\ln\alpha_s$ terms starting from three loops signals the cancellation of contributions coming from the different energy scales $1/r$ and α_s/r:

$$\ln\alpha_s = \ln\frac{\mu}{1/r} + \ln\frac{\alpha_s/r}{\mu}, \qquad (3)$$

where μ is a factorization scale. Indeed in the short range $E_0(r)$ is characterized by a hierarchy of well-separated scales, see figure 2.

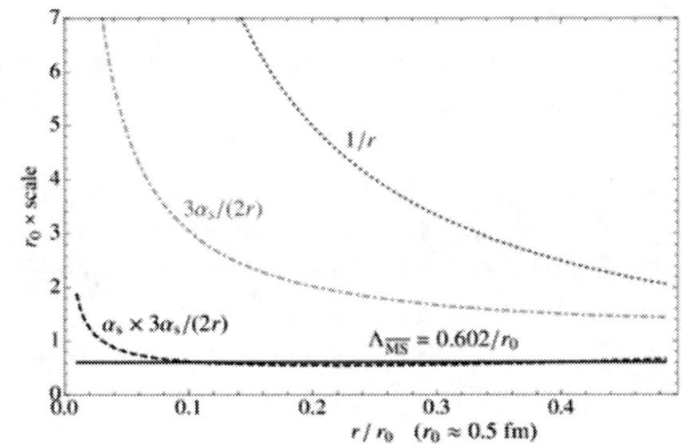

Figure 2. Energy scales for the short-range static energy in the quenched case, from [17].

The separation of contributions coming from gluons scaling like $1/r$ from contributions coming from gluons scaling like α_s/r leads to the factorization formula [11, 12]:

$$E_0(r) \;=\; \Lambda_s \;+\; V_s(r,\mu) - i\frac{V_A^2}{3}\int_0^\infty dt\, e^{-it(V_o-V_s)}\,\langle \mathrm{Tr}\,\{\mathbf{r}\cdot g\mathbf{E}(t,\mathbf{0})\,\mathbf{r}\cdot g\mathbf{E}(0,\mathbf{0})\}\rangle(\mu) + \dots\,, \quad (4)$$

where $V_s(r,\mu) = -4\alpha_s/(3r) + \dots$ is the color-singlet static potential, $V_o(r,\mu) = \alpha_s/(6r) + \dots$ is the color-octet static potential, V_A a Wilson coefficient describing the chromoelectric dipole coupling, and $\mathbf{E}$ the chromoelectric field. The color-singlet static potential encodes the contributions from the scale $1/r$, while the low energy contributions are in the term proportional to the two chromoelectric dipoles.

All the terms in (4) are well known in perturbation theory. The three-loop expression of the color-singlet static potential has been computed in [13, 14]. The Wilson coefficient V_A is 1 up to corrections of order α_s^2 [16]. The chromoelectric field correlator, $\langle \mathrm{Tr}\,\{g\mathbf{E}(t,\mathbf{0})\cdot g\mathbf{E}(0,\mathbf{0})\}\rangle$ (Wilson lines connecting the fields are understood), is known up to two loops [19]. The color-octet static potential is known up to three loops [20].

An advantage of equation (4) is that it allows for an efficient resummation of the logarithmic contributions to the static potentials and eventually to the static energy. This is achieved by evaluating the anomalous dimensions of the potentials through the computation of the ultraviolet divergences in the relevant integrals proportional to the two chromoelectric dipoles and then by solving the renormalization group equations for the static potentials [15, 17]. This is how contributions to the static energy of the type $\alpha_s^{3+n}\ln^n\alpha_s/r$ and $\alpha_s^{4+n}\ln^n\alpha_s/r$ have been computed for all n.

The perturbative expansion of V_s is affected by a renormalon ambiguity of order Λ_{QCD}. The ambiguity reflects in the poor convergence of the perturbative series.

A first method to cure the poor convergence of the perturbative series of V_s consists in subtracting a (constant) series in α_s from V_s and reabsorb it in a redefinition of the normalization constant Λ_s [21]. This is the strategy we followed in the earlier analysis [9].

A second possibility consists in considering the force

$$F(r,\nu) = \frac{d}{dr}E_0(r). \quad (5)$$

β	7.373	7.596	7.825
r_1/a	5.172(34)	6.336(56)	7.690(58)
Volume	$48^3 \times 64$	64^4	64^4

Table 1. Lattice couplings, spacings and volumes used for the extraction of α_s.

It does not depend on Λ_s and is free from the renormalon of order Λ_{QCD}. In (5) we have written explicitly the dependence of F on the renormalization scale for the coupling, ν, which appears when computing F order by order in $\alpha_s(\nu)$. Once integrated upon the distance, the force gives back the static energy

$$E_0(r) = \int_{r_*}^{r} dr'\, F(r', 1/r'), \tag{6}$$

up to an irrelevant constant determined by the arbitrary distance r_*, which can be reabsorbed in the overall normalization when comparing with lattice data. Notice that the integration in (6) can be done (numerically) keeping the strong-coupling constant running at a natural scale of the order of the inverse of the distance. This scheme does not generate therefore potentially large logarithms of the type $\ln \nu r$. Such logarithms are typically generated within the first method. In the newer analysis of [3], we have adopted this second method.

3 Numerical analysis

In order to determine α_s, we match the three-loop expression of $E_0(r)$ in the short range with the determination of the same quantity in 2+1-flavor lattice QCD obtained from the tree-level improved gauge action and the Highly-Improved Staggered Quark (HISQ) action by the HotQCD collaboration [22]. The strange quark mass, m_s, has been fixed to its physical value, while the light quark masses, m_l, have been set equal to $m_s/20$. This corresponds to a pion mass of about 160 MeV in the continuum limit. Several lattice spacings and volumes have been considered, but for our determination of α_s we restricted to the three finest spacings shown in table 1. The largest gauge coupling, $\beta = 7.825$, corresponds to a lattice spacing of $a = 0.041$ fm. The lattice spacing has been fixed using the r_1 scale defined as $r^2 dE_0(r)/dr|_{r=r_1} = 1.0$. In turn, the scale r_1 could be fixed from the pion decay constant [23]: $r_1 = 0.3106 \pm 0.0017$ fm.

Because of the many lattice points available at short distance (see figure 3) we can fit α_s (or equivalently $\Lambda_{\overline{MS}}$) for each lattice spacing separately and eventually average these determinations. This is different from our earlier analysis in [9], where, because of the less data points, we had to fit α_s on data points from all lattice spacings at the same time. Such a procedure introduced a sizeable extra error source due to the different lattice normalizations. This error is absent in the newer analysis in [3].

We use the following procedure: *(i)* for each lattice spacing we fit $\Lambda_{\overline{MS}}$ at different orders of perturbative accuracy; the overall normalization constant that matches the lattice data with the perturbative expression of $E_0(r)$ is typically fixed on the 7th, 8th or 9th lattice point called $N_{\rm ref}$ in some of the following figures (this choice is dictated by the fact that lattice points at short distance are expected to be better described by perturbation theory, but at too short distance may be affected by lattice artefacts); *(ii)* we repeat the fits for each of the following distance ranges: $r < 0.75r_1$, $r < 0.7r_1$, $r < 0.65r_1$, $r < 0.6r_1$, $r < 0.55r_1$, $r < 0.5r_1$, and $r < 0.45r_1$, and use only the ranges where the reduced χ^2 either

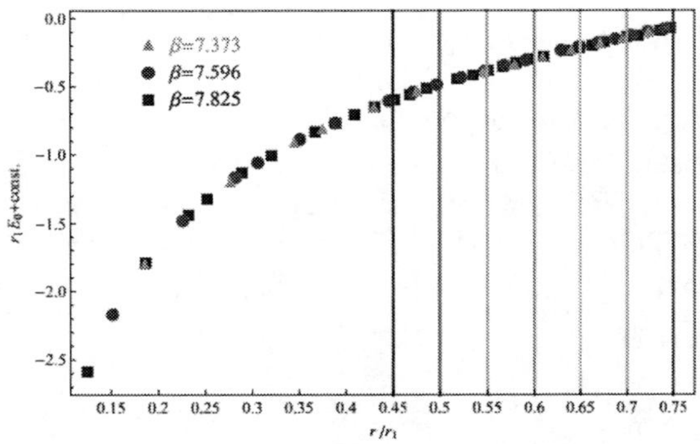

Figure 3. Data sets used in the analysis of [3]. Lattice data are from [22].

decreases or does not increase by more than one unit or is smaller than 1, when increasing the perturbative order; *(iii)* to estimate the perturbative uncertainty of the result, we repeat the fits by varying the renormalization scale in the perturbative expansion from $\nu = 1/r$ to $\nu = \sqrt{2}/r$ and $\nu = 1/(\sqrt{2}r)$, or by adding/subtracting a term $\pm 4\alpha_s^{n+2}/(3r^2)$ to the n-loop expression of the force; we take the largest uncertainty.

As expected, perturbative QCD describes better the data in the shortest distance ranges. In particular, criterion *(ii)* is fulfilled for data in the range $r < 0.6r_1$ or shorter; to be on the most conservative side, in the final result we will use only fits for $r < 0.5r_1$.

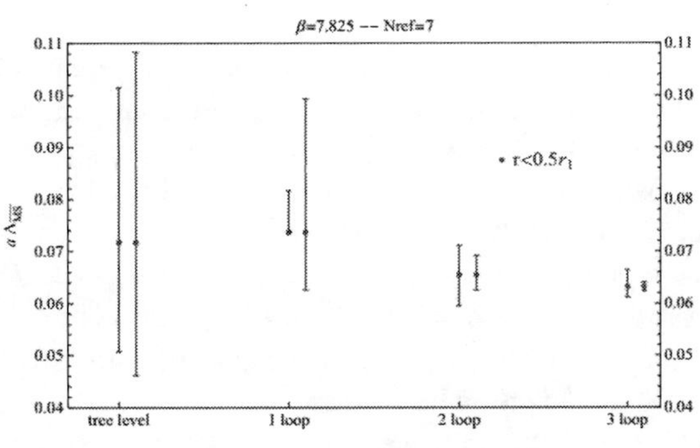

Figure 4. Results for $a\Lambda_{\overline{\rm MS}}$ from the fits to the lattice data at $\beta = 7.825$ for different orders of perturbative accuracy. The first error bar corresponds to the ν variation, whereas the second one to the addition of a generic higher-order term. The static energies have been normalized on the 7th point.

We have performed several checks and estimated several error sources. *(a)* When considering the perturbative errors of *(iii)* we notice that uncertainties due to the higher-order term $\pm 4\alpha_s^{n+2}/(3r^2)$ dominate for the static energy at tree level and one loop, while starting from two loops the dominant source of uncertainties comes from variation of ν. Uncertainties decrease in going to higher orders, and in all cases they fall inside the error bar of the lower-order determination, see figure 4. *(b)* The result is insensitive or very little sensitive to changes in the lattice spacing, in the range of short distances used (see figure 5) and in the point chosen for the normalization with the lattice data. *(c)* We have estimated and added to our determination of $\Lambda_{\overline{\rm MS}}$ a statistical error taken as the variation in

$\Lambda_{\overline{\rm MS}}$ when one allows for the fit to be one χ^2 unit above minimum. Statistical errors are smaller than perturbative errors, but, in particular for the shortest distance ranges, not negligible. *(d)* An analysis done comparing the perturbative expression of the force (5) with the numerical derivative of the lattice data for $E_0(r)$ gives consistent results, but with larger uncertainties associated with the interpolation of the lattice points. *(e)* By repeating the fits adding a monomial term proportional to r^3 and r^2, which could be associated with gluon and quark local condensates, and also a term proportional to r, we could not find evidence for a significant non-perturbative term at short distances and the value of $\Lambda_{\overline{\rm MS}}$ remains unchanged.

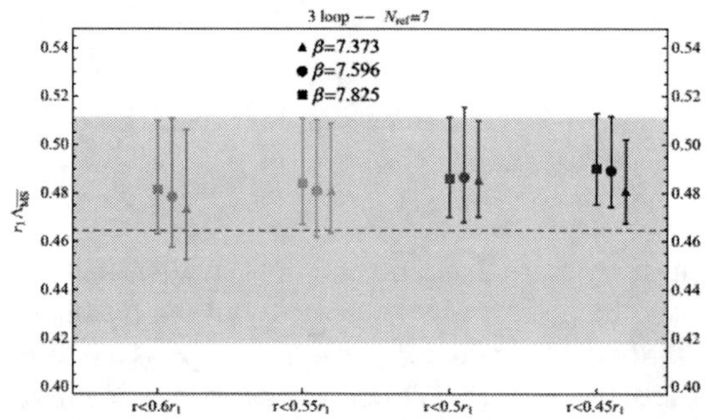

Figure 5. Results for $r_1\Lambda_{\overline{\rm MS}}$ at three-loop accuracy for the different lattice spacings of table 1 and ranges of figure 3. We show the fit for data normalized with respect to the 7th lattice point. Similar analyses with consistent results have been done in [3] also for different normalization points. The band shows our previous result [9].

The analysis revealed that the data are not sensitive to logarithmic corrections of order $\alpha_s^{4+n}\ln^n\alpha_s/r$, moreover these corrections are much smaller than the non-logarithmic contribution of order α_s^4/r, which turns out to be numerically as large as the logarithmic corrections of order $\alpha_s^{3+n}\ln^n\alpha_s/r$, but with opposite sign. Hence we have included in our perturbative expression of the static energy only logarithmic corrections of the type $\alpha_s^{3+n}\ln^n\alpha_s/r$, whereas we have not included logarithmic corrections of the type $\alpha_s^{4+n}\ln^n\alpha_s/r$, although they are known. We chose $\mu = 1.26r_1^{-1} \sim 0.8$ GeV for the factorization scale. Variations of μ only produce small effects on the results.

As shown in figure 5, the newer analysis of [3] is consistent with the older one of [9], but with an error which is roughly half. The reduction in the error can be traced back to two main improvements. First, as already discussed in this section, the newest lattice data provide much more points in the short range so that we can fit for each lattice spacing independently and avoid normalization errors that arise when combining data from different lattices. Second, as discussed at the end of section 2, fitting with the perturbative expression obtained from (6) allows avoiding uncertainties associated with a fixed renormalization scale ν, which is a potential source of large logarithms of the type $\ln\nu r$.

4 Results

The results of the fits of $\Lambda_{\overline{\rm MS}}$ for the different lattice spacings are summarized in table 2. The first error is the perturbative one, the second error is the one from the statistical uncertainties in the fit, and, for the last column, the third error corresponds to the one coming from the conversion from a to r_1. Errors in the right-hand side of the last column are added in quadrature. The final number for $r_1\Lambda_{\overline{\rm MS}}$ with three flavors is obtained as a weighted average of the results for the three different lattices with errors added linearly. Once converted in physical units, it reads

$$\Lambda_{\overline{\rm MS}} = 315^{+18}_{-12}\ \text{MeV}. \tag{7}$$

	$a\Lambda_{\overline{\rm MS}}$; $N_{\rm ref}=7$	$a\Lambda_{\overline{\rm MS}}$; $N_{\rm ref}=8$	$a\Lambda_{\overline{\rm MS}}$; $N_{\rm ref}=9$	$a\Lambda_{\overline{\rm MS}}$; range spanned	$r_1\Lambda_{\overline{\rm MS}}$; range spanned
$\beta=7.373$	$0.0957^{+0.0046}_{-0.0028}$ ± 0.0017	$0.0957^{+0.0046}_{-0.0028}$ ± 0.0017	$0.0957^{+0.0046}_{-0.0028}$ ± 0.0017	$0.0957^{+0.0046}_{-0.0028}$ ± 0.0017	$0.4949^{+0.0240}_{-0.0144} \pm 0.0086 \pm 0.0025$ $= 0.4949^{+0.0256}_{-0.0170}$
$\beta=7.596$	$0.0781^{+0.0046}_{-0.0029}$ ± 0.0007	$0.0784^{+0.0043}_{-0.0027}$ ± 0.0010	$0.0785^{+0.0049}_{-0.0029}$ ± 0.0007	$0.0783^{+0.0048}_{-0.0031}$ ± 0.0010	$0.4961^{+0.0303+0.0066}_{-0.0197-0.0061} \pm 0.0044$ $= 0.4961^{+0.0313}_{-0.0211}$
$\beta=7.825$	$0.0644^{+0.0032}_{-0.0019}$ ± 0.0006	$0.0642^{+0.0033}_{-0.0020}$ ± 0.0008	$0.0643^{+0.0032}_{-0.0020}$ ± 0.0008	$0.0643^{+0.0033}_{-0.0021}$ ± 0.0008	$0.4944^{+0.0256}_{-0.0159} \pm 0.0065 \pm 0.0037$ $= 0.4944^{+0.0267}_{-0.0175}$
Average					$r_1\Lambda_{\overline{\rm MS}} = 0.495^{+0.028}_{-0.018}$

Table 2. $\Lambda_{\overline{\rm MS}}$ determinations from the three lattice spacings and averaging.

With this value of $\Lambda_{\overline{\rm MS}}$ we can compare the static energy as computed in perturbative QCD at three loop accuracy with resummation of the $\alpha_s^{3+n}\ln^n\alpha_s/r$ terms with lattice data. In figure 6 this comparison is done for the $\beta = 7.825$ lattice data. The comparison shows also visually (the numerical quantitative analysis has been summarized in the previous section) that perturbation theory agrees with lattice data up to about 0.2 fm.

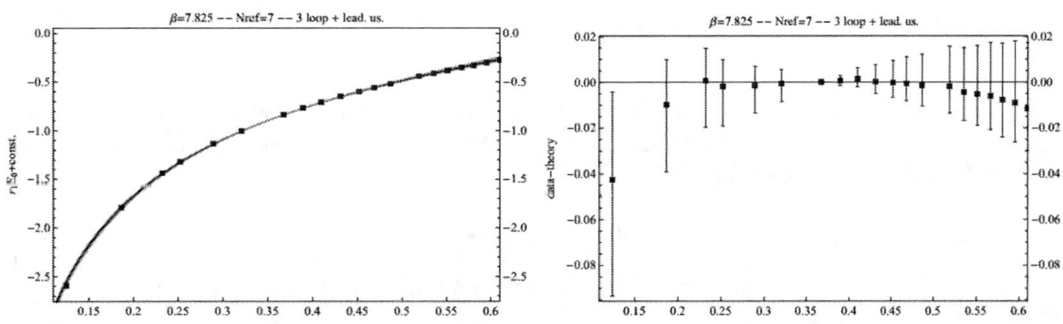

Figure 6. Left panel: comparison of the lattice data for $\beta = 7.825$ with the perturbative expression at three loops plus logarithmic corrections of the type $\alpha_s^{3+n}\ln^n\alpha_s/r$. We take $r_1\Lambda_{\overline{\rm MS}} = 0.495^{+0.028}_{-0.018}$; the grey band reflects the uncertainty in $r_1\Lambda_{\overline{\rm MS}}$. Right panel: result of subtracting the perturbative expression from the lattice data. In both panels the static energy has been normalized on the 7th lattice point.

By converting (7) to α_s at the highest energy scale we used, i.e., 5.1 GeV, and then evolving the value of α_s down to 1.5 GeV with 3 flavors, we obtain

$$\alpha_s(1.5\ \text{GeV}) = 0.336^{+0.012}_{-0.008}, \tag{8}$$

which is one of the most accurate determinations of the strong coupling constant at this low-energy scale. Finally, by evolving α_s at four loops up to the Z mass, M_Z, including the decoupling relations at the quark thresholds ($M_c = 1.6$ GeV and $M_b = 4.7$ GeV), we obtain

$$\alpha_s(M_Z) = 0.1166^{+0.0012}_{-0.0008}. \tag{9}$$

The effects of higher-order terms in the running are negligible with the current accuracies, but may become relevant when and if the precision of $\alpha_s(M_Z)$ will be reduced to the per mil level.

5 Outlook

We have computed the strong coupling constant, α_s, by fitting the static energy, as obtained in perturbation theory with three loop accuracy and with resummation of the $\alpha_s^{3+n} \ln^n \alpha_s/r$ terms, with the short-distance part of the static energy as obtained on physical lattices. The result is a very accurate determination of α_s at low energy. At the Z-mass scale the determination is competitive with the others entering the PDG average. Its central value is lower than the PDG average of 2014 [1], but closer to the expected PDG average of 2016 [7].

There are several ways in which the present determination may be improved with the potential of becoming the most precise determination of α_s. As shown in figure 4, to achieve the present accuracy the inclusion of the three-loop result is crucial but, as discussed at the end of section 3, not all of the presently available perturbative information has been used. More precise lattice data on finer lattices and with more data points at short distances could take advantage of it and improve the determination of α_s. Also, it would be important, in order to reduce possible systematic effects, to perform the same study on Wilson loops computed on different lattices with different actions.

A possible systematic effect is due to the finite lattice spacing. A continuum extrapolation would reduce this effect and allow for a precise determination of the force between static charges along the same lines developed in [24] for the quenched case. As we mentioned in section 3, a study of the force with the present lattice data, although consistent with the result obtained from the static energy, did not provide a comparable accuracy because of finite lattice spacing effects in the calculation of the slope.

Another possibility consists of computing the force directly from the lattice, i.e., not as the slope of the static energy. A matrix element that gives the force, F, between a static quark located in $\mathbf{r}$ and a static antiquark located in $\mathbf{0}$ is [25, 26]

$$F(r) = -\lim_{T\to\infty} \frac{\left\langle \mathrm{Tr\,P}\,\hat{\mathbf{r}}\cdot g\mathbf{E}(t,\mathbf{r}) \exp\left\{ig \oint_{r\times T} dz^\mu A_\mu\right\}\right\rangle}{\left\langle \mathrm{Tr\,P} \exp\left\{ig \oint_{r\times T} dz^\mu A_\mu\right\}\right\rangle}. \tag{10}$$

The chromoelectric field $\mathbf{E}(t,\mathbf{r})$ on the right-hand side is located at the quark line of the Wilson loop. It would be interesting to compute this matrix element and examine if, besides being an alternative way to compute the force, it may also provide a more accurate determination of α_s.

Acknowledgements

I thank Alexei Bazavov, Nora Brambilla, Xavier Garcia i Tormo, Péter Petreczky and Joan Soto for a long going collaboration on the subject. This work is supported in part by DFG and NSFC (CRC110). I also acknowledge financial support from the DFG cluster of excellence "Origin and structure of the universe" (www.universe-cluster.de).

References

[1] K. A. Olive *et al.* [Particle Data Group Collaboration], Chin. Phys. C **38** (2014) 090001.
[2] S. Moch *et al.*, arXiv:1405.4781 [hep-ph].

[3] A. Bazavov, N. Brambilla, X. Garcia i Tormo, P. Petreczky, J. Soto and A. Vairo, Phys. Rev. D **90** (2014) 7, 074038 [arXiv:1407.8437 [hep-ph]].
[4] A. H. Hoang, D. W. Kolodrubetz, V. Mateu and I. W. Stewart, Phys. Rev. D **91** (2015) 9, 094018 [arXiv:1501.04111 [hep-ph]].
[5] D. d'Enterria *et al.*, arXiv:1512.05194 [hep-ph].
[6] D. Boito, M. Golterman, K. Maltman, J. Osborne and S. Peris, Phys. Rev. D **91** (2015) 3, 034003 [arXiv:1410.3528 [hep-ph]].
[7] S. Bethke and G. Dissertori, at this conference.
[8] http://pdg.lbl.gov/2015/html/rpp_archives.html
[9] A. Bazavov, N. Brambilla, X. Garcia i Tormo, P. Petreczky, J. Soto and A. Vairo, Phys. Rev. D **86** (2012) 114031 [arXiv:1205.6155 [hep-ph]].
[10] N. Brambilla, X. Garcia i Tormo, J. Soto and A. Vairo, Phys. Rev. Lett. **105** (2010) 212001 [Phys. Rev. Lett. **108** (2012) 269903] [arXiv:1006.2066 [hep-ph]].
[11] N. Brambilla, A. Pineda, J. Soto and A. Vairo, Phys. Rev. D **60** (1999) 091502 [hep-ph/9903355].
[12] N. Brambilla, A. Pineda, J. Soto and A. Vairo, Nucl. Phys. B **566** (2000) 275 [hep-ph/9907240].
[13] C. Anzai, Y. Kiyo and Y. Sumino, Phys. Rev. Lett. **104** (2010) 112003 [arXiv:0911.4335 [hep-ph]].
[14] A. V. Smirnov, V. A. Smirnov and M. Steinhauser, Phys. Rev. Lett. **104** (2010) 112002 [arXiv:0911.4742 [hep-ph]].
[15] A. Pineda and J. Soto, Phys. Lett. B **495** (2000) 323 [hep-ph/0007197].
[16] N. Brambilla, X. Garcia i Tormo, J. Soto and A. Vairo, Phys. Lett. B **647** (2007) 185 [hep-ph/0610143].
[17] N. Brambilla, A. Vairo, X. Garcia i Tormo and J. Soto, Phys. Rev. D **80** (2009) 034016 [arXiv:0906.1390 [hep-ph]].
[18] X. Garcia i Tormo, Mod. Phys. Lett. A **28** (2013) 1330028 [arXiv:1307.2238].
[19] M. Eidemüller and M. Jamin, Phys. Lett. B **416** (1998) 415 [hep-ph/9709419].
[20] C. Anzai, M. Prausa, A. V. Smirnov, V. A. Smirnov and M. Steinhauser, Phys. Rev. D **88** (2013) 5, 054030 [arXiv:1308.1202 [hep-ph]].
[21] A. Pineda, J. Phys. G **29** (2003) 371 [hep-ph/0208031].
[22] A. Bazavov *et al.* [HotQCD Collaboration], Phys. Rev. D **90** (2014) 9, 094503 [arXiv:1407.6387 [hep-lat]].
[23] A. Bazavov *et al.* [MILC Collaboration], PoS LATTICE **2010** (2010) 074 [arXiv:1012.0868 [hep-lat]].
[24] S. Necco and R. Sommer, Nucl. Phys. B **622** (2002) 328 [hep-lat/0108008].
[25] N. Brambilla, A. Pineda, J. Soto and A. Vairo, Phys. Rev. D **63** (2001) 014023 [hep-ph/0002250].
[26] A. Pineda and A. Vairo, Phys. Rev. D **63** (2001) 054007 [Phys. Rev. D **64** (2001) 039902] [hep-ph/0009145].

EPJ Web of Conferences **120**, 07005 (2016) DOI: 10.1051/epjconf/201612007005
ISMD 2015

World Summary of α_s (2015)

Siegfried Bethke[1,a], Günther Dissertori[2,b], and Gavin Salam[3,c]

[1] *Max-Planck-Institut für Physik, München*
[2] *ETH Zürich*
[3] *CERN, Geneva; on leave from CNRS, LPTHE, Paris*

Abstract. This is a **preliminary** update of the measurements of α_s and the determination of the world average value of $\alpha_s(M_Z^2)$ presented in the 2013/2014 edition of the Review of Particle Properties [1].

A number of studies which became available since late 2013 provide new results for each of the (previously 5, now) 6 subclasses of measurements for which pre-average values of $\alpha_s(M_Z^2)$ are determined. In the following, we list those new results which are used to determine the new average values of α_s, i.e. which are based on at least complete $NNLO$ perturbation theory and are published in peer-reviewed journals, as well as those which are used for demonstrating asymptotic freedom (although being based on NLO only) :

- updated results from τ-decays [2] [3] [4], based on a re-analysis of ALEPH data and on complete N^3LO perturbation theory,
- more results from unquenched lattice calculations, [5][6],
- further results from world data on structure functions, in $NNLO$ QCD [7],
- from e^+e^- hadronic event shapes (C-parameter) in soft collinear effective field theory ($NNLO$) [8],
- α_s determinations at LHC, from data on the ratio of inclusive 3-jet to 2-jet cross sections [9], from inclusive jet production [10], from the 3-jet differential cross section [11], and from energy-correlations [12], all in NLO QCD, plus one determination in complete $NNLO$, from a measurement of the $t\bar{t}$ cross section at $\sqrt{s} = 7$ TeV [13];
- and finally, an update of α_s from a global fit to electroweak precision data [14].

All measurements available in subclasses of τ-decays, lattice results, structure functions and from e^+e^--annihilation are summarized in figure 1.

With the exception of lattice results, most results within their subclass are strongly correlated, however to an unknown degree, as they largely use the same data sets. The large scatter between many of these measurements, sometimes with only marginal or no agreement within the given errors, indicate the presence of additional systematic uncertainties from theory or caused by details of the analyses. In such cases, a pre-average value is determined,

[a] e-mail: bethke@mpp.mpg.de
[b] e-mail: Guenther.Dissertori@cern.ch
[c] e-mail: Gavin.Salam@cern.ch

 DOI: 10.1051/epjconf/201612007005

with a symmetric, overall uncertainty that encompasses the central values of all individual determinations ('*range averaging*'). For the subclass of lattice results, the average value determined in Ref. [5] is taken over. For the subclasses of hadron collider results and electroweak precision fits, only one result each is available in full $NNLO$, so that no pre-averaging can be applied. Note, however, that more measurements of top-quark pair production at LHC are meanwhile available, indicating that on average, a larger value of $\alpha_s(M_Z^2)$ is likely to emerge in the future; see also [15]. The resulting subclass averages are indicated in figure 1, and are summarized in table 1.

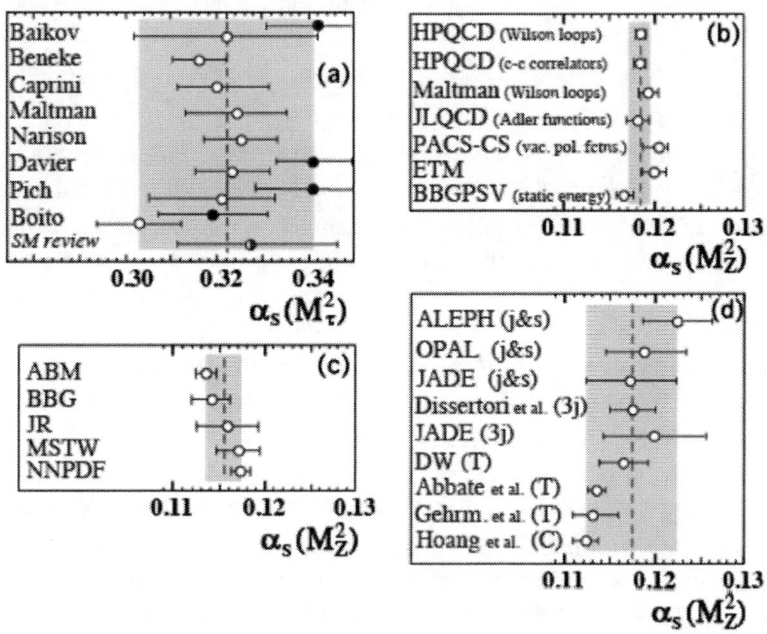

Figure 1. Summary of determinations of α_s from hadronic τ-decays (a), from lattice calculations (b), from DIS structure functions (c) and from e^+e^- annihilation (d). The shaded bands indicate the pre-average values explained in the text, to be included in the determination of the final world average of α_s. In a), full circles are results obtained using CIPT, and open circles using FOPT expansions, see text.

Assuming that the resulting pre-averages are largely independent of each other, the final world average value is determined as the weighted average of the different input values. An initial uncertainty of the central value is calculated treating the uncertainties of all measurements as being uncorrelated and of Gaussian nature, and the overall χ^2 to the central value is determined. If the initial value of χ^2 is smaller than the number of degrees of freedom, an overall, a-priori unknown correlation coefficient is introduced and determined by requiring that the total χ^2/d.o.f. equals unity. Applying this procedure to the values listed in table 1 results in a **preliminary** new world average of

$$\alpha_s(M_Z^2) = 0.1177 \pm 0.0013 \; .$$

Table 1. Pre-average values of subclasses of measurements of $\alpha_s(M_Z^2)$. The value from τ-decays was converted from $\alpha_s(M_\tau^2) = 0.322 \pm 0.019$, using the QCD 4-loop β-function plus 3-loop matching at the charm- and bottom-quark pole masses.

Subclass	$\alpha_s(M_Z^2)$
τ-decays	0.1187 ± 0.0023
lattice QCD	0.1184 ± 0.0012
structure fuctions	0.1154 ± 0.0020
e^+e^- [jets & shps]	0.1174 ± 0.0051
hadron collider	$0.1151^{+0.0028}_{-0.0027}$
ew precision fits	0.1196 ± 0.0030

This value is in reasonable agreement with that from 2013/2014, which was $\alpha_s(M_Z^2) = 0.1185 \pm 0.0006$, however at a somewhat decreased central value and with an overall uncertainty that has doubled. These changes are mainly due to the following reasons:

- the uncertainty of the lattice result, now taken from the estimate made by the FLAG group, is more conservative than that used in the previous review, leading to a larger final uncertainty of the new world average, and to a reduced fixing power towards the central average value;
- the decreased pre-average value of $\alpha_s(M_Z^2)$ from τ-decays, due to the most recent re-evaluations and their unexplained, increased inconsistency with respect to each other;
- the relatively low value of α_s from the new sub-class of hadron collider results, which currently consists of only one measurement of the $t\bar{t}$ cross section at $\sqrt{s} = 7$ TeV, and which appears to be "lowish" if compared to further measurements at higher $\sqrt{s}$ [15].

Note that pending revisions of the procedures applied may still change the results shown in table 1 and the final world average value which therefore are regarded to be **preliminary**.

While there is still room for improved measurements and treatments of systematic uncertainties, the data and results, especially when including measurements which are available at *NLO* only, consistently demonstrate and prove asymptotic freedom and the running of α_s, as predicted by QCD, up to energies beyond 1 TeV, see figure 2.

References

[1] K. A. Olive *et al.* [PDG Collab.], Chin. Phys. C **38** (2014) 090001.
[2] M. Davier *et al.*, Eur. Phys. J. C74, 2803 (2014), [arXiv:1312.1501 [hep-ph]].
[3] D. Boito *et al.*, Phys. Rev. D91, 034003 (2015), [arXiv:1410.3528 [hep-ph]] .
[4] A. Pich, Prog. Part.Nucl. Phys. 75 (2014) 41, [arXiv:1310.7922 [hep-ph]].
[5] S. Aoki *et al.*, Eur. Phys. J. C74, 2890 (2014), [arXiv:1109.1388 [hep-ph]].
[6] A. Bazavov *et al.*, Phys. Rev. D90, 074038 (2014), [arXiv:1407.8437 [hep-ph]].
[7] L. A. Harland-Lang *et al.*, [arXiv:1506.05682 [hep-ph]]
[8] A.H. Hoang *et al.*, Phys. Rev. D91, 094018 (2015), [arXiv:1501.04111 [hep-ph]],
[9] S. Chatrchyan *et al.* [CMS], Eur. Phys. J. C73, 10 (2013), [arXiv:1304.7498 [hep-ex]].
[10] V. Khachatryan *et al.* [CMS], Eur. Phys. J C75, 6 (2015), [arXiv:1410.6765 [hep-ex]].

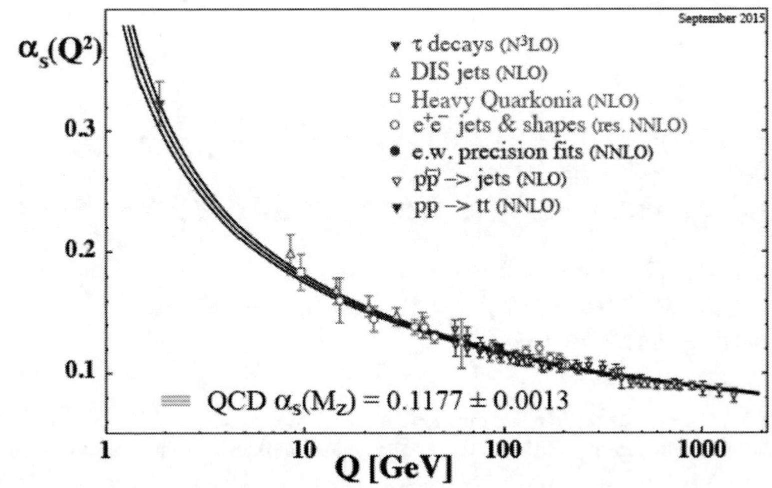

Figure 2. Summary of measurements of α_s as a function of the energy scale Q. The respective degree of QCD perturbation theory used in the extraction of α_s is indicated in brackets (NLO: next-to-leading order; NNLO: next-to-next-to leading order; res. NNLO: NNLO matched with resummed next-to-leading logs; N^{3}LO: next-to-NNLO).

[11] V. Khachatryan *et al.* [CMS], Eur. Phys. J. C75 5, 186 (2015) [arXiv:1412.1633 [hep-ex]].

[12] G. Aad *et al.* [ATLAS], Phys.Lett. B750 (2015) 427, [arXiv:1508.01579 [hep-ex]].

[13] S. Chatrchyan *et al.* [CMS], Phys. Lett. B728, 496 (2014) [arXiv:1307.1907 [hep-ex]]

[14] M. Baak *et al.* [Gfitter], Eur. Phys. J. C74 , 304660 (2014), [arXiv1407.3792 [hep-ph]].

[15] G. Salam, contribution to the Workshop on High Precision α_s Measurements: from LHC to FCC-ee, CERN, Geneva, October 12th-13th, 2015.

Final *ep* DIS cross sections from H1 and ZEUS and HERAPDF

Iris Abt[1,a]

[1] *Max-Planck-Institut für Physik, München*

Abstract. HERA was the first and only *ep* collider. It was in operation from 1992 to 2007 and the two collaborations ZEUS and H1 collected about 0.5 fb^{-1} of integrated luminosity each. Inclusive data from both collaborations were combined and used to determine the set of parton distribution functions called HERAPDF2.0. The history of HERA is presented as well as the long path to the final data combination and HERAPDF2.0. The proton remains full of wonders and parton distribution functions might have nothing to do with them. The legacy of HERA are the precise *ep* data which can be plotted as legacy plots.

1 Introduction

Sometimes it makes no sense to rewrite what was previously written. Please consult the original paper [1] from H1 and ZEUS for the full story of HERA data combination and HERAPDF2.0.

For a discussion on whether parton distribution functions at low Bjorken x have anything to do with the proton, please see [2].

2 Legacy Plots

As I have decided not to rewrite the data combination and HERAPDF2.0 paper, I just provide some HERA legacy plots.

3 Conclusions

The publication of their combined data together with the new PDF set HERAPDF2.0 was a major milestone for the ZEUS and H1 collaborations. If you need more than some plots, please consult the paper [1].

References

[1] ZEUS and H1 Collaborations, H. Abramowicz et al, Eur. Phys. J. **C 75**, 580 (2015)
[2] A. Caldwell and G. Grindhammer, Physik Journal **6**, 39 (2007)

[a] e-mail: isa@mpp.mpg.de

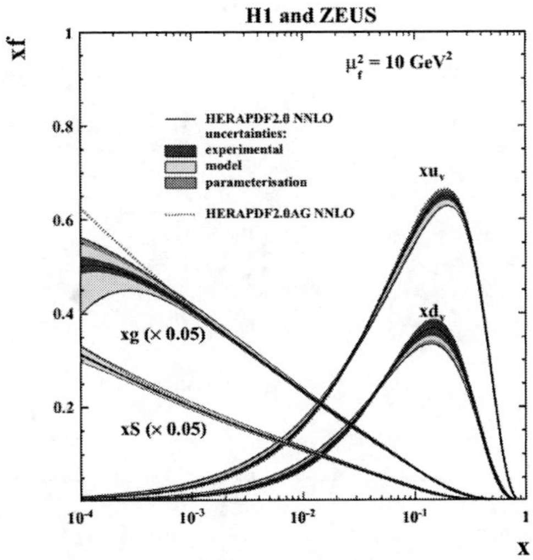

Figure 1. The parton distribution functions xu_v, xd_v, $xS = 2x(\bar{U} + \bar{D})$ and xg of HERAPDF2.0 NNLO at $\mu_f^2 = 10\,\mathrm{GeV}^2$. The gluon and sea distributions are scaled down by a factor 20. The experimental, model and parameterisation uncertainties are shown. The dotted lines represent HERAPDF2.0AG NNLO with an alternative gluon parameterisation preventing a turn-over.

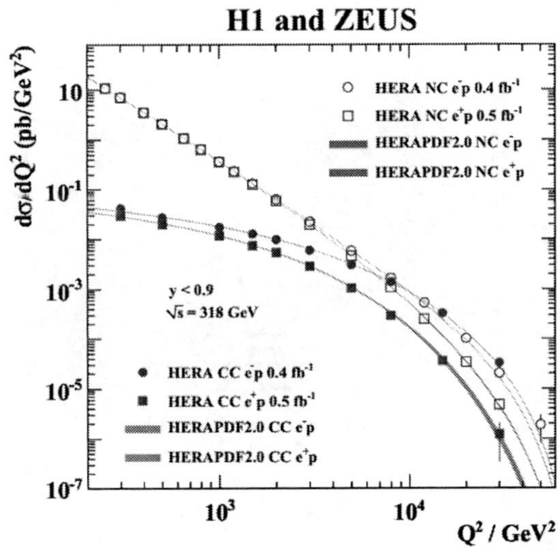

Figure 2. The combined HERA NC and CC e^-p and e^+p cross sections, $d\sigma/dQ^2$, together with predictions from HERAPDF2.0 NLO. The bands represent the total uncertainty on the predictions.

DOI: 10.1051/epjconf/201612008001

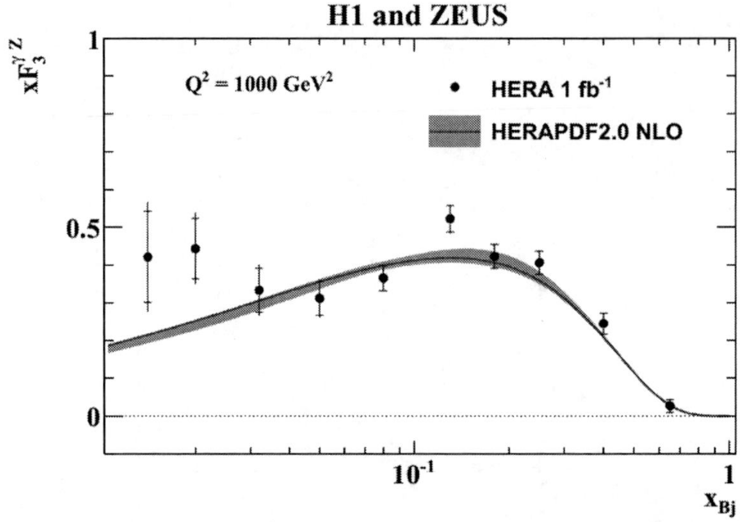

Figure 3. The structure function $xF_3^{\gamma Z}$ connected to the valence quarks averaged over $Q^2 \geq 1000\,\text{GeV}^2$ at the scale $Q^2 = 1000\,\text{GeV}^2$ together with the prediction from HERAPDF2.0 NLO. The band represents the total uncertainty on the prediction.

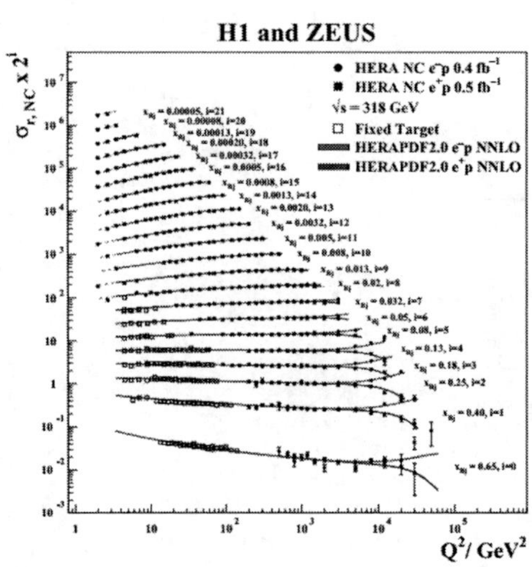

Figure 4. The combined HERA data for the inclusive NC e^+p and e^-p reduced cross sections together with fixed-target dataand the predictions of HERAPDF2.0 NNLO. The bands represent the total uncertainties on the predictions. Dashed lines indicate extrapolation into kinematic regions not included in the fit.

EPJ Web of Conferences **120**, 08002 (2016) DOI: 10.1051/epjconf/201612008002
ISMD 2015

ATLAS Measurements Sensitive to the Proton Structure

P. J. Bussey[1,a]
on behalf of the ATLAS Collaboration

[1] *SUPA, School of Physics and Astronomy, University of Glasgow, Glasgow G12 8QQ, United Kingdom*

Abstract. This talk presents a survey of ATLAS measurements at the LHC that are sensitive in various ways to the parton distribution functions of the proton.

1 Introduction

The parton structure of the proton plays a central role in the study of all particle interactions at the LHC that take place with large momentum transfer. An accurate understanding of such processes therefore requires reliable determination of the parton distribution functions (PDFs) of the proton. These are essentially derived from experiment. In the present talk we survey a number of measurements made with ATLAS [1] that are sensitive to the proton PDFs, which are presented in terms of models that have been constructed in various ways, based on measurements at the LHC but also at HERA, Fermilab and other laboratories. Different processes depend on the proton structure in different ways. The study of inclusive jets, dijets and trijets in high energy events provides insight into the gluon PDF of the proton. Drell-Yan processes are sensitive to sea quark distributions, while processes in which Z and W bosons are emitted can also give information on sea quark distributions.

2 Inclusive jets, dijets and trijet states.

In a recent study of high-momentum inclusive jet production [2], ATLAS first verify that the NLOJET++ model is capable of giving a good general description of the jet production over all accessible rapidity and transverse momentum (p_T) values, using the standard CT10 PDF set. This enables different PDF models to be compared with confidence and gives an extensive coverage of the standard kinematic quantities x and Q^2. In fig. 1, the ratio of the measured and theoretical cross sections is shown for a variety of published PDF sets. Of these, CT10, MSTW 2008, NNPDF 2.1 and HERAPDF 1.5 are found to do well over nearly all the region, but ABM11 (with five flavours) shows significant discrepancies at central rapidity values, where the fit probabilities are low. Specific references to the relevant PDF sets can be found in the cited ATLAS publications.

Dijet [3] and trijet [4] final states have also been studied, and again the general form of the cross section is first shown to be well modelled by the NLOJET Monte Carlo. Figure 2 shows a selection of the comparison of the results with proton structure models, in bands of the mean rapidity difference y^* of the dijets, and an analogous rapidity measure Y^* of the trijets. The same models that succeeded well

[a]e-mail: peter.bussey@glasgow.ac.uk

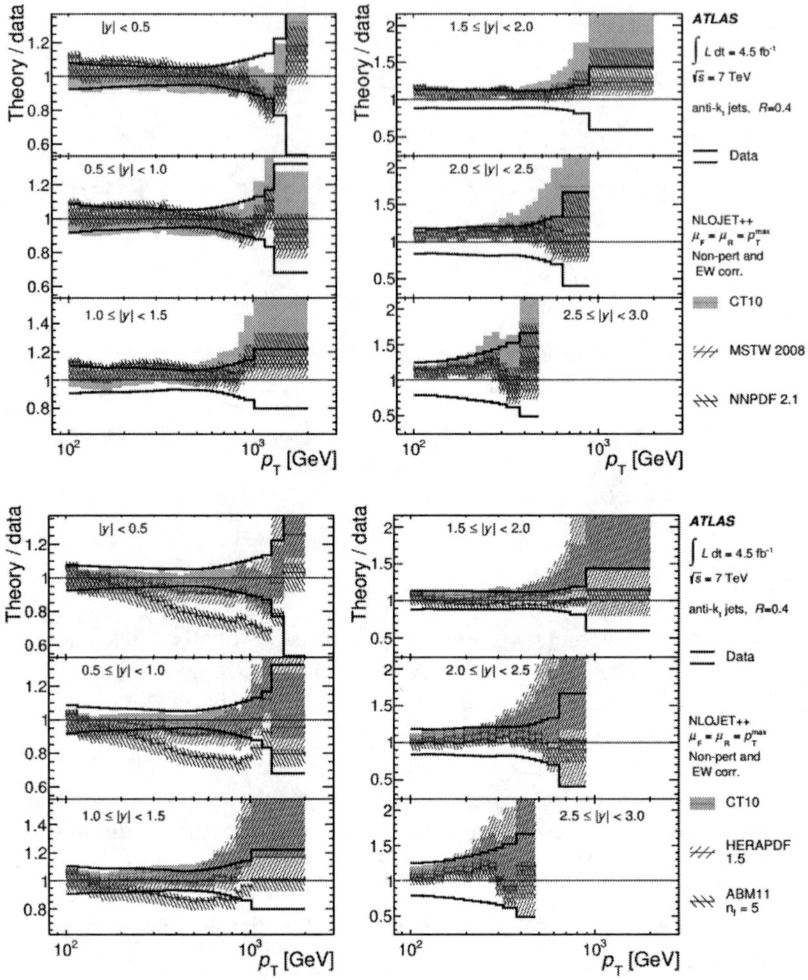

Figure 1. Transverse momentum of inclusive jets at ATLAS [2], compared with various models of the proton structure. Theory uncertainties are shown by coloured bands.

for the inclusive jets succeed well also for these cases, with the addition of epATLjet 13, which was based on a dedicated fit to HERA and ATLAS data. ABM 11 is unsatisfactory, as before. Electroweak corrections were applied to the dijet measurements but were unavailable for the trijets.

In principle, inclusive prompt photon cross sections should also show sensitivity to the proton structure. In practice, this is quite low and although the measured cross sections are all described well by the MCFM model, there is little differentiation between different PDF sets at present.

3 Top quark production

The production of pairs of top quarks provides a different perspective on the proton structure, and high statistics are available at the LHC. There are many parameters that can be measured, and several

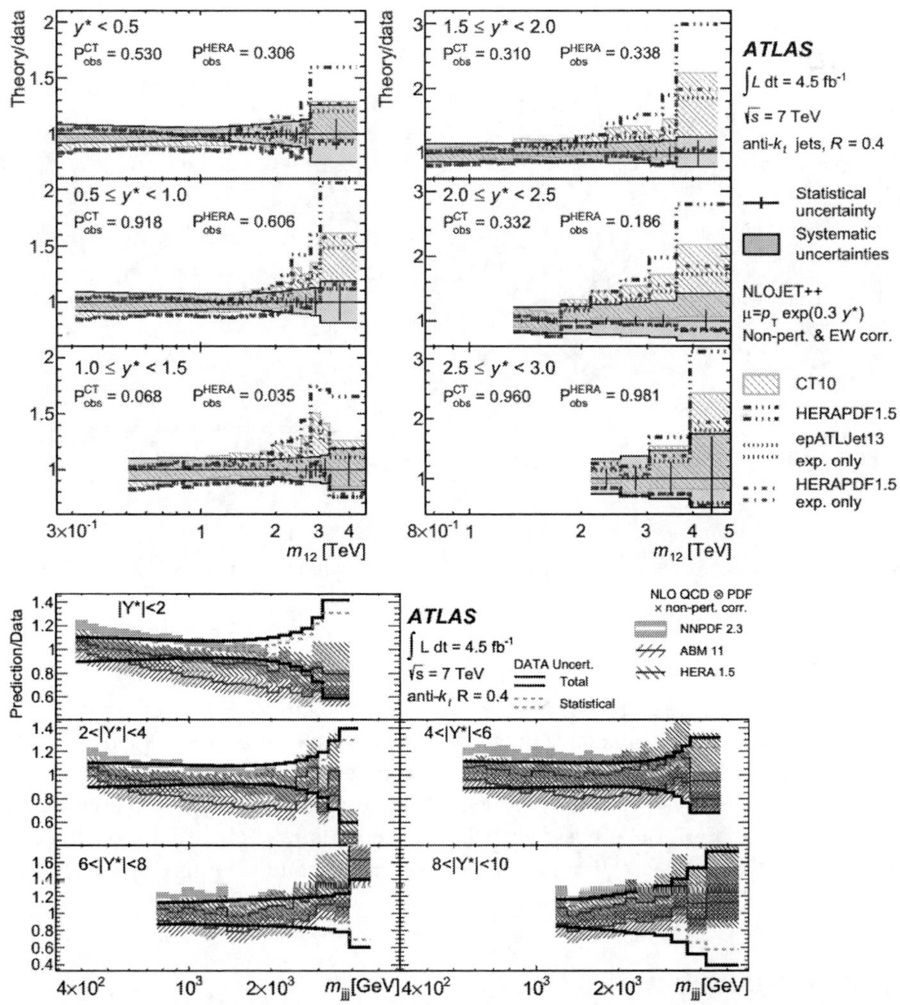

Figure 2. Mass of dijets [3] (upper) and trijets [4] (lower) at ATLAS, compared with various models of the proton structure. Theory uncertainties are shown by coloured bands.

models of the dynamics of the process. Of these, ALPGEN, MC@NLO, POWHEG (all using HERWIG) and POWHEG (using PYTHIA) are all found to be satisfactory, although the agreement with data is not perfect [5]. An example is the mass of the top-antitop system, which can be reconstructed in various decay modes of the *W* bosons in the decays. ATLAS have studied this process using final states defined with a lepton and four jets, with at least one *b*-meson tag. Figure 3 (left) shows that three standard models evaluated using NLO theory (CT10 NLO, MSTW 2008 and NNPDF 2.3) tend to lie above the data at high values of the mass, while HERAPDF 1.5 gives a good description over the whole range. Other variables do not discriminate well between the models.

These studies have been continued using 8-TeV LHC data. A typical example is shown in fig. 3 (right) where the rapidity of single top mesons is compared with theory, again at the NLO level. In

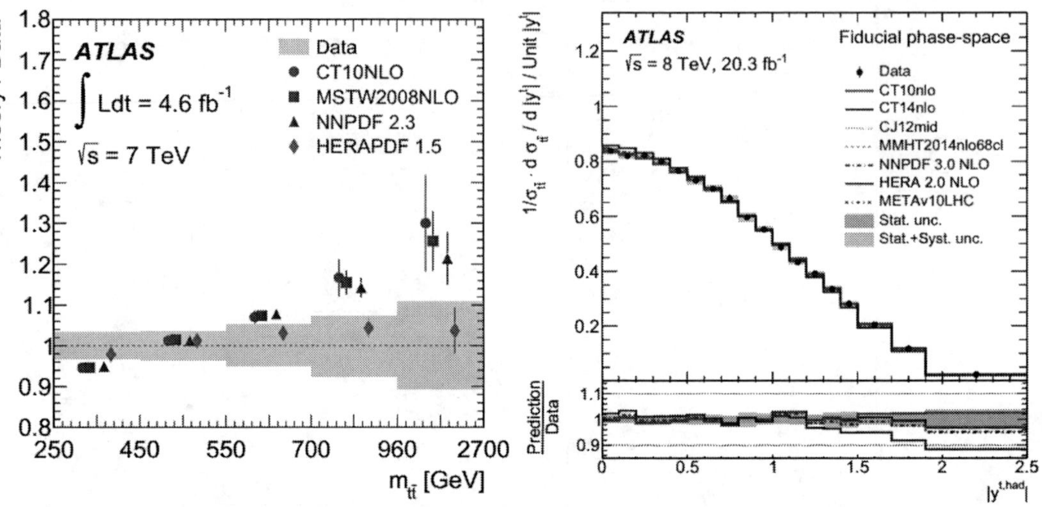

Figure 3. Left: mass of $t\bar{t}$ system produced at ATLAS [5], compared with various models of the proton structure. Right: rapidity of top mesons.

this case, all the theories except HERAPDF 2.0 are satisfactory. When other variables are plotted, slightly different results are obtained; thus CT10nlo is found to be less satisfactory than the others when the rapidity of the top pair and of its boost are considered.

Other pairs of high-energy leptons are available for study. However the production of electron-positron pairs by the Drell-Yan process gives little differentiation between the available proton models at masses above 116 GeV, and all are found to give cross sections lying lower than the data [6]. At lower masses the data and theory are in agreement.

4 *Z* and *W* production

Z bosons are produced copiously at the LHC, but the inclusive *Z* cross section is poorly described by NLO theory as a function of the p_T of the Z [7]. In principle this process could be used to study the proton structure, but in practice the dynamics still need to be better modelled and understood.

The situation improves when further constraints are applied to the process under study, and the production of a *Z* accompanied by *b*-jets is much more favourable [8]. Figure 4 shows (upper) that different models predict significantly different values for the cross section, but that MFCM and aMC-NLO are satisfactory. At present, the scale uncertainties in the models prevent discrimination between different choices of PDF (lower figure), but there is potential for improvement using higher-order calculations. At present the MSTW 2008, NNPDF 2.3 and CT10 PDFs may all be regarded as satisfactory, when used with an appropriate model of the production dynamics.

The measurement of *W*/*Z* ratios provides information that can be used to extract the elusive strange quark density in the proton. In specifying the density of the strange quark relative to the down quark as $r_s = 0.5(s+\bar{s})/d$, a suppression factor of 0.5 has been commonly employed. ATLAS have measured this ratio to be close to 1.0 [9], which is compared to several proton PDF models in fig. 5.

The production of *W* mesons accompanied by a charm particle is capable of giving further information [10]. The charm quark is normally of the opposite sign to the *W*, and to remove backgrounds

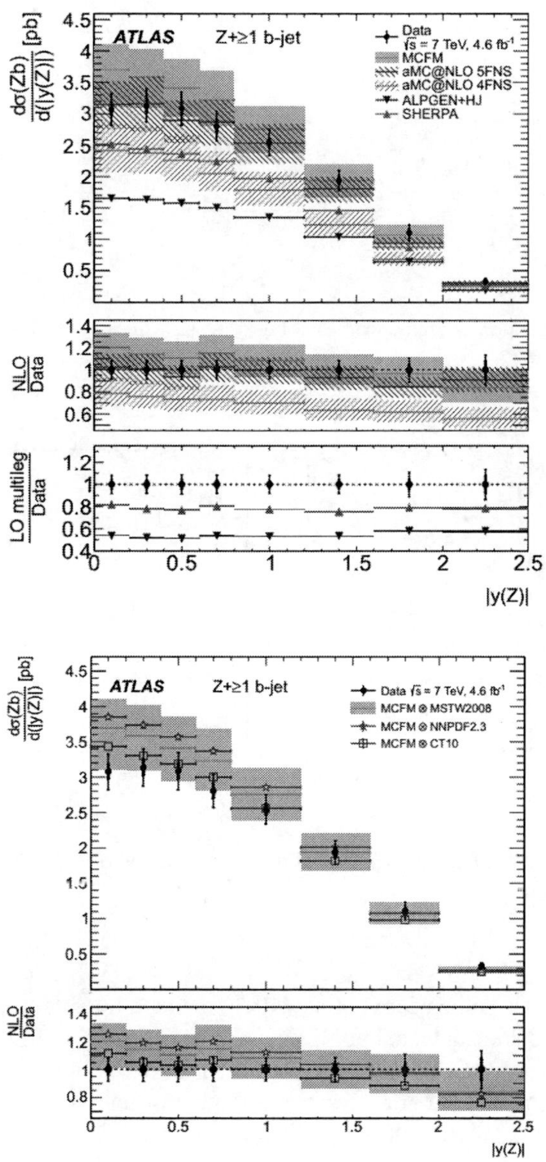

Figure 4. Rapidity of Z accompanied by b-jets [8], compared with various models of the production process (upper) and proton structure (lower). Scale-dominated theoretical uncertainties are shown by coloured bands.

the same-sign backgrounds are subtracted. Results for two channels are shown in fig. 6, giving a certain amount of discrimination between different PDF models. There is scope for improvement by achieving smaller uncertainties both on the data and on the theory. The HERAFITTER program has been run using these data, and favours an unsuppressed strange quark contribution.

The simultaneous measurement of W, Z, and $t\bar{t}$ cross sections is capable of providing an important new set of viewpoints both on the production of these final states and of the proton structure [11]. As

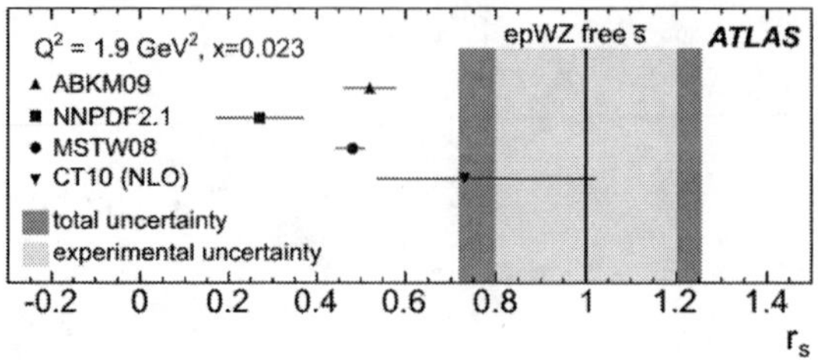

Figure 5. The strange quark suppression factor measured by ATLAS in the W/Z production ratios [9], compared to other determinations.

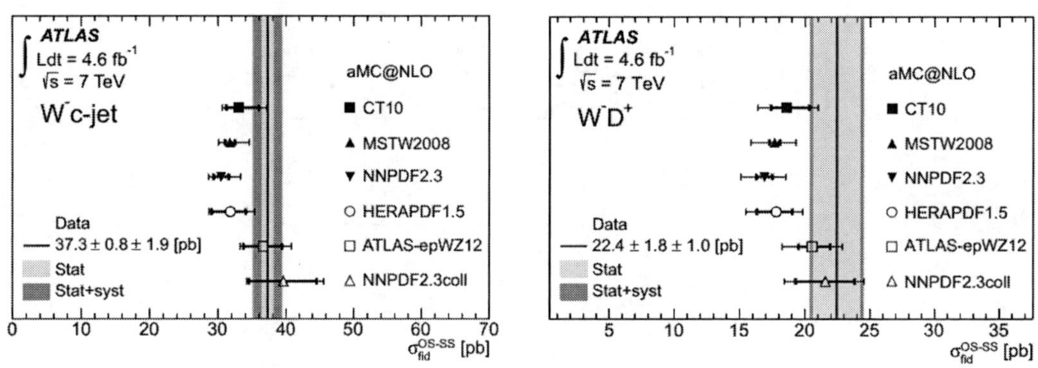

Figure 6. Background-subtracted fiducial cross sections for production of a W boson accompanied by a charm jet and by a D meson [10], compared to predictions.

an example, fig. 7 shows two-dimensional plots with associated uncertainties for Z production and $t\bar{t}$ production. If the theories are calculated at NLO, using corresponding PDF sets, there is substantial disagreement with the data when both channels are considered simultaneously. The theories show some spread, but have large uncertainties. At NNLO, most of the theoretical uncertainties decrease, and the MSTW 2008, CT10, HERAPDF 1.5 (with large uncertainties), NNPDF23 and epWZ models agree well with each other and with the data. The ABM11 and JR09 models now apparently fail.

5 Conclusions

The measurements of ATLAS have enlarged the kinematic region over which the proton structure can be studied, and this will improve further when 13 TeV data are available. A variety of models have been presented, and they have varying success in describing the different channels that have been analysed. The precision of the experimental data is an important factor, as is also the precision of the theoretical preductions, which can benefit considerably by extending the model to next-to-next-to-leading order. Different processes have different sensitivities to different PDFs, and the their simultaneous measurement appears to be a powerful technique. At the same time, in any given case,

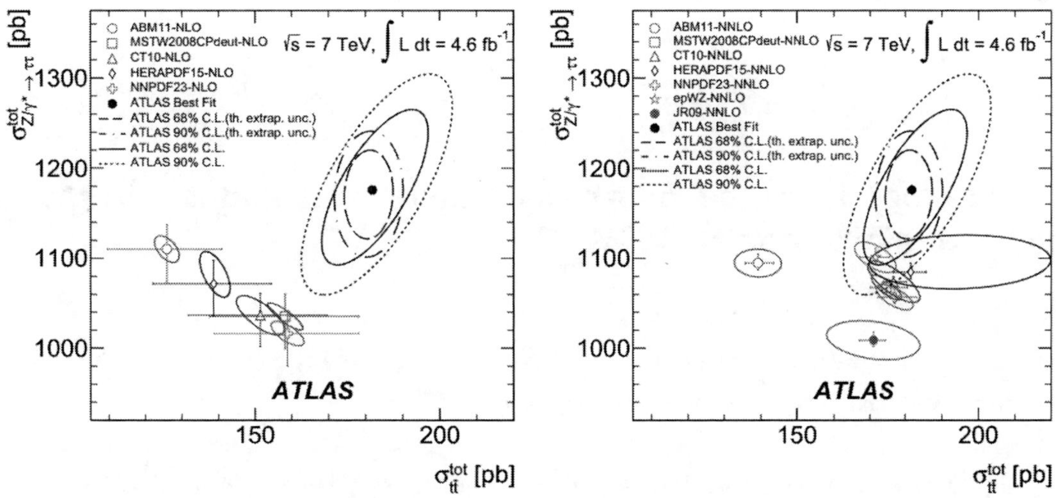

Figure 7. Cross sections for top pair production and Z production plotted simultaneously [11] and compared to theoretical models evaluated at next-to-leading order (left) and next-to-next-to-leading order (right).

it is important to have confidence in the basic dynamical model that is to be used in evaluating the predictions.

With higher energy running with the upgraded ATLAS apparatus, we can expect more precise b-tagging, higher statistics in many channels, and improved developments in reconstruction techniques. At the same time, some channels may also have higher backgrounds and tougher running conditions to cope with.

A good knowledge of the proton structure is essential for the accurate measurement of many physical quantities. ATLAS has well demonstrated the importance of this area for present and future work.

References

[1] ATLAS Collaboration, JINST **3**, S08003 (2008)
[2] ATLAS Collaboration, JHEP bf 15, 153 (2015)
[3] ATLAS Collaboration, JHEP bf 14, 059 (2014)
[4] ATLAS Collaboration, Eur. Phys. J. **C 75**, 228 (2015)
[5] ATLAS Collaboration, Phys. Rev. **D 90**, 072004 (2014)
[6] ATLAS Collaboration, Phys. Lett. **B 725**, 223 (2013)
ATLAS Collaboration, JHEP **06**, 112 (2014)
[7] ATLAS Collaboration, JHEP **09**, 145 (2014)
[8] ATLAS Collaboration, JHEP **10**, 141 (2014)
[9] ATLAS Collaboration, Phys. Rev. Lett. **109**, 012001 (2012)
[10] ATLAS Collaboration, JHEP **05**, 068 (2014)
[11] ATLAS Collaboration, Phys. Rev. **D 91**, 052005 (2015)

EPJ Web of Conferences **120**, 08003 (2016) DOI: 10.1051/epjconf/201612008003
ISMD 2015

Measurements of the Collins asymmetries for kaons and pions in e^+e^- annihilations at *BABAR*

A. Filippi[1,a]
on behalf of the *BABAR* Collaboration

[1] *I.N.F.N. sez. di Torino, via P. Giuria, 1, 10125 Torino, Italy*

Abstract. New measurements of the Collins asymmetries were performed by *BABAR* exploiting inclusive $e^+e^- \to h_1h_2X$ annihilations (with $h_{1,2} = \pi$ and/or K) mainly at the energy of the $\Upsilon(4S)$, which corresponds to a squared transferred momentum Q^2 ~110 GeV$^2/c^4$. For the first time asymmetries following strange quarks fragmentation could be derived as a function of the fractional energy carried out by inclusively emitted hadron pairs.

1 Introduction

The inclusive cross section of e^+e^- annihilation in hadrons, produced following the hadronization of a $q\bar{q}$ pair, may be parameterized via phenomenologial fragmentation functions (FF's), which are universal, non-perturbative and have to be determined from experimental data. According to the factorization theorem this cross section may be written as:

$$\frac{d\sigma}{dz}(e^+e^- \to hX) = \sum_q \sigma(e^+e^- \to q\bar{q}) \cdot [D_q^h(z, Q^2) + D_{\bar{q}}^h(z, Q^2)] \quad (1)$$

where $z = 2E_h/\sqrt{s}$ is the fractional energy carried by the emitted hadron h, $D_{q(\bar{q})}$ is the FF for q ($\bar{q}$), Q is the transferred momentum, σ is the cross section for the $e^+e^- \to q\bar{q}$ elementary process and the sum runs over the quark flavors allowed by the center-of-mass energy of the reaction.

The description of the hadronization stage can largely benefit from the information related to the parton spins and transverse degrees of freedom of the parton motion. The Collins FF's were introduced [1] to describe this contribution, and are related to the probability that a transverse polarized quark will fragment into a spinless hadron. The fragmentation function of a polarized quark $q^\uparrow$ hadronizing into spinless hadron, $D_1^{q\uparrow}$, reported in eq. (2), is given by the sum of the D_1^q unpolarized contribution, and a polarized term. The latter is proportional to a spin-orbit coupling, expressed through a mixed product of the quark spin $\mathbf{s}_q$, its direction $\mathbf{k}_q$ and the transverse momentum $\mathbf{P}_\perp$ of the emitted hadron, times the polarized fragmentation function $H_1^{\perp q}$:

$$D_1^{q\uparrow}(z, \mathbf{P}_\perp; s_q) = D_1^q(z, \mathbf{P}_\perp; s_q) + \frac{P_\perp}{zM_h} H_1^{\perp q} \mathbf{s_q} \cdot (\mathbf{k_q} \times \mathbf{P}_\perp). \quad (2)$$

[a] e-mail: filippi@to.infn.it

This function is known as Collins FF and basically describes the strength of the cosine dependence of the polarized part of $D_1^{q\uparrow}$. The so-called *Collins effect* stands in the azimuthal modulation of the production cross section of spinless hadrons in a e^+e^- annihilation process, and can be measured experimentally from the azimuthal distribution of hadron pairs emitted in opposite hemispheres in the reaction center of mass. Differences in the amplitude of this modulation for hadron pairs of different charges and flavors determine the presence of measurable asymmetries.

First measurements of the Collins effect were performed in Semi-Inclusive Deep Inelastic Scattering (SIDIS) experiments, like HERMES [2] and COMPASS [3], with unpolarized lepton beams off polarized targets. In this case the direction of the quark spin is known, but the cross section of the hadron production depends on the convolution of the Collins FF and another chiral-odd probability distribution function, known as transversity. Therefore, one of the two needs to be determined independently. This can be done resorting to e^+e^- annihilations in $q\bar{q}$, with the inclusive production of hadron pairs (h_1, h_2): $e^+e^- \to q\bar{q} \to h_1^{\pm}h_2^{\mp(\pm)}X$. In this case, the reaction cross section is proportional, through a cosine term expressing an azimuthal modulation with respect to the quark spins, to the product of the Collins FF's for the fragmenting q and $\bar{q}$. The e^+e^- annihilation reaction is a golden channel for the derivation of Collins FF's since no hadron is present in the initial state; moreover, even if the directions of the q and $\bar{q}$ spins are unknown, they have to be parallel and this introduces a correlation in the emission of hadron pairs. The first measurements of the Collins asymmetries, as a function of z of each inclusively produced hadron, were performed by Belle [4] for pion pairs, at $Q^2 \sim 110$ GeV$^2/c^4$. Later, BABAR confirmed these results, at the same energy, with additional information on the dependence of the Collins asymmetries also on fractional pion momenta [6]. New results on inclusive pion pairs production have been recently obtained also by BESIII [5], and are in agreement with the previous findings (even though data were taken at a lower Q^2, $\sim$13 GeV$^2/c^4$).

A new analysis performed by BABAR extends also to strange quarks the investigation of spin related effects in the hadronization process [7], thanks to the inclusion of kaons in the set of identified pairs. This is the first attempt to infer Collins FF's for the fragmentation of strange quarks both to strange and non-strange mesons, known respectively as "favored" and "disfavored" FF's, depending on the preservation of the parent quark flavor in the hadronization process. According to the flavor of the measured particles and their charge, different combinations of Collins favored and disfavored FF's can occur and therefore be measured.

In the following a summary of the results obtained in the latest BABAR analysis will be reported. In this analysis, 468 fb^{-1} of e^+e^- annihilations, taken at the $\Upsilon(4S)$ resonance mass and 40 MeV below, were used. More details are available in Ref. [7].

2 Method and analysis overview

The extraction of Collins asymmetries is based on the determination of the yields of particle pairs as a function of the azimuthal angles of their emission. These angles, and consequently the cross section for the particles production, can be defined in different reference systems. The fundamental step is to identify an axis that, together with the e^+e^- direction in the reaction rest frame, forms the plane with respect to which the directions of the emitted hadrons and azimuthal angles can be defined. A natural choice for this axis is the $q\bar{q}$ emission direction, but since this is unknown, approximations have to be applied. The most used reference systems are the "Thrust Reference Frame" (RF12), in which this preferential direction is taken by the thrust axis, and the transverse momenta of both the hadrons are being measured, and the "Second Hadron Reference Frame" (RF0), in which the axis used to define the reaction plane coincides with the direction of one of the two emitted hadrons [6]. In both the cases the cross sections exhibit an azimuthal cosine modulation; however, the dependence on Collins FF's of the parent q and $\bar{q}$ is different. Even if from the experimental point of view in RF0 the particle distributions can be obtained in a more direct way, the theoretical interpretation of the results is much more difficult as compared to RF12 due to the more complicated entanglement of the involved FF's.

The analysis strategy for the extraction of the Collins effect is based on the simultaneous measurements of hadron pair yields ($\pi\pi$, πK and KK in whichever charge combination) and their fit through $(b + a\cos\phi)$ functions. Care must be taken as the modulation due to a genuine Collins effect is to be distinguished by similar effects due to apparatus acceptance and other asymmetry sources. The evaluation of asymmetries is performed in both the mentioned reference systems, which deliver complementary results (even if strongly correlated, as obtained from the same experimental sample).

Events are selected with a minimum of three hadrons, belonging to two jets emitted in opposite hemispheres in the reaction center of mass and well contained within the apparatus acceptance. The requirement of a total measured energy larger than 11 GeV allows to reject most of the 2τ and 2γ background, and photon/gluon radiation events with the γ or the jets emitted almost collinear to the beam line. *Ah hoc* cuts are moreover applied on the total energy and the event thrust in order to remove most of the leptonic events (Bhabha, $\mu^+\mu^-$, $\mu^+\mu^-\gamma$).

The particle identification efficiency, performed through the energy loss and Cherenkov radiation information from, respectively, the BABAR drift chamber and DIRC [8], is required to be as large as ~80% for kaons and ~90% for pions, with a minimal misidentification of electrons and muons (smaller than 2 and 4%, respectively). Pion/kaon pairs misidentification probabilities must also be known with good accuracy, as they must be used to weight the function fitting the modulated yield distributions.

2.1 Raw asymmetries

All the distributions of the normalized yields of $\pi\pi$, πK, KK pairs of the same (L), opposite (U) or all charge combinations (C) exhibit a cosine dependence as a function of the azimuthal angle, in both the considered reference frames. The difference in the amplitude of the distributions for particles of different charge is the signal of the Collins effect. As an example, the normalized yield distributions for KK pairs of the same (L, open crosses), opposite (U, open squares) and all charges (C, full points) are reported, for RF12, in Fig. 1 (left).

Asymmetries not dependent on the charge are generated by other sources, which have to be identified and isolated to extract the information on the pure Collins effect. The use of (uds) simulated data samples helps identifying these sources: visible modulations in simulated data where no Collins effect has been injected are mainly due to the apparatus acceptance, and to the emission of gluon radiation from the final state (FSR) and/or photon radiation from the initial state (ISR). In order to eliminate these contributions, which occur for all charge combinations, a common procedure consists on performing the ratios of the normalized raw yields, known as *double ratios* method.

2.2 Double ratios

The ratio of unlike-to-like (A^{UL}) and unlike-to-all sign hadron pairs (A^{UC}) carry the same information of raw yield distributions and still present a cosine modulation, also after the factorization of acceptance effects. Residual modulations not due to the Collins effect and mainly related to ISR emission can still be observed in the double ratios of simulated distributions, and they can be retained as systematic errors. The double ratio (DR) distribution for KK pairs in RF12, is reported in Fig. 1 (right): the UL ratio is represented by open crosses, the UC ratio by full points.

The DR distributions for each of the three particle combinations ($j = \pi\pi$, πK, KK), may again be fitted by a $(B + A\cos\phi)$ function, in which the A coefficients contain the Collins and residual radiative

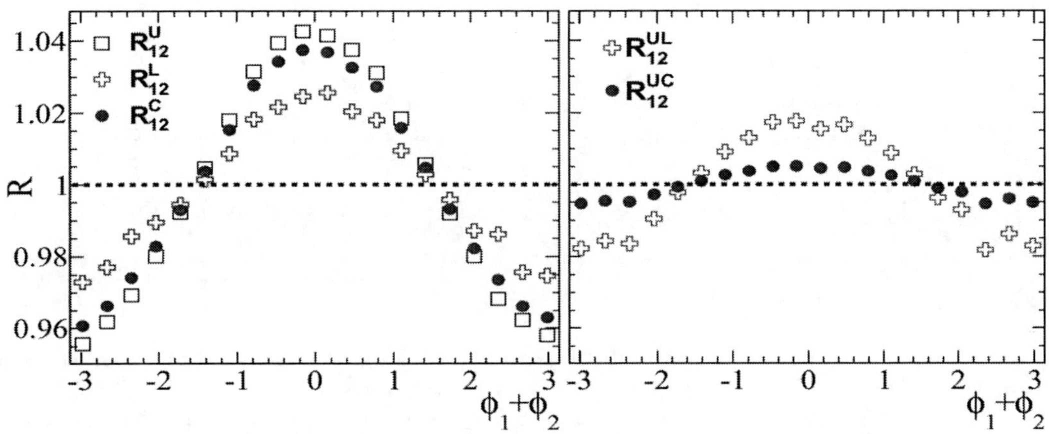

Figure 1. Distribution of normalized yields (left) of KK pairs for unlike (U), like (L) and all charge combinations (C), and their double ratios in RF12 (right).

effects. On the other hand, they may also be expressed as a sum of a pure (uds) Collins-related part, and a non-Collins contribution carried by background channels:

$$A_j^{meas} = F_{uds}^j A_j^{Collins} + \sum_i F_i^j A_j^i. \tag{3}$$

In this equation, the F^j coefficients express the fraction of j-type pairs coming from (uds) (signal) or from the i-th background channel, constrained by the requirement $F_{uds} + \sum_i F_i = 1$, and may be determined by means of proper Monte Carlo simulations. Eq. (3) is a system of linear equations whose unknowns are the three Collins asymmetries $A_j^{Collins}$ (corrected by the proper pair misidentification efficiency mentioned above) and the A_j^i contributions to the DR asymmetry given by the background channels.

Dedicated Monte Carlo simulations show that just a few background reactions really bring an important contribution to the yields. While the selection cuts are effective enough to reduce the contribution by $B\overline{B}$ and $\tau^+\tau^-$ events, which are, respectively, mainly important at low and high z (therefore $A^{B\overline{B}} = 0$ and $A^{\tau^+\tau^-} = 0$ may be set), the contribution from the charm source is sizeable and larger than 30%, so it cannot be neglected. A charm-enhanced control data sample, in which the presence of at least one D^* per event is required, is used to perform an analogous evaluation of the asymmetry, and provides additional equations to the linear system. In this way eq. (3) can be solved to extract the relevant (uds) Collins asymmetries for the $\pi\pi$, $K\pi$ and KK channels.

2.3 Corrections and systematic effects

The obtained results need to be corrected for a number of effects, and an accurate estimation of systematic errors is needed as well. One of the most important corrections mainly affects the evaluations performed in RF12, and is related to the approximation of taking the thrust axis in place of the unknown $q\overline{q}$ direction. By means of Monte Carlo simulations one sees that the opening angle between the two axes has a distribution with the maximum around 100 mrad. By means of the same simula-

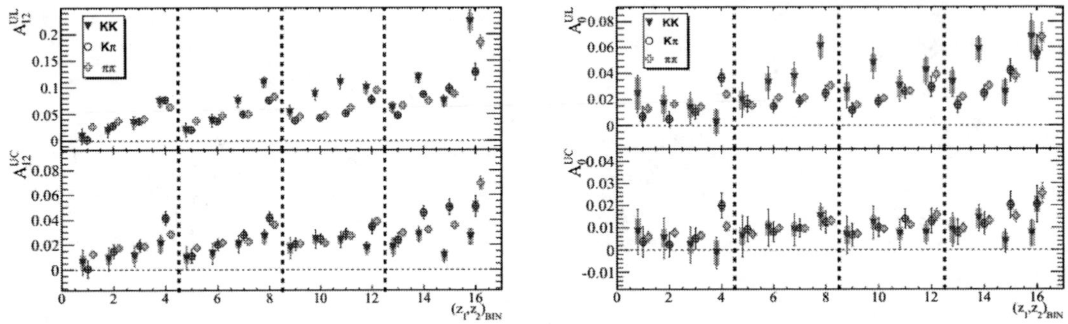

Figure 2. Collins asymmetries measured by *BABAR* [7] for KK (triangles), πK (circles) and $\pi\pi$ (crosses) pairs, in RF12 (left) and RF0 (right). Vertical dashed lines defines subplots for ranges of z_1: [0.15,0.2], [0.2,0.3], [0.3,0.5], [0.5,0.9]. In each subplot z_2 varies in the same ranges. The bars and bands report the statistical and systematic uncertainty, respectively.

tions a dilution factor can determined, to be applied to the measured asymmetries; it depends on the fractional energy and increases with it.

Several sources of systematic uncertainties have to be taken into account; they act differently in the two considered reference systems and depend sizeably on the fractional energies of both the emitted hadrons. The larger effects are played by uncertainties in the Monte Carlo simulations of background channels, in the particle identification efficiencies and in the fit procedures, and by fluctuations due to different choices for the selection cuts. Minor effects are added by beam polarization, fluctuations of asymmetries in different data taking periods, second order couplings between the Collins asymmetries and detector acceptance effects, and higher order harmonics neglected in the fits.

The absolute value of the systematic uncertainty is at most at the 2% level, mostly relevant in higher z bins; the maximum relative total systematic uncertainty is about 10%.

3 Results

The experimental results are expressed in terms of double ratio asymmetries for unlike-to-like (A^{UL}) and unlike-to-all sign particles (A^{UC}), in four bins of fractional energy z for both the hadrons, and in both RF12 and RF0. The measured asymmetries for KK, $K\pi$ and $\pi\pi$ pairs, corrected for the above mentioned effects, are reported in Fig. 2, for RF12 (left) and RF0 (right). The total systematic uncertainties are depicted in the figure as bands.

Some traits may be singled out in the trends of the measured asymmetries, common to both reference systems:

- most of the asymmetries are significantly different from zero;
- the asymmetries show a similar magnitude for all channels;
- all A^{UL} asymmetries increase with z; the same is approximately valid for A^{UC} except for the KK channel, for which the rising trend is less remarkable;
- A^{UC} is always smaller than A^{UL};
- the asymmetries for the KK pairs are almost consistent with zero at low z;
- A^{UL} for KK and $K\pi$ pairs is larger than A^{UL} for $\pi\pi$ pairs especially at large z values; this observation is important as directly related to the magnitude of the favored Collins FF for the strange quark.

The results obtained in the two reference systems, even if correlated, are complementary as referring to different combinations of the Collins FF's. Global fits based on each of them are expected to deliver consistent outcomes.

4 Conclusions

The new results for $\pi\pi$ pairs have been checked for consistency with the previous $\pi\pi$ results published by *BABAR* [6]. After a proper rescaling through a factor depending on the polar angle, needed since the two sets of results are referred to two different kinematic regions, such consistency emerges clearly, which fosters more confidence on the reliability of the novel results involving strangeness.

The new measurements deliver important information to the present knowledge of strange quark fragmentation, since this is the first time they are provided for KK and $K\pi$ pairs. Therefore, they are valuable inputs for global fits, which include SIDIS data and are aimed to the simultaneous extraction of both the Collins FF and the transversity PDF. First global fits were performed on the results from pion pairs [9] and are currently being updated using the new experimental findings. The trend found for asymmetries of strange hadron pairs are in agreement with the existing theoretical expectations [10]. New results on this subject are awaited from Belle and the coming Belle-II experiment.

References

[1] J.C. Collins, Nucl. Phys. **B396**, 161 (1993);
J.C. Collins and D.E. Soper, Nucl. Phys. **B193**, 381 (1981)

[2] HERMES Collaboration, A. Airapetian *et al.*, Phys. Lett. **B693** (2010), 11

[3] COMPASS Collaboration, M. Alekseev *et al.*, Phys. Rev. Lett. **B692**, 240 (2010)

[4] Belle Collaboration, R. Seidl *et al.*, Phys. Rev. **D78**, 032011 (2008)

[5] BESIII Collaboration, M. Ablikin *et al.*, ArXiv:1507.06824 (2015) [hep-ex]

[6] *BABAR* Collaboration, J.P. Lees *et al.*, Phys. Rev. **D90**, 052003 (2014)

[7] *BABAR* Collaboration, B. Aubert *et al.*, ArXiv:1506.05864 (2015) [hep-ex], accepted for publication in Phys. Rev. **D**

[8] *BABAR* Collaboration, B. Aubert *et al.*, Nucl. Instr. Meth. **A729**, 615 (2013)

[9] A. Bacchetta *et al.*, Phys. Lett. **B659**, 234 (2007);
M. Anselmino *et al.*, Phys. Rev. **D75**, 054032 (2007);
M. Anselmino *et al.*, Phys. Rev. **D87**, 094019 (2013)

[10] D. Boer, Nucl. Phys. **B306**, 23 (2009)

EPJ Web of Conferences **120**, 09001 (2016) DOI: 10.1051/epjconf/201612009001
ISMD 2015

First 13 TeV results from CMS

Ferenc Siklér[1,a] for the CMS Collaboration

[1] *Wigner RCP, Budapest, Hungary*

Abstract. An overview of the first 13 TeV results from CMS is presented, including measurements of pseudorapidity-density of charged hadrons, two-particle angular correlations, inclusive $t\bar{t}$ and single top production, as well as exotica.

1 Pseudorapidity-density of charged hadrons

The pseudorapidity-density of charged hadrons ($dN_{ch}/d\eta$) is one of the most basic physical observables in high-energy particle collisions. It presents the essential first step in exploring the physics of a new energy regime, since it reflects the combination of perturbative and nonperturbative QCD effects: saturation of parton densities, multiparton interactions, parton hadronization, and soft diffractive scattering. In CMS the measurement was performed under special circumstances. Because of a cryogenic problem, no magnetic field could be maintained during the first stable collisions. Hence it was impossible to measure particle charge or momentum, but only direction.

Inelastic collision events were selected online, where a coincidence of signals from both beam-sensing devices was required, indicating the presence of both proton bunches crossing the interaction point; and offline, where at least one reconstructed interaction vertex is needed, according to the tracklet or track vertex reconstruction methods described below. While only the reconstructed collision vertex with the highest multiplicity is used in the tracklet analysis, all reconstructed vertices in a given bunch crossing are processed for the track analysis.

The results are corrected to correspond to a sample of inelastic collisions [1]. The corrections from the detector-level offline event selection to the hadron-level event definitions are derived from MC simulations with Pythia8 v208 [2, 3] (several tunes [4]) and Epos LHC [5] (LHC tune [6]), which cover a wide range of possible model ingredients. The detector response is simulated with Geant4 [7] and processed through the same event reconstruction chain as collision data. The MC simulations are produced with the location and shape of the interaction region as extracted from data.

The analysis of the recorded data is performed with two reconstruction techniques based on hits of charged particles detected by the CMS pixel detector. (A very preliminary result from June, using hits only from the strip tracker, is displayed in Fig. 1-left.) While both start by searching for hit pairs in different layers at similar azimuthal angles ϕ (Fig. 1-right), the tracklet method (using hit pairs) performs background subtraction based on control samples in data, while the track method (using hit triplets) minimizes background contributions by requiring an additional hit in the detector. This results in a slightly narrower accessible η range for the track method as compared to the tracklet method.

[a] e-mail: sikler.ferenc@wigner.mta.hu

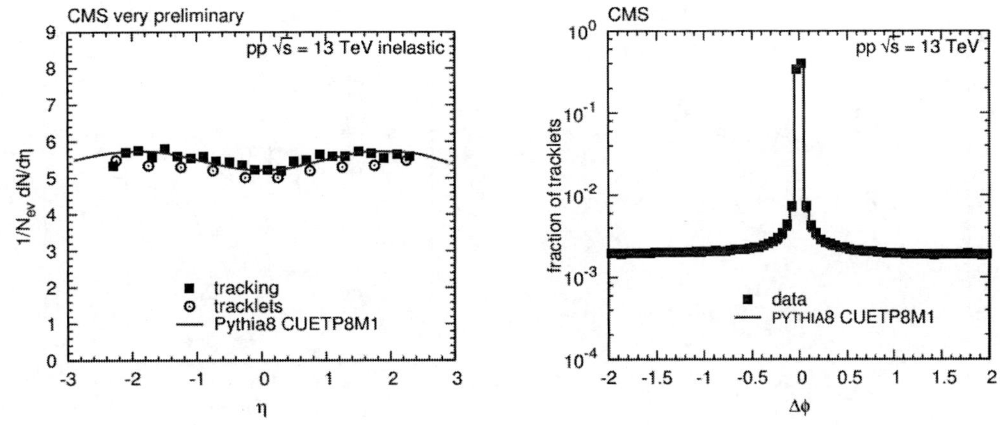

Figure 1. Left: Distributions of the pseudorapidity density of charged hadrons in inelastic pp collisions at 13 TeV measured in *strip-only data* (solid markers, tracking and tracklet results), and predicted by the PYTHIA8 CUETP8M1 event generator (curve) [8]. Right: The $\Delta\phi$ distributions of hit pairs on tracklets in the pixel data (squares) and from MC simulation (histogram) [1].

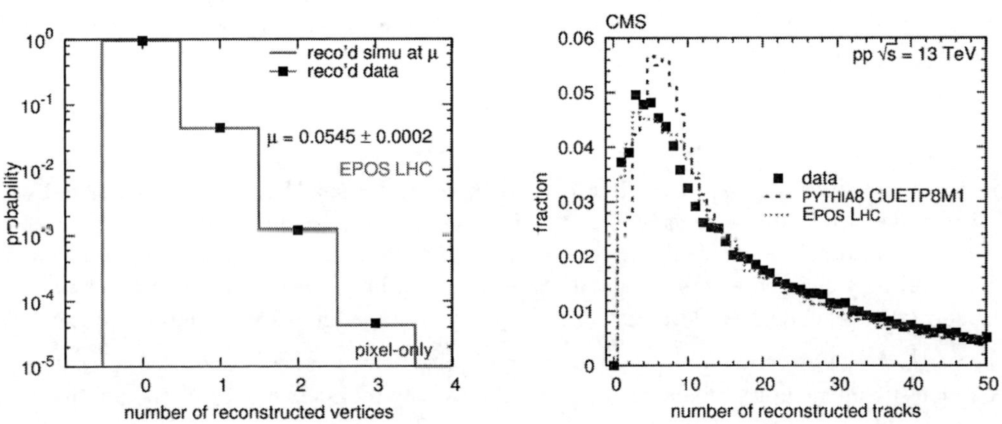

Figure 2. Left: Distribution of the number of reconstructed vertices from data and that based on the EPOS LHC model including simulations. The extracted mean (μ) of the corresponding Poissonian is also indicated. Right: Distributions of the number of reconstructed tracks per event from data and from simulations, for inelastic pp collisions [1].

Since various factors, such as detector alignment, material, detector response, and dependence on MC event generators, influence the two techniques somewhat differently, their final combination into a single measurement provides a more robust result.

For the tracklet analysis, all recorded bunch crossings are processed (about 170 000 collision events), while for the track analysis only 1 million of them are used (about 55 000 collision events). After corrections, the agreement between the tracklet and track $dN_{ch}/d\eta$ results is better than 2% at

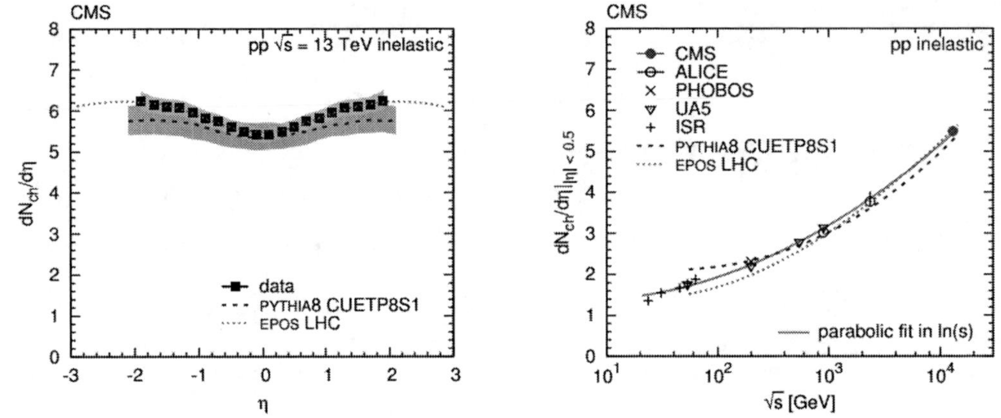

Figure 3. Left: Distributions of the pseudorapidity density of charged hadrons in the region $|\eta| < 2$ in inelastic pp collisions at 13 TeV measured in data (solid markers, combined track and tracklet results, symmetrized in η), and predicted by the PYTHIA8 CUETP8S1 and the EPOS LHC event generators (curves) [1]. The grey shaded area encompassing the data points indicates their correlated systematic uncertainties. The blue band corresponds to the envelope of the CUETP8S1 tune parametric uncertainties. Right: Center-of-mass energy dependence of $dN_{ch}/d\eta|_{|\eta|<0.5}$ including lower energy data [1]. The solid curve shows a second-order polynomial in $\ln(s)$ fit to the data points, including the new result at $\sqrt{s} = 13$ TeV. The dashed and dotted curves show the PYTHIA8 CUETP8S1 and EPOS LHC predictions, respectively.

central pseudorapidity and better than 3% at forward pseudorapidities. Hence, averaging their $dN_{ch}/d\eta$ values is justified. Since the systematic uncertainties dominate and are mostly correlated between the two analyses, the simple mean of the central values and of systematic uncertainties are taken. The fraction of primary charged leptons is about 1% of the total long-lived charged particles produced, and the correction from detector-level tracklets and tracks is done for charged hadrons only.

The performance of vertexing in simulation and in data was compared and used to estimate the average inelastic interaction probability μ per bunch crossing by assuming a Poissonian distribution (Fig. 2-left). While PYTHIA8 CUETP8M1 gives $\mu = 0.0525 \pm 0.0003$, EPOS LHC leads to 0.0545 ± 0.0002. The product of the number of analyzed bunch crossings and μ gives the number of inelastic collision events, to be used for the calculation of the final $dN_{ch}/d\eta$ values. As Fig. 2-right shows, the distribution of the number of reconstructed tracks in data is in fair agreement with the MC event generator predictions in particular at the higher multiplicities.

Pseudorapidity density distributions of charged hadrons in the region $|\eta| < 2$ for inelastic pp events are shown in Fig. 3-left. The data points and uncertainties are symmetrized in $\pm\eta$. While the predictions of both PYTHIA8 (with CUETP8S1 and CUETP8M1) and EPOS LHC agree with the measured central value, the measured $dN_{ch}/d\eta$ distribution in the full η range is better described by the latter. The uncertainty band of PYTHIA8 corresponds to the envelope of the uncertainties of the tune parameters of CUETP8S1; the EPOS LHC predictions have no uncertainty associated with its parameter settings.

For central pseudorapidities ($|\eta| < 0.5$), the charged-hadron multiplicity density is $dN_{ch}/d\eta|_{|\eta|<0.5} = 5.49 \pm 0.01(\text{stat}) \pm 0.17(\text{syst})$, a value obtained by combining the two methods.

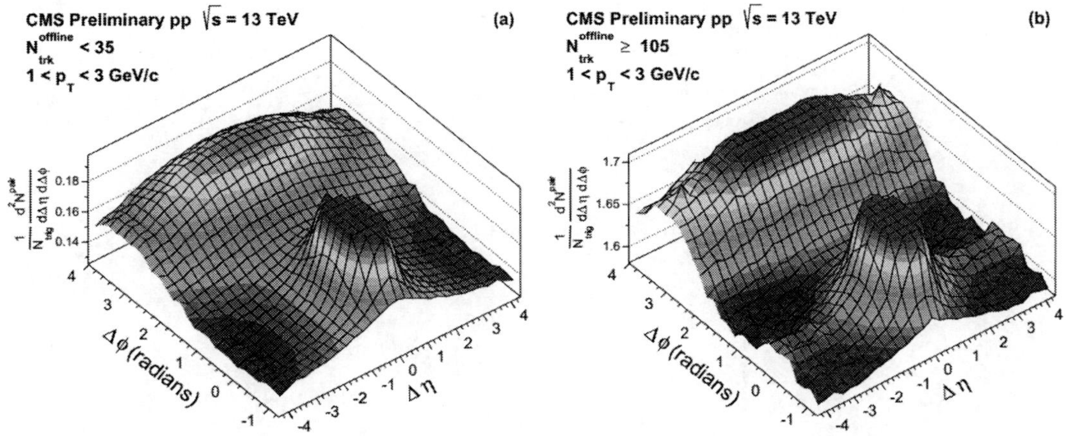

Figure 4. The 2D $(\Delta\eta, \Delta\phi)$ two-particle correlation functions in pp collisions at $\sqrt{s}$ = 13 TeV for pairs of charged particles both in the range $1 < p_T < 3$ GeV/c. Results are shown for (a) low-multiplicity events $N_{\text{track}}^{\text{offline}} < 35$) and for (b) a high-multiplicity sample ($N_{\text{track}}^{\text{offline}} \geq 105$). The sharp peaks from jet correlations around $(\Delta\eta, \Delta\phi) = (0, 0)$ are truncated to better illustrate the long-range correlations.

The center-of-mass energy dependence of $\mathrm{d}N_{\text{ch}}/\mathrm{d}\eta$ is shown in Fig. 3-right. For comparison, inelastic pp measurements at lower energies (ISR [9, 10], UA5 [11, 12], PHOBOS [13], and ALICE [14]) are also plotted. The measured values are empirically fitted using a second-order polynomial in $\ln(s)$ as $1.55 - 0.113\ln(s) + 0.0168\ln(s)^2$ which provides a good description of the available data over the full energy range. The Pythia8 and Epos LHC event generators globally reproduce the collision-energy dependence of hadron production in inelastic pp collisions.

These results constitute the first CMS measurement of hadron production at the new center-of-mass energy frontier, and provide new constraints for the improvement of perturbative and nonperturbative QCD aspects implemented in hadronic event generators.

2 Angular correlation of charged hadrons

CMS has performed studies of two-particle angular correlations for charged particles produced in pp collisions at a center-of-mass energy of 13 TeV [15], with a special emphasis on the long-range near-side two-particle structure seen in lower energy pp and pPb collisions. The data correspond to an integrated luminosity of about 270 nb^{-1}.

The correlations are studied over a broad range of pseudorapidity ($|\eta| < 2.4$) and over the full azimuth (ϕ) as a function of charged particle multiplicity and transverse momentum. In high-multiplicity events, a long-range ($|\Delta\eta| > 2.0$), near-side ($\Delta\phi \approx 0$) structure emerges in the two-particle $\Delta\eta - \Delta\phi$ correlation functions (Fig. 4). The magnitude of the correlation exhibits a pronounced maximum in the range $1.0 < p_T < 2.0$ GeV/c and an approximately linear increase with the charged particle multiplicity (Fig. 5). The overall correlation strength at $\sqrt{s}$ = 13 TeV is similar to that found in earlier pp data at $\sqrt{s}$ = 7 TeV, but is measured up to much higher multiplicity values.

The near-side long-range yield obtained with the zero-yield-at-minimum (ZYAM) procedure is found to be consistent with zero in the low-multiplicity region, with an approximately linear increase

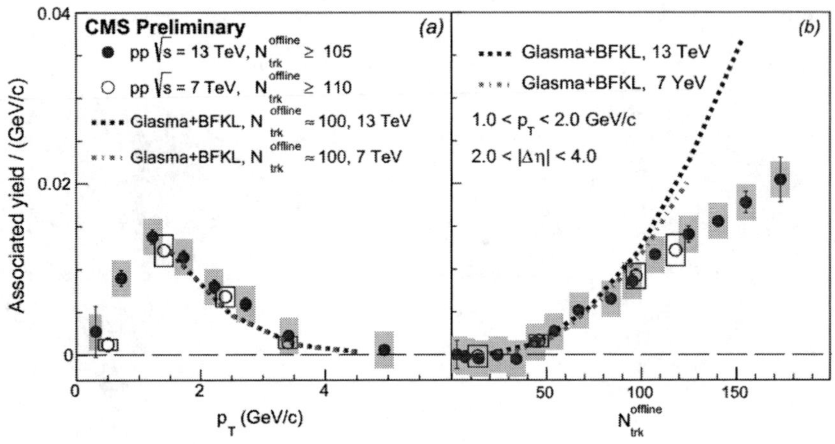

Figure 5. Associated yield for the near side of the correlation function averaged over $2 < |\Delta\eta| < 4$ and integrated over the region $|\Delta\phi| < \Delta\phi_{ZYAM}$ for pp data at $\sqrt{s} = 13\,\text{TeV}$ (filled circles) and 7 TeV (open circles). Panel (a) shows the associated yield as a function of p_T for $N_{\text{track}}^{\text{offline}} \geq 105$. In panel (b) the associated yield for $1 < p_T < 2$ GeV/c is shown as a function of multiplicity $N_{\text{track}}^{\text{offline}}$. The p_T selection applies to both particles in each pair. The error bars correspond to the statistical uncertainties, while the shaded areas and boxes denote the systematic uncertainties. Curves represent calculations from the gluon saturation model [16].

with multiplicity $N_{\text{trk}}^{\text{offline}} \gtrsim 40$. A strong collision system size dependence is observed when comparing data from pp, pPb, and PbPb collisions. Comparing the pp data at $\sqrt{s} = 7\,\text{TeV}$ and 13 TeV, no collision energy dependence of the near-side associated yields is observed over the overlapping event multiplicity and p_T ranges.

3 Top production

The study top production has a great discovery potential for physics beyond the standard model. It helps to test the production mechanisms, to check the validity of QCD, and to provide an important source of background in searches for physics beyond the standard model.

Top quarks can decay leptonically t → W b → (l ν) (b-jet), and hadronically t → hadrons (three jets). In case of top pair production, di-leptonic decays of both quarks can be selected by requiring one electron and one muon of opposite charge, and at least two jets in the final state. The case when one decay is leptonic and the other one is hadronic is selected by asking for one electron or one muon, and at least four jets in the final state. The measurements detailed below all use data corresponding to an integrated luminosity of 42 pb^{-1}, taken at the beginning of the 2015 pp run.

The top-antitop quark ($t\bar{t}$) production cross section has been measured for the first time by the CMS in proton-proton collisions at $\sqrt{s} = 13\,\text{TeV}$ [17]. The measurement is performed by analyzing events with one electron and one muon and at least two jets. The results are obtained through an event-counting analysis. The measured cross section is $\sigma_{t\bar{t}} = 772 \pm 60(\text{stat}) \pm 62(\text{syst}) \pm 93(\text{lumi})$ pb, resulting in a total relative uncertainty of 16.4% (Fig. 6). This measurement is consistent with the standard model prediction of $\sigma_{t\bar{t}}^{\text{NNLO + NNLL}} = 832^{+40}_{-46}$ pb for a top quark mass of 172.5 GeV, and also with a recent measurement from the ATLAS Collaboration [18].

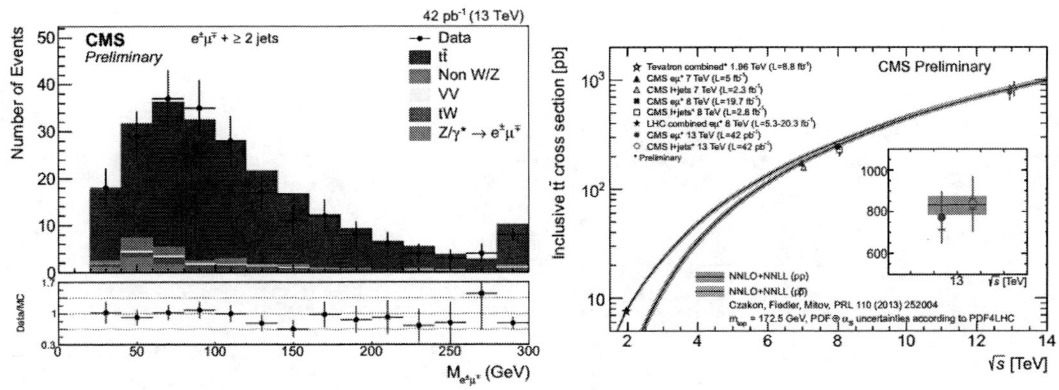

Figure 6. Left: Distribution of the dilepton invariant mass after all selections [17]. The last bin contains the overflow events. The ratios of data to the sum of the expected yields are given at the bottom. Right: Top quark pair production cross section in $p\bar{p}$ and pp collisions as a function of center-of-mass energy.

The total inclusive and the normalized differential cross sections for the production of top quark pairs have been determined in proton-proton collisions at 13 TeV [19, 20]. The measurements are performed in the l+jets decay channels with an electron or a muon in the final state. The $t\bar{t}$ production cross section is measured as a function of p_T and y of the top quarks and as a function of p_T, y, and invariant mass of the $t\bar{t}$ system, as well as of the jet multiplicity. Due to the limited available integrated luminosity the precision of the measurement is mostly dominated by the statistical uncertainties. However, in some kinematic regions the uncertainties in the parton shower and hadronization model are higher. The results are compared to several standard model calculations (Fig. 7-left). All calculations are compatible with the measured results though the POWHEG simulation shows the best agreement in the description of $p_T(t\bar{t})$. In the l+jets decay channels we measure a total inclusive $t\bar{t}$ production cross section of $836 \pm 27(\text{stat}) \pm 88(\text{syst}) \pm 100(\text{lumi})$ pb. This is in good agreement with the NNLO standard model prediction, hence no significant deviation from the standard model prediction is observed.

The measurement of single top production helps testing QCD and electroweak processes, specifically the tWb vertex, and the CKM matrix element V_{tb}. In the analysis, events with one muon (from $t \to W \to \mu$ or via $t \to W \to \tau \to \mu$ decays) in the final state are selected. The signal can be most easily got from a fit to the η distribution of the (light) recoil jet.

A measurement of the t-channel single top-quark cross section has been performed using the very first proton-proton collisions at 13 TeV [21]. Events are selected with one muon in the final state. The signal is extracted from a fit to the pseudorapidity distribution of the recoiling jet. The inclusive cross section is measured to be $\sigma_{\text{t-ch}} = 274 \pm 98(\text{stat}) \pm 52(\text{syst}) \pm 33(\text{lumi})$. The observed (expected) significance of the signal contribution is 3.5 (2.7) standard deviations with respect to the background-only hypothesis. The magnitude of the CKM matrix element V_{tb} is also calculated, yielding $|V_{tb}| = 1.12 \pm 0.24(\text{exp}) \pm 0.02(\text{theo})$. The results are found to be in agreement with predictions by the standard model (Fig 7-right).

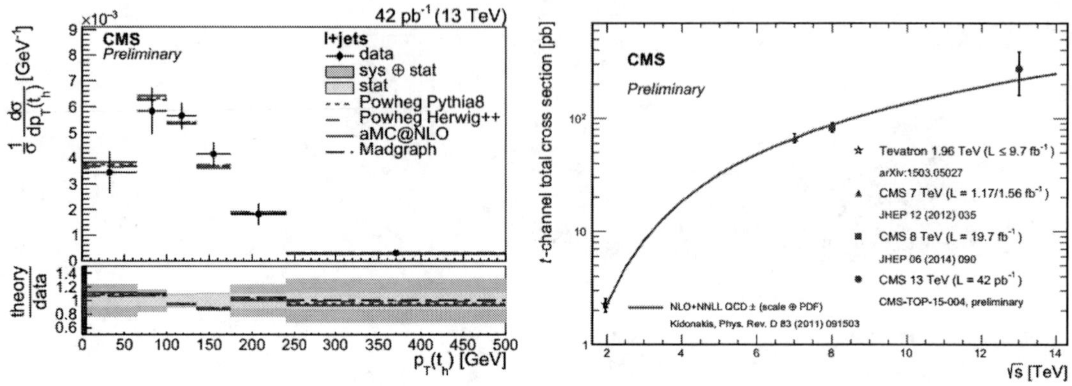

Figure 7. Left: Normalized differential cross sections as a function of p_T(t) compared to the predictions of several MC event generators [19]. Right: The t-channel single top-quark cross section summary of the most precise CMS measurements in comparison with QCD calculations [21].

4 Exotica

A search for narrow resonances decaying into a pair of jets has been performed using 42 pb^{-1} of pp collisions at $\sqrt{s} = 13$ TeV. The dijet invariant mass distribution is measured to be a smoothly falling distribution (Fig. 8-left), as expected within the standard model. In the analyzed data sample there is no significant evidence for new particle production. We present generic upper limits that are applicable to any model of narrow dijet resonance production. A list specific lower limits is set on the mass of string resonances, excited quarks, axigluons, colorons, color octet scalars, and scalar diquarks (Fig. 8-right). These give the most stringent constraint to date on the mass of string resonances, a lower mass limit of 5.1 TeV. This dijet resonance search is more sensitive than the previous Run 1 searches for dijet resonances beyond 5 TeV.

Acknowledgments

The author wishes to thank to the Hungarian Scientific Research Fund (K 109703), and the Swiss National Science Foundation (SCOPES 152601) for their support.

References

[1] V. Khachatryan et al. (CMS), Phys. Lett. B **751**, 143 (2015), 1507.05915
[2] T. Sjöstrand, S. Mrenna, P. Skands, JHEP **05**, 026 (2006), hep-ph/0603175
[3] T. Sjöstrand, S. Mrenna, P.Z. Skands, Comput. Phys. Commun. **178**, 852 (2008), 0710.3820
[4] CMS Collaboration (2014), CMS-PAS-GEN-14-001
[5] K. Werner, F.M. Liu, T. Pierog, Phys. Rev. C **74**, 044902 (2006), hep-ph/0506232
[6] T. Pierog, I. Karpenko, J.M. Katzy, E. Yatsenko, K. Werner (2013), dESY-13-125, 1306.0121
[7] S. Agostinelli et al. (GEANT4), Nucl. Instrum. Meth. A **506**, 250 (2003)
[8] CMS Collaboration (2015), CMS-DP-2015-011

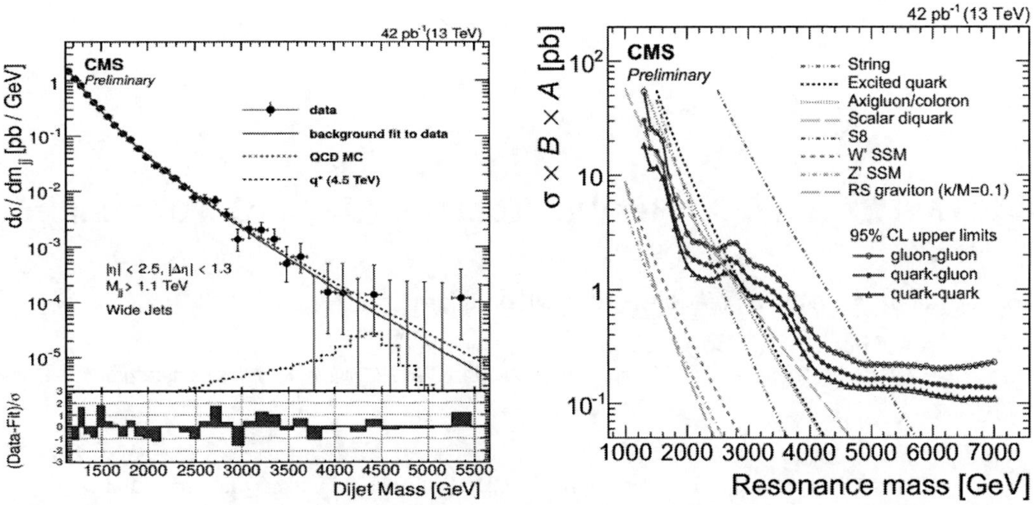

Figure 8. Left: Dijet mass spectrum from wide jets (points) compared to a smooth fit (solid) and to predictions including detector simulation of QCD (dashed) and an excited quark signal (histogram) [22]. The QCD prediction has been normalized to the data. Right: The observed 95% confidence level upper limits for dijet resonances of the type gluon-gluon, quark-gluon, and quark-quark, compared to theoretical predictions [22].

[9] A.M. Rossi, G. Vannini, A. Bussière, E. Albini, D. D'Alessandro, G. Giacomelli, Nucl. Phys. B **84**, 269 (1975)

[10] W. Thome et al. (Aachen-CERN-Heidelberg-Munich), Nucl. Phys. B **129**, 365 (1977)

[11] G.J. Alner et al. (UA5), Z. Phys. C **33**, 1 (1986)

[12] G.J. Alner et al. (UA5), Phys. Rept. **154**, 247 (1987)

[13] B.B. Back et al. (PHOBOS), Phys. Rev. Lett. **93**, 082301 (2004), `nucl-ex/0311009`

[14] K. Aamodt et al. (ALICE), Eur. Phys. J. C **68**, 89 (2010), `1004.3034`

[15] V. Khachatryan et al. (CMS) (2015), `1510.03068`

[16] K. Dusling, R. Venugopalan, Phys. Rev. D **87**, 094034 (2013), `1302.7018`

[17] V. Khachatryan et al. (CMS) (2015), `1510.05302`

[18] ATLAS Collaboration (2015), ATLAS-CONF-2015-033

[19] CMS Collaboration (2015), CMS-PAS-TOP-15-005

[20] CMS Collaboration (2015), CMS-PAS-TOP-15-010

[21] CMS Collaboration (2015), CMS-PAS-TOP-15-004

[22] CMS Collaboration (2015), CMS-PAS-EXO-15-001

EPJ Web of Conferences **120**, 09002 (2016) DOI: 10.1051/epjconf/201612009002
ISMD 2015

Early Run 2 Hard QCD Results from the ATLAS Collaboration

Nicola Orlando[1,2,a] on behalf of the ATLAS Collaboration

[1] *The University of Hong Kong, Department of Physics, Hong Kong*
[2] *Aristotle University of Thessaloniki, Laboratory of Nuclear and Particle Physics, Thessaloniki, Greece, 54124*

Abstract. We provide an overview of hard QCD results based on data collected with the ATLAS detector in proton-proton collision at $\sqrt{s}$ = 13 TeV at the Large Hadron Collider. The production of high transverse momentum jets, photons and photon-pairs were studied; the inclusive jet cross section is found to agree well with the prediction of perturbative QCD calculations performed at next-to-leading accuracy. The production cross sections for W and Z bosons in their e and μ decays was measured; in general, agreement is found with the expectation of next-to-next-to leading order QCD calculations and interesting sensitivities to the proton structure functions are already observed. The top production cross sections were measured in different top decay channels and found to agree with the state of the art QCD predictions.

1 Introduction

Since June 2015 the Large Hadron Collider (LHC) has started to deliver proton-proton (pp) collisions at the unprecedented centre-of-mass energy $\sqrt{s}$ = 13 TeV. The ATLAS [1] collaboration has analysed this very early 13 TeV data taken in both special conditions with very low pileup and those dedicated to high p_T physics.

The 13 TeV ATLAS physics program benefits from the installment of the Insertable B-Layer (IBL) [2], a fourth layer added to the present Pixel detector between a new beam pipe and the current inner Pixel layer (B-layer) and many other improvements affecting the detector, data-taking as well as analysis software-framework.

The measurements presented in this paper use between 6.4 pb^{-1} up to 85 pb^{-1} of integrated luminosity collected at LHC in proton-proton (pp) collisions at $\sqrt{s}$ = 13 TeV. The precision of many measurements is limited by the early luminosity calibration is based on pair of x-y beam-separation scans performed in June 2015, leading to about 9% uncertainty on the luminosity scale determination.

2 Jets and photons measurements

The measurement of cross-sections of jets produced at high transverse momentum provides a test of the validity of perturbative QCD (pQCD) at the TeV scale.

[a] e-mail: nicola.orlando@cern.ch

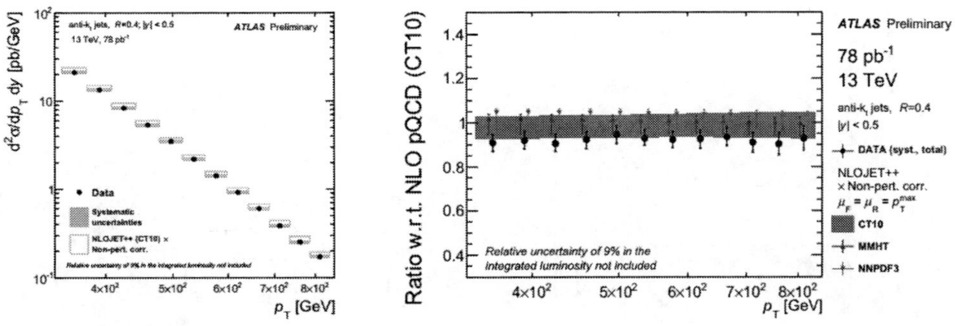

(a) (b)

Figure 1. Inclusive-jet cross sections (a) as a function of the jet p_T in $|y| < 0.5$; the error bars show the statistical uncertainties and the filled area shows the experimental systematic uncertainties. Ratio of the measured inclusive-jet cross section to the NLO pQCD prediction (b), shown as a function of the jet p_T in $|y| < 0.5$. The figures are taken from Ref. [3].

The inclusive-jet cross section is measured differentially as a function of the jet transverse-momentum, p_T, in a jet rapidity range of $|y| < 0.5$ [3][1]. Theoretical predictions at next-to-leading order (NLO) in pQCD are available from the NLOJET++ generator [4]; the predictions are corrected for non-perturbative effects. As shown in Fig. 1, the pQCD calculations are consistent with the measurement; the 9% uncertainty on the luminosity calibration is not included in the figures.

The production of prompt photons provides a test for pQCD in a cleaner environment compared to jet production, since it is less affected by hadronisation effects. The measurement of two prompt isolated photons is a tool to probe pQCD predictions relevant to understand the irreducible background to physics processes involving photons in the final state, like the Higgs boson decay in photon-pair.

The kinematic distributions of the signal yield for inclusive photon production [5], presented as functions of photon transverse-energy, E_T^{γ}, and pseudorapidity, $|\eta^{\gamma}|$, agree with the predictions of the SHERPA generator [6], as shown in Fig. 2 (a-b).

Fig. 3 shows the invariant mass spectrum, $m_{\gamma\gamma}$, and the transverse-momentum distribution, $p_T^{\gamma\gamma}$, of the selected photon-pairs [7]. An enhancement in the $p_T^{\gamma\gamma}$ spectrum is observed between 20 GeV and 40 GeV, about the sum of the E_T^{γ} thresholds of the two photons, as expected from beyond leading-order (LO) QCD corrections.

3 Heavy flavour production

The study of heavy quarkonium is a challenge for the understanding of perturbative and non-perturbative QCD regimes. Although many aspects of quarkonium production and decay can be described reasonably well, the literature lack of a coherent theoretical picture which can explain all quarkonium measurements simultaneously. A first analysis of the non-prompt fraction of J/ψ production at $\sqrt{s} = 13$ TeV is performed in the di-muon decay channel [8]. The non-prompt fraction is found

[1]ATLAS uses a right-handed coordinate system with its origin at the nominal interaction point (IP) in the centre of the detector and the z-axis along the beam pipe. The x-axis points from the IP to the centre of the LHC ring, and the y-axis points upward. Cylindrical coordinates (r,φ) are used in the transverse plane, φ being the azimuthal angle around the beam pipe. The pseudorapidity is defined in terms of the polar angle θ as $\eta = -\ln[\tan(\theta/2)]$. Transverse momentum and energy are defined as $p_T = p\sin\theta$ and $E_T = E\sin\theta$, respectively.

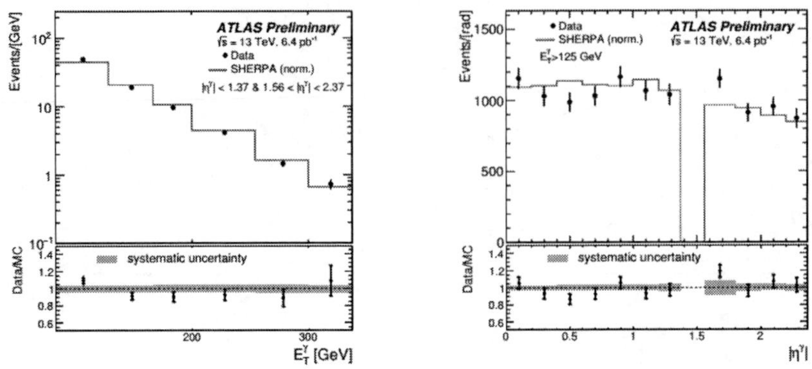

(a) (b)

Figure 2. Measured isolated-photon yield as a function of E_T^γ (a) and $|\eta^\gamma|$ (b) for $|\eta^\gamma| < 2.47$. The figures are taken from Ref. [5].

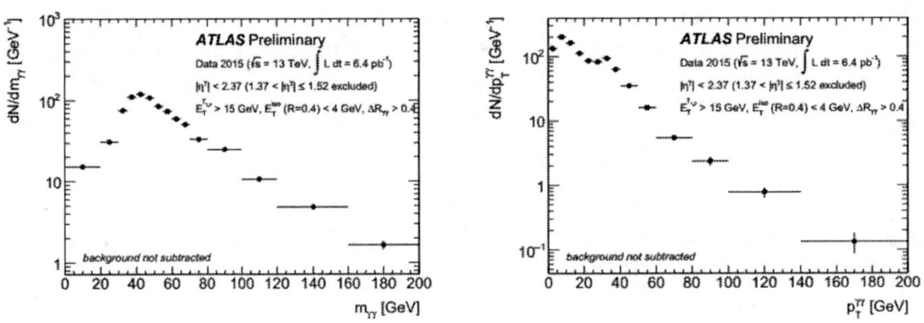

(a) (b)

Figure 3. Number of isolated photon-pair candidates shown as a function of $m_{\gamma\gamma}$ (a) and $p_T^{\gamma\gamma}$ (b). The figures are taken from Ref. [7].

to increase steadily from 0.25 at the J/ψ transverse momentum of 8 GeV to 0.65 at 40 GeV, with no significant variation with rapidity observed within the precision of the measurement. The centre-of-mass energy (and initial-state) dependence of the non-prompt fraction is studied by comparing these results to previous ATLAS measurements at $\sqrt{s} = 2.76$ TeV and at $\sqrt{s} = 7$ TeV, and to CDF measurements $\sqrt{s} = 1.96$ TeV. The comparison between the various measurements is illustrated in Fig. 4; no significant change in the non-prompt fraction is observed from $\sqrt{s} = 7$ TeV to $\sqrt{s} = 13$ TeV, contrary to the significant difference observed between the $\sqrt{s} = 7$ TeV measurement and the measurements at lower energies.

4 W/Z cross sections

The measurement of weak vector bosons, W/Z, production at hadron colliders provides a benchmark of our understanding of QCD and electroweak (EW) phenomenology. The large production cross

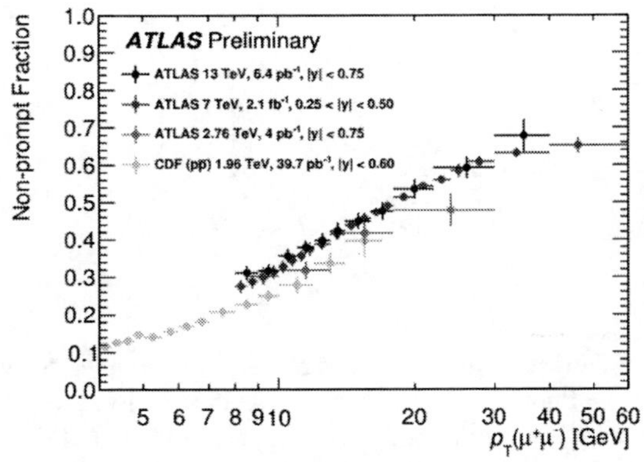

Figure 4. Non-prompt J/ψ production fraction measured in the rapidity $|y| < 0.75$ compared to previous measurements from ATLAS in pp collisions at $\sqrt{s} = 2.76$ TeV and $\sqrt{s} = 7$ TeV, and from CDF in $p\bar{p}$ collisions at $\sqrt{s} = 1.96$ TeV. The error bars represent the total uncertainty on the measurements (statistical and systematic). The figures are taken from Ref [8].

sections and easily identifiable W/Z decays to leptonic final states offer a clean experimental signature which can be measured to high precision. In addition, theoretical predictions for inclusive W/Z production are available at next-to-next-to-leading-order (NNLO) accuracy in QCD, and include EW corrections at NLO accuracy [9]; the cross-section predictions are dependent on the parton distribution functions (PDFs), therefore, measurements of W and Z bosons production offer a unique opportunity to test modeling of PDFs at the new centre-of-mass energy of $\sqrt{s} = 13$ TeV.

Fig. 5 (a-b) shows the measured fiducial W/Z cross sections [10]. The data are compared to predictions calculated using different PDF sets. The measurements agree well with the predictions. The experimental precision is already comparable to PDF uncertainties, however, the current experimental precision is limited by the preliminary luminosity calibration.

The measurement of the ratio $R_{W^+/W^-} = \sigma^{fid}_{W^+}/\sigma^{fid}_{W^-}$ [10] is shown in Fig 5 (c); there is a significant scatter for different PDF predictions and the accuracy of the experimental result is comparable to the spread among them. The data favours results from the PDFs which include LHC measurements lower centre-of-mass energy (NNPDF3.0 [11], MMHT14nnlo68cl [12], and ABM12LHC [13]).

Measurements of vector bosons production in association with jets represent an important test of pQCD and constitute a relevant background for studies of the Higgs boson and searches for new phenomena. The fiducial production cross sections for Z bosons plus inclusive one to four jet multiplicities have been measured [15] with a precision ranging from about ~10% to ~20%. The measurements have been compared to predictions from Sherpa and Madgraph [16], normalised to the NNLO inclusive cross-section for Z boson production; as can be observed from Fig. 6, the predictions show a reasonable agreement with the data.

5 Top quark pair cross sections

The top quark is the heaviest known elementary particle, with a mass close to the scale of electroweak symmetry breaking. Precise measurements of the top-pair cross section, $\sigma_{t\bar{t}}$, are sensitive to the gluon parton distribution function (PDF) of the proton, the strong coupling constant α_s, the top-quark mass m_t, and potential enhancements due to physics beyond the Standard Model (SM).

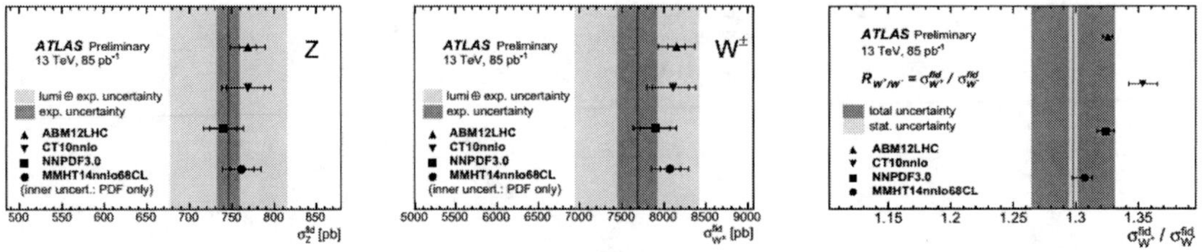

(a) (b) (c)

Figure 5. Measured fiducial W/Z cross sections (a,b) and R_{W^+/W^-} (c) (red line); the green (cyan) band corresponds to the experimental uncertainty without (with) the luminosity uncertainty; the yellow band in (c) represent the statistical uncertainty on the measurement. The data are compared to NNLO QCD predictions derived with four PDF sets; the inner error bar of the predictions represents the PDF uncertainty while the outer error bar includes the sum of all other systematics. The figures are taken from Ref [10].

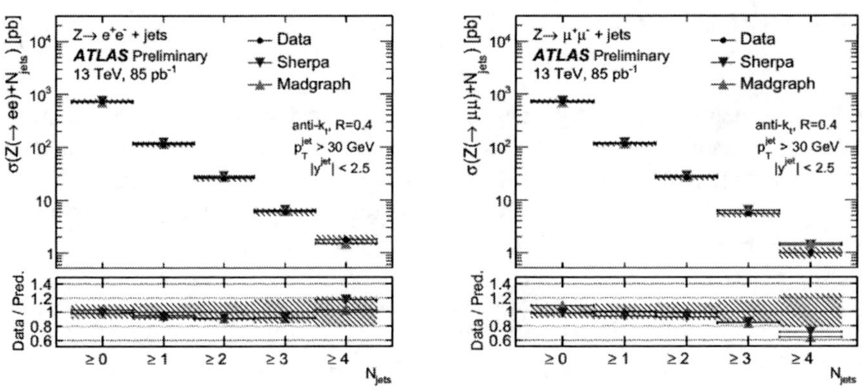

Figure 6. Fiducial cross sections as a function of the inclusive jet multiplicity in the Z→ ee channel (left) and the Z→ $\mu\mu$ channel (right). The measured cross section is compared to predictions from SHERPA and MADGRAPH. The hatched error band corresponds to the total systematic and statistical uncertainty of the cross section, together with the luminosity uncertainty. The figures are taken from Ref [15].

The inclusive $t\bar{t}$ production cross-section was measured in several decay channels, assuming m_t = 175.5 GeV [17, 18]. As shown in Fig. 7 (a), the data agree well with the state of the art SM prediction based on NNLO QCD perturbative expansion supplemented with next-to-next-to-leading logarithm (NNLL) QCD resummation [19].

A measurement of the ratio between the $t\bar{t}$ and Z-boson production cross-sections is also presented, using on the measurement of the $t\bar{t}$ cross-section in the $e\mu$ final state and of the Z-boson cross-section in the ee and $\mu\mu$ decay modes. The measurement of the $t\bar{t}$ cross-section in the $e\mu$ final state is used to calculate the ratio since it is the most precise ATLAS measurement of the $t\bar{t}$ cross-section at $\sqrt{s}$ = 13 TeV. This ratio benefits from significant cancellations in systematic uncertainties between the two cross-section measurements, and is sensitive to ratios of the gluon and sea-quark PDFs.

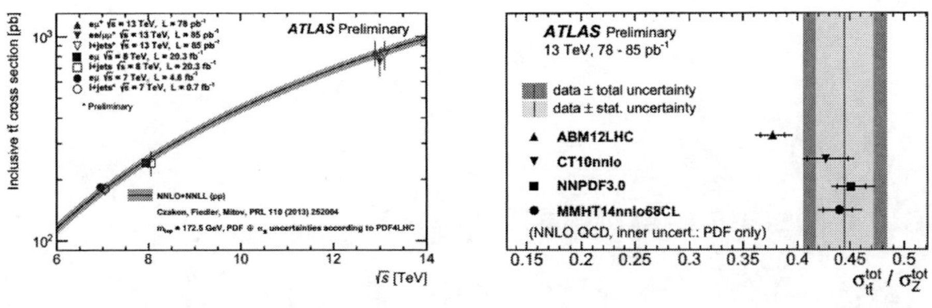

Figure 7. Cross-section for $t\bar{t}$ pair production in pp collisions as a function of centre-of-mass energy (a); the ATLAS measurements at $\sqrt{s}$ = 7, 8 and 13 TeV are compared to the NNLO+NNLL theory predictions. Measured cross-section ratio $R_{t\bar{t}/Z}$ compared to NNLO predictions at $\sqrt{s}$ = 13 TeV based on several PDF sets; the inner shaded band corresponds to the statistical uncertainty on the measurement, whilst the outer shaded band includes both statistical and systematic uncertainties; the inner error bars on the predictions correspond to PDF uncertainties only, while the outer error bars also include QCD scales and α_s uncertainties. The figures are taken from Ref [17, 18].

The measured $t\bar{t}/Z$ cross-section ratio, $R_{t\bar{t}/Z}$ [18], agree with predictions obtained with CT10NNLO, NNPDF3.0 and MMHT14NNLO68CL, as can be seen from Fig. 7 (b). However, data are only marginally consistent with the prediction derived with the ABM12LHC PDF set, which gives a 12% smaller $t\bar{t}$ cross-section than CT10NNLO.

References

[1] ATLAS Collaboration, JINST **3** (2008) S08003.
[2] ATLAS collaboration, CERN-LHCC-2010-013. ATLAS-TDR-19 (2010).
[3] ATLAS collaboration, ATLAS-CONF-2015-034 (2015).
[4] Z. Nagy, Phys. Rev. D **68** (2003) 094002 [hep-ph/0307268].
[5] ATLAS collaboration, ATL-PHYS-PUB-2015-016 (2015).
[6] T. Gleisberg *et al.* JHEP **0902** (2009) 007 [arXiv:0811.4622 [hep-ph]].
[7] ATLAS collaboration, ATL-PHYS-PUB-2015-020 (2015).
[8] ATLAS collaboration, ATLAS-CONF-2015-030 (2015).
[9] Y. Li and F. Petriello, Phys. Rev. D **86** (2012) 094034 [arXiv:1208.5967 [hep-ph]].
[10] ATLAS collaboration, ATLAS-CONF-2015-039 (2015).
[11] R. D. Ball *et al.* [NNPDF Collaboration], JHEP **1504** (2015) 040 [arXiv:1410.8849 [hep-ph]].
[12] L. A. Harland-Lang *et al.* Eur. Phys. J. C **75** (2015) 5, 204 [arXiv:1412.3989 [hep-ph]].
[13] S. Alekhin *et al.* Phys. Rev. D **89** (2014) 5, 054028 [arXiv:1310.3059 [hep-ph]].
[14] J. Gao *et al.*, Phys. Rev. D **89** (2014) 3, 033009 [arXiv:1302.6246 [hep-ph]].
[15] ATLAS collaboration, ATLAS-CONF-2015-041 (2015).
[16] J. Alwall *et al.*, JHEP **1407** (2014) 079 [arXiv:1405.0301 [hep-ph]].
[17] ATLAS collaboration, ATLAS-CONF-2015-033 (2015).
[18] ATLAS collaboration, ATLAS-CONF-2015-049 (2015).
[19] M. Czakon and A. Mitov, Comput. Phys. Commun. **185** (2014) 2930 [arXiv:1112.5675 [hep-ph]].

EPJ Web of Conferences **120**, 09003 (2016) DOI: 10.1051/epjconf/201612009003
ISMD 2015

First Run II results from ALICE

Alberica Toia[1,2,a]
on behalf of the ALICE Collaboration

[1] *Goethe University of Frankfurt*
[2] *GSI*

Abstract. The ALICE Collaboration is collecting data with both Minimum Bias and Muon triggers with pp collisions at $\sqrt{s}$ = 13 TeV in the ongoing LHC Run II. An excellent performance of tracking and PID in the central barrel and in the muon spectrometer has been obtained. First results on the charged-particle pseudorapidity density and on identified particle transverse momentum spectra at $\sqrt{s}$ = 13 TeV is presented.

1 Introduction

After a two-year long shutdown, the Large Hadron Collider (LHC) at CERN restarted its physics programme in June 2015 with proton-proton collisions at $\sqrt{s}$ = 13 TeV, the highest centre-of-mass energy reached so far in laboratory.

The charged particle multiplicity density is a measure of the initial energy density and provides information on the dynamics of the soft particle production and its relation to the initial collision geometry. Particle production at collider energies originates from the interplay of perturbative (hard) and non-perturbative (soft) QCD processes. Soft scattering processes and parton hadronisation dominate the bulk of particle production at low transverse momenta and can only be modelled phenomenologically. Hence, these measurements provide constraints for a better tuning of models and event generators for hadron-collider and cosmic-ray physics. In addition, they are valuable reference data to study nuclear effects in nucleus-nucleus and proton-nucleus collisions, in particular allowing one to measure the nuclear modification factors R_{AA} and R_{pA}, respectively, of the transverse-momentum spectra.

To investigate particle production mechanisms, ALICE has extensively studied the first moment, $\langle p_T \rangle$, of the charged-particle transverse momentum spectrum and its correlation with the charged-particle multiplicity N_{ch}. This has been studied by many experiments at hadron colliders covering collision energies from $\sqrt{s}$ = 31 GeV up to 7 TeV. All experiments observed an increase of $\langle p_T \rangle$ with N_{ch} in pp collisions, which could be reproduced in PYTHIA only including color reconnections in the hadronization and which was recognized as a hint of coherence in a high multiplicity environment [1].

2 ALICE in Run II

The ALICE apparatus and its performance in Run I are presented in detail elsewhere [2]. Here we only briefly mention the upgrades set up for Run II. More details are presented in [3]. During the

[a] e-mail: alberica.toia@cern.ch

shut-down the construction and installation of the TRD was completed, so that it now has full central barrel acceptance. The full TRD coverage provides an improved momentum resolution for tracking and a uniform electron and hadron identification.

The detectors located outside TOF used by two electromagnetic calorimeters with thicknesses of 1 nuclear interaction length, the high-resolution lead-tungsten PHOS and the large-acceptance lead-scintillator EMCal, have been upgraded. The EMCal was extended with the installation of eight (six full-size and two one-third-size) super modules, forming the so-called dijet calorimeter (DCal) back-to-back to the existing EMcal. The PHOS trigger electronics was modified so that the PHOS and Dcal essentially act as a single calorimeter, able to measure both single shower, as well as jet triggers.

A new subsystem, the ALICE diffractive (AD) detector, has been installed during LS1 to enhance the capabilities of the experiment to tag diffractive events. The AD detector consists of two double layers of scintillation counters, sensitive to charged particles with p_T down to 20 MeV/c, placed far away from the interaction region on both sides, one in the ALICE cavern (at z = 17.0 m) and one in the LHC tunnel (at z = -19.5 m), with a pseudorapidity coverage of $4.9 < \eta < 6.3$ and $-7.0 < \eta < -4.8$, respectively.

The TPC has a 90 m^3 drift volume and is divided into two parts by the central cathode, which is kept at -100 kV. The end plates are equipped with multiwire proportional chambers (MWPC). The gas mixture of the TPC was changed from $Ne - CO_2$ (90:10) to $Ar - CO_2$ (88:12) for Run II to ensure a more stable response for the high particle flux generated during p–Pb and Pb–Pb running without significant degradation of momentum resolution at the low p_T.

3 Data Analysis

We present the pseudorapidity (η) and transverse-momentum (p_T) distributions of primary charged particles measured in proton-proton collisions at the centre-of-mass energy $\sqrt{s}$ = 13 TeV with the ALICE detector at the LHC [4]. Primary particles are defined as prompt particles produced in the collisions, including all decay products, with the exception of those from weak decays of strange particles. The pseudorapidity distribution is measured at central rapidity in $|\eta| < 1.8$. The measurements reported here have been obtained for inelastic events (INEL) and events having at least one charged particle in the pseudorapidity interval $|\eta| < 1$ (INEL>0). The transverse-momentum distribution of charged particles is measured in the range $0.15 < p_T < 20$ GeV/c and $|\eta| < 0.8$ for INEL>0 events. The correlation between transverse momentum and particle multiplicity is also investigated by studying the evolution of the spectra with event multiplicity. The data have been compared to calculations from models commonly used at the LHC.

The data were collected after the startup of LHC Run II in June 2015, with beam conditions leading to a luminosity of about $5x10^{27}cm^{-2}s^{-1}$, which corresponds to a rate of about 350 Hz for inelastic proton-proton collisions, and a small pile-up probability (10^{-3}). The data were collected using a minimum-bias trigger requiring a hit in either the V0 scintillators or in the AD arrays. After removing background events with an offline selection based on the timing information from the V0 and the AD detectors, and the correlation between the number of clusters of pixel hits and the number of tracklets the remaining contamination is estimated with control triggers taken for various combinations of beam and empty buckets, to be negligible. The final minimum-bias sample used for the data analysis consist of about 1.5 million, which are further required to have a valid reconstructed vertex within $|z| < 10$ cm, to ensure uniform acceptance.

For the corrections we use Monte Carlo events from the PYTHIA 6 (Perugia-2011) event generator with particle transport performed via a GEANT3 simulation of the ALICE detector.

3.1 Pseudorapidity Distribution

The measurement of the charged-particle pseudorapidity distribution is performed using short track segments, termed "tracklets". Tracklet candidates are formed using the position of the primary vertex and a pair of hits, one in each SPD layer. The primary vertex position is extracted by correlating hits in the two SPD layers. This technique effectively allows to reconstruct charged particles with p_T above the 50 MeV/c cut-off determined by particle absorption in the material. The vertex position requirement results in an effective $|\eta| < 1.8$ coverage. The charged particle pseudorapidity density is obtained from the measured distribution of tracklets, corrected for the acceptance and efficiency for a primary particle to produce a tracklet, and the contamination of reconstructed tracklets from combinations of hits not produced by the same primary particle. Both correction factors are determined as a function of the z position of the primary vertex and the pseudorapidity of the tracklet from detector simulations. Differences in strange-particle content between data and simulations, observed at lower beam energies, are taken into account by scaling the strangeness production in the Monte Carlo event sample.

3.2 Momentum Distribution

The transverse-momentum distribution is measured from tracks reconstructed using the information from the Inner Tracking System (ITS) and the Time Projection Chamber (TPC) detectors. Candidate tracks are selected with cuts on the number of space points used for tracking and on the quality of the track fit, as well as on the distance of closest approach to the reconstructed vertex, which result in an effective aceptance of $|\eta| < 0.8$. The efficiency for track reconstruction is calculated with PYTHIA 6, and, since it depends on the particle type, a correction is applied to reproduce the spectra of strange particles measured by ALICE in minimum-bias pp collisions at $\sqrt{s} = 7$ TeV. Finally, the residual contamination from secondary particles, estimated from Monte Carlo simulations, is subtracted from the spectrum.

Various are the contributions to the systematic uncertainties of these analyses. One of the main contributions to the normalisation of the results comes from the limited knowledge of cross-sections and kinematics of diffractive processes, not yet measured at $\sqrt{s} = 13$ TeV, and therefore solely based on previous experimental data at lower collision energies and simulations with Monte Carlo event generators. For the transverse-momentum distribution the dominant sources of uncertainty are the track selections, the efficiency corrections and, for low p_T, the contamination from weak decays of strange hadrons. For the pseudorapidity distribution the dominant sources of uncertainty are the detector acceptance and efficiency, the correction due to p_T extrapolation down to zero, and also the contamination from weak decays of strange hadrons. Other minor systematic uncertainties also arise from the knowledge of the material budget, and of the particle composition.

4 Results

Figure 1, left panel, shows the average charged-particle density distribution $dN_{ch}/d\eta$ measured in INEL and INEL>0 events in the pseudorapidity range $|\eta| < 1.8$. Also shown in the Figure are the results recently published by the CMS Collaboration for inelastic collisions, which agree, within the uncertainties, with the ALICE measurement. We compared our measurement to Monte Carlo calculations performed with PYTHIA 6 (Perugia-2011), PYTHIA 8 (Monash-2013) and EPOS LHC in both the INEL and INEL>0 event classes. PYTHIA 6 calculations are in better agreement with the data than PYTHIA 8 in both classes. EPOS LHC calculations are sligthly higher than the data both in INEL and INEL>0 events.

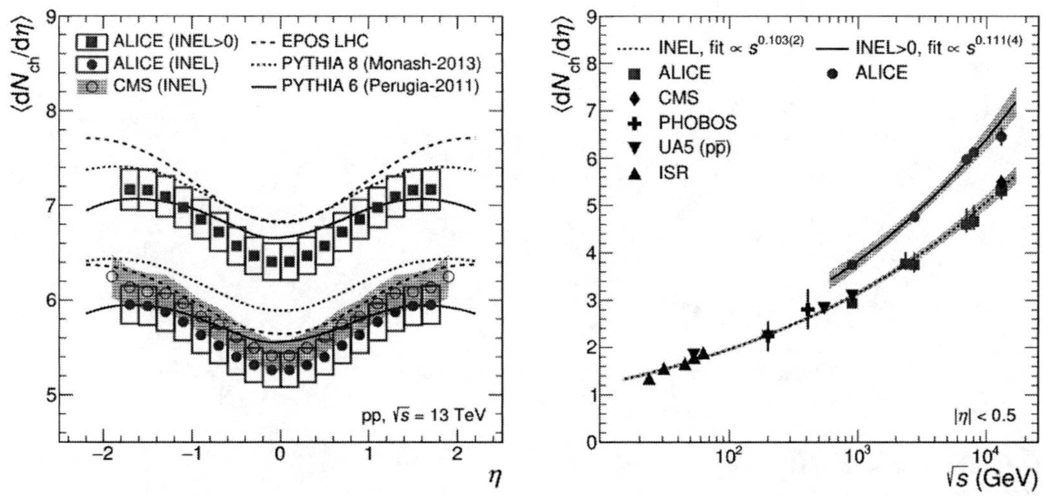

Figure 1. left: Average pseudorapidity density of charged particles as a function of eta produced in pp collisions at $\sqrt{s}$ = 13 TeV. The ALICE results are shown in the normalisation classes INEL and INEL>0 and compared to Monte Carlo calculations and to the results from the CMS Collaboration. The uncertainties are the quadratic sum of statistical and systematic contributions. Right: Charged-particle pseudorapidity density measured in the central pseudorapidity region $|\eta|<0.5$ for INEL and INEL>0 events. The uncertainties are the quadratic sum of statistical and systematic contributions. The lines are power-law fits of the energy dependence of the data and the grey bands represent the standard deviation of the fits.

In the right panel of the Figure we show a compilation of results on pseudorapidity density of charged particles measured at central rapidity ($|\eta| < 0.5$) at different proton-proton collider energies. The energy dependence of $dN_{ch}/d\eta$ is parametrised by the power law function fitted to data. Although the fit results assume that uncertainties at different centre-of-mass energies are independent, which is not strictly the case, the new measuremnt of Run II seems to be in line with the expectations from the data measured at lower energy.

Figure 2 presents the measured p_T spectrum and its comparison with calculations with PYTHIA 6 (Perugia-2011), PYTHIA 8 (Monash-2013) and EPOS LHC. For bulk particle production, the mechanism of colour reconnection is an important one in the PYTHIA models. EPOS is a model based on the Gribov-Regge theory at parton level. Collective (flow-like) effects are incorporated in the EPOS3 version and treated via parametrisations in the EPOS LHC version. These event generators, benefitting from the tuning performed on the LHC data in Run I, describe the p_T spectrum reasonably well, although not in detail. It is interesting to note that both PYTHIA 8 and EPOS LHC models show a similar pattern in the ratio to data with discrepancies up to 20% and that PYTHIA 6 overestimates particle production at high p_T.

The right panel of the Figure shows the ratio of transverse-momentum spectra of charged particles at $\sqrt{s}$ = 13 TeV and 7 TeV [5], properly re-normalized in the INEL>0 event class. As expected, the spectrum is significantly harder at $\sqrt{s}$ = 13 TeV than at $\sqrt{s}$ = 7 TeV. PYTHIA 6, PYTHIA 8 and EPOS LHC reproduce the trend observed in the data, but exhibit a slightly more pronounced hardening with energy in the region of a few GeV/c. The effect appears to be more significant in PYTHIA 8 than in PYTHIA 6 and EPOS LHC.

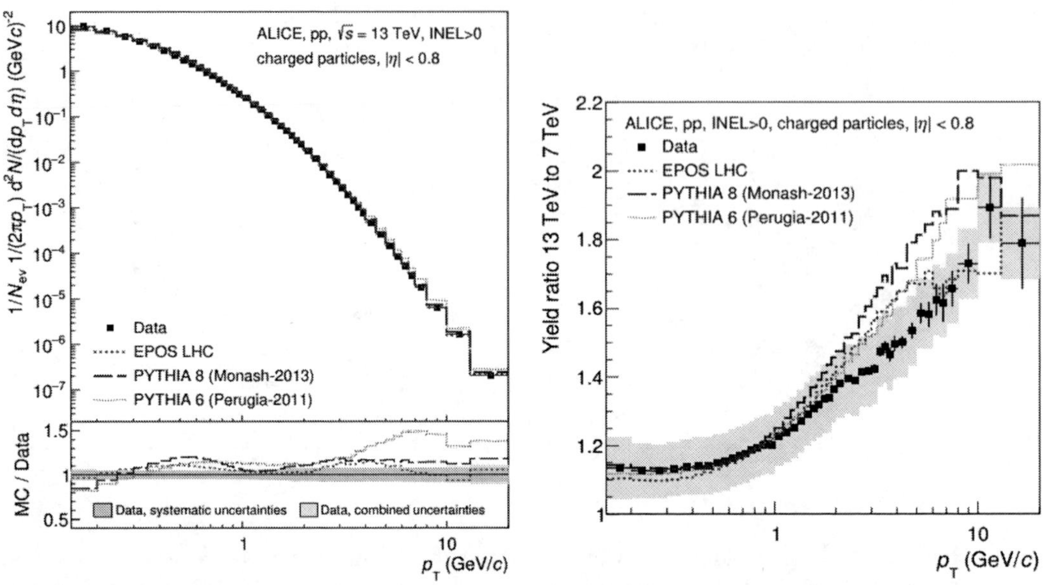

Figure 2. Left:Invariant charged-particle yield as a function of p_T normalised in the INEL>0 event class. The data are compared to Monte Carlo calculations . For the ratio of models (MC) and data (lower panel) the systematic and total uncertainties of the data are shown as grey bands. Right: Ratio of transverse-momentum spectra in INEL>0 events at $\sqrt{s}$ = 13 and 7 TeV. The boxes represent the systematic uncertainties, given by the quadratic sum of uncertainties at the two energies. The data are compared to Monte Carlo calculations.

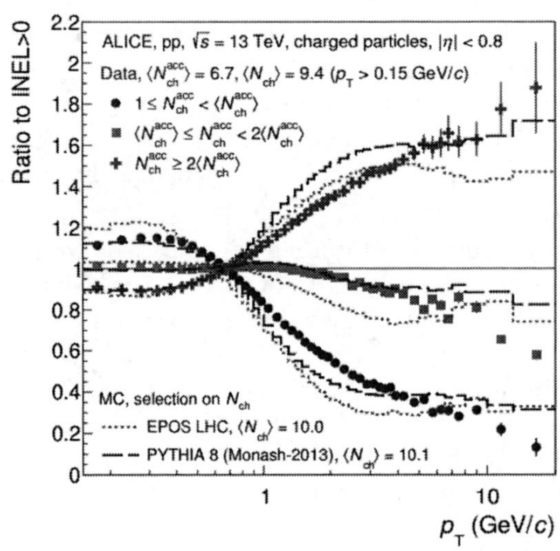

Figure 3. Ratios of transverse-momentum distributions of charged particles in three intervals of multiplicities to the respective one for inclusive (INEL>0) collisions. The spectra were normalized by the integral prior to division. The data are compared to Monte Carlo calculations.

DOI: 10.1051/epjconf/201612009003

The correlation of the particle mean transverse momentum ($\langle p_T \rangle$) with the multiplicity of the event (N_{ch}) first observed at the SppS collider, has been studied by many experiments at hadron colliders in pp($\bar{p}$) covering collision energies from $\sqrt{s}$ = 31 GeV up to 7 TeV. The increase of $\langle p_T \rangle$ with N_{ch} in the central rapidity region observed in all experiments could be reproduced in the PYTHIA event generator only if a mechanism of hadronisation with colour reconnections (CR) is considered. Whether a connection exists between CR and flow in the sense of a genuine space-time evolution of the system is a subject of current investigation. A mechanism involving collective string hadronisation is also used in the EPOS model.

To investigate closely the changes of the spectral shape with multiplicity, we calculate ratio of spectra measured in three intervals of multiplicity to the inclusive (INEL>0) spectrum, shown in Figure 3. In this way, the systematic uncertainties of the measured spectra cancel out completely in the ratios. For this ratio, the spectra were normalised by the integral prior to dividing. The selection is performed on the multiplicity measured in the same kinematic region as the spectrum, $|\eta| < 0.8$ and $0.15 < p_T < 20$ GeV/c, using the uncorrected measured track multiplicity for data and the true value of N_{ch} known in Monte Carlo events. The multiplicity cuts correspond to the average and twice the average value.

The data in the Figure show that the correlation of the spectrum with multiplicity is prominent for the whole p_T range and particularly stronger at high p_T. This correlation is influenced by hard parton production, which, through fragmentation (leading to jets), contributes to charged-particle multiplicity and this contribution is increasing with the parton energy. The increase of $\langle p_T \rangle$ with multiplicity, typically moderate, is most enhanced in the present analysis of spectra and multiplicity, measured in the same kinematic region, where jets presumably dominate the high-multiplicity domain.

The general features seen in the data, which are similar to those first seen $\sqrt{s}$ = 0.9 TeV, are reproduced by PYTHIA 8 and EPOS LHC fairly well. Some discrepancies, of no more than 20%, are noticeable in particular in the high p_T region. This is more prominent for EPOS LHC. It was shown earlier that both EPOS LHC and PYTHIA 8 reproduce well, although slightly overpredicting, the correlation of $\langle p_T \rangle$ with N_{ch}. The present data on spectral shape highlight some deficiencies in both models concerning the description of spectral shapes as a function of multiplicity.

5 Conclusions

In conclusions, measurement of the pseudorapidity and tranverse-momentum distribution of charged particles produced in proton-proton collisions at $\sqrt{s}$ = 13 TeV were presented. The p_T spectrum is significantly harder than at $\sqrt{s}$ = 7 TeV and its shape depends strongly on the multiplicity (in the same kinematic region). The correlation of the spectrum with multiplicity is prominent for all p_T and is observed to be stronger at high p_T, influenced by hard parton production, which, through fragmentation (leading to jets), contributes to charged-particle multiplicity. All in all, there is a fair agreement of the data with expectations from lower energy extrapolations and with the Monte Carlo models used for comparison (PYTHIA6, PYTHIA8, EPOS-LHC).

References

[1] B. Abelev et al. (ALICE Collaboration), Phys. Lett. B 727 (2013) 371–380
[2] B. Abelev et al. (ALICE Collaboration), Int. J. Mod. Phys. A 29 (2014) 1430044
[3] J. Adam et al. (ALICE Collaboration), ALICE-PUBLIC-2015-004 (2015)
[4] J. Adam et al. (ALICE Collaboration), Phys. Lett. B 753 (2016) 319-329
[5] B. Abelev et al. (ALICE Collaboration), Eur. Phys. J. C73 no. 12 (2013)

AUTHOR INDEX

EDP Sciences – Web of Conferences
17, Avenue du Hoggar
Parc d'Activité de Courtabouf
BP 112
F-91944 Les Ulis Cedex A
France

ISBN 978-1-5108-2603-8

9 781510 826038

CURRAN ASSOCIATES INC.
proceedings.com